Optics for Engineers

Optics for Engineers

Edited by **Jason Penn**

Clanrye International
New Jersey

Published by Clanrye International,
55 Van Reypen Street,
Jersey City, NJ 07306, USA
www.clanryeinternational.com

Optics for Engineers
Edited by Jason Penn

International Standard Book Number: 978-1-63240-564-7 (Hardback)

Printed in the United States of America.

Contents

Preface

Optics is the branch of physics that studies the properties of light and its interaction with matter. It also involves design and construction of instruments required for studying the various phenomena related to light. This discipline usually includes ultraviolet, visible and infrared radiation. Light is an electromagnetic wave and it exhibits the behavior of wave as well as particle. To understand this dual nature optics has been further divided into ray optics and wave optics. It includes daily use objects and new technologies to study the properties of light such as lenses, mirrors, microscopes, fibre optics and lasers. This book provides significant information of this discipline to help develop a good understanding of optics and related fields. It includes contributions of experts and scientists which will provide innovative insights into this field. The aim of this text is to present researches that have transformed this discipline and aided its advancement. It is appropriate for students seeking detailed information in the field of optics as well as experts.

Various studies have approached the subject by analyzing it with a single perspective, but the present book provides diverse methodologies and techniques to address this field. This book contains theories and applications needed for understanding the subject from different perspectives. The aim is to keep the readers informed about the progress in the field; therefore, the contributions were carefully examined to compile novel researches by specialists from across the globe.

Indeed, the job of the editor is the most crucial and challenging in compiling all chapters into a single book. In the end, I would extend my sincere thanks to the chapter authors for their profound work. I am also thankful for the support provided by my family and colleagues during the compilation of this book.

Editor

Third Order Optical Nonlinearities of C_{450} Doped Polymer Thin Film Investigated by the Z-Scan

Zainab S. Sadik[1], Dhia H. Al-Amiedy[2], Amal F. Jaffar[3]
[1]Department of Physics, College of Science, University of Baghdad, Baghdad, Iraq
[2]Department of Physics, College of Science for Women, University of Baghdad, Baghdad, Iraq
[3]Ministry of High Education & Scientific Research, Foundation of Technical Education, Institute of Medical Technology, Mansour, Iraq
Email: zainab_ss2004@yahoo.com

ABSTRACT

In the present work, z-scan technique was used to study the nonlinear properties, represented by nonlinear refractive index and nonlinear absorption coefficient for Coumarin 450 (C_{450}) doped PMMA as a function of concentration for two different solvents. The results show change of the effect from self-focusing to self-defocusing in closed aperture z-scan (or change of the effect from two photon absorption to saturable absorption in open aperture z-scan), when the sample dissolved in low viscosity and more polar solvent.

Keywords: Nonlinear Optics; Solvent Effect; Z-Scan Technique

1. Introduction

There is considerable interest in understanding the optical nonlinearities of dyes for widespread applications. Dye molecules are used mostly to generate tunable laser sources and optical shutters, optical signal-processing devices [1-4], two-photon microscopy [5], up conversion lasers [6,7], optical limiting [8,9], optical data storage [10,11] and three-dimensional microfabrication [12]. The basic absorption processes in dyes could be divided into 1) linear absorption; 2) saturation of absorption (SA) and 3) reverse saturable absorption (RSA). Saturation of absorption is vital for use of the dyes in mode-locking. The most important application of RSA is for optical limiting devices [13,14] that protect sensitive optical components, including human eye, from laser-induced damage.

In the present work we studied the nonlinear properties of PMMA thin films doped Coumarin 450 (C_{450}) dye used as an active media of solid state dye laser with two kinds of excitation, CW diode laser at 650 nm and pulsed Nd:YAG at 532 nm.

2. Experimental Materials

All materials used without further purification. Coumarin 450 (C_{450}) supplied from Lambda Physik and used without further purification, Polymethacrylate (PMMA) from ICI company. The solvents used were Tetrahydrofloran (THF) from (BDH Chemicals Ltd. Poole England, Chloroform: From Lab-Scan Ltd. Analitical Sciences HPLC, Dublin Ireland and Methanol from Riedel_dehaen.

All solvents are spectroscopic grade. The choose of the solvents was based on the requirement of solubility of polymers.

2.1. Samples Preparation

Solution of concentrations (10^{-5}, 10^{-4} and 10^{-3}) M/L with different solvents where prepared by weighting amount of the material by using a matter balance having a sensitivity of 10^{-4} gm. To enhance the solubility of the dye in each solvent, the dye firstly dissolved in Methanol with ratio (20%) for each solvent.

Dye doped polymer films were fabricated by the free casting technique (FC), the solution of the polymer was prepared by dissolving the required amount of polymer (7 gm in 100 ml of the used solvent).

The free casting (FC), involves casting a polymer solution on to a flat bottomed glass cup (Petri dish) without imposing hydrodynamic stress on the liquid .The dishes were arranged on a glass plate in order that the dishes have a plan situation, then the dishes were covered with a heavy paper box to be protected from from li light. Solvent is allowed to evaporate under ambient conditions (30˚C) until the films harden. The harden films were then removed from the Petri dish by washing them off with distilled water and placed in the oven (50˚C) for 10 minutes to dry. The thickness of the films produced by this method is dependent on straight forward way of concentration and the volume of the liquid in the dish. In this study the liquid volumes are ranged from (6 - 14) ml, yielding film thickness of (90 - 210) μm. The thickness of the films was measured with an electrical device (Mini-test 3000 microprocessor coating thickness) from electro, Phyisk, Germany (ERICHSEN).

2.2. Z-Scan System

The experimental setup of closed and open aperture z-scan is shown in **Figure 1**. The laser that is used in the present work is a home built CW Diode laser of 650 nm (maximum power 50 mw, beam diameter 1.5 mm and beam divergence 1.5 mrad).

3. Results & Discussion

3.1. Closed Aperture Z-Scan Using CW Diode Laser

Closed aperture z-scan was used to investigate the nonlinear refractive index, where a circular aperture with transmissivity S < 1 is placed behind the sample in the far field to control the cross section of the beam coming out of the sample. As the sample is scanned through the beam, the far field profile shows intensity variation across the beam profile, which is recorded through the aperture and the transmission is recorded as a function of z position. **Figures 2** and **3** show the closed aperture z-scan trace for the dye C_{450} in PMMA doped films in Chloroform and THF.

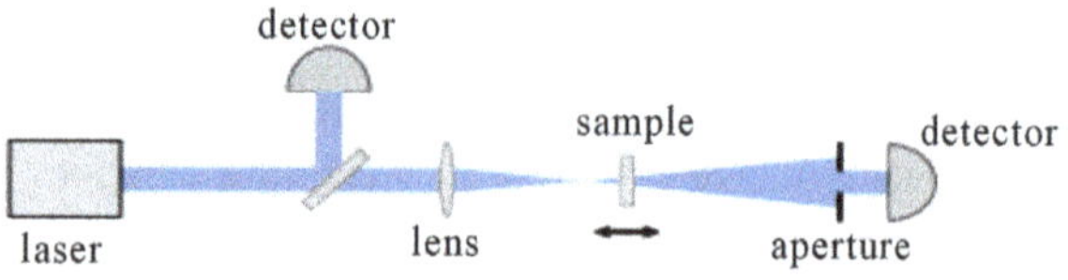

Figure 1. The setup of z-scan system.

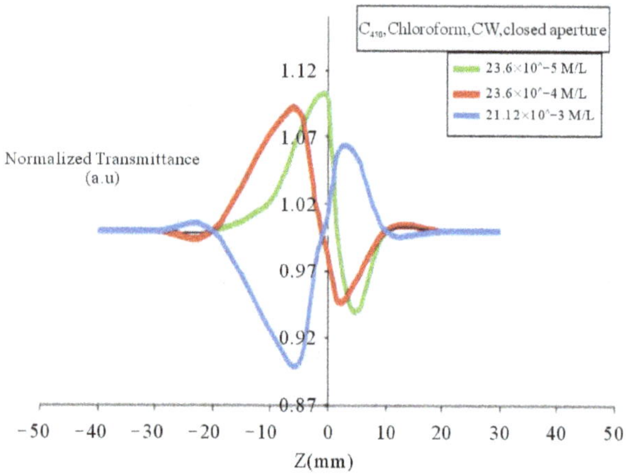

Figure 2. Closed aperture z-scan with CW laser for C_{450} in PMMA doped films in chloroform.

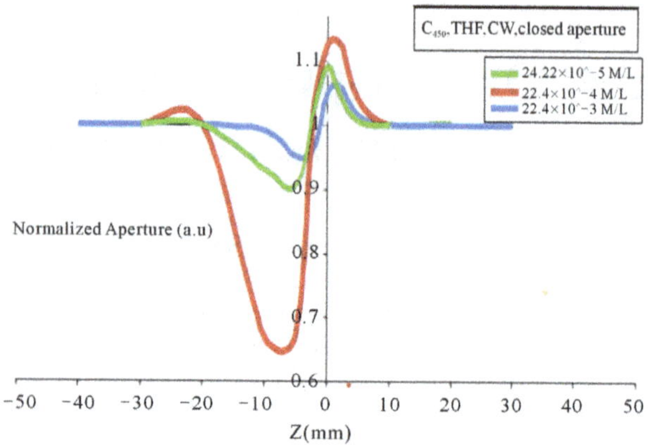

Figure 3. Closed aperture z-scan with CW laser for C_{450} in PMMA doped films in THF.

Figure 2 indicates negative refractive index for films of concentrations (23.6×10^{-5}, 23.5×10^{-4}) M/L and positive refractive index for film of concentration (21.12×10^{-3}) M/L.

Figure 3 indicates positive refractive index for the three concentrations.

It is clear from the figures above that the z-scan trace varies for each films as the dye concentration is changed and this is because the difference of dye concentration in each film this causes variation in films thickness which causes a variation in the nonlinear phase shift which in turn changes the value of the nonlinear refractive index and the third order electric susceptibility. It is seen that, under the condition of small phase shift, z-scan curves are approximately symmetric, where the increase and decrease of normalized intensity at the aperture are basically the same. z-scan curves broaden and their peak-to-valley separation increases as the nonlinear phase shift increases so that the z-scan curves becomes gradually asymmetric. This can be easily explained by the beam intensity distribution on the aperture plane.

Figure 3 and for film of C_{450} with concentration (22.4×10^{-4}) M/L in THF shows that the valley has amplitude which is much larger than its peak and this is because of the large phase shift in the laser beam. In the case of a large nonlinear phase shift, at the transmittance peak of the z-scan curves, the normalized intensity at the aperture changes rapidly and then arrives at zero. In contrast, the change of normalized intensity is slow at the transmittance valley of the z-scan curves. Accordingly, the change of the peak of z-scan is quicker than the valley due to the difference of the radial distribution of normalized intensity on the aperture plane, so it is seen that, the shape of the z-scan curves becomes gradually asymmetric and broaden. Moreover, the peak almost disappears while the valley still exists. Our results are in good agreement with the results of S.-Qi Chen *et al.* [15].

In a closed-aperture z-scan the nonlinear refractive index n_2 can be calculated from the following formula [15]:

$$n_2 = \Delta\Phi_0 / k I_0 L_{eff} \quad (1)$$

where

$$k = 2\pi/\lambda$$

$\Delta\Phi_0$: nonlinear phase shift;

λ is the wavelength of the beam.

L_{eff}: the effective length of the sample which can be determined from the following formula [16]

$$L_{eff} = \left(1 - e^{-\alpha_0 L} / \alpha_0\right) \quad (2)$$

where

L: the sample length;

α_0: linear absorption coefficient.

In Equation (1), I_0 is the intensity at the focal spot given by [17]

$$L_0 = 2P_{peak} / \pi\omega_0 \quad (3)$$

where

ω_0: the beam radius at the focal spot;

P_{peak}: the peak power given by [17]

$$P_{peak} = E/\Delta t \quad (4)$$

where

E: the energy of the pulsed laser;

Δt: the time duration.

The closed-aperture z-scan defines variable transmittance values, which used to determine the nonlinear phase shift $\Delta\Phi_0$ and the nonlinear refractive index n_2 using the above equations. This can be shown in **Table 1**, which represent the $\Delta\Phi_0$, n_2 and other linear and nonlinear parameters of the C_{450} doped with PMMA films in the two solvents with CW diode laser at 650 nm.

Figure 4 below represent the variation of the nonlinear refractive index with the nonlinear phase shift of C_{450} doped films in the two solvents.

It is clear from **Figure 4** that the relation between the nonlinear phase shift and the nonlinear refractive index is linearly increasing relation. As the dye concentration is increased, the film's thickness of the doped films is increased and it causes the increase the laser beam phase shift.

Films of C_{450} doped with THF shift laser beam are more than doped films in chloroform .This indicates: the different of the nonlinear response of each film causing a different third order nonlinear susceptibility and different nonlinear refractive index.

3.2. Closed Aperture Z-Scan with Pulsed Nd:YAG Laser

Closed aperture z-scan with pulsed Nd:YAG laser at 532 nm (second harmonic output of the Nd:YAG laser) with

Table 1. The linear and nonlinear parameters of the C_{450} doped with PMMA films in the solvents chloroform and THF by using closed aperture z-scan with CW diode laser at 650 nm.

C_{450} Doped Films in Chloroform			
C (M/L)	23.6×10^{-5}	23.5×10^{-4}	21.12×10^{-3}
α_0 cm^{-1}	4.4961	4.854	7.661
L_{eff} (cm)	0.019	0.0195	0.0194
T_{max}	1.099	1.163	1.111
T_{min}	0.879	0.929	0.852
ΔT	0.22	0.234	0.258
$\Delta\phi_o$	0.541	0.528	0.636
Re$\chi^{(3)}$ cm^2/watt	7.27×10^{-10}	7.573×10^{-10}	1.038×10^{-9}
n_2 type	−ve	−ve	+ve
n_2 cm^2/watt (CW)	1.157×10^{-6}	1.204×10^{-6}	1.337×10^{-6}
C_{450} Doped Films in THF			
C (M/L)	24.2×10^{-5}	22.4×10^{-4}	22.4×10^{-3}
α_0 cm^{-1}	4.831	5.355	5.049
L_{eff} (cm)	0.0175	0.0174	0.0176
T_{max}	1.087	1.167	1.106
T_{min}	0.917	0.439	0.864
ΔT	0.172	0.736	0.243
$\Delta\phi_o$	0.423	1.813	0.598
Re$\chi^{(3)}$ cm^2/watt	5.883×10^{-10}	2.65×10^{-9}	8.435×10^{-10}
n_2 type	+ve	+ve	+ve
n_2 cm^2/watt (CW)	9.925×10^{-7}	4.425×10^{-6}	1.383×10^{-6}

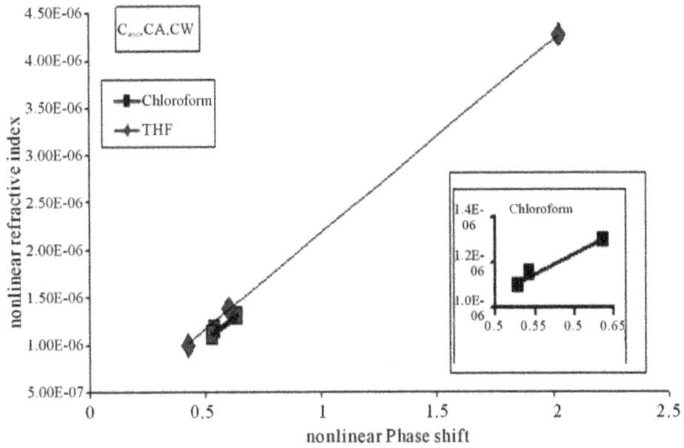

Figure 4. Variation of the nonlinear refractive index as a function of the phase shift of the laser beam for the dye C_{450} in PMMA doped films in the solvents chloroform and THF by using diode laser at 650 nm. The inset figures represent the variation of the nonlinear refractive index with nonlinear phase shift for the film of C_{450} in chloroform alone.

nanosecond laser pulses were used to investigate the nonlinear refractive index of the dye C_{450} in PMMA doped films in chloroform only. The following **Figures 5** and **6** show the closed aperture z-scan results:

Figure 5 shows z-scan for C_{450} in chloroform.

Figure 5 shows that the behavior of C_{450} doped films in chloroform of concentrations (23.6×10^{-5}, 21.12×10^{-3}) M/L, have negative refractive index while film of concentration (23.5×10^{-4}) M/L has positive refractive index. We noticed also that the peak of films with concentrations (23.6×10^{-5}, 23.5×10^{-4}) M/L was enhanced and the valley was suppressed and this is because of large nonlinear phase shift in these films.

Figure 6 shows the variation of the nonlinear refracttive index with the nonlinear phase shift of the dye C_{450} doped films in chloroform.

The first observation forms all the results of closed aperture z-scan by using CW and pulsed laser is that: the behavior of nonlinear refractive index n_2 changes between negative and positive behavior as the magnitude of laser beam nonlinear phase shift is changed. Nonlinear phase shift was changed with films thickness, the intensity at the focus and film's linear absorption coefficient. Since the dyes concentration in each film limits the value of the linear absorption coefficients α_0, so α_0 is another factor which limits the value of n_2, (**Table 2**). All these factors limit the amount of beam's energy which is transferred to each sample through the linear absorption. These indicate that the general shape of curve as well as the nonlinear refractive index sign is medium dependant. F. F. Alonso and coworkers consider the measured nonlinear refractive index as being only parametric in nature *i.e.*, "off-resonance", that is not involving a net change of population from the ground state to other excited states and the refractive index is positive for frequencies below resonance and negative at frequencies above it [18].

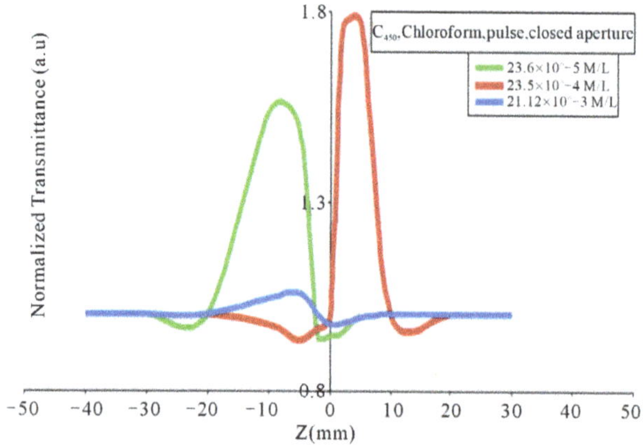

Figure 5. Closed aperture z-scan with pulsed Nd:YAG laser for C_{450} in PMMA doped films in chloroform.

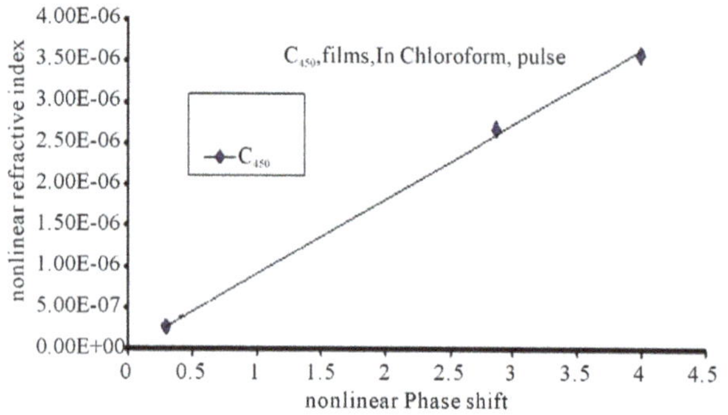

Figure 6. Variation of the nonlinear refractive index as a function of the phase shift of the laser beam for the dye C_{450} in PMMA doped films in chloroform, by using Nd:YAG pulsed laser.

The difference between the experimental results of CW and pulsed-laser indicates that the third order nonlinearities in the two cases have different origins. With CW pumping major the contribution for the observed third-order nonlinearities is to be thermal in nature. The energy from the focused laser beam is transferred to sample through linear absorption and is manifested in terms of heating the medium leading to a temperature gradient and there by the refractive index changes across the sample which then acts as a lens (Kerr effect). The phase of propagating beam will be distorted due to the presence of this thermal lens. Since low intensity of the CW diode laser rules out purely electronic contributions (253.97 watt/cm^2) compared with the intensity of pulsed laser (285.68×10^8 watt/cm^2), the thermal effect for CW laser should be considered, while the electronic polarization contributes to the change of the index of refraction in pulsed laser excitation so it was mainly electronic in origin.

We observed from the above tables also that the values of n_2 as a function of the dye films thickness and concentration are changed with the kind of used solvents where

Table 2. The linear and nonlinear parameters of the dye C_{450} in PMMA doped films in the solvent chloroform by using ND:YAG laser.

	C_{450} Doped Films in Chloroform		
C (M/L)	23.6×10^{-5}	23.5×10^{-4}	21.12×10^{-3}
α_0 cm^{-1}	4.96	4.854	7.66
L_{eff} (cm)	0.0189	0.0195	0.0199
T_{max}	2.073	2.537	1.088
T_{min}	0.909	0.914	0.97
ΔT	1.163	1.622	0.118
$\Delta\phi_0$	2.865	3.995	0.290
$Re\chi^{(3)}$ cm^2/watt	1.673×10^{-9}	2.25876×10^{-9}	2.074×10^{-10}
n_2 type	−ve	+ve	−ve
n_2 cm^2/watt (CW)	2.66×10^{-6}	3.592×10^{-6}	2.074×10^{-7}

the two solvents which we used are different in their dielectric constant, refractive index, vapor pressure and viscosity. The choice of solvent has a major influence on the film thickness: For solvents with low vapor pressure (*i.e.* the surrounding air will take up only a small amount of solvent before being saturated), the evaporation process will take longer than for solvents with a high vapor pressure, such process effects the films' thickness [19].

The kind of used solvent effects also the polymer chains where polymer chains dissolved in "good" solvents tend to have an open conformation, allowing easy access for chromophores to come into interchain contact. Polymer chains in "poor" solvents, on the other hand, tend to form tight coils, and make it difficult for chromophores to become physically adjacent even through the chains that tend to clump together [18,20,21]. For these reasons we notice a large difference in the nonlinear behavior of the films by changing the solvents.

3.3. Open Aperture Z-Scan with CW Diode Laser

Open aperture z-scan was used to investigate the nonlinear absorption coefficient by removing the aperture. This case corresponds to collecting all the transmitted light and therefore it is insensitive to any nonlinear beam distortion due to nonlinear refraction [22].

The coefficients of nonlinear absorption can be easily calculated from such transmittance curves [13]. The total transmittance is given by [16]:

$$T(z)=\sum_{m=0}^{\infty}\frac{\left[\frac{\beta I_0 L_{eff}}{1+(Z/Z_o)}\right]^m}{(m+1)^{3/2}} \quad (5)$$

where, Z: is the sample position at the minimum transmittance;

Z_o: the diffraction length;

m: integer;

$T(z)$: the minimum transmittance.

The two terms in the summation are generally sufficient to determine the nonlinear absorption coefficient β.

Figures 7 and **8** show the open aperture z-scan results for C_{450} doped films in the two solvents:

Figure 7 indicates that the behavior of C_{450} in PMMA doped films in chloroform: films of concentrations (23.6 × 10^{-5}) and (21.12 × 10^{-3}) M/L have saturation behavior while film of concentration (23.5 × 10^{-4}) M/L has TPA behavior.

Figure 8 indicates that the behavior of C_{450} in PMMA doped films in THF: film concentrations (24.2 × 10^{-5}) and (22.4 × 10^{-4}) M/L have saturation behavior while (22.4 × 10^{-3}) M/L has TPA behavior.

The open-aperture z-scan defines variable transmittance values, which used to determine β at CW Diode laser. This can be reported in **Table 3**.

3.4. Open Aperture Z-Scan with Nd:YAG Pulsed Laser

Figures 9 and **10** show the open aperture results of C_{450} doped films in PMMA, chloroform and THF.

Figure 9 indicates that the (TPA) behavior for the

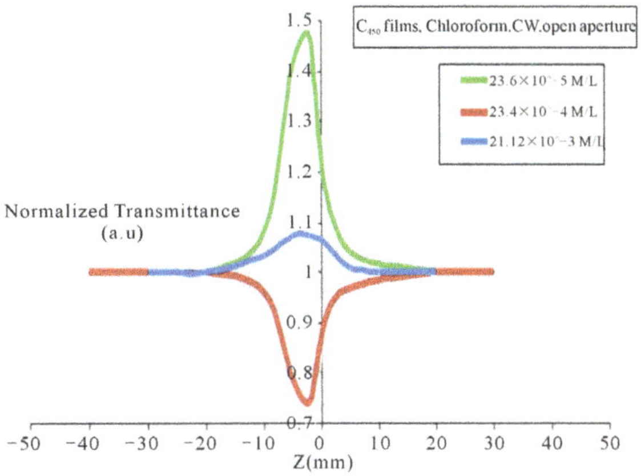

Figure 7. Open aperture z-scan for C_{450} doped in PMMA films in chloroform using CW diode laser.

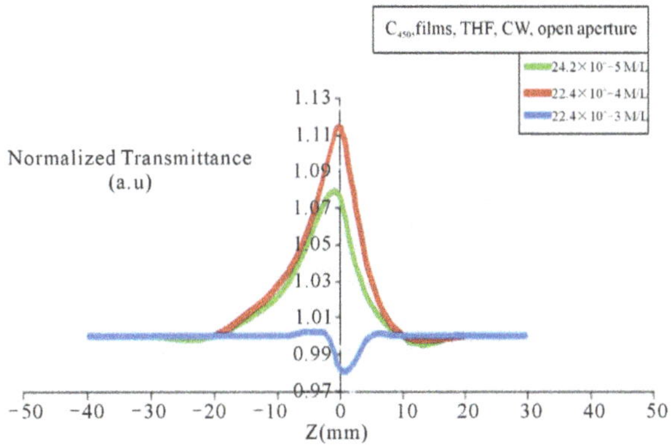

Figure 8. Open aperture z-scan for C_{450} in PMMA doped flms in THF using CW diode laser.

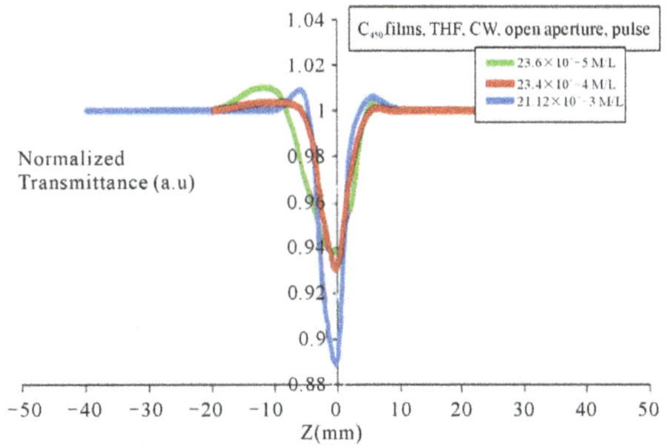

Figure 9. Open aperture z-scan for C_{450} doped with PMMA films in chloroform using Nd:YAG laser.

Table 3. Calculated of the nonlinear parameters of C_{450} doped films in the solvents chloroform and THF, by using open aperture z-scan with CW Diode laser.

C_{450} Doped Films in Chloroform			
C (M/L)	23.7×10^{-5}	23.4×10^{-4}	21.1×10^{-3}
L_{eff}. (cm)	0.019	0.0195	0.0194
The behavior	SA	TPA	SA
$Im\chi^{(3)}$ (cm/watt)	9.07×10^{-19}	1.95	6.68×10^{-21}
β (cm/watt), CW	2.153×10^{-19}	0.60	2.054×10^{-21}
C_{450} Doped Films in THF			
C (M/L)	24.2×10^{-5}	22.5×10^{-4}	22.4×10^{-3}
L_{eff}. (cm)	0.0175	0.0174	0.0176
The behavior	SA	SA	TPA
$Im\chi^{(3)}$ (cm/watt)	5.06×10^{-19}	2.29×10^{-19}	2.32
β (cm/watt), CW	1.65×10^{-19}	7.12×10^{-20}	0.74

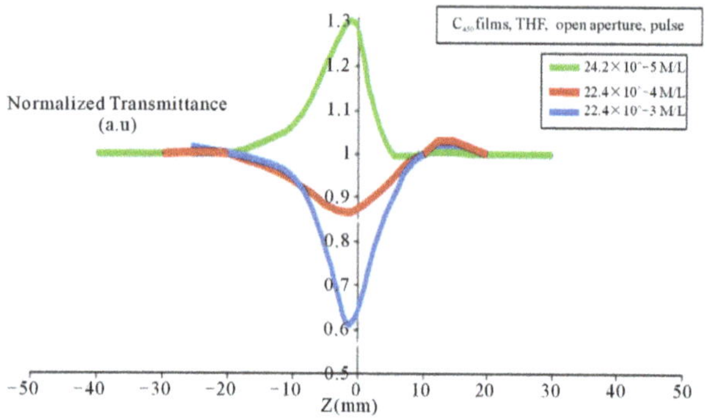

Figure 10. Open aperture z-scan for C_{450} doped with PMMA films in THF using Nd:YAG laser.

three concentrations.

Figure 10 indicates that the behavior of C_{450} in PMMA doped films in THF: film with concentration (24.22×10^{-5}) M/L has SA behavior while films with concentrations (22.5×10^{-4} and 22.4×10^{-3}) M/L have (TPA) behaviors.

In open-aperture, z-scan with Nd:YAG laser defines variable transmittance value. **Table 4** can be used to determine β.

We notice from the results above that: for open aperture z-scan, the nonlinear absorption behavior switched over between saturable absorption (SA) and reversed saturable absorption (TPA) for all the concentrations and the behavior is TPA for the films of concentration 10^{-3} M/L (high concentration) by using CW and pulsed laser.

TPA occrued when the excited state cross section is greater than the ground state cross section where enhancement of TPA is attributed to intermediate state resonance that makes the cross section of the excited

Table 4. Calculated nonlinear parameters of C_{450} doped films in chloroform and THF, by using open aperture z-scan with Nd:YAG laser.

C_{450} Doped Films in Chloroform			
C (M/L)	23.7×10^{-5}	23.5×10^{-4}	21.1×10^{-3}
L_{eff}. (cm)	0.019	0.0195	0.0194
The behavior	TPA	TPA	TPA
$Im\chi^{(3)}$ (cm/watt)	1.52×10^{-8}	1.54×10^{-8}	1.11×10^{-8}
β (cm/watt), CW	4.4×10^{-9}	4.55×10^{-9}	4.19×10^{-9}
C_{450} Doped Films in THF			
C (M/L)	24.2×10^{-5}	22.4×10^{-4}	22.4×10^{-3}
L_{eff}. (cm)	0.0175	0.0174	0.0176
The behavior	SA	TPA	TPA
$Im\chi^{(3)}$ (cm/watt)	1.52×10^{-26}	1.2238×10^{-8}	8.47453×10^{-9}
β (cm/watt), CW	6.05×10^{-27}	4.638×10^{-9}	3.28×10^{-9}

state greater than that of the ground state. Hence the intensity at the focus is increased and exhibited significant transmittance drop .While the behavior of the saturation absorption is due to the accumulation of molecules in the singlet excited state leading to depletion of the ground state, so the intensity at the focus is reduced and exhibited high transmittance. The reason for this behavior is the occurrence of a photon-assisted off-resonance energy transfer and that was demonstrated by L. C. Olivier and S. C. Zhilio [23].

β in case of TPA was calculated from Equation (5). It is clear from this equation that the factors which limit the nonlinear absorption coefficient are the samples L_{eff}, transmittance at $Z = 0$ and I_0 the peak intensity at the focus. For this we notice that the β's values and the imaginary part of the nonlinear susceptibility values measured by using Nd:YAG laser is too much lower than that measured by CW laser and that is because the intensity of the Nd:YAG laser (258.6×10^8 watt/cm^2) is too much greater compared with the CW laser intensity (253.97 watt/cm^2). In case of SA, β was calculated from equation below

$$\beta = \sigma_o / I_s \qquad (6)$$

where

σ_o: is the ground state cross section;

I_s: is the saturation intensity.

It is clear from the above results that the values of β and the imaginary part of the nonlinear susceptibility ($Im\chi^{(3)}$) for the dyes C_{450} in the two cases: TPA and SA are changed with the kind of used solvents.

Naga Srinivas and workers [24] attributed the switch over from SA to RSA behavior with increasing the dye concentration to localization of energy which leads to resonant TPA, and the change over from REA to SA is attributed to the aggregation and fast decay times in molecules which get populated through energy transfer.

4. Conclusion

The values of nonlinear refractive index coefficient for C_{450} doped PMMA thin films change with the kind of laser used for excitation, and that is because the different of the origin of the third-order nonlinearity caused by each kinds of laser

REFERENCES

[1] F. P. Schafer, "Dye Lasers," Springer-Verlag, Berlin, 1973.

[2] R. L. Sutherland, "Handbook of Nonlinear Optics," Marcel Dekker, New York, 1996.

[3] H. S. Nalwa and S. Miyata, "Nonlinear Optics of Organic Molecules and Polymers," CRC Press, Boca Raton, 1997.

[4] F. P. Schafer, "Organic Dyes in Laser Technology," *Angewandte Chemie International Edition in English*, Vol. 9, No. 1, 1970, pp. 9-25. doi:10.1002/anie.197000091

[5] W. Denk, J. H. Strickler and W. W. Webb, "Two-Photon Laser Scanning Fluorescence Microscopy," *Science*, Vol. 24, No. 4951, 1990, pp. 73-76. doi:10.1126/science.2321027

[6] M. Anandi, "Two-Photon Pumped Unconverted Lasing in Dye Doped Polymer Waveguides," *Applied Physics Letters*, Vol. 62, No. 26, 1993, pp. 3423-3425. doi:10.1063/1.109036

[7] G. S. He, C. F. Zhao, J. D. Bhawalkar and P. N. Prasad, "Two-Photon Pumped Cavity Lasing in Novel Dye Doped Bulk Matrix Rods,'' *Applied Physics Letters*, Vol. 67, No. 3703, 1995, pp. 3703-3705. doi:10.1063/1.115355

[8] J. E. Ehrilich, X. L. Wu, I. Y. S. Lee, Z. Y. Hu, H. Rockel, S. R. Marder and J. W. Perry, "Two-Photon Absorption and Broadband Optical Limiting with Bis-Donor Stilbenes," *Optics Letters*, Vol. 22, No. 24, 1997, pp. 1843-1845.

[9] G. S. He, G. Xu, P. N. Prasad, B. A. Reinhardt, J. C. Bhatt and A. G. Dillard, "Two-Photon Absorption of Novel Organic Compounds," *Optics Letters*, Vol. 20, No. 5, 1995, pp. 435-437. doi:10.1364/OL.20.000435

[10] D. A. Parthenopoulos and P. M. Rentzepis, "Three-Dimensional Optical Storage Memory," *Science*, Vol. 245, No. 4920, 1989, pp. 843-845. doi:10.1126/science.245.4920.843

[11] H. Strickler and W. W. Webb, "Three-Dimensional Optical Data Storage in Refractive Media by Two-Photon Excitation," *Optics Letters*, Vol. 16, No. 22, 1991, pp. 1780-1782. doi:10.1364/OL.16.001780

[12] B. H. Cumpston, S. P. Ananthavel, S. Barlow, D. L. Dyer, J. E. Ehrlich, L. L. Erskine, A. A. Heikal, S. M. Kuebler, I. Y. S. Lee, D. M. Maughon, J. Quin, H. Rockel, M. Rumi and J. W. Perry, "Two-Photon Polymerization Initiators for Three-Dimensionaloptical Data Storage and Microfabrication," *Science*, Vol. 398, No. 4, 1999, pp. 51-54.

[13] L. W. Tutt and T. F. Boggess, "A Review of Optical Limiting Mechanisms and Devices Using Organics, Fullerenes, Semiconductors and Other Materials," *Progress in Quantum Electronics*, Vol. 17, No. 4, 1993, pp. 299-338. doi:10.1016/0079-6727(93)90004-S

[14] S. VenugopalRao, N. K. M. Naga Srinivas and D. Narayana Rao, "Nonlinear Absorption and Excited State Dynamics in Rhodamine B Studied Using Z-Scan and Degenerate Four Wave Mixing Techniques," *Chemical Physics Letters*, Vol. 361, No. 5-6, 2002, pp. 439-445. doi:10.1016/S0009-2614(02)00928-4

[15] S.-Q. Chen, Z.-B. Liu, W.-P. Zang, J.-G. Tian, W.-Y. Zhou, F. Song and C.-P. Zhang, "Study on Z-Scan Characteristics for a Large Nonlinear Phase Shift," *Journal of the Optical Society of American B*, Vol. 22, No. 9, 2005, pp. 1911-1916. doi:10.1364/JOSAB.22.001911

[16] M. Sheik-Bahae, A. A. Said, T. H. Wei, D. J. Hagan and E. W. Van Stryland, "Sensitive Measurement of Optical Nonlinearities Using a Single Beam," *IEEE Journal of Quantum Electronics*, Vol. 26, No. 4, 1990, pp. 760-769. doi:10.1109/3.53394

[17] J. F. Ready, "Industrial Application of Lasers," 2nd Edition, Academic Press, San Diego, 1978.

[18] F. F. Alonso, P. Marovino, A. M. Paoletti, M. Righini and G. Rossi, "Third-Order Optical Non-Linearities in Titanium Bis-Phthalocyanine/Toluene Solutions," *Chemical Physics Letters*, Vol. 356, No. 5-6, 2002, pp. 607-613. doi:10.1016/S0009-2614(02)00428-1

[19] F. Pschenitzka, "Patterning Techniques for Polymer Light-Emitting Devices," Ph.D. Thesis, Princeton University, Princeton, 2002.

[20] A. Yavrian, T. V. Galstian and M. Piche, "Photoinduced Absorption and Refraction in Azo Dye Doped PMMA Films: The Aging Effect," *Optical Materials*, Vol. 26, No. 3, 2004, pp. 261-265.

[21] T. Q. Nguyena, R. Y. Yee and B. J. Schwartz, "Solution Processing of Conjugated Polymers: The Effects of Polymer Solubility on the Morphologyand Electronic Properties of Semiconducting Polymer Films," *Journal of Photochemistry and Photobiology A: Chemistry*, Vol. 144, No. 1, 2001, pp. 21-30. doi:10.1016/S1010-6030(01)00377-X

[22] M. Sheik-Bahae and M. P. Hassaibeak, "Handbook of Optics IV," McGraw-Hill, Boston, 2000.

[23] L. C. Oliveira and S. C. Zilio, "Chromium-Doped Saturable Absorbers Investigated by the Z-Scan," *Brazilian Journal of Physics*, Vol. 24, No. 2, 1994, pp. 498-501.

[24] N. K. M. Naga Srinivas, S. Venugopal Rao and D. Narayana Rao, "Saturable and Reverse Saturable Absorption of Rhodamine B in Methanol and Water," *Journal of Optical Society American B*, Vol. 20, No. 12, 2003, pp. 2470-2479. doi:10.1364/JOSAB.20.002470

2

Fiber Optic Displacement Sensor with New Reflectivity Compensation Method

Ansgar Wego[1*], Gundolf Geske[2]
[1]Department of Electrical Engineering, Hochschule Wismar, Wismar, Germany
[2]ASTECH Angewandte Sensortechnik GmbH, Rostock, Germany
Email: [*]ansgar.wego@hs-wismar.de

ABSTRACT

In this paper, a fiber optic displacement sensor with a new reflectivity compensation method is presented. The proposed compensation method is based on two light receiving channels with characteristic displacement sensitivities. The sensitivity characteristic for each channel is achieved by using fibers with different numerical apertures. The ratio of the intensity values of the two receiving channels is a function of the object displacement and fairly independent from the reflectivity of the measured object. The sensor is characterized by a well-defined measurement spot. By use of a focus lens mounted onto the fiber optics probe head, the object displacement range can be extended. The sensor is suitable for measurements with changing object reflectivity and demanding distance ranges.

Keywords: Fiber Optics; Displacement; Sensor; Reflectivity Compensation

1. Introduction

Fiber optic displacement sensors are known for decades. The simplest fiber optical sensor types use a single transmitting fiber for object illumination and a single receiving fiber for receiving the reflected light [1,2]. These types are not able to compensate for changes of the object reflectivity. So the practical use of these simple types is limited. Because of this limitation, the desire for fiber optic displacement sensors, which deliver a signal fairly independent of the object's surface reflectivity, in recent years, led to the development of several compensation methods.

A method described by Kissinger [3] exploits a measurement situation where the object performs periodic movements and where the mean gap between the sensor probe and the object remains constant. Due to the alternating sensor signal, the mean value can be eliminated by filter circuitry. The sensitivity of the sensor, which is independent from the object surface reflectivity, delivers a measure for the object displacement. The described method is limited to special applications and is therefore not of much interest.

Further known methods for reflectivity compensation of fiber optic displacement sensors are based on the difference of the displacement sensitivity of two separate light receiving channels. In all cases, the sensitivity difference of the receiving channels is achieved by a distinctive geometrical arrangement of the involved optical fibers in the probe head of the sensor [4-10]. For this class of reflectivity compensated sensors, the intensity ratio of the two channels is fairly independent from the object reflectivity and is therefore a measure of the probe displacement. The disadvantages of the geometrical methods are a complex fiber assembly as well as limitations in measurement ranges, in spot sizes and in spot characteristics.

2. Setup of the Sensor System

An overview of the sensor system with the new compensation method is shown in [11]. **Figure 1** illustrates the setup of the proposed fiber optic displacement sensor. The new method is based on one light sending channel and two light receiving channels too, as it has been described by Kissinger *et al*. But in contrast to these known methods, now light receiving fibers with different numerical apertures are used to achieve characteristic displacement intensity curves for the two light receiving channels. The preferred arrangement of the fibers in the probe head of the sensor now is "statistically mixed" (cf. **Figure 2**) in contrast to the previously known well-ordered geometrical arrangements.

[*]Corresponding author.

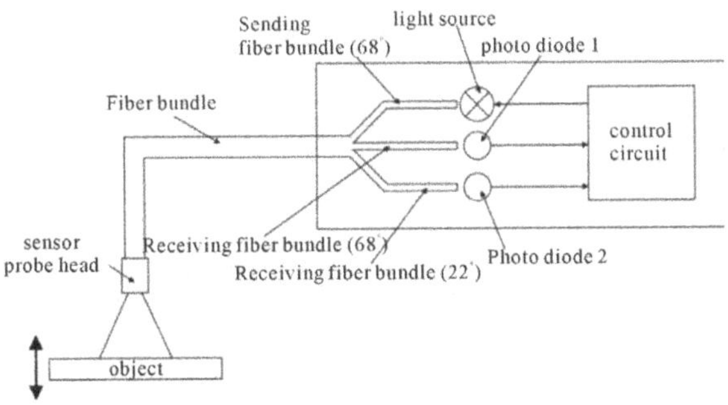

Figure 1. Reflectivity compensated fiber optical sensor.

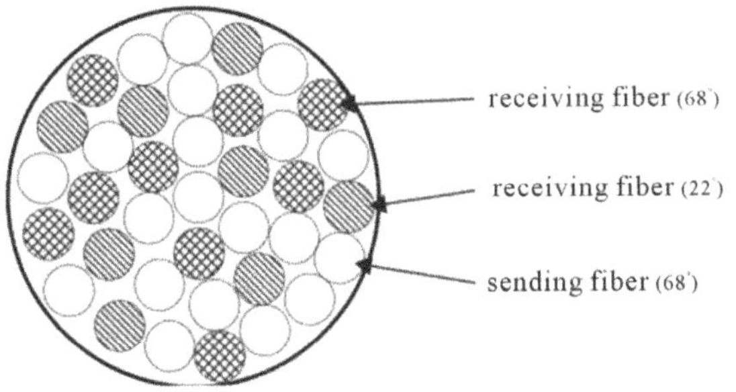

Figure 2. "Statistically mixed" fiber arrangement in the sensor probe head.

The "statistically mixed" arrangement of the fibers provides for a uniform light and measurement spot and compensates for slightly inhomogeneities of the object's surface. As a light source, a white light emitting diode was used to provide a broad band sending light. For the light detection, two equal silicon PIN photo diodes were employed as broad band receivers. The sensor system includes all necessary electronic circuitry for the analog and digital signal processing (*i.e.* transimpedance amplifiers, analog to digital converters, LED driver circuit, microcontroller, interface circuits, etc.).

3. Working Principle

3.1. System without Focus Lens

The numerical aperture *NA* of the fibers determines the opening angle α of the measurement light cone according to Equation (1).

$$\alpha = 2 \cdot \arcsin(NA) \tag{1}$$

Furthermore, the opening angle influences the intensity curves of the two light receiving channels. **Figure 3** shows the measured intensities of the two light receiving channels of the proposed sensor. The fibers of channel 1 and 2 have an opening angle of 22° and 68° respectively. The measurement was done for several objects with different reflectivity. **Figure 3** exemplarily shows the curves for a gray and a bluish object.

As it can be seen, the corresponding curves are almost identical. Thus, the ratio of the intensities is an explicit measure of the object displacement. The intensity curves were recorded as a function of the object displacement d in the range of (1 to 6) mm using a triangulation sensor for the distance measurement. The normalization of the curves was done for a displacement of $d = 1.6$ mm. This displacement defines the nominal working distance of the sensor. At this displacement, the absolute intensity of channel 2 is decreased by approximately 20% from its maximum (maximum is given at $d = 1$ mm). Clipping of the sensor signal in the range of (1 to 1.6) mm is thereby avoided. This normalization method is a tradeoff for enabling a maximum bidirectional displacement variation and a moderate intensity loss at the nominal working distance.

The sensitivity S of a receiving channel is given by the negative slope of the corresponding intensity curve according to Equation (2).

$$S = -\frac{\mathrm{d}I}{\mathrm{d}d} \tag{2}$$

The sensitivity curves of the two sensor channels in the displacement range of (1 to 6) mm are illustrated in **Figure 4**. As it can be seen, the sensitivity curves for the

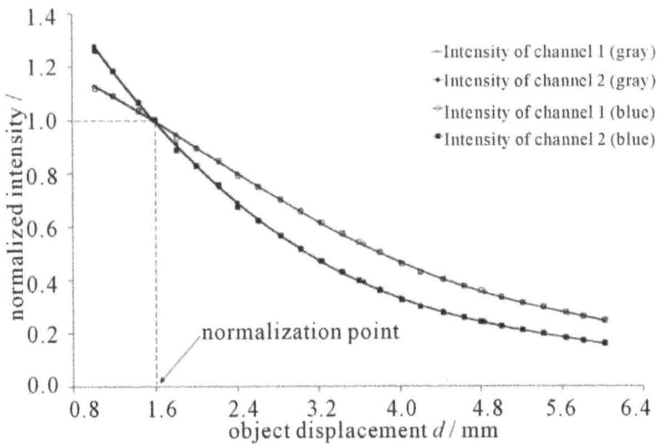

Figure 3. Normalized intensity curves vs. object displacement for different object reflectivities.

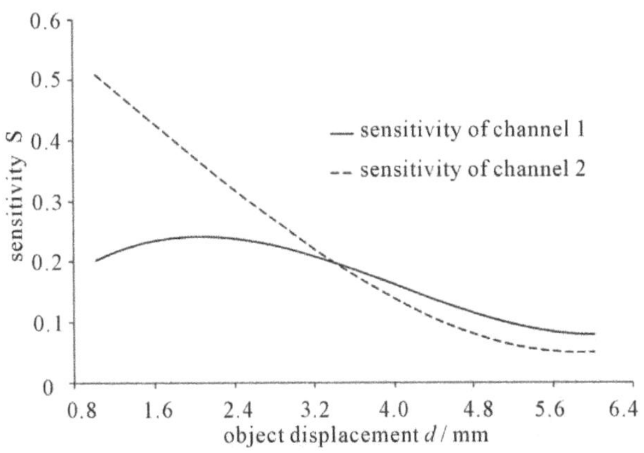

Figure 4. Sensitivity curves of the two sensor channels.

two channels are different. This is caused by the different opening angles of the used fibers for the two receiving channels. For such an arrangement, the ratio R of the intensity signals I_1 and I_2 of these two channels is a function of the displacement d of a reflective object (cf. Equation (3)).

$$R = \frac{I_1}{I_2} = f(d) \tag{3}$$

Because the sensitivity of a fiber optic displacement sensor is generally fairly independent from the surface reflectivity of the measurement object, the ratio R of the two intensity signals is independent from the reflective properties of the object too. The effect can be seen in **Figure 5**. The ratio R is shown in the range of (1 to 6) mm. The figure shows an almost linear range from (1 to 2.8) mm. But using a look-up-table (LUT) and a linear interpolation method, the non-linear range from approximately (2.8 to 6) mm may also be used to extend the working range. Alternatively to the LUT, a polynomial fitting function can be applied to linearize the sensor output.

3.2. System with Focus Lens

Figure 6 shows the arrangement with a focus lens. The lens is mounted on top of the sensor probe head and focuses the light of the fiber optics onto the object. By using a lens, the nominal working distance of the sensor can by extended significantly. For the used lens arrangement, the intensity curve of the sensor system has a distinctive maximum at 45 mm (cf. **Figure 7**).

In the proximity of the maximum, the sensitivity curves of the two receiving channels show a different trend. Thus, the ratio R of the two intensities again is a measure of the object displacement as described in the section above. **Figure 8** depicts the two intensity curves and their ratio as a function of object displacement d in the range of (40 to 56) mm. In this figure, the ratio values are scaled for a better presentation.

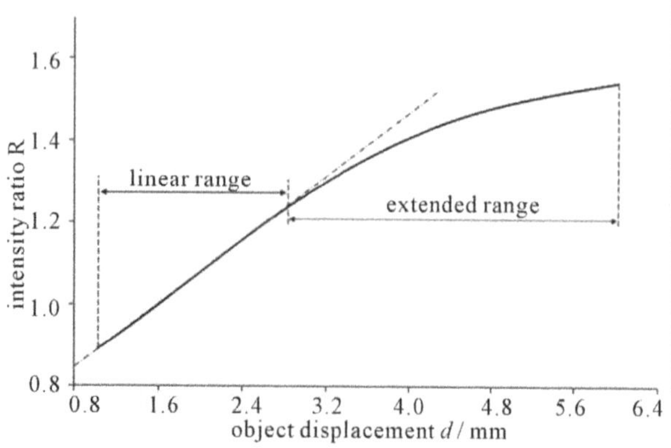

Figure 5. Intensity ratio vs. object displacement.

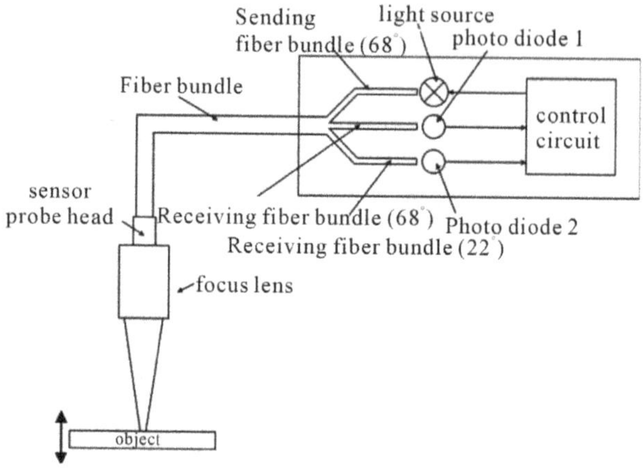

Figure 6. Sensor setup with mounted focus lens.

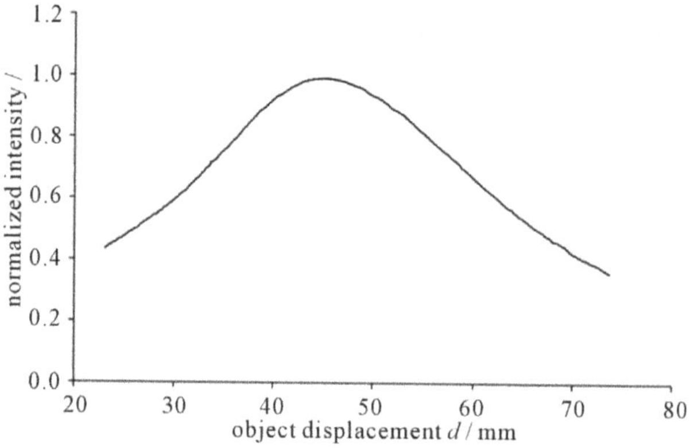

Figure 7. Normalized intensity vs. object displacement of the sensor with focus lens.

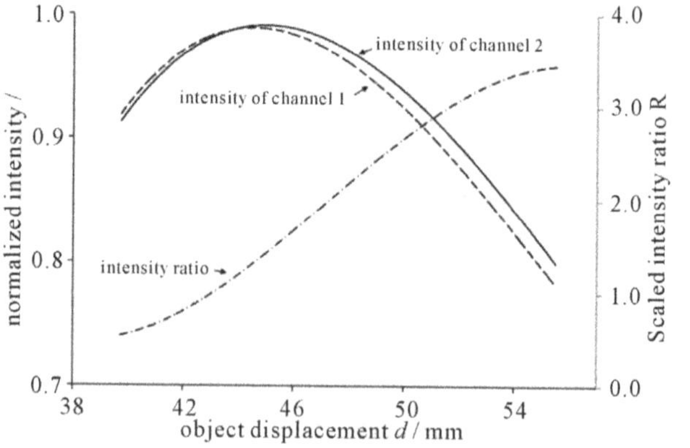

Figure 8. Normalized intensities and scaled intensity ratio vs. object displacement of the sensor with focus lens.

4. Conclusions

In comparison to the conventional methods, the proposed reflectivity compensation method for the fiber optic displacement sensor has several advantages. First, it is easy to implement and easy to handle. Due to the arrangement of the fibers in the sensor probe head, the light and measurement spot is homogeneous and well defined. The achievable measurement range of (1 to 6) mm is comparatively large and almost the half of the range is linear.

Second, by using a focus lens on top of the sensor probe head, the working range can be easily extended. This enables various applications with different object properties.

5. Acknowledgements

The authors would like to thank WINGS—Wismar International Graduation Services GmbH for support.

REFERENCES

[1] W. E. Frank, "Detection and Measurement Device Having a Small Flexible Fiber Transmission Line," US Patent No. 3273447, 1966.

[2] C. D. Kissinger, "Fiber Optic Proximity Probe," US Patent No. 3327584, 1967.

[3] C. D. Kissinger, "Fiber Optic Proximity Instrument Having Automatic Surface Reflectivity Compensation," US Patent No. 4247764, 1981.

[4] C. D. Kissinger and R. Dormann, "Reflectivity Compensating System for Fiber Optic Sensor Employing Dual Probes at a Fixed Gap Differential," US Patent No. 4488813, 1984

[5] L. Hoogenboom, "Fiber Optic Proximity Sensors for Narrow Targets with Reflectivity Compensation," US Patent No. 4701610, 1987.

[6] C. D. Kissinger, "Reflectivity Compensated Fiber Optic Sensor," US Patent No. 4701611, 1987.

[7] R. Hafle, "Fiber Optic Probe Sensor for Measuring Target Displacement," US Patent No. 5017772, 1991.

[8] F. Suganuma, A. Shimamoto and K. Tanaka, "Development of a Differential Optical-Fiber Displacement Sensor," *Applied Optics*, Vol. 38, No. 7, 1999, pp. 1103-1109. doi:10.1364/AO.38.001103

[9] X. Li, K. Nakamura and S. Ueha, "Reflectivity and Illuminating Power Compensation for Optical Fibre Vibrometer," *Measurement Science and Technology*, Vol. 15, No. 9, 2004, pp. 1773-1778. doi:10.1088/0957-0233/15/9/014

[10] M. L. Casalicchio, G. Perrone, D. Tosi, A. Vallan and A. Neri, "Non-Contact Low-Cost Fiber Distance Sensor with Compensation of Target Reflectivity," *IEEE International Instrumentation and Measurement Technology Conference*, Singapore, 5-7 May 2009, pp. 1671-1675. doi:10.1109/IMTC.2009.5168724

[11] A. Wego and G. Geske, "Reflexionskompensierter Faseroptischer Abstandssensor," *Photonik*, Vol. 17, No. 5, 2012, pp. 62-64.

3

Flexible Optical Waveguide Bent Loss Attenuation Effects Analysis and Modeling Application to an Intrinsic Optical Fiber Temperature Sensor

Mustapha Remouche, Francis Georges, Patrick Meyrueis
Laboratoire des Systèmes Photoniques, École Nationale Supérieure de Physique de Strasbourg,
Université de Strasbourg, Strasbourg, France
Email: remouche@lsp.u-strasbg.fr

ABSTRACT

The temperature dependence of the bending loss light energy in multimode optical fibers is reported and analyzed. The work described in this paper aims to extend an initial previous analysis concerning planar optical waveguides, light energy loss, to circular optical waveguides. The paper also presents a novel intrinsic fiber optic sensing device base on this study allowing to measure temperatures parameters. The simulation results are validated theoretically in the case of silica/silicone optical fiber. A comparison is done between results obtained with an optical fiber and the results obtained from the previous curved optical planar waveguide study. It is showed that the bending losses and the temperature measurement range depend on the curvature radius of an optical fiber or waveguide and the kind of the optical waveguide on which the sensing process is implemented.

Keywords: Optical Fiber; Losses; Curvature; Sensor; Temperature; Micro Technology

1. Introduction

A curvature effect is easily reached with optical fibers; therefore many laboratories have investigated the effects of curvatures on optical fiber measuring responses. Gratings implemented in optical fibers provide measuring performances related to bending that were also investigated [1-12]. Therefore, the power attenuation coefficient of bent fibers is one of the parameters that must be determined for using the fiber as a transducer.

In previous paper [1], we described a geometrical method to determine the local numerical aperture and the light output power attenuation with the bending of an optical waveguide. We showed that this method can be applied only when the optical waveguide is curved during the manufacturing at a temperature close to glass melting. In this case, the core and the cladding refractive index are modified by temperature effects independently of the curvature radius.

In this a previous work [1], we described a set of methods for using a flat optical waveguide as a transducer for measuring temperature by bending, in which only some transmitted light intensity effects are involved. We also investigated previously some photonic effects allowing using a multimode integrated optical waveguide as an intrinsic temperature sensor operating by light intensity modulation at the output of the sensor.

The purpose of this new work is to extend the analysis of the planar optical waveguide response to the temperature [1], to a circular optical waveguide.

An optical fiber is a good example of cylindrical optical waveguide. The optical fiber bending loss phenomena is used as a transduction effect in some types of intrinsic optical fiber sensors (temperature, displacement, strain…) [1,2,13-20].

The step-index optical fiber that we use is curved during its manufacturing at a high temperature. A geometrical modeling is used to describe the light propagation in the optical fiber and to determine the light power attenuation at the output of the fiber due to the fiber bending. In this case the geometrical approach is similar to the one presented for the planar waveguide sensing modeling.

2. Analysis If the Power Attenuation with the Bending of an Optical Fiber

The guidance of the core rays in a straigth step-index optical fiber is achieved by ensuring that the propagation angle θ, satisfies the condition: $0 \le \theta \le \theta_c$, where the critical angle, θ_c, and the critical angle, α_c, are given respectively, at room-temperature ($T_0 = 20^\circ\text{C}$) by:

$\theta_c = \cos^{-1}(n_2/n_1)$ and $\alpha_c = \sin^{-1}(n_2/n_1)$. The numerical aperture (*NA*) is given at T_0 by:

$NA = n_1 \sin\theta \le (n_1^2 - n_2^2)^{1/2}$. n_1 and n_2 are the core and the cladding refractive index respectively [1].

When the optical fiber is bent with a curvature radius *R* (**Figure 1**), the local numerical aperture at a given location of the optical fiber curved part will be changed. In a meridional-plane of the optical fiber, when the position angle at the beginning of the bend is $\phi = 0°$ or 180°, the optical fiber behavior becomes identical to the one an optical planar waveguide [1], and the local numerical aperture NA_1 is given, at T_0, by:

$$NA_l(R, \rho_0) = n_1 \sin\theta = n_1^2\left[1 - \frac{n_2^2}{n_1^2}\left[\frac{R+\rho}{R+\rho_0}\right]^2\right]^{1/2} \quad (1)$$

where ρ is the fiber core radius and $r(= R + \rho_0)$ is the abscissa on the input optical fiber aperture where the origin is "*O*". "ρ_0" will satisfy the relation: $-\rho \le \rho_0 \le \rho$. However, in all others optical fiber planes, when $\phi \ne 0°$ and 180°, the local numerical aperture, NA_l, can be calculated by using Equation (1), where the quantity $(R - \rho_0 \cdot \cos\phi)$ replaces $(R + \rho_0)$ and "ρ_0" to satisfy the relation: $0 \le \rho_0 \le \rho$. Finally, the local numerical aperture of a curved step-index optical fiber is given by:

$$NA_l(R, \rho_0, \phi) = n_1^2\left[1 - \frac{n_2^2}{n_1^2}\left[\frac{R+\rho}{R-\rho_0 \cdot \cos\phi}\right]^2\right]^{1/2} \quad (2)$$

where ϕ is required to satisfy the relation: $0 \le \phi \le 2\pi$.

Equation (2) becomes identical to equation: $NA = (n_1^2 - n_2^2)^{1/2}$ when *R* is infinite (straight optical fiber case) and becomes identical to Equation (1) when ϕ = 0° or 180° (bent step-index optical planar waveguide case). We used the silica/silicone step-index optical fiber with the following characteristics at T_0, for a wavelength λ = 633 nm: n_1 = 1.4570, n_2 = 1401, 2ρ = 200 µm and *NA* = 0.4000.

In **Figure 2**, we plot the local numerical aperture, at T_0 for various values of *R*, as a function of the position "ρ_0" with a value of ϕ. It is clearly seen that the local numerical aperture increase with the increase of the values of *R*

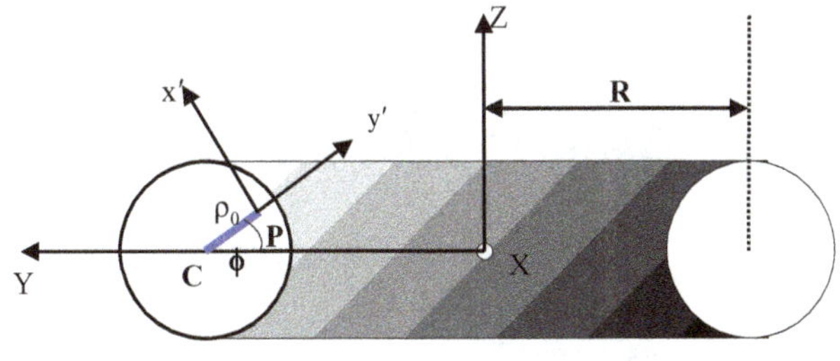

Figure 1. Schematic presentation of a section of a fiber bent with an arc of radius *R*.

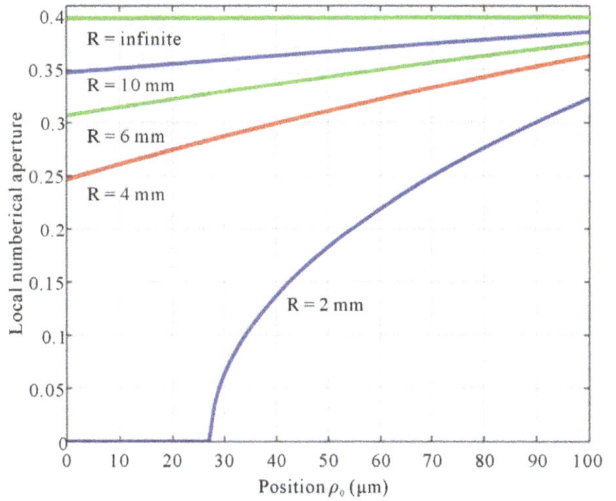

Figure 2. Local numerical aperture according to the position "ρ_0" for several values of *R* with ϕ = 135°.

and ρ_0. The maximum of the local numerical aperture is obtained when the angle ϕ is equal to 180°.

3. An Approach of the Description of the Light Propagation in a Bent Optical Fiber

The ray paths in the core of a step index fiber are straight lines, but the geometrical description is more complicated than in a planar waveguide, that we analyzed previously [1], due to the presence of the skew rays and of the depending on when the optical fiber is curved. In the bent planar waveguide [1], the angle of incidence of a ray on the outer core-cladding interface remains the same along a given ray path, and this property is true for all the ray paths. However, in the bent optical fiber, only the rays entering the bent part of the fiber in the meridional-plane containing this bent part behave in this manner. For the skew rays entering this plane, the subsequent reflections within the core do not follow a simple repeatable pattern because of the asymmetry introduced by bending the fiber. The complicated form of the ray invariants [3] and the differential equations describing the ray path offer no possibility of a simplification. We therefore propose a numerical technique for tracing each ray separately along the bent fiber.

In the **Figure 3**, we consider an arbitrary incident ray entering the bent section of the optical fiber from the cross section *X'X* at point "*P*" at the beginning of the bent part (ξ = 0). The bent part of optical fiber forms a torus portion by the revolution of the circular cross section about the origin "*O*" with a radius *R*. The axis system *OXYZ* has an axis *OX* pointing directly into the paper. A local coordinate system *ox'y'z'* with *oz'* pointing into the paper is also indicated at *P*. An arbitrary ray, incident at "*P*", has its direction cosines $(\cos\theta_{x'}, \cos\cos\alpha_i, \cos\cos\theta_{z'})$ relative to *ox'y'z'*. The

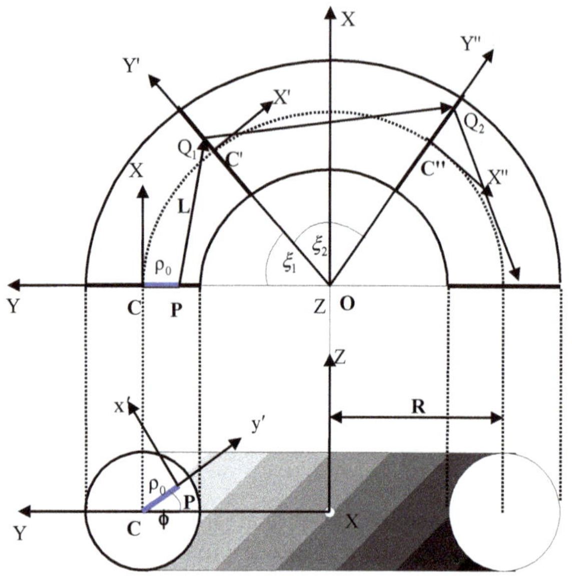

Figure 3. Schematic presentation of a section of a fiber bent with the curvature following an arc having a radius *R*.

vector ***OP*** is given by:

$$\boldsymbol{OP} = (0, R - \rho_0 \cos\phi, \rho_0 \sin\phi) \tag{3}$$

where the distance of the point "*P*" from the fiber center is ρ_0, if we let r be a point along the ray distant "*L*" from "*P*" then:

$$\begin{aligned} \boldsymbol{r} &= L\cos\cos\theta_{z'}\hat{X} \\ &+\left[R-\rho_0\cos\phi + L(\cos\cos\theta_{x'}\sin\phi - \cos\alpha_i\cos\phi)\right]\hat{Y} \\ &+\left[\rho_0\sin\phi + L(\cos\phi+\cos\alpha_i\sin\phi)\right]\hat{Z} \end{aligned} \tag{4}$$

If the ray meets the torus at "Q_1" then we can write $|\boldsymbol{L}| = \overline{PQ_1}$, where "$Q_1$" is the position of the first reflection. At "Q_1", we have:

$$\left|\boldsymbol{r} - R\sin\xi\hat{X} - R\cos\xi\hat{Y}\right| = \rho \tag{5}$$

And this gives the angular distance ξ around the axis of the bent optical fiber curvature:

$$\tan\xi = \boldsymbol{r}\hat{X}/\boldsymbol{r}\hat{Y} \tag{6}$$

ξ_1 corresponds to $\overline{POQ_1}$ angle. From Equation (6), we deduce an equation which will provide, as a function of the position "*L*", the intersection of the optical ray with the bent optical fiber core-cladding interface. The smallest real positive solution of the equation represents the distance "L_1".

From the point of the first reflection "Q_1" on the optical fiber torus, the optical ray is reflected to another point on the surface of the torus (core-cladding interface) around the bend.

After the first reflection all solutions for the distance "*L*" to the next reflection are given as solutions of the cubic problem. For obtaining the cubic equation, we must replace ρ_0 by ρ in the quadratic equation.

Some simple coordinates rotations and translations simplify the calculation of the incident and the reflected ray angles at each reflection point [2]. The reflected ray direction cosines at "Q_1" are given by multiplying a matrix transformation "*M*" by the direction cosines a "*P*" where:

$$\begin{bmatrix} \text{Direction} \\ \text{cosines} \\ \text{at } Q_1 \end{bmatrix} = [M]\cdot\begin{bmatrix} \cos\theta_{x'} \\ \cos\alpha_i \\ \cos\theta_{z'} \end{bmatrix} \tag{7}$$

With: (see Equation (8)).

A rotation around "Q_1" will bring the local x-axis tangential to the surface of the torus at Q_1. This rotation is ζ. All the others parameters were defined previously.

Having determined the geometry of the ray path, we can then calculate the fractional power loss at each reflection point along a given path by using the following equation:

$$P(\xi) = P(0)\exp(-\gamma\xi) \tag{9}$$

where γ is the attenuation coefficient of each ray, which varies from one reflection to the next one along the bent optical fiber, it is given by:

$$\gamma = \sum_{i=1}^{N} T_i \Big/ \sum_{i=1}^{N} \Delta\xi_i \tag{10}$$

where $\Delta\xi$ is the angular separation between two successive reflections and N is the total number of reflections. We can then use the Generalized Fresnel's Law to calculate the transmission coefficient T_i at each reflection point along a given path.

An algebraic expression for the transmission coefficient of the refracted rays ($V \ll 1$), is given by [3,11,12]:

$$T_i = \frac{4K_{y_1'}K_{y_2'}}{\left(K_{y_1'} + K_{y_2'}\right)^2} \tag{11}$$

where $K_{y_1'}$ and $K_{y_2'}$ are given by: $K_{y_1'} = k_1\cos\alpha_i$ and $K_{y_2'} = k_1\left(\sin^2\alpha_c - \sin^2\alpha_i\right)^{1/2}$ and: $k_1 = 2\pi n_1/\lambda$.

$$M = \begin{bmatrix} \cos\zeta\cos\phi + \sin\zeta\cos\xi\sin\phi & \sin\phi\cos\zeta - \sin\zeta\cos\xi\cos\phi & \sin\xi\sin\zeta \\ \sin\zeta\cos\phi - \cos\zeta\cos\xi\sin\phi & \sin\zeta\cos\phi + \cos\zeta\cos\xi\cos\phi & -\sin\xi\cos\zeta \\ -\sin\phi\sin\xi & \sin\xi\,\mathrm{coss}\,\phi & \cos\xi \end{bmatrix} \tag{8}$$

For tunneling rays when $V \gg 1$, the transmission coefficient at a reflection point of the radius ρ_c is given by:

$$T_i = \frac{4K_{y_1'}K_{y_2'}}{\left(K_{y_1'} + K_{y_2'}\right)^2}\exp\left\{-\frac{4}{3}V^{3/2}\right\} \tag{12}$$

where the parameter "V" is given by:

$$V = \frac{\rho_c}{2}\left(1 - \frac{\sin^2\alpha_c}{\sin^2\alpha_i}\right)\left(\frac{2k_1^2\sin^2\alpha_i}{\rho_c}\right) \tag{13}$$

The parameter ρ_c is the radius of curvature of the core-cladding interface in the incidence plane at "0". It is defined by the normal to the interface and the incident ray direction. It is given by [3,11,12]:

$$\rho_c = \frac{\rho_{x'}\rho_{z'}\sin^2\alpha_i}{\rho_{x'}\cos^2\theta_{z'} + \rho_{z'}\cos^2\theta_{x'}} \tag{14}$$

where $\rho_{x'} = \rho$ and $\rho_{z'} = \rho + R/\cos\phi$.

When $V \cong 0$, the transmission coefficient is given by:

$$T_i = 3.182\left(\frac{\lambda\cos^2\theta_c}{2\pi n_1\rho_c}\right)\frac{1}{\sin\theta_c} \tag{15}$$

But there are some optical rays that reach the interface with incidence angles close to θ_c.

Finally, the total intensity at the end of the angular length "ξ" of the bent part of the optical waveguide is found by a quadruple summation of Equation (9) across the cross-sectional area (r,ϕ) of the optical fiber at $X'X$ and the distribution of the ray angle (θ,ψ) [3,11,12]:

$$P_s(\xi) = P_0\int_0^{\varrho} r\mathrm{d}r\int_0^{2\pi}\mathrm{d}\psi\int_0^{\theta_c}\cos\theta\sin\theta\mathrm{d}\theta\int_0^{2\pi}\mathrm{d}\phi\exp(-\gamma\xi) \tag{16}$$

We plot in **Figure 4** the normalized power attenuation against the normalized distance (z/ρ) along the bent fiber axis for a step index optical fiber. Where "z" is the curvature length ($z = R\cdot\xi$). Each curve shows the same characteristics. Initially, there is the transition region, having a rapid power loss. This region dominated by the refracted rays and tunnel rays with the largest attenuation coefficient. The behavior of the bent part of the fiber is similar to the planar waveguide bent part [1], but the loss of power in the transition region is not as important, since only a few rays close the meridional plane have large losses.

In **Figure 5**, we plot the normalized power attenuation against normalized distance (z/ρ) along the bent planar waveguide and along the bent optical fiber for the same curvature radius (R = 2 mm). Beyond the transient region, the loss of power for a bent optical fiber case is higher than for the bent planar waveguide. This is because the skews rays can pass through the regions of high attenuation, although they may initially have a low attenuation.

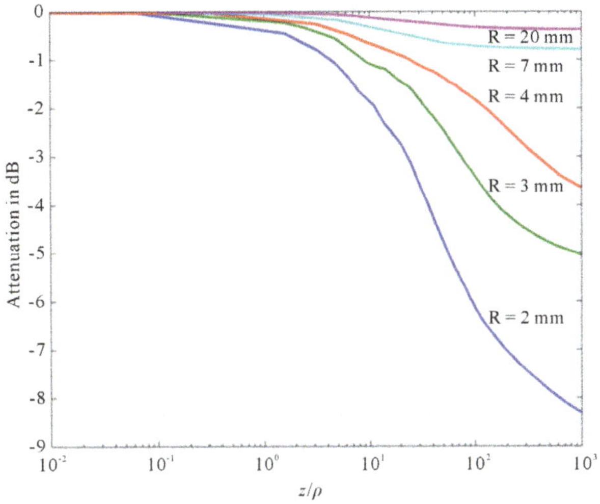

Figure 4. Intensity attenuation according to the ratio (z/ρ) in the step profile optical fiber for several values of R.

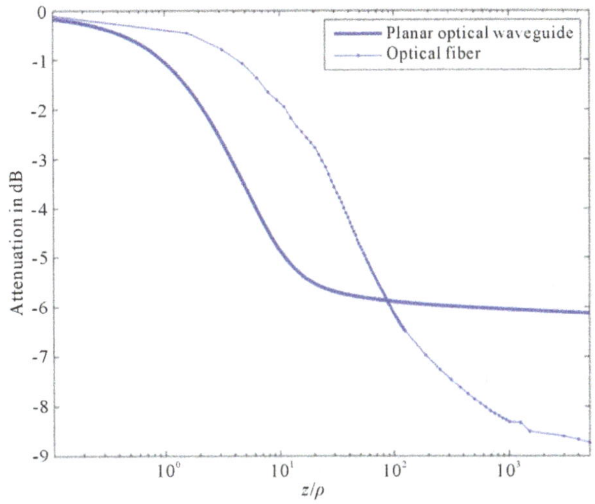

Figure 5. Intensity attenuation comparison between a bent planar waveguide and the bent optical fiber for the same curvature radius (R = 2 mm).

4. Temperatures Measurements Principle by Using a Bent Optical Fiber as a Transducer

The optical fiber sensor that we propose is an intrinsic optical fiber sensor; the sensitive element is the curved part of the fiber (**Figure 6**). This sensor principle is based on the variations of the sensing fiber output light intensity according to the fiber temperature variations. The light energy losses induced by the bending effect of the fiber will be compensated dynamically by the thermal effect; consequently, the output signal intensity is modulated only by the temperature variations.

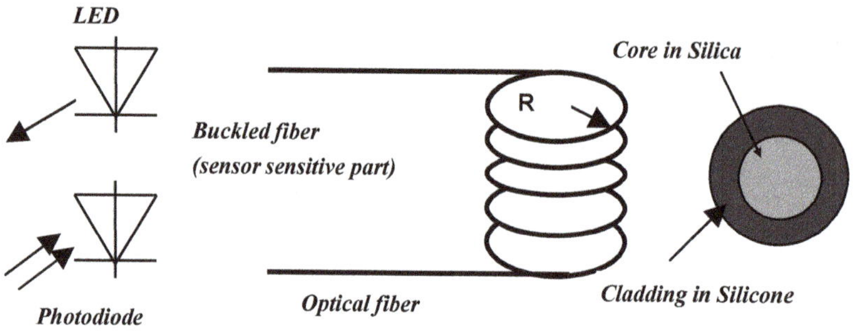

Figure 6. Temperature sensor general diagram.

5. The Temperature Effect on a Local Numerical Aperture inside a Bent Optical Fiber

For a given radius ($R = R_0$) of the optical fiber curvature and from Equation (2), the local numerical aperture inside this bent portion of the fiber according to the temperature is written in the following way:

$$NA_l\left(R_0,T,\rho_0,\phi\right) = n_1^2\left(R_0,T\right)\left[1-\frac{n_2^2\left(R_0,T\right)}{n_1^2\left(R_0,T\right)}\left[\frac{R_0+\rho}{R_0-\rho_0\cos\phi}\right]^2\right]^{1/2} \quad (17)$$

where the core refractive index and the cladding refractive index are written according to the temperature (T) in the following way [1]

$$\begin{aligned} n_1\left(R_0,T\right) &= n_1\left(T_0\right)+K_1\left(T-T_0\right), \\ n_2\left(R_0,T\right) &= n_2\left(T_0\right)+K_2\left(T-T_0\right) \end{aligned} \quad (18)$$

The coefficient $K_1 = \mathrm{d}n_1/\mathrm{d}T$ and $K_2 = \mathrm{d}n_2/\mathrm{d}T$ are respectively the thermo-optic coefficient of core the refractive index and the thermo-optic coefficient of cladding the refractive index. $K_1 = -3.78 \times 10^{-4}$/°C and $K_2 = 1.7744 \times 10^{-5}$/°C.

Silicone-based polymers possess a unique set of properties that makes them highly suitable for optical applications. The excellent thermal stability (−115°C to 260°C) allows this material to be useful for high temperature sensing applications [22-26].

In this optical fiber the small positive thermo-optic coefficient effect in inorganic glass waveguides used as the core is canceled out by using the negative thermo-optic coefficient of polymers used to constitute the cladding.

For an applied temperature T, we plot in **Figure 7** the local numerical apertures in an optical fiber according to the value of "ρ_0". We observe that the local numerical aperture increases when the applied temperature increases. Consequently, an optical ray unguided at room temperature becomes guided at temperature greater than T_0.

6. Bent Optical Fiber Temperature Response as a Sensor

In this section, we analyze the effect of the temperature variations on the light propagation in a curved optical fiber. We present the effects of the curvature radius on the optical fiber temperature response. For this analysis, the curvature radius R and the length of the bent part of the optical fiber (ξ) are given. The refractive index of the core and the cladding of the optical fiber depend on the temperature. The geometrical model is used to evaluate the light output power according to the temperature. The output power at the end of the transducer bent part of the fiber is given by:

$$\begin{aligned} P_s\left(R_0,T\right) = P_0\int_0^{\rho} r\mathrm{d}r\int_0^{2\pi}\mathrm{d}\psi\int_0^{\theta_c}\cos\theta\sin\theta\mathrm{d}\theta \\ \int_0^{2\pi}\mathrm{d}\phi\exp\left(-\gamma\left(R_0,T\right)\xi\right) \end{aligned} \quad (19)$$

For a given curvature length ($\xi = 2\pi R_0$), we plot in **Figure 8** the normalized output light intensity response of the sensor to the temperature variations for several values of R_0.

The response curve of the optical fiber operating as a

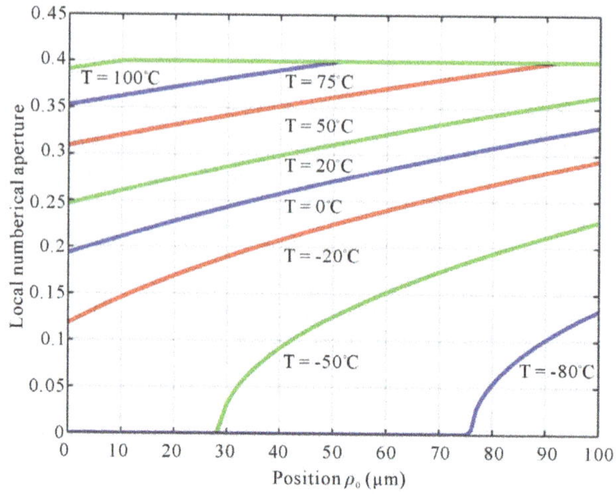

Figure 7. Local numerical aperture according to the position "ρ_0" for several applied temperature, with R_0 = 4 mm and ϕ = 135°.

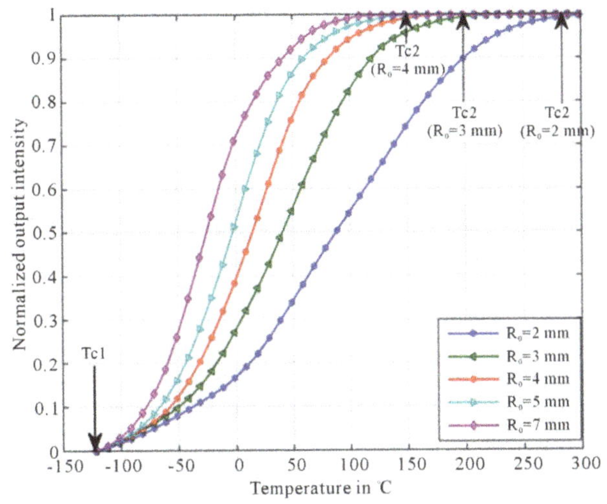

Figure 8. Transmission rate according to the temperature for various curvature radiuses.

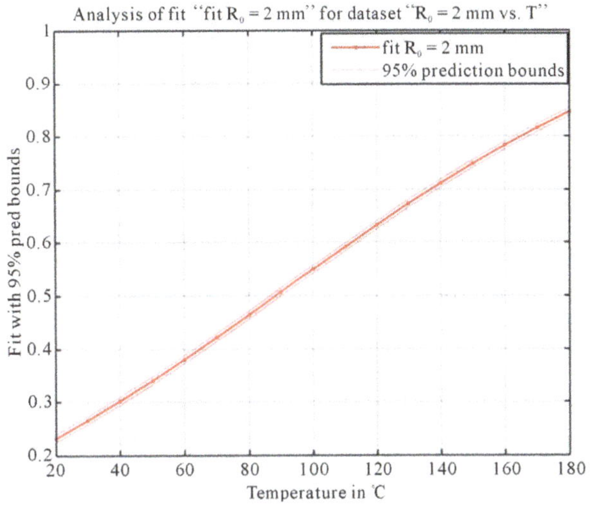

Figure 9. The temperature response curve fitting for R_0 = 2 mm between 20˚C and 180˚C.

temperature sensor that we propose in **Figure 8** is similar to the one of a planar waveguide temperature sensor [1]. For this first approach we have not considered errors measuring protocols. We note that the transmission rate (P/P_0) increases when the local numerical aperture is increasing up to a saturation level. Each curve is corresponding to the recovery of the losses induced by the refracted optical rays and the recovery of the losses induced by the tunnel effect, up to saturation. We can say that the temperature sensing range of the temperature optical fiber that we propose depends on the curvature radius of the optical fiber.

We deduce from **Figure 8** the following statements:

- The part of the response curve between $Tc1$ and T_0 is corresponding to intensity losses caused only by the temperature effect.
- The part of the response curve between T_0 and $Tc2$ is corresponding to intensity losses caused only by the optical fiber curvature.
- The linear region of the optical fiber temperature sensor response is increasing when the curvature radius of the bent fiber is decreasing.

The sensor sensitivity is given by:

$$S = \frac{\mathrm{d}P(R_0,T)}{\mathrm{d}T} \tag{20}$$

where $P(R_0,T)$ is the output intensity. We deduce from **Figure 8** that the sensitivity of the temperature sensor is *inversely* proportional to the curvature radius.

Figure 9 shows the *temperature response* curve *fitting* for R_0 = 2 mm. In the liner zone between 20˚C and 180˚C, this *sensor* has a *sensitivity* of 0.004˚C^{-1}.

7. Conclusion

The response to temperature variations on the bending light power loss of a multimode optical fiber with different bent fiber curvature radii have been analyzed. It has been found that the bending losses due to the internal optical fiber numerical aperture variations increase when the fiber bending angle increase. The more important losses are caused by refraction and tunnel effects. We have shown that a bent optical fiber can be used as a temperature transducer. The use of an optical fiber curved during its manufacturing at high temperature allows to minimize some residual mechanical effects, and allows to use rigorously the geometrical approach to describe the light propagation, to evaluate the losses in light power output values and to calibrate the temperature sensor. We have shown that if we use, for example a silica/silicone fiber as a transducer, we can obtain good performances with an excellent sensitivity and an excellent linearity associated to a large temperature measurement range, mainly because the thermo-optic effect value on the silicone is negative and important.

REFERENCES

[1] M. Remouche, R. Mokdad and A. Chakari, "Intrinsic Integrated Optical Temperature Sensor Based on Waveguide Bend Loss," *Optics & Laser Technology*, Vol. 39, No. 7, 2007, pp. 1454-1460. doi:10.1016/j.optlastec.2006.09.015

[2] M. Remouche, R. Mokdad and M. Lahrashe, "Intrinsic Optical Fiber Temperature Sensor Operating by Modulation of the Local Numerical Aperture," *Optical Engineering*, Vol. 46, No. 2, 2007, p. 024401. doi:10.1117/1.2709854

[3] A. W. Snyder and J. D. Love, "Optical Waveguide Theory," Chapman & Hall, Upper Saddle River, 1983.

[4] A. Zendehnam, M. Mirzaei1 and A. Farashiani, "Investigation of Bending Loss in a Single-Mode Optical Fibre,"

Pramana Journal of Physics-Indian Academy of Sciences, Vol. 74, No. 4, 2010, pp. 591-603.

[5] R. Ulrich, S. C. Rashleigh and W. Eickhoff, "Bending-Induced Birefringence in Single-Mode Fibers," *Optics Letters*, Vol. 5, No. 6, 1980, pp. 273-275. doi:10.1364/OL.5.000273

[6] E. A. J. Marcatilli, "Bends in Optical Dielectric Guides," *Bell System Technical Journal*, Vol. 48, No. 7, 1969, pp. 2103-2132.

[7] D. Gloge, "Bending Loss in Multimodes Fibers with Graded and Upgraded Core Index," *Applied Optics*, Vol. 11, No. 11, 1972, pp. 2187-2506. doi:10.1364/AO.11.002506

[8] D. Marcuse, "Influence of Curvature on the Losses of Doubly Clad Fibers," *Applied Optics*, Vol. 21, No. 23, 1982, pp. 4208-4213. doi:10.1364/AO.21.004208

[9] M. Semenkoff, "Bending Effect on Light Propagation in an Optical Fiber: Application to a Temperature Sensor," *Optics and Lasers in Engineering*, Vol. 17, No. 3-5, 1992, pp. 179-186. doi:10.1016/0143-8166(92)90035-6

[10] D. Marcuse, "Bending Losses of the Asymmetric Slab Waveguide," *Bell System Technical Journal*, Vol. 50, No. 8, 1969, pp. 2551-2563.

[11] A. W. Snyder and D. J. Mitchell, "Generalized Fresnel's Laws for Determining Radiation Loss from Optical Waveguides and Curved Dielectric Structures," *Optik*, Vol. 40, No. 4, 1974, pp. 438-459.

[12] A. W. Snyder and D. J. Mitchell, "Bending Losses of Multimode Optical Fibers," *Electron Letter*, Vol. 10, No. 1, 1974, pp. 11-13. doi:10.1049/el:19740008

[13] J. Dakin, "Optical Fiber Sensors," Artech House, Norwood, 1988.

[14] T. S. Yu Francis, "Fiber Optical Sensors," Marcel Dekker, Inc., New York, 2002.

[15] F. Pang; W. Liang and W. Xiang, "Temperature-Insensitivity Bending Sensor Based on Cladding-Mode Resonance of Special Optical Fiber," *IEEE Photonics Technology Letters*, Vol. 21, No. 2, 2009, pp. 76-78. doi:10.1109/LPT.2008.2008657

[16] R. P. Hu and X. G. Huang, "A Simple Fiber-Optic Flowmeter Based on Bending Loss," *IEEE Sensors Journal*, Vol. 9, No. 12, 2009, pp. 1952-1955. doi:10.1109/JSEN.2009.2031845

[17] R. M. Gavalis, P. Y. Wong and J. A. Eisenstein, "Localized Active-Cladding Optical Fiber Bend Sensor," *Optical Engineering*, Vol. 49, No. 6, 2010, p. 064401. doi:10.1117/1.3449110

[18] X. Chen, C. Zhang and D. J. Webb, "Bragg Grating in a Polymer Optical Fibre for Strain, Bend and Temperature Sensing," *Measurement Science and Technology*, Vol. 21, No. 9, 2010, p. 094005.

[19] J. H. Kuang, P. C. Chen and Y. C. Chen, "Plastic Optical Fiber Displacement Sensor Based on Dual Cycling Bending," *Sensors*, Vol. 10, 2010, pp. 10198-10210. doi:10.3390/s101110198

[20] J. Zhang, H. Liu and X. Wu, "Curvature Optical Fiber Sensor by Using Bend Enhanced Method," *Frontiers of Optoelectronics in China*, Vol. 2, No. 2, 2009, pp. 204-209. doi:10.1007/s12200-009-0032-x

[21] N. singgh, V. Mishra and S. C. Jain, "Enhanced Sensitivity Refractive Index Sensor Based on Segmented Fiber with Bending," *Indian Journal of Pure & Applied Physics*, Vol. 47, No. 9, 2009, pp. 655-657.

[22] Z. Zhang, G. Z. Xiao and P. Zhao, "Planar Wave Guide-Based Silica-Polymer Hybrid Variable Optical Attenuator and Its Associated Polymers," *Applied Optics*, Vol. 44, No. 20, 2005, pp. 2402-2408. doi:10.1364/AO.44.002402

[23] H. Ma, A. K.-Y. Jen and L. R. Dalton, "Polymer-Based Optical Waveguides: Material, Processing, and Devices," Adv Mater, Vol. 14, No. 19, 2002, pp. 1339-1365.

[24] M. Zhou, "Low-Loss Polymeric Material for Waveguide Components in Fiber Optical Telecommunication," *Optical Engineering*, Vol. 41, No. 7, 2002, pp. 1631-1643. doi:10.1117/1.1481895

[25] L. Eldada, "Advance in Polymer Integrated Optics," *IEEE Journal of Selected Topics in Quantum Electronics*, Vol. 6, No. 1, 2000, pp. 54-68.

[26] T. Watanabe, N. Ooba and S. Hayashida, "Polymeric Optical Waveguide Circuits Formed Using Silicone Resin," *Journal of Lightwave Technology*, Vol. 16, No. 6, 1998, pp. 1049-1055. doi:10.1109/50.681462

4

Behavioral Variations of Gain and NF Owing to Configurations and Pumping Powers

Belloui Bouzid
Hafr Al-Batin Community College (HBCC), King Fahd University of Petroleum and Minerals (KFUPM), Dhahran, Saudi Arabia
Email: Bouzid@hbcc.edu.sa

ABSTRACT

Six configurations are proposed in this paper to explore the gain and noise figure (NF) variations under the pumping power effect. I propose a new investigation of gain and NF at different EDFA configurations. Configurations such as SPSS, DPSS, DPSSF, TPDS, TPDSF, and QPDSF are designed, investigated and compared. A continuous progress of gain values is observed from SPSS to QPDSF, and a change of NF values related to configurations is recorded. The NF variations show different behaviors at different configurations. High gain of 59.49 dB and low NF value of 4.22 dB are recorded for the QPDSF configuration and low gain and low NF are recorded for the SPSS configuration.

Keywords: Double Pass; Single Pass; Erbium Doped Fiber; Configuration; Pumping Power

1. Introduction

EDFA is a crucial milestone in optical communication systems and wide growing internet. The progressive development of EDFA since its early stage at the end of eightieth is showing a continuous progress at the two levels of EDFA knowledge, theoretical and experimental. EDFA is an important part in the long haul optical fiber communication, and it is considered as another interesting research topic for laser phenomena of spontaneous and stimulated emission, where interaction of matter-light-matter still in its early stage.

Fiber to the home will guide the future changes of communication in the near future. Laser with the stimulated and spontaneous emission are considered to be the main factor in the next quantum revolution. Let us think how terabits transmission can be carried out without fiber optics or fiber amplifier? How the narrow band of electronics can be handling the wide broad band of communication systems? At the beneath of the laser phenomena there are many promising future development and new discoveries that can be conditionally achieved. Optical amplifier and lasers are used nearly in all the wide spectrum of science such as medicine, military, education and manufacturing.

Research in optical communication systems is growing daily. In addition, the obligation for efficient systems to fulfill the practical need of communication for high capacity and high speed is extremely demanded due to the fast growing of high speed and high capacity transmission. Highly efficient and stable EDFAs is playing a milestone role in this crucial era.

The lack of entire description of behavioral study of gain and NF at different configurations in the published papers will affect the fast growing of optical amplifier and reduce its effectiveness. Most of the published papers are focusing on NF and efficiency [1], gain and gain flattening [2-5], and optimization of pumping power [6] without any research papers which, focuses on the description of the complete behavioral variations and trends of EDFA gain and NF at different configurations.

All factors affecting the EDFA are needed to be elaborated and investigated to enhance the amplification outputs. Focusing on improvement of gain and noise figure, the devised and optimized configuration is to establish a wide and flat EDFA gain [2-5], and to discover an efficient EDFA at all levels. The obligation for an efficient EDFA is to fulfill the practical need of communication, with high capacity and high speed.

The EDFA description at different configurations is rarely found in the published papers [7], and the use of the entire physical phenomena to describe, illustrate, and interpret the variations of gain and NF at different configurations is highly required to understand their trends at large-scale.

In this paper general descriptions with illustrations and analysis are performed based on the studies of six different configurations: single pass single stage (SPSS), double pass single stage (DPSS), double pass single stage with filter (DPSSF), triple pass double stage (TPDS), triple pass double stage with filter (TPDSF) and quadruple pass double stage with filter (QPDSF) [8].

2. Experiment Setup and Discussion of Results

The used erbium-doped fiber in this experiment is characterized by: NA of 0.27 cutoff wavelength of 840 nm, peak absorption at the signal 1527 nm wavelength of 6 dB/m, erbium concentration of 440 ppm, and Er^{3+} core doped in silica/germania. The New port, tunable band pass filter (TBF) is mechanically tuned with a pass-band of 1 nm, an insertion loss of 1.5 dB at the tuned wavelength, and a tuning range limited to 45 nm from 1520 to 1565 nm. The Tunable Laser Source (TLS) is a continuous wavelength source of the 1550 nm input signal power, and the wavelength division multiplexer (WDM) is to merge both 980 nm pump and 1550 nm signal in EDF. The signal will be reacted with stimulated emission and amplified spontaneous emission (ASE) where at the output the ASE and the stimulated signal will be displayed. The filter will eliminate almost all ASE; the OSA displays the amplified signal with small portion of ASE at the bottom of the signal.

The six configurations were shown in **Figure 1(a)** SPSS: single pass single stage, **Figure 1(b)** DPSS: dou- ble pass single stage, **Figure 1(c)** DPSSF: double pass single stage with filter, **Figure 1(d)** TPDS: triple pass double

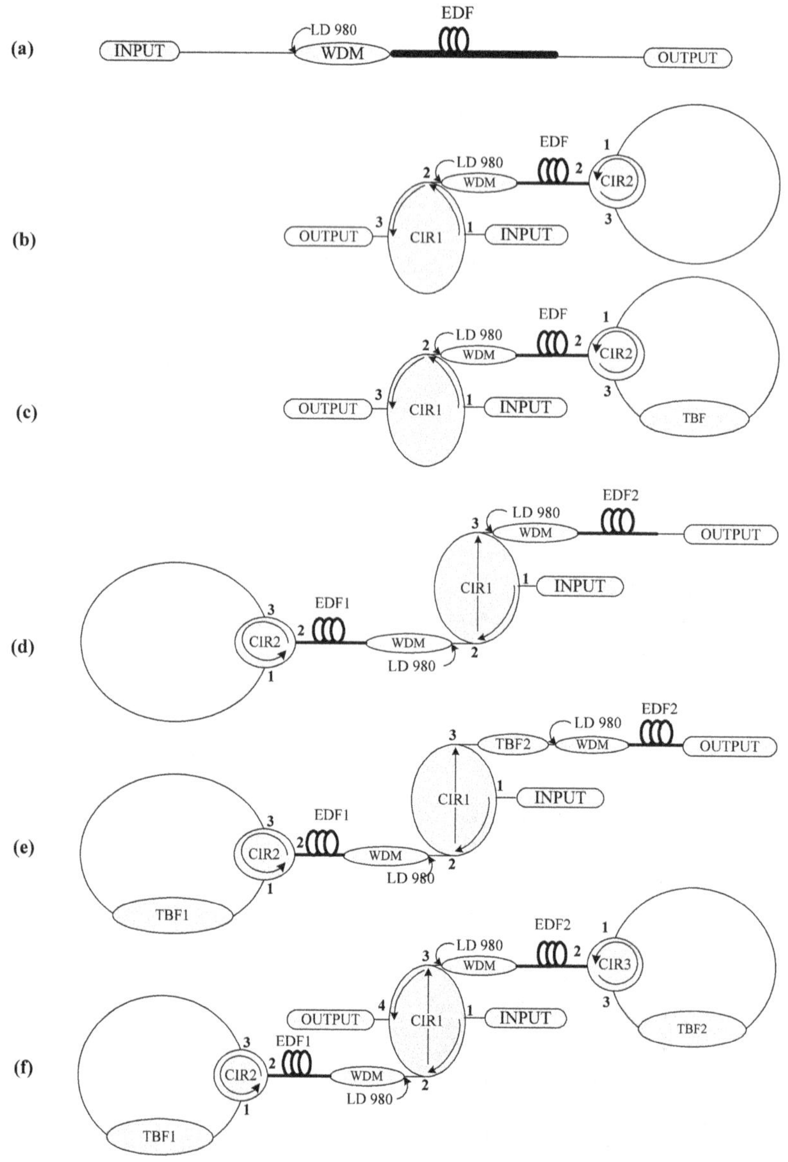

Figure 1. Experimental configurations of EDFA: (a) single pass single stage (SPSS); (b) double pass single stage (DPSS); (c) double pass single stage with filter (DPSSF); (d) triple pass double stage (TPDS); (e) triple pass double stage with filter (TPDSF) and (f) quadruple pass double stage with filter (QPDSF). TBF: tunable bandpass filter, CIR: circulator, EDF: erbium-doped Fiber, LD: laser diode, and WDM: wavelength division multiplexing, INPUT: tunable laser source, and OUTPUT: optical spectrum analyzer.

stage, **Figure 1(d)** TPDSF: triple pass double stage with filter, and QPDSF: quadruple pass double stage with filter. The difference between these configurations is owing to the additions of TBF and the second stage which can be single pass or double pass. The circulators are used as loop back where port1 and 3 are spliced and the TBF is incorporated between these ports to suppress and eliminate the unwanted ASE. The Key role of TBF in this continuous increase of gain is impressive and crucial where stimulated emission will strongly amplify the signal.

Due the large number of configurations used in these experiments, we will describe only the signal in the QPDSF configuration. The configuration in **Figure 1(f)** showed that each turn consistently gave a signal attenuation of 12 dB, this loss is due to three circulators, two TBF filters, and two WDMs. Thus, the amplified signal will propagate through the CIR1 from port1 to port2 then travel through EDF1; the signal will be affected by the first amplification from EDF1, through port2 into port3 of CIR2, passing through the first TBF1 filter into port1 and back to port2 to be amplified during the second pass by EDF1 into port2 of CIR1, and therefore will propagate again in the second stage through EDF2, CIR3, and TBF2 for the third and fourth passes. The output signal power was displayed through the OSA from port4 of CIR1. Traveling from port1 to port4 of CIR1, the signal will be affected by four amplifications during the four passes, or, as we mentioned, the quadruple pass double stage with filter configuration [8].

Figure 2 shows gain on dB versus configurations and pumping power. The input signal power is at 1550 nm wavelength and –50 dBm only the SPSS. From the 3D graph, all the six configurations show an increase of gain at the changing of configurations from SPSS, DPSS, DPSSF, TPDSF and QPDSF. Except for the TPDS without filter, which shows a lower gain compared to DPSSF and TPDSF.

It can be seen clearly, by following the variations of gain values versus configurations, at lower pump power of 10 mW, the gain is varied at different configurations where a shift between 9.65 and 45 dB of the gain values is recorded for SPSS and QPDSF respectively. At higher pumping power of 90 mW the gain is shifted owing to configurations change from 20.04 to 59.49 dB. All these results are at 1550 nm input signal power and –50 dBm except the SPSS. This good result shows clearly the varied configurations, the filter, and the double pass impact on the gain values. So, with the change of configuration from SPSS to QPDSF, the gain is increased to 45 dB at low pumping power. The gain gap between the SPSS and QPDSF reach to 39.45 dB at high pump power.

Go in details for the shown results in the figure. The SPSS records the lowest gain where the QPDSF records the highest one. At this level of explanation, the effect of the configurations type is very crucial where the gain gap between the SPSS and QPDSF reach approximately 35.83 dB at 10 mW pump power. It is evident that the effect of the pumping power and configuration with filter are effective for the gain enhancement. It is also observed, that the DPSSF has a higher gain compared with TPDS where single stage has higher gain compared to double stage. I think, the role of filter in the design is certainly crucial and principal in this reversed phenomenon. In this case it is recorded that adding single stage single pass to the double pass configuration will reduce the gain.

Figure 3 shows experimental NF versus configura-

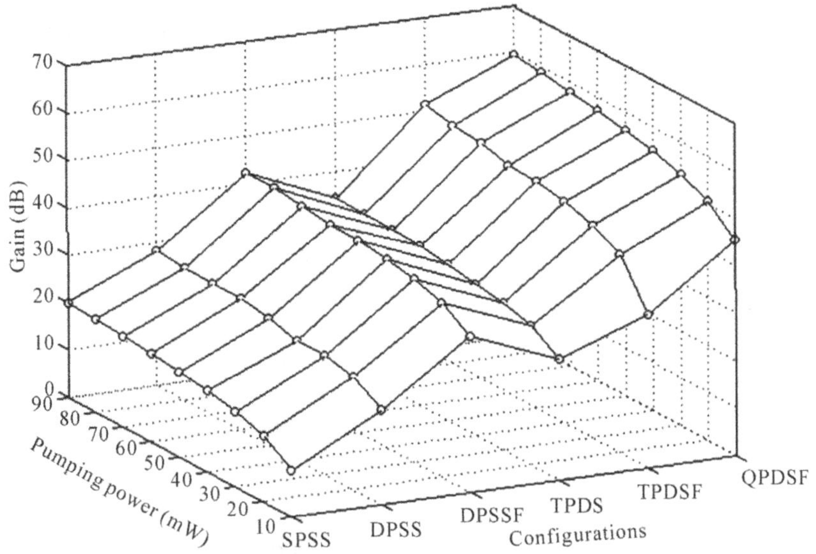

Figure 2. Experimental gain versus configurations and pumping power at 1550 nm wavelength at –50 dBm input signal power. SPSS: single pass single stage, DPSS: double pass single stage, DPSSF: double pass single stage with filter, TPDS: triple pass double stage, TPDSF: triple pass double stage with filter, and QPDSF: quadruple pass double stage with filter.

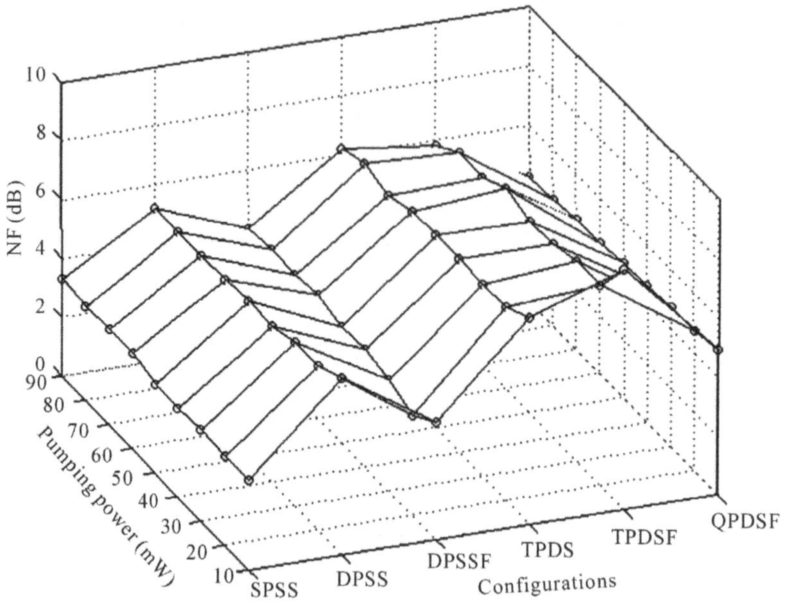

Figure 3. Experimental NF versus configurations and pumping power at 1550 nm wavelength at −50 dBm input signal power. SPSS: single pass single stage, DPSS: double pass single stage, DPSSF: double pass single stage with filter, TPDS: triple pass double stage, TPDSF: triple pass double stage with filter, and QPDSF: quadruple pass double stage with filter.

Table 1. Gain and NF variations due to configuration.

Configurations	Gain	NF
SPSS	LOW	LOWEST
DPSS	LOW	HIGH
DPSSF	HIGH	LOW
TPDS	LOW	HIGH
TPDSF	HIGH	HIGH
QPDSF	HIGHEST	LOW

tions and pumping power at input signal power of 1550 nm wavelengths and −50 dBm except SPSS. It is evident that the NF is decreased at specific configurations and increased at others. The 3D graph shows clearly that the NF is decreased at three configurations. It is also evident that the NF is bellow 5 dB for the SPSS, DPSSF, and QPDSF configurations, and above 5 dB for the DPSS, TPDS, and TPDSF configurations. The TPDS and TPDSF show the highest NF compared with the other configurations.

It can be noted the effect of adding the single pass stage to the double pass on increasing of noise figure. In the double stage double pass the TBF filter is affecting the QPDSF and the DPSSF and enhance their NF. The configuration SPSS records the lowest NF with the lowest gain, QPDSF records the highest gain. This can be related to the multi factors that have been included such as: double pass, double stage, and the filter positioned between the two ports of the circulator. It is clearly shown in this paper, that the configuration structure and the filter are crucial in the variation of the gain and NF. In particular, high gain can be generated simply with simple modification related to the design structure. QPDSF records 59.49 dB gain and 4.22 dB NF. This result can be increased higher with fifth or sixth passes and higher pump power. **Table 1** shows the description and the comparison of gain and NF at different configurations.

3. Conclusion

The configurations SPSS, DPSS, DPSSF, TPDS, TPDSF, and QPDSF were exposed to the same conditions of input signal power wavelength and different pumping power. The full investigation of gain and NF shows high gain of 59.49 dB and low NF of 4.22 dB values of QPDSF configuration. Gain and NF were profoundly described and clearly investigated using both 3D and 2D graphs display. A continuous increasing of gain values was demonstrated. The gain progress was recorded from SPSS, DPSS, DPSSF, TPDS, TPDSF, to QPDSF configurations. A variation of NF related to different configurations was proved and demonstrated. The TBF positioned in the between the circulator ports, configurations structure, and pumping power are a principal factors for gain and NF control.

4. Acknowledgments

The author wishes to acknowledge UPM, MMU (Malaysia) and HBCC/KFUPM (Saudi Arabia) for their support in providing the various facilities utilized in the presentation of this paper.

REFERENCES

[1] A. C. Çokrak and A. Altuncu, "Gain and Noise Figure Performance of Erbium-Doped Fiber Amplifiers," *Journal of Electrical & Electronics Engineering*, Vol. 4, No. 2, 2004, pp. 1111-1122.

[2] B. Bouzid, B. M. Ali and M. K. Abdullah. "A High Gain EDFA Design Using Double Pass Amplification with a Band-Pass Filter," *Photonics Technology Letters*, Vol. 15, No. 9, 2003, pp. 1195-1197. doi:10.1109/LPT.2003.814901

[3] Y. B. Lu and P. L. Chu, "Gain Flattening by Using Dual-Core Fiber in Erbium-Doped Fiber Amplifier," *IEEE Photonics Technology Letters*, Vol. 12, No. 12, 2000, pp. 1616-1617 .

[4] A. Mori, T. Sakamoto, K. Shikano, K. Kobayashi, K. Hoshino and M. Shimizu, "Gain Flattened Er^{3+}-Doped Tellurite Fiber Amplifier for WDM Signals in the 1581 - 1616 nm Wavelength Region," *Electronics Letters*, Vol. 36, No. 36, 2000, pp. 621-622.

[5] C. Yang, "Design and Simulation of Gain-Flattened Ultra Wideband Fiber Amplifiers Covering S-, C-, and L-bands," Ph. D Thesis, The University of North Carolina at Charlotte, Charlotte, 2003.

[6] M. A. Mahdi, K. A. Khairi, B. Bouzid and M. K. Abdullah "Optimum Pumping Scheme of Dual-Stage Triple-Pass Erbium-Doped Fiber Amplifier," *IEEE Photonics Technology Letters*, Vol. 16, No. 2, 2004, p. 419. doi:10.1109/LPT.2003.821059

[7] A. Sellami, K. Al-Khateeb and B. Belloui, "The Influence of EDFA's Configuration on the Behavioral Trends of Gain," *International Conference on Computer and Management*, Kuala Lumpur, 9-11 May 2006, pp. 853-856.

[8] B. Bouzid, "High-Gain and Low-Noise-Figure Erbium-Doped Fiber Amplifier Employing Dual Stage Quadruple Pass Technique," *Optical Review*, Vol. 17, No. 3, 2010, pp. 100-102.

Bragg-Angle Diffraction in Slant Gratings Fabricated by Single-Beam Interference Lithography

Xinping Zhang*, Shengfei Feng, Tianrui Zhai
College of Applied Sciences, Beijing University of Technology, Beijing, China
Email: *zhangxinping@bjut.edu.cn

ABSTRACT

A single-beam interference-lithography scheme is demonstrated for the fabrication of large-area slant gratings, which requires exposure of the photoresist thin film spin-coated on a glass plate with polished side-walls to a single laser beam in the ultraviolet and requires small coherence length of the laser. No additional beam splitting scheme and no adjustments for laser-beam overlapping and for optical path-length balancing are needed. Bragg-angle diffractions are observed as strong optical extinction that is tunable with changing the angle of incidence. This device is important for the design of efficient filters, beam splitters, and photonic devices.

Keywords: Slant Grating; Single-Beam Interference Lithography; Bragg-Angle Diffraction

1. Introduction

Interference lithography [1,2] is a conventional technique for producing large-area one- and two-dimensional grating structures in the micro- or nano-scale. In particular, this technique can be used to fabricate the master gratings for the construction of metallic photonic crystals [3], which are important for the development of new lasers,[4] polarizers [5], filters [6], and other photonic devices [7,8]. High flexibility in the interference-lithography scheme using different arrangements of the laser beams enables realization of a variety of photonic structures [9-11]. Conventionally, more than two beams are required in the optical design for interference lithography, so that the interference pattern is recorded directly by the medium within the overlapping area of multiple laser beams. In this work, we demonstrate a simple interference lithography scheme using a single laser beam in the ultraviolet to fabricate slant gratings, where the slant interference pattern forms in the photoresist between the directly transmitted part of the incident laser beam through the substrate and the other part that is refracted into the substrate through the side wall. Slant gratings may enable optical functions that cannot be easily realized in conventional grating structures, where the incident angle with respect to the plane of the grating sidewall may exceed 90 degrees. Using slant gratings, investigations based on large incident angles become feasible and properties of photonic band gaps can be studied using the off-plane incidence scheme, where the Bragg-angle diffractions can be achieved. This is important for practical applications in high-contrast filters, beam splitters, and in the design of photonic crystal devices.

*Corresponding author.

2. Fabrication of the Slant Gratings

Figure 1 demonstrates the basic principles of this single-beam interference lithography technique. The UV laser beam is incident at an angle of θ_i onto the sample consisting of a glass-plate substrate with polished sidewalls and the photoresist film spin-coated on the backside. The size of the laser spot is much larger than that of the sample and part of the incident light has the chance to be coupled into the substrate through the side wall at an angle of 90° - θ_i. This part of light will be incident on the glass-photoresist interface at a large angle of 90° - $\sin^{-1}(\cos\theta_i/n_S)$, whereas, the other part of the incident light will enter the photoresist layer at an angle of $\sin^{-1}(\sin\theta_i/n_S)$, where n_S is the refractive index of the glass substrate. The interference pattern forms bisecting the propagation directions of these two parts of the laser beam when they are refracted into the layer of photoresist, which actually determines the slant angle of the photoresist grating after the exposure and the development processes.

Thus, the slant angle of the grating can be calculated as follows:

$$\theta_{Slant} = \frac{1}{2}\sin^{-1}\left\{\frac{n_S}{n_{PR}}\cos\left[\sin^{-1}\left(\frac{\cos\theta_i}{n_S}\right)\right]\right\} + \frac{1}{2}\sin^{-1}\left(\frac{\sin\theta_i}{n_S}\right), \quad (1)$$

where n_{PR} is the refractive index of photoresist at the laser wavelength. The slant period of the grating that can be measured directly on the top surface is evaluated by:

$$\Lambda_{Slant} = \frac{\lambda}{2n_{PR}\sin\left[\theta_{Slant} - \sin^{-1}\left(\frac{\sin\theta_i}{n_{PR}}\right)\right]\cos\left(\theta_{Slant}\right)} = \frac{\Lambda}{\cos\theta_{Slant}} \quad (2)$$

where Λ is correspondingly the "true" grating period if it is measured in the plane perpendicular to the side wall of the grating, as illustrated schematically in **Figure 3**. According to **Figure 1**, the effective area of the slant grating may be evaluated by the value of L with:

$$L = H\left\{1/\tan\left[\sin^{-1}\left(\cos\theta_i/n_S\right)\right] - \tan\left[\sin^{-1}\left(\sin\theta_i/n_S\right)\right]\right\}, \quad (3)$$

where H is the thickness of the glass substrate.

In the practical fabrication, a single 355 nm laser beam is used as the UV light source, which is incident at an angle of θ_i = 34° onto the sample that is prepared by spin-coating S1805 photoresist from onto a fused-silica glass plate with a thickness of about 1.5 mm and an area of 10 × 10 mm^2 at a speed of 2000 rpm. Furthermore, the glass substrate and the photoresist have a refractive index of n_S = 1.476 and n_{PR} = 1.74 at 355 nm, respectively. Thus, we can easily obtain Λ_{Slant} = 31.66°, Λ_{Slant} = 536 nm, Λ = 456 nm, and L = 1.6 mm.

The atomic force microscopic (AFM) image of the fabricated slant grating is shown in **Figure 2(a)** and a three-dimensional (3D) re-drawing of **Figure 2(a)** is given in **Figure 2(b)**. Clearly, the grating structures have a period of about 532 nm, or $\Lambda_{Slant} \approx$ 532 nm and Λ = 453 nm, which agrees very well the designed or the calculated values of Λ_{Slant} = 536 nm and Λ = 456 nm. **Figure 2(a)** also shows a modulation depth smaller than 600 nm of the grating and **Figure 2(b)** indicates convincingly the slant features of the grating. The fabrication experiments

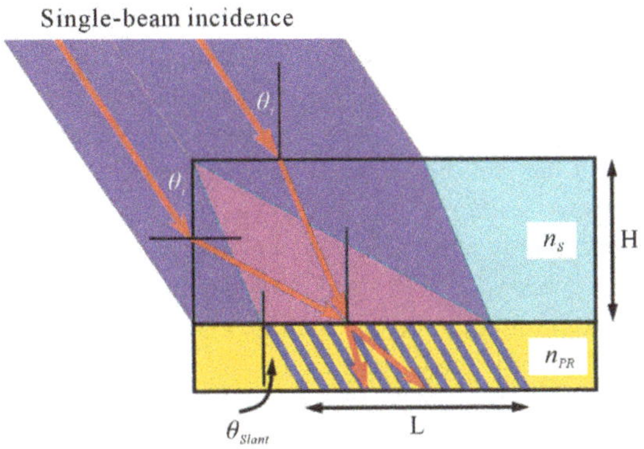

Figure 1. Schematic illustration of the single-beam interference lithography.

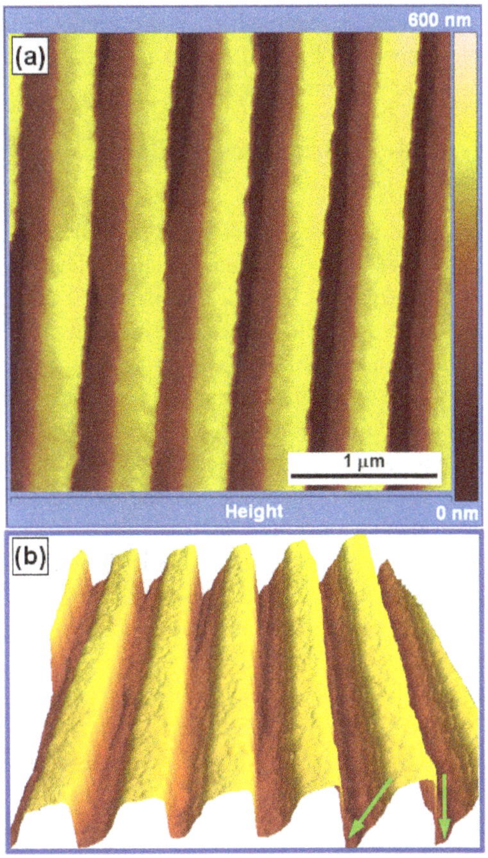

Figure 2. (a) AFM image of the photoresist slant grating structures; (b) A three-dimensional re-drawing of the AFM image in (a).

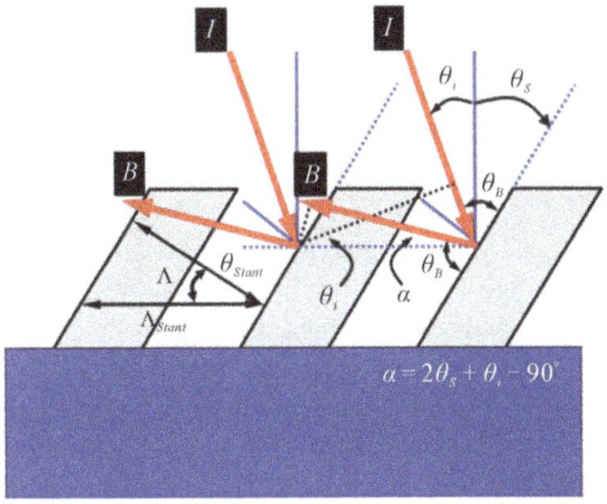

Figure 3. The schematic illustration of the Bragg-angle diffractions in the slant gratings.

showed that the period and the slant angle of the grating can be tuned by changing the incident angle of the UV laser. It should be noted that the laser beam coupled into the substrate through the side-wall will totally reflected by the top and bottom surfaces of the sample and propagates within the substrate, so that it interacts with the directly transmitted light multiple times within the photoresist layer. Thus, multiple domains of the interference patterns can be recorded as slant gratings in the photoresist, as can be observed in the inset of **Figure 2(a)**, which actually extend the applicable grating area.

3. Bragg-Angle Diffractions in Slant Gratings

This kind of slant grating enables easily the Bragg-like diffractions even at normal incidence of light. The basic principles are illustrated schematically in **Figure 3**. The incident light at an angle of θ_i is actually incident at $\theta_B = \theta_i + \theta_{Slant}$ with respect to the side surface of the slant grating lines. If we look at the diffraction in the direction of beam B that is symmetric with the incident light beam *I* about the normal to the slant surface, which can be defined as a kind of Bragg-angle diffraction, the diffraction condition may be written as:
$n_{eff}\Lambda_{Slant}(\sin\theta_i + \cos\alpha) = \lambda_B$, where n_{eff} is the effective refractive index of the grating layer consisting of the air and photoresist, θ_B is the resonance wavelength of the Bragg-angle diffraction. Obviously, the value of n_{eff} is dependent not only on the duty cycle of the slant grating and the wavelength of the incident light, but also on the shape of the grating profile and how the light is incident onto the grating. Considering $\alpha = 2\theta_{slant} + \theta_i - 90°$, we have $n_{eff}\Lambda_{Slant}\left[\sin\theta_i + \sin(\theta_i + 2\theta_{Slant})\right] = \lambda$, which can actually be rewritten as:
$2n_{eff}\Lambda_{Slant}\sin(\theta_i + \theta_{Slant})\cos\theta_{Slant} = \lambda$. However, we already obtained $\Lambda_{Slant}\cos\theta_{Slant} = \Lambda$ and $\theta_B = \theta_i + \theta_{Slant}$, thus, the diffraction condition becomes:

$$2n_{eff}\Lambda\sin\theta_B = \lambda. \qquad (4)$$

This is exactly the condition for Bragg diffraction and defines the Bragg-angle diffraction in the slant grating. On this basis, the slant gratings may be taken as a kind of one-dimensional photonic crystal structures that enables direct incidence from the side surface of the grating. It is this kind of Bragg-angle diffraction that induces strong spectroscopic response of the slant gratings.

Figure 4 presents the optical characterization of the device shown in **Figure 2**, which is demonstrated by the angle-resolved tuning properties of the optical extinction spectrum. The incident angle θ_i is changed from −44 to +36 degrees. The definitions of the positive and negative values of θ_i are illustrated in the inset of **Figure 4(a)**. For a positive value of θ_i, a strong extinction signal with a bandwidth of about 90 nm at full width at half maximum (FWHM) can be observed, which is tuned from about 550 to 810 nm as the angle of incidence is increased from 0 to 28 degrees in steps of 4 degrees, as shown in **Figure 4(a)**. For $\theta_i = 0$, the spectral peak of the Bragg diffraction or the Bragg resonance mode is observed at about 550 nm. Using the Bragg diffraction condition in (4), a grating period of $\Lambda_{Slant} = 532$ nm or $\Lambda \approx 453$ nm, a slant angle of $\Lambda_{Slant} \approx 31.66°$, we can obtain $n_{eff} \approx 1.16$ at about 550 nm for normal incidence.

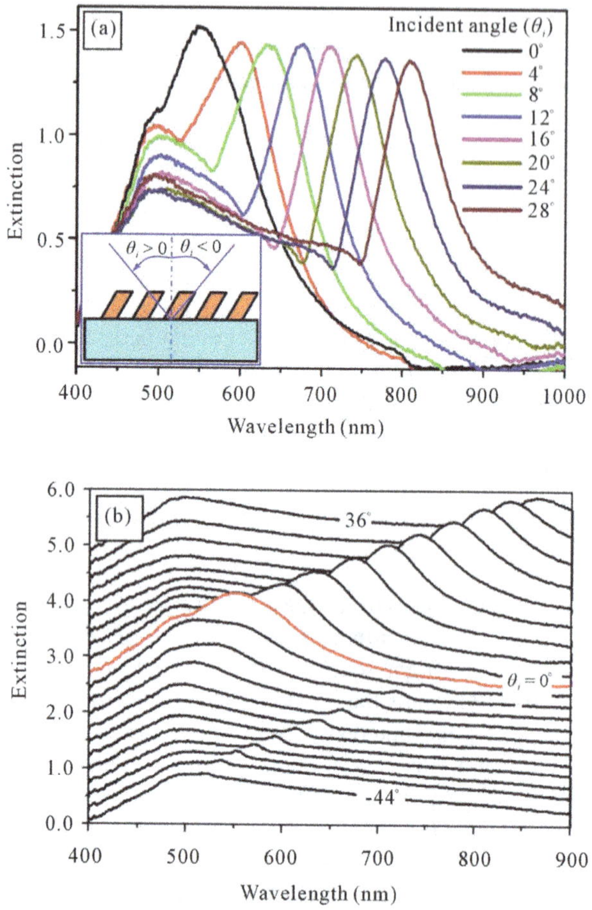

Figure 4. (a) Angle-resolved tuning properties of the Bragg resonance mode of the slant gratings for a positive incident angle increased from 0 to 28 degrees; (b) Angle-resolved tuning properties of the optical extinction spectrum with the incident angle changed from −44 to +36 degrees.

As shown in **Figure 4**, the amplitude of the extinction signal due to the Bragg-angle diffraction is ranging from 1.4 to 1.6 (at $\theta_i = 0$), meaning a reduction of 75% - 80% in the transmission spectrum. Furthermore, the resonance spectrum is tuned to longer wavelengths as the angle of incidence onto the slant surface of the grating is reduced while the incidence angle with respect to the normal of the substrate is increased. This is the typical feature of band-gaps of the one-dimensional photonic crystals.

For negative values of θ_i, a relatively small resonant mode with a bandwidth narrower than 10 nm at FWHM can be observed, which is tuned from about 790 to 520 nm as the incident angle is changed from 0 to −44°. This is the shorter-wavelength branch of the waveguide resonance mode of the WGS structures [12], where a thin layer of photoresist remained at the bottom of the grating structures after the development process and it acts as the waveguide. Furthermore, a broad-band feature can be observed in addition to the above two tunable resonance modes, which is peaked at about 495 nm and extends to longer wavelengths with increasing the incident angle. This actually results from the normal grating diffraction and the absorption by the photoresist.

4. Conclusion

In conclusion, we demonstrated fabrication of slant gratings using a simple single-beam interference lithography scheme. The period and the slant angle of the grating structures may be tuned by changing the incident angle of the UV laser beam. Bragg-angle diffraction interprets the strong optical response of this kind of nanostructures with well-established theoretical model. The tunable resonance mode with a bandwidth of about 90 nm at FWHM may be taken as the band-gap of a kind of one dimensional photonic crystal structures. This introduces a new approach for the realization of photonic devices, which may be applied in filters and beam splitters, or used as master gratings for the realization of plasmonic devices.

5. Acknowledgements

The authors acknowledge the financial support by the National Science Foundation of China (11074018), the Program for New Century Excellent Talents in University (NCET), and the Research Fund for the Doctoral Program of Higher Education of China (RFDP, 20091103110012).

REFERENCES

[1] W. W. Ng, C. S. Hong and A. Yariv, "Holographic Interference Lithography for Integrated-Optics," *IEEE Transactions on Electron Devices*, Vol. 25, No. 10, 1978, pp. 1193-1200. doi:10.1109/T-ED.1978.19251

[2] V. Berger, O. Gauthier-Lafaye and E. Costard, "Photonic Band Gaps and Holography," *Journal of Applied Physics*, Vol. 82, No. 1, 1997, pp. 60-64. doi:10.1063/1.365849

[3] A. Christ, S. G. Tikhodeev, N. A. Gippius, J. Kuhl and H. Giessen, "Waveguide-Plasmon Polaritons: Strong Coupling of Photonic and Electronic Resonances in A Metallic Photonic Crystal Slab," *Physical Review Letters*, Vol. 91, No. 18, 2003, p. 183901. doi:10.1103/PhysRevLett.91.183901

[4] J. Stehr, J. Crewett, F. Schindler, R. Sperling, G. von Plessen, U. Lemmer, J. M. Lupton, T. A. Klar, J. Feldmann, A. W. Holleitner, M. Forster and U. Scherf, "A Low Threshold Polymer Laser Based on Metallic Nanoparticle Gratings," *Advanced Materials*, Vol. 15, No. 20, 2003, pp. 1726-1729. doi:10.1002/adma.200305221

[5] X. P. Zhang, H. M. Liu, J. R. Tian, Y. R. Song and L. Wang, "Band-Selective Optical Polarizer Based on Gold-Nanowire Plasmonic Diffraction Gratings," *Nano Letters*, Vol. 8, No. 9, 2008, pp. 2653-2658. doi:10.1021/nl0808435

[6] X. P. Zhang, H. M. Liu, J. R. Tian, Y. R. Song, L. Wang, J. Y. Song, and G. Z. Zhang, "Optical Polarizers Based on Gold Nanowires Fabricated Using Colloidal Gold Nanoparticles," *Nanotechnology*, Vol. 19, No. 28, 2008, p. 285202. doi:10.1088/0957-4484/19/28/285202

[7] X. P. Zhang, B. Q. Sun, J. M. Hodgkiss and R. H. Friend, "Tunable Ultrafast Optical Switching via Waveguided Gold Nanowires" *Advanced Materials*, Vol. 20, No. 23, 2008, pp. 4455-4459. doi:10.1002/adma.200801162

[8] D. Nau, R. P. Bertram, K. Buse, T. Zentgraf, J. Kuhl, S. G. Tikhodeev, N. Gippius and H. Giessen, "Optical Switching in Metallic Photonic Crystal Slabs with Photoaddressable Polymers," *Applied Physics B—Lasers and Optics*, Vol. 82, No. 4, 2006, pp. 543-547. doi:10.1007/s00340-005-2103-z

[9] X. L. Yang, L. Z. Cai and Y. R. Wang, "Larger Bandgaps of Two-Dimensional Triangular Photonic Crystals Fabricated by Holographic Lithography Can Be Realized by Recording Geometry Design," *Optics Express*, Vol. 12, No. 24, 2004, pp. 5850-5856. doi:10.1364/OPEX.12.005850

[10] Y. Yang, Q. Li and G. P. Wang, "Design and Fabrication of Diverse Metamaterial Structures by Holographic Lithography," *Optics Express*, Vol. 16, No. 15, 2008, pp. 11275-11280. doi:10.1364/OE.16.011275

[11] D. C. Meisel, M. Diem, M. Deubel, F. Pérez-Willard, S. Linden, D. Gerthsen, K. Busch and M. Wegener, "Shrinkage Precompensation of Holographic Three-Dimensional Photonic-Crystal Templates," *Advanced Materials*, Vol. 18, No. 22, 2006, pp. 2964-2968. doi:10.1002/adma.200600412

[12] D. Rosenblatt, A. Sharon and A. A. Friesem, "Resonant Grating Waveguide Structures," *IEEE Journal of Quantum Electronics*, Vol. 33, No. 11, 1997, pp. 2038-2059. doi:10.1109/3.641320

6

Speckle Reduction in Imaging Projection Systems

Weston Thomas, Christopher Middlebrook
Electrical and Computer Engineering Department, Michigan Technological University, Houghton, USA
Email: whthomas@mtu.edu

ABSTRACT

Diffractive diffusers (phase gratings) are routinely used for homogenizing and beam shaping for laser beam applications. Another use for diffractive diffusers is in the reduction of speckle for pico-projection systems. While diffusers are unable to completely eliminate speckle they can be utilized to decrease the resultant contrast to provide a more visually acceptable image. Research has been conducted to quantify and measure the diffusers overall ability in speckle reduction. A theoretical Fourier optics model is used to provide the diffuser's stationary and in-motion performance in terms of the resultant contrast level. Contrast measurements of two diffractive diffusers are calculated theoretically and compared with experimental results. Having a working theoretical model to accurately predict the performance of the diffractive diffuser allows for the verification of new diffuser designs specifically for pico-projection system applications.

Keywords: Diffractive Diffusers; Speckle Contrast Reduction; Laser Pico-Projectors

1. Introduction

The observance of speckle in laser images is caused by the interference of the coherent source. Diffusers reduce speckle by decreasing the temporal and spatial coherence of the source. A diffuser has multiple cells each with different phase values. By rotating or vibrating the diffuser over a discrete time period the phase levels alter the coherence, thereby reducing the speckle contrast. The time-averaging of the speckle must occur over a discrete time smaller than the integration time of the human eye. Preceding investigations into the reduction of speckle contrast using rotational diffusers are well known and have had limited success [1-5]. The criterion for an effective diffuser is based on the calculation of the speckle contrast. Equation (1) gives the contrast, calculated by the standard deviation divided by the average intensity.

$$C = \frac{\sqrt{\langle I^2\rangle - \langle I\rangle^2}}{\langle I\rangle} = \frac{\sqrt{\frac{\sum_{i=1}^{N}\left(I_i - \langle I\rangle\right)^2}{N-1}}}{\langle I\rangle} \quad (1)$$

Figure 1 shows an example of a binary diffuser that is currently being used for reducing speckle in laser projections systems and will be discussed in detail later. This specific diffuser has circular symmetry for rotation purposes.

2. Theory and Model

This diffuser is modeled as a single scattering phase screen, written in the form of a transmission aperture, given by Equation (2) [6]. **Figure 2(a)** illustrates the placement of the diffuser in line with the initial beam and followed by the imaging screen in the (x, y) plane which has the varying phase pattern across the diffuser surface as shown in **Figure 3**.

$$t_A = \mathrm{e}^{\varphi(x,y)} \quad (2)$$

The simulation will revolve around mimicking the operation of two distinct diffuser types: binary and grayscale. The binary diffuser is based on a hadamard matrix and thus contains only two distinct phase level sections. This generates sharp edges in the output image. The grayscale diffuser takes advantage of 64 discrete phase levels creating a smoother granular profile [7]. Both types are designed to reduce the visual degradation

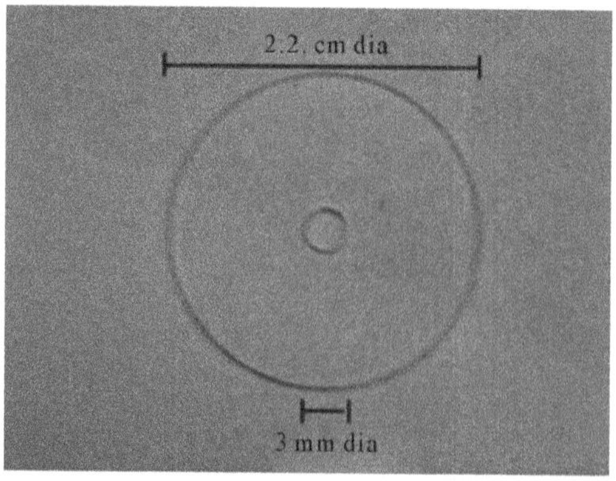

Figure 1. Binary diffractive diffuser.

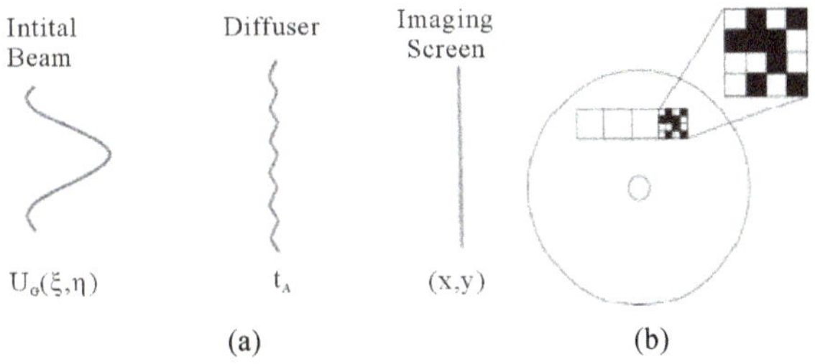

Figure 2. (a) Diffuser model layout; (b) Diffuser unit cell layout.

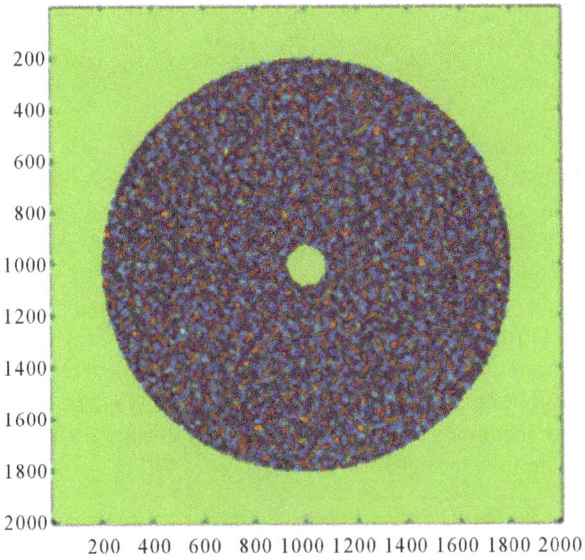

(a)

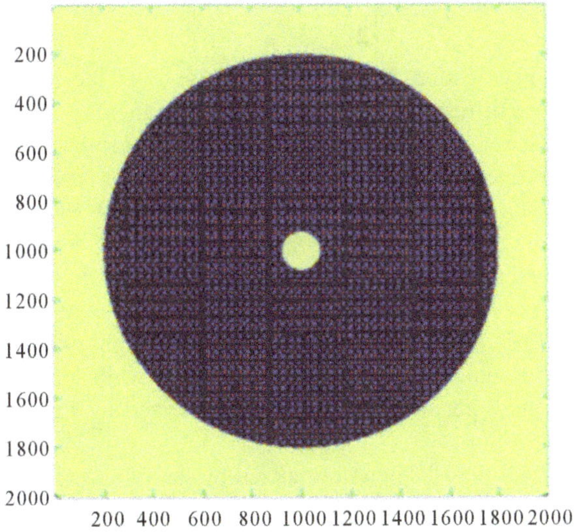

(b)

Figure 3. (a) Grayscale diffuser array; (b) Binary diffuser array.

caused by speckle through rotation. They are both modeled with an array of 2000 × 2000 pixels. The phase screen arrays are demonstrated in **Figures 3(a)** and **(b)**. Each diffuser is composed of individual cells replicated across the surface as demonstrated in **Figure 2(b)**. The binary diffuser's unit cell is an order 16 Hadamard matrix while each Grayscale unit cell is a random organization of the 64 phase levels. These two unit cells are approximations of the physical diffusers that were used for experimental measurements but are not exact due to proprietary information.

Equation (3) demonstrates the linear size limitation given this array size and a z depth of 2×10^4 pixels. Every individual phase element will consist of 25 pixels for each diffuser model. The minimum phase element size is 2.8 pixels for accurate sampling. This has been increased to be arranged evenly and symmetrically across the diffuser surface [8,9]. A distance mapping along the optical axis can be characterized along the z-axis. Every square pixel is related to a physical length by 2.5 μm. This mapping can be established for the propagation axis, z. This framing scale is maintained for all arrays unless distinctly noted on the axis of the image.

$$\Delta x > \frac{\lambda z}{L} \quad (3)$$

For the discrete version, Δx is the linear phase element size, λ is the wavelength, z is the propagation distance in pixels and L represents the total physical side length of the array. The output from the aperture can be found by using a Fourier approximation for distance propagation such as the Fresnel equation [10]. Treating the mathematical computation as a linear system allows the separation of code to flow freely with the functional partitions. The system can be broken up into multiple parts as well based on the needs of the simulation. In this instance it is helpful to separate the Fourier optics propagation and the diffuser rotational program from the individual diffuser models allowing the diffuser models to be interchanged without rewriting large amounts of the simulation.

The initial beam is considered to be a collimated monochromatic Gaussian beam, as demonstrated in **Figure 4**, which is defined as [11]

$$U_G(\xi,\eta) = \mathrm{e}^{\frac{[(\xi+a)\times\Delta\xi]^2+[(\eta+b)\times\Delta\eta]^2}{w_0^2}} \quad (4)$$

where the position (a, b) is the center of the Gaussian beam, is the beam waist, and $(\Delta\xi,\Delta\eta)$ represents the physical size divided by the number of pixels in each direction. The laser wave front will be the size reference from which the rest of the simulation is measured. The xy-plane is the plane of incidence of the wave-front. The Gaussian beam is propagated through the diffuser in two steps. First the wave is multiplied by the transmission aperture, t_A adding a phase displacement to the initial beam. The output of this calculation, shown in Equation (5), represents the field directly after the diffuser [12].

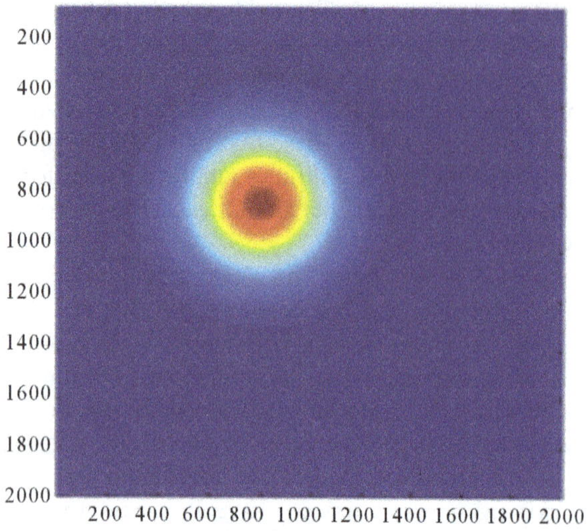

Figure 4. Gaussian beam profile.

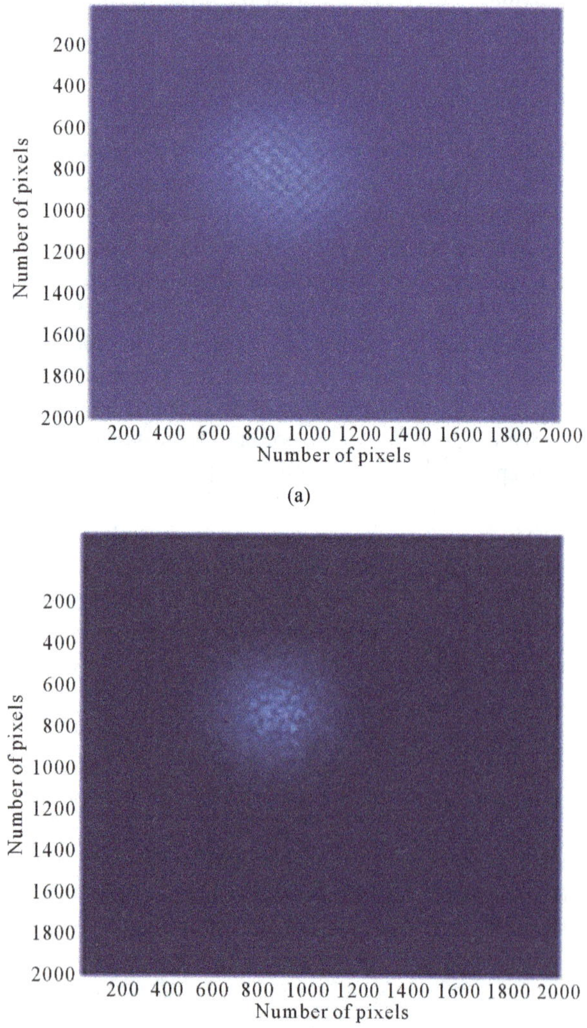

(a)

(b)

Figure 5. (a) Stationary binary diffuser; (b) Stationary grayscale diffuser.

$$U_1(\xi,\eta) = U_G(\xi,\eta) \times t_A \tag{5}$$

The second part involves the Fourier propagation of onto the (x, y) image plane. This is accomplished through the Fresnel approximation of the Rayleigh-Sommerfield equation [8,13]. The Fresnel approximationis given in Equation (6) where λ is the wavelength, k is the wavenumber, z is the propagation distance along the optical axis and (ξ,η) is the aperture plane. The wavelength, once again, is chosen to be 535 nm, representing a generic green laser diode as well as the center wavelength of the visible spectrum.

$$U_2(x,y) = \frac{e^{jkz}}{j\lambda z} \iint U_1(\xi,\eta) e^{j\frac{k}{2z}\left[(x-\xi)^2+(y-\eta)^2\right]} d\xi d\eta \tag{6}$$

The final field image cannot be generated too close to the diffuser without causing quadratic errors in the exponential component. Using previous values for the diffuser aperture size and imaging array size, 1600 and 2000 pixels respectively, a proximity limit of 4500 pixels is calculated using Equation (7). This is considered the near-field region of measurement [13]. While the Fresnel integral does work in the far-field, the loss of image clarity due to beam expansion makes it less convenient to use and the model will refrain from approaching that region, defined by Equation (8) [13].

$$z^3 \gg \frac{\pi}{4\lambda}\left[(x-\xi)^2 + (y-\eta)^2\right]^2_{\max} \tag{7}$$

$$z \gg \frac{k\left(\xi^2+\eta^2\right)^2_{\max}}{2} \tag{8}$$

The final section of the simulation involves the rotation of the diffuser and the modeling of a camera or capture device. Using previous data, the diffuser can be sufficiently operated at 60 (+/−0.5) revolutions per second [3]. In addition, the camera will operate at 30 Hz, to mimic the eye's refresh rate or sampling time of around 23 frames per second [14]. To accomplish this the diffuser array will be rotated prior to propagation of the field $U_1(\xi,\eta)$. The formation of the final image is completed by addition of the individual fields while maintaining pixel position (j,k) as characterized by Equation (9).

The final field is then normalized by the total number of images, h. The intensity image is then found by squaring the absolute value of the field, $U_F(x,y)$. A capture device integrates on the order of 10^{20} photons for a single image during a predetermined exposure time. In order to represent this, an approximation is resolved. This was completed through a series of Monte Carlo simulations. A generic noise pattern was created and then rotated a single revolution. The images were rotated and integrated at various degree increments ranging from

0.01 degrees to 60 degrees. The resolution of the various images was compared against one another until the individual noise parameters were indistinguishable. This is the degree resolution from which the simulation will be completed. With the degree resolution defined at 0.75 the total number of images integrated to create a single frame is 480. Comparing the speed of the diffuser at 60 Hz to the camera integration speed of 30 Hz it is identified that 2 rotations are completed for every single image frame. This will require at least 144,000 distinct images to represent a 10 second video capture.

3. Results

Calculations were made for the speckle contrast of stationary and moving diffuser images using Equation (1). The images for the stationary diffusers are shown below in **Figures 5(a)** and **(b)** for the binary and grayscale versions, respectively. The central spot of the diffuser is the zero-order diffractive mode. Most diffractive optical elements will consist of some form of zero-order mode and is considered to be the DC portion of the element. An optimum diffuser will minimize this zero-order and smooth out the overall output profile of the speckle. The first order of the element contains the majority of the incident power, 90% (+/−5%) as measured experimentally, as well as the highest speckle reductive properties. The speckle contrast is calculated for this first order spot of the diffuser image. The binary version has a contrast of 0.76 and the grayscale version has a contrast value of 0.74. These values correspond to data taken from intensity images of the simulation.

Close up versions of the central order of the diffuser image are shown in **Figures 6(a)** and **(b)**. This is the physical representation of a focused image from the diffuser. They are 300 × 300 pixel array snapshots of the total diffuser image from above. This can be directly compared with bench top experiments taken using binary and grayscale diffusers with a 532 nm DPSS laser [3]. The camera used for the physical experiments had a 480 × 640 pixel array. The stationary, focused diffuser images are shown in **Figures 6(c)** and **(d)**. The contrast values for the bench top experiment are 0.77 for the binary diffuser and 0.68 for the grayscale version. The contrast values themselves are accurate to within 10%. The images themselves, however, are not as comparable. The simulated versions have a more rigid structure while the bench top images are more fluid and organic, specifically related to the binary diffuser image. It is believed that this is a result of the inputs for the creation of the individual diffusers. The contrast results alone do provide enough accuracy for the simulation to be acceptable.

The simulation is finished by creating rotating variations of the diffuser patterns and propagating them onto the image plane. The parameters were set to mimic the rotational speed of the diffuser as 60 Hz. The speckle was only measured for the central order as it was for the stationary diffuser images. The simulated images are shown in **Figures 7(a)** and **(b)**. The calculated contrast values for the binary and grayscale images are 0.297 and 0.276, respectively. The contrast values found in the bench top experiments were slightly less at 0.18 for the binary and 0.14 for the grayscale versions. This relates to a contrast difference of around 10% - 13% contrast reduction between simulation and actual results.

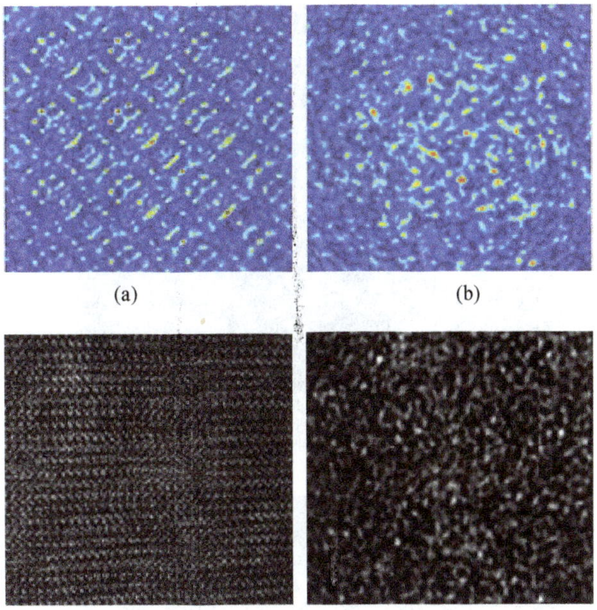

(a) (b) (c) (d)

Figure 6. (a) Focused, stationary binary diffuser; (b) Focused, stationary grayscale diffuser; (c) Physical binary diffuser output intensity image; (d) Physical grayscale diffuser output intensity image.

The current version of this model focuses on a singlescattering approach [4]. While this approach provides a close approximation it does not fully agree with the physical diffuser scattering profile. This will be the main concentration for future work with the project. It is also assumed that a single polarization is incident on the diffusers and that the diffusers are polarization maintaining. Previous research showed that the polarization incident on the diffuser did not have an impact on the final contrast results [3]. This is an approximation that can be changed in the simulation to ensure the theoretical and experimental results match up. The simulation is still a good approximation as it does keep the contrast reduction rate between stationary and rotational the same at around 50% speckle decrease. This allows the simulation to become a useful tool for preliminary design work with theoretical diffuser shapes.

As mentioned earlier, the diffusers were illuminated

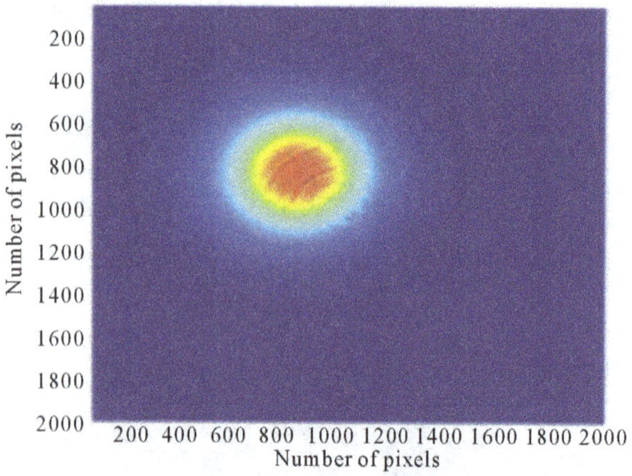

(a)

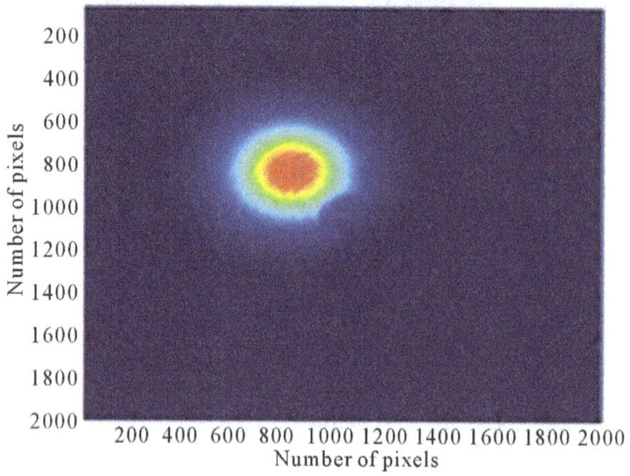

(b)

Figure 7. (a) Stationary binary diffuser; (b) Stationary grayscale diffuser.

with a coherent Gaussian source. All contrast values were measured without subtracting the Gaussian beam difference in the speckle. While uniform illumination speckle statistics are generally considered proper, it is by no means correct as standard laser diodes will have some sort of Gaussian shell and any speckle created from such a diode will have statistics correlating to the Gaussian laser beam. Verification was completed using the simulation model and comparisons were made between uniform illumination and a Gaussian source. Minor differences can be seen within **Figure 6** but the primary result of the Gaussian illumination is the slight spreading of the speckle pattern and lower intensity within the center. Based on the speckle numbers this result is trivial and evens out across the final image plane, as demonstrated through **Figures 8(a)** and **(b)**.

4. RGB Integration

In order to make full use of this model for a projector system more than one color must be combined for an

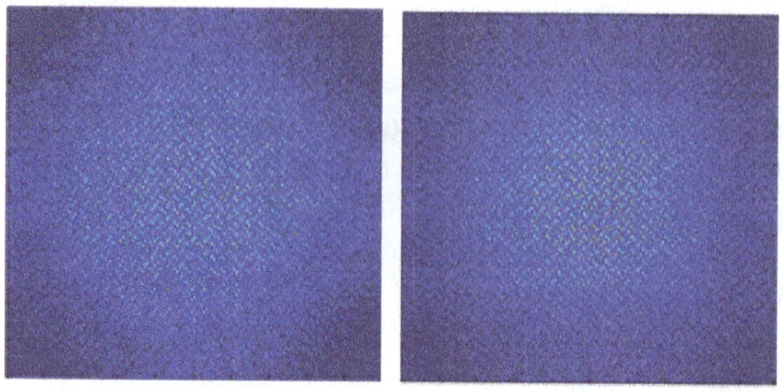

(a) (b)

Figure 8. (a) Diffuser model with uniform illumination; (b) Diffuser model with Gaussian illumination.

image. Most diffusers are created for a single wavelength and errors result from using polychromatic light or multiple wavelengths. This problem will need to be solved before full implementation of diffusers into pico-projectors can be accomplished. In addition, integration of multiple colors into a single image can be difficult on its own due to the variation in the human eye's perception of colors [15]. Simulation will help to reduce problems in the creation and testing of multi-wavelength diffusers.

Beginning experiments were conducted to measure the wavelength dependence of the diffuser designs. The diffuser arrays were tested at wavelengths 635 and 450 nm to compare with the 535 nm beam results. The three wavelengths individual images for the grayscale diffuser simulation are shown in **Figures 9(a)-(c)**. All three wavelength simulations were completed at the same distance from the diffuser along the optical axis. The obvious comparison between them is the expansion of the overall wave front. Speckle contrast measurements were taken for the central order of the diffuser. The contrast for the 450 nm image is 0.711 and is 0.65 for the 635 nm image.

The simulated images also allow the ability to combine the three wavelengths together into a single RGB image. **Figure 9(d)** illustrates this process and shows how the three distinct grayscale speckle patterns overlay onto one another. Current speckle calculations do not allow for 3 dimensional measurements and thus any calculation would simply be an average of the results for reach individual wavelength. In this case such an average gives a speckle contrast of 0.70 for the simulated grayscale diffuser. Also the colored images had to be normalized based on the visual response of the eye. Otherwise, the colors would appear out of sync and any speckle measurements would not align with visual representation of the speckle image [16]. A new measurement protocol will be needed to fully realize the speckle reduction ability of a diffuser design. Current investigations have led to comparisons between correlation times of diffusers and their resultant speckle contrast.

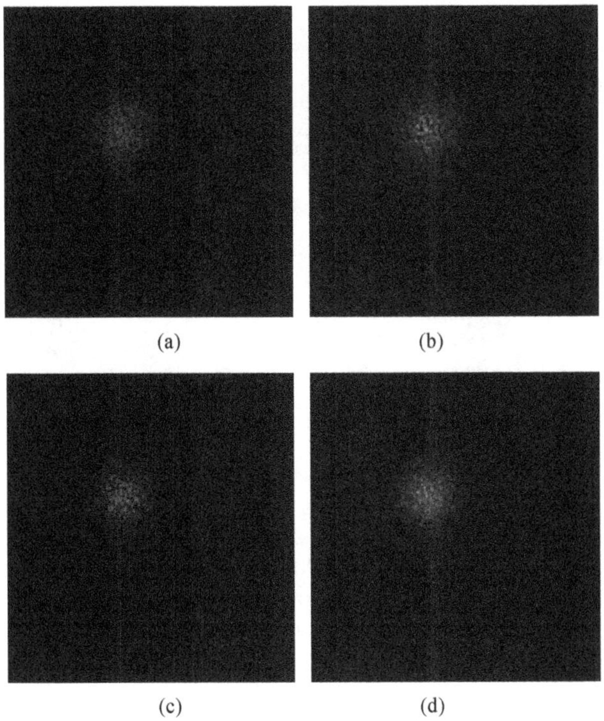

Figure 9. Simulated images of grayscale diffuser at (a) 450 nm; (b) 535 nm; (c) 635 nm; (d) Multi-wavelength combination.

5. Conclusion

A working mathematical model was developed to model diffractive diffusers for speckle reduction in imaging projection systems. The diffuser simulation was verified by comparing the results to experimental measured values for two distinct diffusers. The diffusers modeled have been shown to have contrast values ranging from 65% - 77% and are accurate to within 10% of experimental results. Having the ability to perform multi-wavelength analysis of diffractive diffuser performance has also been shown with the model and verified with experimental results. The model will be used to aid in the design and performance of various diffractive diffuser designs for speckle reduction applications. This work provides the ability to quickly analyze multiple diffuser designs solely focused on creating a reduced speckle laser imaging projector system.

REFERENCES

[1] J. W. Goodman, "Speckle Phenomena in Optics: Theory and Applications," Roberts & Company, Englewood, 2007, p. 387.

[2] J. I. Trisnadi, "Hadamard Speckle Contrast Reduction," *Optics Letters*, Vol. 29, No. 1, 2004, pp. 11-13. doi:10.1364/OL.29.000011

[3] W. Thomas, C. Middlebrook and J. Smith, "Laser Speckle Contrast Reduction Measurement Using Diffractive Diffusers," *Proceedings of SPIE, Emerging Liquid Crystal Technological IV*, Vol. 7232, 2009, 11 p.

[4] D. D. Duncan, S. J. Kirkpatrick and R. K. Wang, "Statistics of Local Speckle Contrast," *Journal of the Optical Society of America*, Vol. 25, No. 1, 2008, pp. 9-15. doi:10.1364/JOSAA.25.000009

[5] C. N. Kurtz, H. O. Hoadley and J. J. Depalma, "Design and Synthesis of Random Phase Diffuser," *Journal of the Optical Society of America*, Vol. 63, No. 9, 1973, pp. 1080-1092. doi:10.1364/JOSA.63.001080

[6] C. N. Kurtz, "Transmittance Characteristics of Surface Diffusers and the Design of Nearly Band-Limited Binary Diffusers," *Journal of the Optical Society of America*, Vol. 62, No. 8, 1972, pp. 982-989. doi:10.1364/JOSA.62.000982

[7] J. W. Goodman, "Statistical Optics," Wiley Classics Library, John Wiley & Sons, Hoboken, 2000, p. 550.

[8] Y. Nakayama and M. Kato, "Diffuser with Pseudorandom Phase Sequence," *Journal of the Optical Society of America*, Vol. 69, No. 10, 1979, pp. 1367-1372. doi:10.1364/JOSA.69.001367

[9] D. Voelz, "Computational Fourier Optics: A MATLAB Tutorial," SPIE Tutorial Texts, SPIE Press, Bellingham, 2011, p. 250.

[10] D. A. Gremaux, "Limits of Scalar Diffraction Theory for Conducting Gratings," *Applied Optics*, Vol. 32, No. 11, 1993, pp. 1948-1953. doi:10.1364/AO.32.001948

[11] J. E. Harvey, "Fourier Treatment of Near-Field Diffraction Theory," *American Journal of Physics*, Vol. 47, No. 11, 1979, pp. 974-980. doi:10.1119/1.11600

[12] J. T. Verdeyen, "Laser Electronics," In: J. N. Holonyak, Ed., *Prentice Hall Series in Solid State Physical Electronics*, 3rd Edition, Prentice Hall, Saddle River, 2000, p. 778.

[13] H. Loui, "Fourier Propagation, in Numerical Methods in Photonics Project 2004," University of Colorado, Boulder, pp. 1-5.

[14] J. W. Goodman, "Introduction to Fourier Optics," 3rd Edition, Roberts & Company, Englewood, 2005.

[15] M. Livingstone, "Vision & Art: The Biology of Seeing," Harry N. Abrams, New York, 2002.

[16] J. W. Tom, A. Ponticorvo and A. K. Dunn, "Efficient Processing of Laser Speckle Contrast Images," *IEEE Transactions on Medical Imaging*, Vol. 27, No. 12, 2008, pp. 1728-1738. doi:10.1109/TMI.2008.925081

Photon Correlation Spectroscopy and SAXS Study of Mixture of NaCl with AOT Microemulsion at X = 6.7

Nahid Karimi, Soheil Sharifi*, Mousa Aliahmad
Department of Physics, University of Sistan and Baluchestan, Zahedan, Iran
Email: *soheil.sharifi@gmail.com, sharifi@df.unipi.it

ABSTRACT

Photon Correlation Spectroscopy is used to study the AOT microemulsion with and without NaCl. Collective diffusion coefficient was investigated by Photon Correlation Spectroscopy technique. We have studied effect of charge on dynamic of water-in-oil microemulsion (nano-droplet of water to the oil), which stabilized by AOT and dispersed in n-Decane at water/AOT with 6.7 molar ratio. The small angle X-ray scattering technique and hard sphere model were used to study the structural information of AOT microemulsion with and without NaCl. The structural investigation of samples shows a decrease of length scale of cylindrical droplets with increasing of NaCl concentration in AOT microemulsion.

Keywords: Nano-Size; Droplet; Microemulsions; Photon Correlation Spectroscopy; SAXS; Diffusion

1. Introduction

The microemulsions are consisting nano-meter size of water droplets in the oil [1-3]. Microemulsion have attracted interest for the delivery of single drug substances with low water solubility and stabilization of drugs in combination due to their preferential solubility in either the water or oil phases [4,5]. Microemulsions can be regarded as reverse-micellar solutions that have solubilized water into the polar surfactant tail region (L_2 phase). The anionic surfactant Aerosol OT (sodium bis-(2-ethylhexyl) sulfosuccinate, AOT) together with water and oil readily forms ternary microemulsions. The aqueous part of microemulsion is usually water or a solution of water with salts. The dynamic properties of microemulsions and colloidal systems are studied by photon correlation spectroscopy (PCS) [6-8] and the structure of AOT microemulsions are well investigated at water to the surfactant molar ration 6.7 [9]. A study shown, for AOT/H_2O/Decane microemulsion, at the low droplet mass fractions ($0.01< m_f<0.1$ and molar ratio 40) the collective diffusion coefficient has a linear behavior as function of the mass fraction [10]. It is well known that at high water concentration with a water to surfactant molar ratio of X = 40 the collective diffusion coefficient D_c depends on the droplet mass fraction (droplet concentration) due to inter-particle interactions [10]. This study is an attempt to understand the collective diffusion coefficient of AOT/H_2O/Decane microemulsion at low water concentration X = 6.7 construct with photon correlation spectroscopy. Our previous work on AOT microemulsion at 6.7 shows a spherical to cylindrical transition [9]. In the present work, we studied the behavior of the collective diffusion coefficient (D_c) at the fix water to oil ratio (X = 6.7) and low mass fraction, by means of photon correlation spectroscopy and SAXS techniques. Moreover, we study the effect of NaCl on the D_c and structure of the AOT/H_2O/Decane microemulsion.

2. Experimental

Photon Correlation Spectroscopy measurements were performed using an ALV single-detector version compact goniometer system, from ALV-GmbH, Langen, Germany. The light source is a He-Ne laser, operating at a wavelength of 632.8 nm with vertically polarized light. The beam was focused on the sample cell through a temperature-controlled cylindrical quartz container (with two plane-parallel windows), which is filled with a refractive index matching liquid (toluene). All the correlation functions in this work were fitted by a single stretched exponential function [10-13].

$$g_1(t,q) = \exp\left[-\left(\frac{t}{\tau}\right)^{\beta}\right] \quad (1)$$

The stretched exponential function describes the decay processes that have a distribution of relaxation times (τ). The parameter β ($0 \le \beta \le 1$) shows width of the distribution function. Small-angle X-ray scattering (SAXS) measurements were performed using the pinhole SAXS in-

*Corresponding author.

strument at the Aarhus [14]. The instrument consists of an X-ray camera (NanoSTAR, Bruker AXS) with a rotating anode X-ray (Cu Kα radiation) source, cross-coupled Göbel mirrors, collimation using three pinholes, an evacuated beam path, and a 2D position-sensitive gas detector (HiSTAR). The experiments were done at a fixed wavelength of λ = 1.54 Å and two different sample-detector distances. In the current experiments small pinholes were used, giving a range of scattering vectors as $0.004 < q(1/\text{Å}) < 0.2$, ($q = 4\pi\sin\theta/\lambda$ where θ is half the scattering angle.

n-Decane, NaCl and Sodium-2-diethylhexyl sulfosuccinate, or AOT 99% were obtained from Sigma-Aldrich. Chemicals were used as received and MilliQ water was used in preparing all samples. The bottles of AOT, which are hermetically sealed, have always been stored in a refrigerator well below the melting point (23°C). After a container was opened, the surfactant was stored in a nitrogen environment; it was still found that the surfactant did not remain stable for more than a month. The samples were prepared by mixing the components directly in glass ampoule in order to minimize the number of transfers and the glass ampoules were tightly sealed with a gas flame. The composition of each system is determined by the molar ratio X of water to surfactant molecules, $X = [H_2O]/[AOT]$ where $[H_2O]$ and [AOT] are molarities of water and surfactant. The droplet mass fraction $m_f = (m_{AOT} + m_{H_2O})/(m_{AOT} + m_{H_2O} + m_{Dec})$ Which varies by the respective mass of the components water (m_{H_2O}), decane (m_{Dec}), and AOT(m_{AOT}). The microemulsions were prepared by weight, in terms of X = 6.7 and the different mass fraction of droplets ($m_{f,drop} = (m_{H_2O} + m_{AOT})/(m_{Dec} + m_{H_2O} + m_{AOT})$), which varies by the respective mass of n-decane (m_{Dec}), AOT(m_{AOT}). The samples were thoroughly shaken to ensure homogenization and then kept at the temperature 20°C in a water bath for several days before the experiment. We observed that all samples were transparent at 20°C. The mixing of NaCl with microemulsions is described by molar ratio of NaCl to AOT, Y = [NaCl]/[AOT], for this experiments, Y = 0.005.

3. Results and Discussion

Microemulsions were formulated by mixing AOT with water and n-Decane at constant molar ratio of water to AOT (X = 6.7) at the different mass fraction ($0.01 < m_f < 0.3$ dilute regime).

Dynamic behavior of the AOT/H_2O/Decane microemulsion was probed with photon correlation spectroscopy. The correlation function shows a single stretch exponential decay at all concentrations, **Figure 1**. The AOT/H_2O/Decane/NaCl microemulsion at X = 6.7 and [NaCl]/[AOT] = 0.005 and different mass fraction is studied by photon correlation spectroscopy, **Figure 2**.

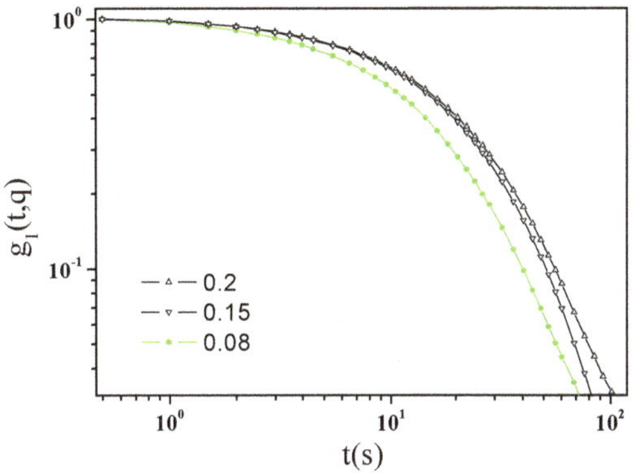

Figure 1. Correlation function as function delay time for AOT/H_2O/Decane microemulsion at X = 6.7 and different mass fraction at 20°C.

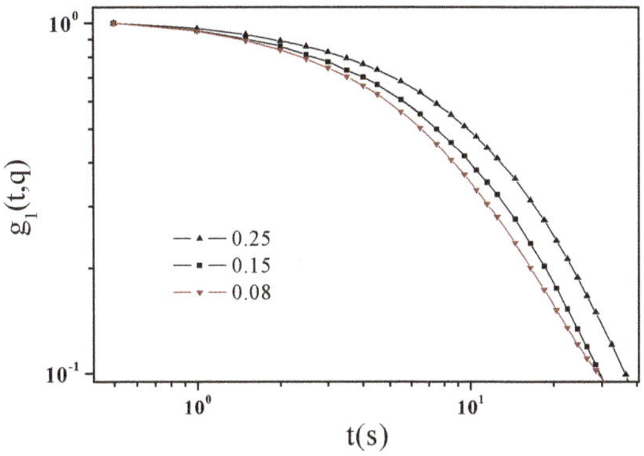

Figure 2. Correlation function as function delay time for AOT/H_2O/Decane/NaCl microemulsion at X = 6.7 and [NaCl]/[AOT] = 0.005, and different mass fraction at 20°C.

All the correlation functions in this work were fitted by a single stretched exponential function, Equation (1). The collective diffusion coefficient D_c were extracted from $D_c = 1/q^2\langle\tau\rangle$ as the function of the mass fraction for the AOT/H_2O/Decane and AOT/H_2O/Decane/NaCl microemulsion illustrated in **Figure 3**. The normalized collective diffusion coefficient show a liner behavior with negative slop between $0.01 < m_f < 0.75$. Our results show after adding NaCl to the AOT/H_2O/Decane microemulsion, diffusion increase clearly, **Figure 3**.

In this work, SAXS is used for the determination of the structure of AOT/H_2O/Decane microemulsions at 293.15 K. The SAXS method was used because the scattering of X-rays of wavelengths of a few angstroms through small angles provides a q range which is particularly appropriate for the determination of both the size of discrete reverse micelles and their interactions. The scattered intensity as a function of q for AOT/H_2O/Decane and AOT/H_2O/Decane/NaCl microemulsion is presented in the **Figure 4**, which the lines are fits to a power law, $I(q) \approx q^{-1}$, with properties of a cylindrical object. At small q, the scattering is only sensitive to the

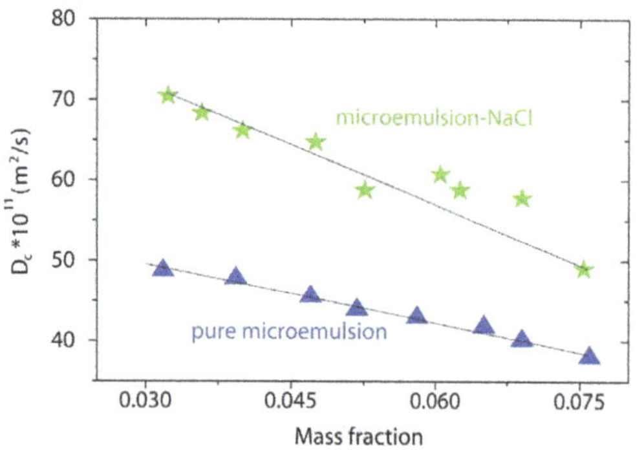

Figure 3. The normalized collective diffusion coefficient as function of mass fraction for AOT/H_2O/Decane with X = 6.7 (up triangle points) and AOT/H_2O/Decane/NaCl with X = 6.7, [NaCl]/[AOT] = 0.005 molar/l (Star points) at the temperature 20°C.

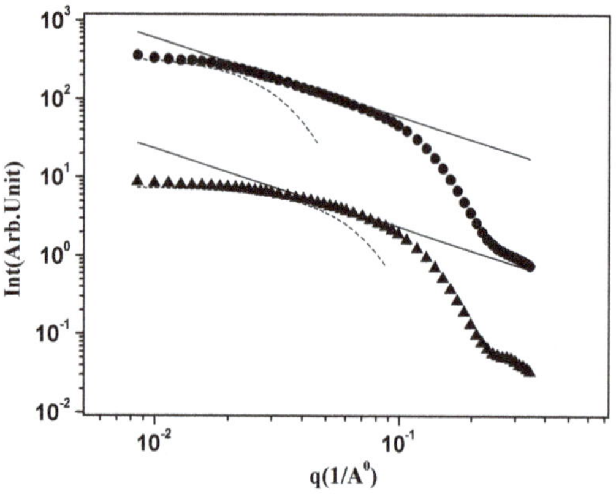

Figure 4. SAXS intensity $I(q)$ of a AOT/H_2O/Decane microemulsion (▲up triangle points) with X = 6.7 and droplet mass fraction (m_f = 0.1) and AOT/H_2O/Decane/NaCl microemulsion (● circle points) with X = 6.7 at constant droplet mass fraction (m_f= 0.1) and [NaCl]/[AOT] = 0.005. The red line is the line with slop (−1) that shown cylinder behavior of the SAXS experiments and dot line at low q is the fit of Guinier's law, $I(q) \approx \exp(-(q\xi)^2/3)$ and ξ change from 50 Å to 30 Å with adding NaCl at 20°C.

overall dimension of the scattering particles, and we analyzing with Guinier's law, $I(q) \approx \exp(-(q\xi)^2/3)$, where ξ is a correlation length.

The correlation length was changed from 60 Å to 30 Å with adding NaCl to the AOT/H_2O/Decane microemulsion, **Figure 4**. The change in the correlation length of microemulsion can increase the collective diffusion coefficient of droplets. For analyzing the data, we applied a model for a mixture of core-shell spheres with an added depletion attraction due to dissolved non-adsorbing polymer. The scattering intensity as a function of scattering vector $I(q)$ of spherical Monodisperse particles can be describe with a form factor component $F(q)$, which is proportional to the scattering of a single particle, and a structure factor $S(q)$, which describes the interaction effect [15,16]:

$$I(q) = cF^2(q)S(q) \tag{2}$$

c being a prefactor, which contains density of scattering particles. For the general case of n shells around a spherical droplet core the form factor reads.

$$F(q) = 4\pi\sum_{i=0}^{n}\Delta\rho_i\left(\frac{\sin(qR_i) - qR_i\cos(qR_i)}{q^3}\right) \tag{3}$$

Where R_i is the radius of the ith shell or, respectively, the core R_0 and $\Delta\rho_i$ is the electron density contrast between the shells i and I + 1 with ρ_{n+1} and ρ_0 being the electron density of the solvent and the core, respectively. The structure factor is the Fourier transform of the pair correlation function $g(r)$.

$$S(q) = 1 + 4\pi n\int_0^\infty (g(r)-1)\cdot r^2\frac{\sin(qr)}{qr}dr \tag{4}$$

The pair correlation function gives the probability to find another particle at a distance r from the center of a given particle, relative to the probability to find a particle at this distance in an ideal gas.

It is closely related to the total correlation function $h(r) = g(r) - 1$ and it can by means of the Ornstein-Zernike equation [17,18]. The **Figure 5** shows the compare of the SAXS experiment with the results of the hard sphere model, the points are experiments data of SAXS and the red lines are hard sphere model with core-shell radius 30nm. This results show that AOT/H_2O/Decane/NaCl microemulsion at X = 6.7 and m_f = 0.1 isn't hard sphere and more similar to cylinder shape. The increase of the NaCl to the microemulsion decrease the length scale of

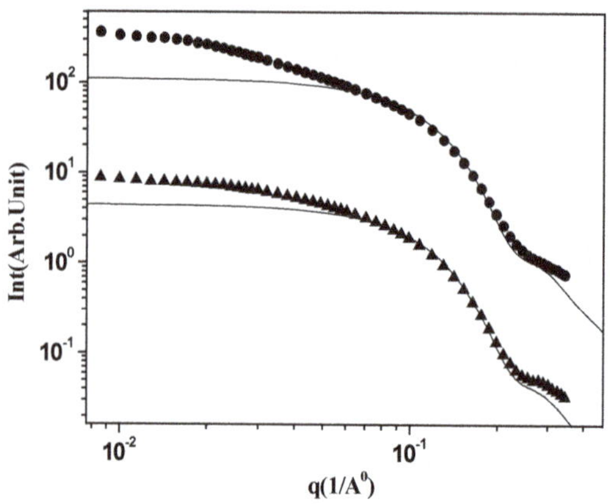

Figure 5. SAXS intensity $I(q)$ of a AOT/H_2O/Decane microemulsion (▲up triangle points) with X = 6.7 and droplet mass fraction (m_f = 0.1) and AOT/H_2O/Decane/NaCl microemulsion (■cubic points) with X = 6.7 at constant droplet mass fraction (m_f = 0.1) and [NaCl]/[AOT] = 0.005 at temperature 20°C. The red line is the hard sphere with core shell model with the core radius 25 nm, core-shell radius 30 nm.

the cylinder and droplets become more similar to the sphere. So, the NaCl can change the cylindrical AOT/H_2O/Decane microemulsion to the spherical microemulsion.

Our results show that the length scale of the droplet changes from 60 Å to 30 Å and shape of droplets changed from spherical to cylindrical by increasing the mass fraction. The study of the $C_{12}E_5$ microemulsion at different droplet mass fraction shows a repulsive behavior that with increase of concentration of PEG on the $C_{12}E_5$ microemulsion, system become attractive [19,20]. In the AOT microemulsion, the collective diffusion coefficient shows a negative slop at mass fraction and so shows a attractive interaction that with add NaCl the interaction stay attractive.

4. Conclusion

The study of the collective diffusion coefficients of AOT/H_2O/decane microemulsions showed a negative slope as a function of droplet mass fraction at the low mass fraction. Small-angle X-ray scattering measurements revealed that AOT microemulsions show cylinder behavior at X = 6.7 and mass fraction 0.1. The NaCl can change the shape of nano-droplets from cylindrical to the spherical and the shape changing of nano-droplets can describe the increasing of collective diffusion coefficient as a function of the mass fractions.

5. Acknowledgements

This work was supported by the University of Sistan and Baluchestan.

REFERENCES

[1] P. Kumar and K. L. Mittal, "Handbook of Microemulsion Science and Technology," Marcel Dekker, New York.

[2] L. K. Pershing, L. D. Lambert and K. Knutson, "Mechanism of Ethanol-Enhanced Estradiol Permeation across Human Skin *in Vivo*," *Pharmaceutical Research*, Vol. 7, No. 2, 1990, pp. 170-175. doi:10.1023/A:1015832903398

[3] P. Liu, T. Kurihara-Bergstrom and W. R. Good, "Cotransport of Estradiol and Ethanol through Human Skin *in Vitro*: Understanding the Permeant/Enhancer Flux Relationship," *Pharmaceutical Research*, Vol. 8, No. 7, 1991, pp. 938-944. doi:10.1023/A:1015876117627

[4] Y.-H. Kim, A.-H. Ghanem, H. Mahmoud and W. I. Higuchi, "Short Chain Alkanols as Transport Enhancers for Lipophilic and Polar/Ionic Permeants in Hairless Mouse Skin: Mechanism(s) of Action," *International Journal of Pharmaceutics*, Vol. 80, No. 1-3, 1992, pp. 17-31. doi:10.1016/0378-5173(92)90258-4

[5] S. Sharifi, M. Amirkhani, J. M. Asla and M. R. Mohammadi and O. Marti, "Light Scattering and SAXS Study of AOT Microemulsion at Low Size Droplet," *Soft Nanoscience Letters*, Vol. 2, No. 1, 2012, pp. 76-80. doi:10.4236/snl.2012.21002

[6] S. N. Tenjarla, "Microemulsions: An Overview and Pharmaceutical Applications," *Critical Reviews in Therapeutic Drug Carrier System*, Vol. 16, 1999, pp. 461-521.

[7] W. Brown, (Ed.), "Dynamic Light Scattering: The Method and Some Applications," Clarendon, Oxford, 1993.

[8] J. Bergenholtz, A. Romagnoli and N. J. Wagner, "Viscosity, Microstructure, and Interparticle Potential of AOT/H_2O/n-Decane Inverse Microemulsions," *Langmuir*, Vol. 11, No. 5, 1995, pp. 1559-1570. doi:10.1021/la00005a025

[9] S. Sharifi, P. Kudla, C. L. P. Oliveira, J. S. Pedersen and J. Bergenholtz, "Variations in Structure Explain the Vicometric Behavior of AOT Microemulsions at Low Water/AOT Molar Ratios," *Zeitschrift für Physikalische Chemie*, Vol. 226, No. 3, 2012, pp. 1-18. doi:10.1524/zpch.2012.0173

[10] T. Blochowicz, C. Gögelein, T. Spehr, M. Müller and B. Stühn, "Polymer-Induced Transient Networks in Water-in-Oil Microemulsions Studied by Small-Angle X-Ray and Dynamic Light Scattering," *Physical Review E*, Vol. 76, No. 4, 2007, p. 041505. doi:10.1103/PhysRevE.76.041505

[11] P. N. Pusay, "Intensity Fluctuation Spectroscopy of Charged Brownian Particles: The Coherent Scattering Function," *Journal of Physics A*, Vol. 11, No. 1, 1978, p. 119.

[12] T. Nose and B. Chu, "Static and Dynamical Properties of Polystyrene in Trans-Decalin. 3. Polymer Dimensions in Dilute Solution in the Transition Region," *Macromolecules*, Vol. 13, No. 1, 1979, pp. 122-132. doi:10.1021/ma60073a024

[13] M. Kotlarchyk, S.-H. Chen, J. S. Huang and M. W. Kim, "Structure of Three-Component Microemulsions in the Critical Region Determined by Small-Angle Neutron Scattering," *Physical Review A*, Vol. 29, No. 4, 1984, p. 2054. doi:10.1103/PhysRevA.29.2054

[14] G. R. Deen, C. L. P. Oliveira and J. S. Pedersen, "Phase Behavior and Kinetics of Phase Separation of a Nonionic Microemulsion of C12E5/Water/1-Chlorotetradecane upon a Temperature Quench," *Journal of Physical Chemistry B*, Vol. 113, No. 20, 2009, pp. 7138-7146. doi:10.1021/jp808268m

[15] M. Schwab and B. Stuhn, "Relaxation Phenomena and Development of Structure in a Physically Crosslinked Nonionic Microemulsion Studied by Photon Correlation Spectroscopy and Small Angle X-Ray Scattering," *Journal of Chemical Physics*, Vol. 12, No. 14, 2000, pp. 6461-6471. doi:10.1063/1.481207

[16] M. Nayeri, M. Zackrisson and J. Bergenholtz, "Scattering Functions of Core-Shell-Structured Hard Spheres with Schulz-Distributed Radii," *Journal of Physical Chemistry B*, Vol. 113, No. 14, 2009, pp. 8296-8302. doi:10.1021/jp811482w

[17] G. Fritz-Popovski, "Determination of Colloidal Interaction Potentials from Small Angle Scattering Data," *Journal of Chemical Physics*, Vol. 131, No. 24, 2009, pp. 8296-8302. doi:10.1063/1.3231606

[18] B. Weyerich, J. Brunner-Popela and O. Glatter, "Small-Angle Scattering of Interacting Particles. II. Generalized

Indirect Fourier Transformation under Consideration of the Effective Structure Factor for Polydisperse Systems," *Journal of Chemical Physics*, Vol. 32, No. 11, 1999, pp. 197-209. doi:10.1107/S0021889898011790

[19] S. Sharifi and M. Amirkhani, "Light Scattering Study of Mixture of Polyethylene Glycol with C12E5 Microemulsion," *Soft Nanoscience Letters*, Vol. 1, No. 3, 2011, pp. 76-80. doi:10.4236/snl.2011.130144

[20] S. Sharifi and A. Alavi, "Dynamic Light Scattering Study of Microemulsion," *Proceedings of the SPIE*, Vol. 8001, 2011, pp. 80012R-80012R-8. doi:10.1117/12.892990

Solvatochromism and Molecular Selectivity of C-(4-chlorophenyl)-N-phenylnitrone: A Photophysical Study

Sneha Salampuria[1], Tandrima Chaudhuri[2*], Manas Banerjee[1*]
[1]Department of Chemistry, University of Burdwan, Burdwan, India
[2]Department of Chemistry, Dr. Bhupendra Nath Dutta Smriti Mahavidyalaya, Hatgobindapur, India
Email: *tanchem_bu@yahoo.co.in, *manasban@rediffmail.com

ABSTRACT

The ground state interaction of C-(4-chlorophenyl)-N-phenylnitrone (N1) with three different α,β-unsaturated ketones (K1 - K3) in very dilute solution (10^{-6} mol·dm^{-3}) has been noticed through charge transfer band formation in the visible region. The experimentally measured transition dipole, ground state resonance energy and formation constants of the complexes indicate interaction selectivity of the acyclic nitrone (N1) for the ketones. Molar absorptivity of the absorbing complexes were determined for all the three N1/K (1:1) interacting systems in toluene. Experimental findings were well rationalized with the help of electron density based global electrophilicity and nucleophilicity indices as well as with frontier molecular orbital calculations.

Keywords: 1,3-DC Efficiency; Binding Constant; Nitrone Selectivity; DFT

1. Introduction

Chemists are increasingly interested in the cooperative effect in non-covalent interactions, which can be successfully employed for the construction of quite complex structures [1]. Multiple hydrogen bonds, coordination bonds, as well as several other forms of homogeneous and heterogeneous weak interactions were thought to have been effective to bind two different entities strongly. In both chemistry and biology non-covalent electron donor-acceptor (EDA) interactions have received much attention in recent years [2-4]. These EDA complex compounds can be utilized as photocatalysts [5], organic semiconductors [6] and dendrimers [7]. They are important in studying redox process [8], nonlinear optical activity [9] and in microemulsion activity [10]. EDA interactions also play a vital role in the field of drug-receptor binding mechanisms [11].

The polar heterocyclic compounds show strong solvatochromism [12]. Nitrones, an important category of 1,3-dipolar species [13-17], undergo facile concerted [$_{\pi}4_s$ + $_{\pi}2_s$]-cycloaddition to different dipolarophiles producing isoxazolidines which serve as a key step in a number of natural product syntheses [13,18-20]. Nitrones are a family of such heterocyclic compounds that have been studied extensively [13-22] due to their stability.

The 1,3-dipolar cycloaddition reaction between a nitrone and an unsaturated dipolarophile is a well-known process in the ground state leading to cyclic adducts. Much computational [23,24] investigation of these reactions were done, mostly following the optimization of transition state (TS) which is energetically an activated complex lying between the reactants and products. The existence of ground state activated complex of these reactions has prompted us to see whether there exists some weak complexation of the reactants in ground state in the solution phase. Any such possibility could be monitored and studied experimentally by common techniques [25] through measuring binding property and formation of such weak complexes [26].

A link is therefore sought for predicting the cycloaddition efficiency in terms of tendency towards formation of weak complexes in solution, despite that many other dominant factors prevailing in solution might affect the prediction in gas phase. However, we have explored this avenue in conjunction with reactivity based on DFT computed electrophilicity/nucleophilicity indices in order to predict the efficiency of 1,3-DC reactions. The success of our model seems to be revealed through the results and discussion section.

Here, we have reported the solvatochromic properties of C-(4-chlorophenyl)-N-phenylnitrone (N1 in **Figure 1**). The objective of this present work is to investigate systematically the intermolecular interactions associated

*Corresponding authors.

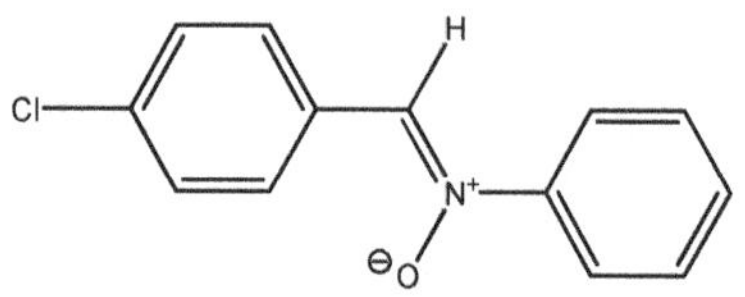

C-(4-chlorophenyl)-N-phenylnitrone (N1)

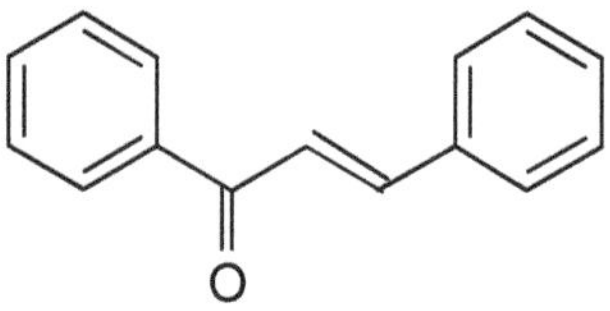

Benzal acetophenone (K1)

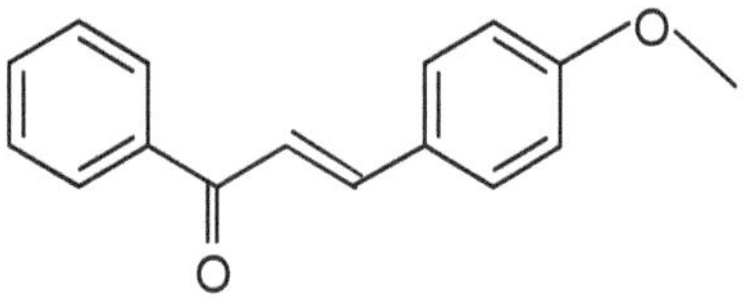

Anisalacetophenone (K2)

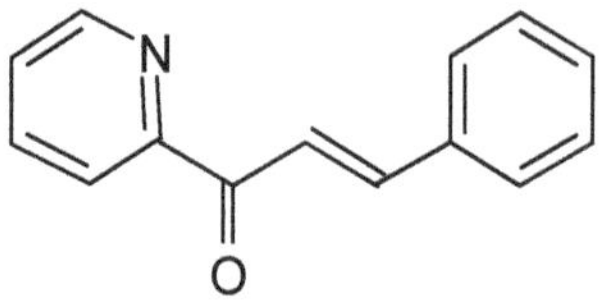

Benzylidene 2-acetylpyridine (K3)

Figure 1. Structures of the nitrone (N1) and the three α,β-unsaturated ketones (K1 - K3) used here.

with the nitrone molecules (**Figure 1**) in different solvent environments as well as in the presence of different α,β-unsaturated ketones. On these grounds, we decided to observe the non-covalent interaction occurring mainly through charge transfer between the nitrone N1 and three different unsaturated ketones K1, K2 and K3 (shown in **Figure 1**) which could predict the molecular selectivity of the nitrone. When we observe the electronic spectra of the complex and the individual compounds involved in the complex, we find a new electronic absorption band attributed to neither the nitrone (N1) nor the ketones (K). In our previous study [27] the molecular selectivity of [60]-fullerene among different porphyrin molecules was checked by using the fluorescence probe method. The present paper reports the trend of forming charge transfer complexes by the nitrone (N1) with the unsaturated ketones (K1 - K3) in terms of experimental transition dipole strengths, resonance energies and formation constants of the complexes in non-polar toluene medium. Experimental findings of the trend in non-covalent interaction were well substantiated theoretically with the help of DFT calculated electrophilicity and nucleophilicity indices [23,24,28-30] as well as in terms of frontier molecular orbital energies [24].

2. Experimental Section

2.1. Materials

Toluene of HPLC grade (for spectroscopic study from Merck Chemical Co.) was used as solvent without further purification. C-(4-chlorophenyl)-N-phenylnitrone (N1) was synthesized and purified according to the existing procedure [18,19]. Benzalacetophenone (K1), anisalacetophenone (K2) and benzylidene-2-acetylpyridine (K3) were synthesized according to literature [31].

2.2. Apparatus and Methods

UV Spectra were recorded using a Shimazdu UV-2450 spectrophotometer. The steady state fluorescence emission spectra were recorded on a Hitachi F-4500 spectrofluorimeter equipped with a temperature controlled cell holder.

The geometries of molecular structures were optimized using the hybrid density functional B3LYP method using 6-31G(d) basis set. All the calculations were performed on a IBM-HS21 server cluster/LAN running parallel version of Linda, employing the LINUX version of Gaussian09 [32] together with Gaussview05.

3. Results and Discussion

3.1. Photophysical Study/Solvent Interaction

The photophysical parameters [electronic absorption maxima (λ_{abs}), emission maxima (λ_{em}) and fluorescence quantum yield (Φ_{fl})] of the nitrone (N1) were determined in a few polar, nonpolar, both protic and aprotic solvents. The energy of maximum absorption [$E(A)$], fluorescence [$E(F)$] and Stoke's shift [$E(St)$] were reported by using the conversion:

$$E/\text{kcal}\cdot\text{mol}^{-1} = 28590/(\lambda/\text{nm})$$

The measured $E(A)$ and $E(F)$ values for the nitrone N1 alone in solvent were given in **Table 1**. The nitrone shows intense $S_0 \rightarrow S_1$ absorption at 86.37 (kcal·mol^{-1}) in carbon tetrachloride (CCl_4) and the absorption maxima were seen to be shifted hypsochromically by (1 - 3 kcal·mol^{-1}) in other solvents. For the steady state fluorescence, solvent-dependent shifts of the nearest fluorescence band were more or less similar (1 - 5 kcal·mol^{-1}) in the respective solvents. In aprotic solvents the fluorescence band shows almost similar type of structure which looses their prominence as the polarity of the solvent is increased (in **Figure 2**). The possible origin of structure of the band is presumably the nitrone vibrations.

The maximum absorption energies $E(A)$ of N1 expressed in kcal·mol^{-1} for aprotic and protic solvents, show a linear correlation with the Reichardt parameter E_T (30) [12], as was shown in **Figure 3**. A linear correlation with the $E_T(30)$ scale of solvent polarity is indicative of

Table 1. Photophysical parameters of N1 in different solvent media.

S_l No.	Solvent	(kcal/mol)			$\varepsilon \times 10^{-4}$ ($dm^3 \cdot mol^{-1} \cdot cm^{-1}$)
		E (A)	E (F)	E (St)	
1	DMSO	87.64	69.09	18.55	2.63
2	ACN	88.59	70.73	17.86	2.40
3	MeOH	89.62	73.19	16.42	2.55
4	Bzn	86.56	73.12	13.44	2.21
5	EtOH	88.95	73.08	15.90	2.02
6	2-PrOH	88.98	72.64	16.34	2.81
7	1-BuOH	88.51	68.04	20.47	3.41
8	THF	87.51	70.42	17.09	2.65
9	Acetic acid	89.68	73.12	16.56	2.58
10	EtOAc	88.51	72.67	15.83	2.37
11	CCl_4	86.37	68.19	18.18	2.43
12	Toluene	86.58	73.04	13.53	1.68
13	n-Hexane	87.43	72.86	14.57	1.11

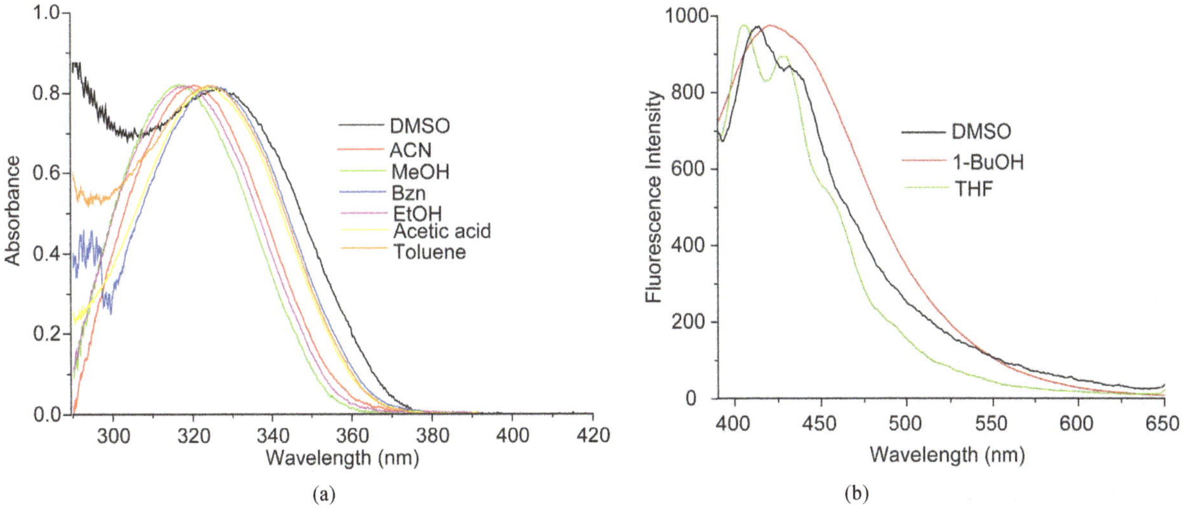

Figure 2. (a) Absorption and (b) emission spectra of N1 (4.75 × 10^{-6} mol·dm^{-3}) in different solvents.

intramolecular charge transfer (ICT) accompanying the $S_0 \rightarrow S_1$ transition. The positive slope of $E(A)$ vs $E_T(30)$ plot in **Figure 3** indicated that the transition is accompanied by a decrease in the dipole moment for the nitrone N1, as is also true for the indicator dye of the $E_T(30)$ scale. The poorer sensitivity of $E(F)$ compared to $E(A)$ towards change in the solvent polarity was observed. This could be intelligible in terms of decreased solute-solvent dipolar interaction in the excited state.

Besides the polarity aspect, Gutmann's donor number (DN) [33] and modified acceptor number (AN^*) of solvents [34] can be used as these are measures of the strength of solvents as Lewis bases and acids respectively. On their basis the effects on $E(A)$ and $E(F)$ could be explained. The maximum absorption energies $E(A)$ of N1 in aprotic and protic solvents, show non-linear growth with the acceptor number of solvents, as has been shown in **Figure 4(a)**. The exponential curvature of the plot indicated that with increasing Lewis acidity, hypsochromic shift in absorption energy is observed and an opposite effect will be expected with increasing basicity of the solvent. It was found that acidity shows better correlation with $E(A)$ than shown by basicity. The opposite type sensitivity of $E(F)$ compared to $E(A)$ towards varia-

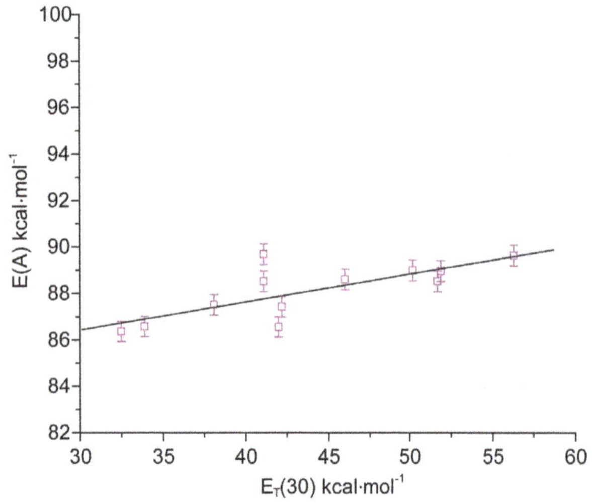

Figure 3. ***E(A)*** **vs** ***E$_T$*****(30) plot for N1.**

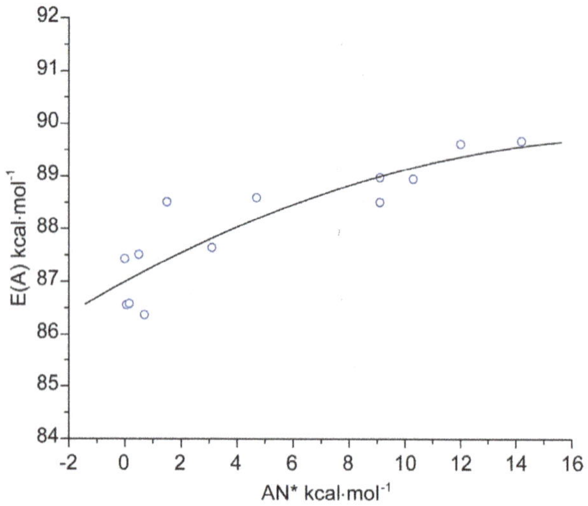

(a)

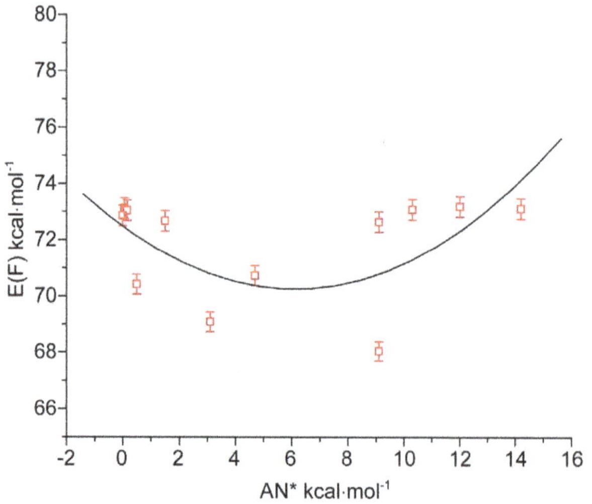

(b)

Figure 4. (a) Plot of ***E(A)*** **of N1 against acceptor number (AN*) of solvents; (b) Plot of** ***E(F)*** **of N1 against acceptor number (AN*) of solvents.**

tion in the acceptor number (AN*) of solvents may be seen from the curvature of the plot in **Figure 4(b)**. It may be noted that with increasing acidity of solvent media energy of fluorescence *E*(*F*) passes through a minimum, just opposite to that in the case of absorption and that the correlation was also poor. The difference is obviously due to different characteristics of the ground and excited states.

Multiple linear regression analysis has been done where correlation of *E*(*A*) or *E*(*F*) was sought with respect to the solvatochromic parameters representing dipolarity (π^*), hydrogen bond donating (α, HBD) and hydrogen bond accepting (β, HBA) ability of solvents (following the Taft's scale [35]: π^*, α and β). This was done with the help of the expression (1)

$$(E_x) = (E_x)_0 + A\alpha + B\beta + C\pi^* + \cdots \quad (1)$$

where $(E_x)_0$ is the value of (E_x) in a hypothetical inert solvent, and A, B and C are the adjusted coefficients reflecting the dependence of E_x on the α, β and π^* parameters. The following regression equations were obtained:

$$E(A) = 86.65 + 2.23\alpha + 0.42\beta + 0.81\pi^* + \cdots; \quad (n = 13,\ r = 0.67) \quad (2)$$

$$E(F) = 74.68 - 0.26\alpha + 1.98\beta - 5.64\pi^* + \cdots; \quad (n = 13,\ r = 0.34) \quad (3)$$

$$[E(St)] = 12.79 + 2.35\alpha - 1.86\beta + 5.57\pi^* + \cdots; \quad (n = 13,\ r = 0.20) \quad (4)$$

Although N1 showed better correlation (r) of absorption energy *E*(*A*) in terms of the Taft's π^*-, α- and β-parameters, the correlation coefficient (r = 0.34) was far from unity for *E*(*F*) due to the low solvatochromic shifts observed in the emission (**Table 1**), leading to a higher dispersion of the experimental values. In our previous studies we have seen that pyromethene dyes [36] or saturated β-diketones [37] also showed similar features. The adjusted coefficient (A) describing the HBD ability of solvent was the major coefficient in Equation (2) only. For the energies of absorption as well as the Stoke's shift, the A coefficients had positive values, corroborating the observed hypsochromic shift with respect to increasing solvent acidity, while for emission, the shift was bathochromic in nature and the coefficient (*A*) had negative sign. The opposite effect was observed as the basicity of solvent was increased. The coefficient (*C*) reflecting the dipolarity/polarizability of solvent, assumed positive value in absorption and negative in the case of emission, suggesting that the absorption bands shift slightly towards higher energies while the emission bands towards lower energies as the polarity of the solvent was increased. The basicity of solvent (β-scale) and dipolarity (π^*-scale) do not evidently affect the absorption while acidity or basicity (α and β-scale) do not much affect the fluorescence minimum since the corresponding adjusted coefficients (B and C in Equation (2), A and B in Equation (3) and B in Equation (4)) were within their standard deviations.

It appear that solvation is mainly controlled here through hydrogen bond donation (α) and dipolar interaction (π^*) with the solvent. The ratio of the regression coefficients of α and π^* indicates the relative importance of HBD over dipolarity interaction in the ground state. The ratio is ca. 2.75 in respect of $E(A)$, whereas the trend is just reciprocal for $E(F)$, implying that HBD plays a significantly greater role in the absorption while dipolarity does in emission.

3.2. Appearance of CT Bands

It was observed that absorption band maximum of N1 in the nonpolar toluene solvent shifted towards red as some dipolarophile compound was added to the solution. **Figure 5(A)** shows the electronic absorption spectra of mixtures containing (N1 and K) in toluene and the shifted band is attributed to CT owing to its characteristic nature. In order to obtain these CT bands, spectra of above solutions were recorded against the pristine nitrone (N1) solution in the respective solvent media as reference to cancel out its own absorbance. It was observed that the new absorption peaks appeared in the visible region. **Figure 5(B)** shows the complete absorption spectra of N1 in the absence and in the presence of increasing concentrations of the ketone K3. The CT absorption peaks were well characterized by fitting to the gaussian function

$$y = y_0 + A\Big/\left[w\sqrt{\left(\pi/2\exp\left[-2(x-x_c)^2/w^2\right]\right)}\right]$$

where x and y denote wavenumber and molar extinction coefficient respectively. When such a Gaussian fit (approximated by the above approximation) is applied over a wide range of (x, y) data points, y_0 obviously represents the lowest bound of the y-data values in the observation range and the remaining term in the expression represents a positive definite quantity. At the value $x = x_c$, the derivative dy/dx disappears giving

$y_c = y_0 + A\Big/\left[w\sqrt{(\pi/2)}\right]$ which provides a numerical significance of the band centre concerning the plot. However, as $x \to \infty$, $y \to y_0$ which implies a significance to the hypothetical absorbance value in the limit of infinite wavelength. One such fitted plot has been shown in **Figure 6**. The results of the Gaussian fit for all the three N1/K systems were presented in **Table 2**. The wavelengths at these new absorption maxima ($\lambda_{max} = x_c$) and the corresponding transition energies ($h\nu$) were summarized in **Table 3**. The gaussian analysis of fitting was done in accordance to the method developed by I. R. Gould *et al.* [38]. One important point to mention here is that the gaussian analysis of a curve generally provided a meaningful result near the maximum of the curve spread over a very small region.

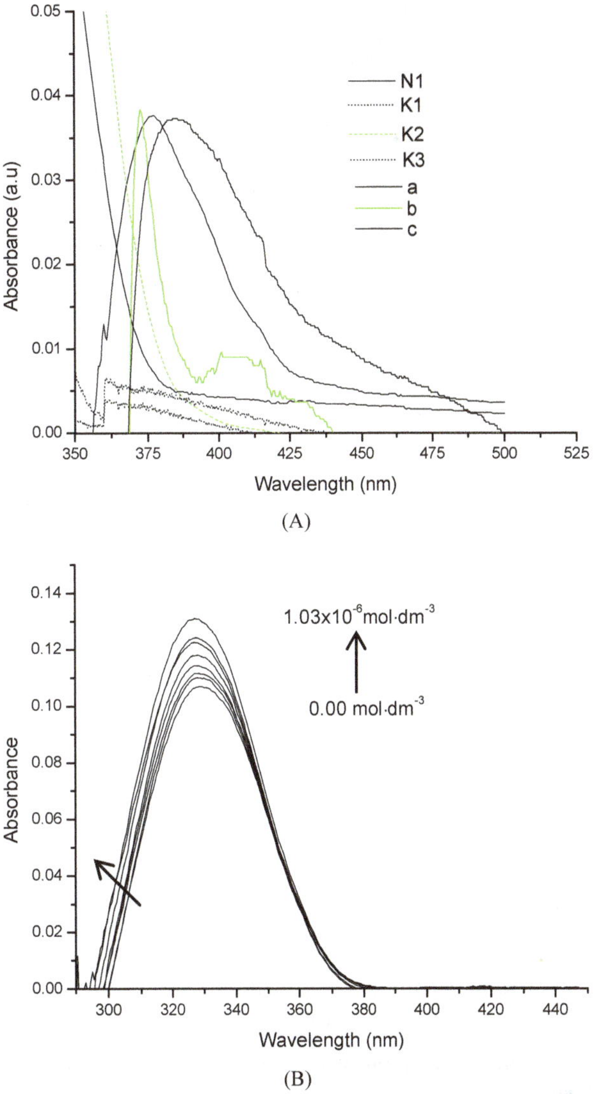

Figure 5. (A) Charge transfer absorption bands for all the three interacting systems: Band (a) was of N1/K1, (b) was of N1/K2 and (c) was of N1/K3 interacting systems in toluene medium; (B) The absorption spectra of N1 (5.18 × 10^{-6} mol·dm^{-3}) in the absence and in the presence of increasing concentrations of the ketone K3.

3.3. Determination of Oscillator Strength (f), Transition Dipole (μ_{EN}), Resonance Energy (R_N)

From the CT absorption spectra, we could estimate os-

Table 2. Gaussian curves analysis for the CT spectra of all the three complexes of N1.

System	Area under the curve $(A) \times 10^2$ ($dm^3 \cdot mol^{-1}$)	Width of the curve $(W) \times 10^3$ (cm^{-1})	Center of the curve $(x_c) \times 10^3$ (cm^{-1})	y_0 ($dm^3 \cdot mol^{-1} \cdot cm^{-1}$)
N1/K1	9.261 ± 0.525	0.19 ± 0.008	2.62 ± 0.002	109.002 ± 9.82
N1/K2	0.568 ± 0.053	0.06 ± 0.004	2.68 ± 0.001	32.69 ± 2.84
N1/K3	20.31 ± 5.148	0.29 ± 0.04	2.58 ± 0.004	122.82 ± 78.40

Table 3. CT absorption maxima and transition energies of the complexes, oscillator strengths (*f*), transition dipole strengths (μ_{EN}), resonance energies (R_N) and theoretically calculated heat of formation of the three complexes of N1.

System	λ_{CT} (nm)	$h\nu_{CT}$ (eV)	$10^9 \times f$ ($dm^3 \cdot mol^{-1} \cdot cm^{-2}$)	μ_{EN} (D)	R_N (eV)	ΔH_f ($kcal \cdot mol^{-1}$)
N1/K1	380.93	3.25	0.400 ± 0.023	0.565 ± 0.134	3.52	−4.08
N1/K2	373.27	3.32	0.025 ± 0.002	0.138 ± 0.042	3.50	−3.58
N1/K3	387.95	3.19	0.877 ± 0.222	0.842 ± 0.424	3.53	−4.20

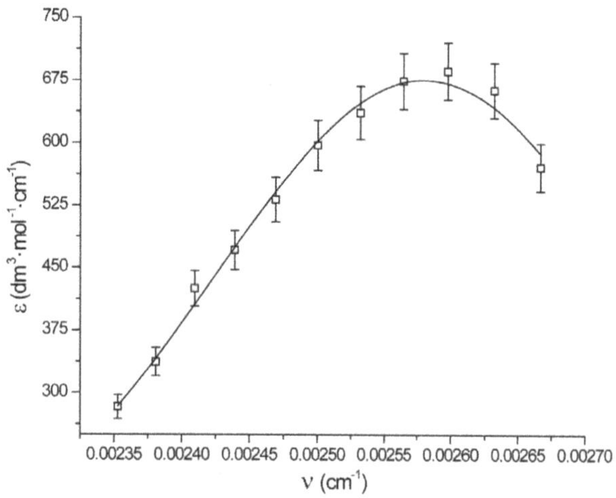

Figure 6. Gaussian analysis of CT spectra of N1/K3 system in toluene.

cillator strength [37] (*f*) for the transition. The oscillator strength is proportional to the square of transition dipole $|\mu_{EN}|$. The observed oscillator strengths of the CT bands are summarized in **Table 3**.

It has been observed that transition dipole strength (μ_{EN}) for the N1/K3 complex was slightly higher than those for the other N1/K complexes.

Again, the resonance energy of the complex in the ground state (R_N) [39] is a contributing factor (a ground state property) to the stability of the complex. The values of R_N for the complexes under study have been provided in **Table 3**. It appears from the trend in R_N values that N1 forms stronger complexes with K1 and K3 as compared to the N1/K2 complex in the non-polar toluene solvent.

3.4. Determination of Formation Constants (K)

The formation constants of the **N1/K** complexes were determined using the Benesi-Hildebrand (BH) [25] equation in the form

$$\frac{1}{d} = \frac{1}{\varepsilon[\text{N1}]_0} + \frac{1}{\text{K}\varepsilon[\text{N1}]_0[\text{K}]_0}. \quad (5)$$

Here $[\text{N1}]_0$ and $[\text{K}]_0$ are the initial concentrations of the nitrone and dipolarophile respectively, d is the absorbance of the donor–acceptor complex at λ_{CT} measured against solvent as reference. Here $d = \left[d_{mix} - d^0_{\text{N1}} - d^0_K\right]$ where d_{mix}, d^0_{N1} and d^0_K are the absorbances of the donor-acceptor mixture, pure N1 and K solutions respectively at the same molar concentrations as are present in the mixture and at the same wavelength against the solvent as reference. The molar extinction coefficient ε is not quite that of the complex. Equation is valid [25] under the approximation $[\text{N1}]_0 \gg [\text{K}]_0$ for 1:1 donor-acceptor complexes. The intensity or the corrected absorbance (d) in the visible portion of the absorption band, increases systematically with gradual addition of the dipolarophile solution as shown in **Table 4**. Thus, it is established in this work that the substantial red shift in the broad 300 - 400 nm absorption band of N1 is due to formation of 1:1 molecular complex between K and N1. The equilibrium constant values were calculated using the Benesi-Hildebrand (BH) model. In all the cases very good linearity were obtained. One such plot has been shown in **Figure 7**.

The following linear regression equations have been obtained from the procured data in toluene at 303 K. For N1/K1 system:

$$\frac{1}{d} = (31.95 \pm 3.78) + \frac{(8.49 \pm 1.26) \times 10^{-5}}{[K1]_0}, \quad (6)$$
$$(\text{correlation coefficient}: r = 0.92)$$

It was observed that N1 binds most strongly with K3 ($[\text{K3}] = 5.31 \times 10^5$ $dm^3 \cdot mol^{-1}$) than with K1 ([K1] = 3.76

Table 4. Data for spectroscopic determination of stoichiometry, formation constant (K) and molar absorptivity (ε) for the N1/K3 complex in toluene at temperature 298 K.

Sl. No.	[K3] × 10^6 (mol·dm^{-3})	[N1] ×10^5 (mol·dm^{-3})	d (at λ_{CT})	K × 10^{-5} dm^3·mol^{-1}	$\varepsilon \times 10^{-3}$ (dm^3·mol^{-1}·cm^{-1})
1	0.115		0.0018		
2	0.229		0.0015		
3	0.344		0.0054		
4	0.459		0.0056		
5	0.574	0.5184	0.0069	5.308 ± 2.189	5.65
6	0.689		0.0077		
7	0.804		0.0085		
8	0.919		0.0095		
9	1.033		0.0106		

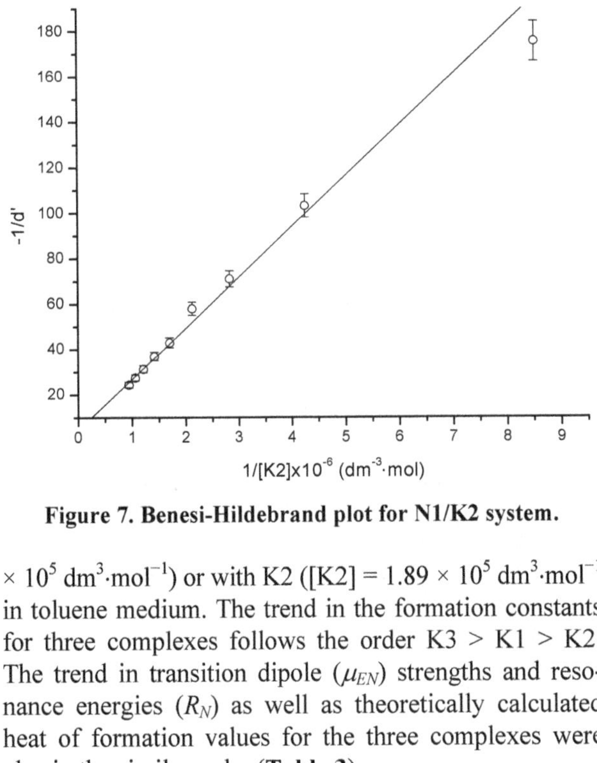

Figure 7. Benesi-Hildebrand plot for N1/K2 system.

× 10^5 dm^3·mol^{-1}) or with K2 ([K2] = 1.89 × 10^5 dm^3·mol^{-1}) in toluene medium. The trend in the formation constants for three complexes follows the order K3 > K1 > K2. The trend in transition dipole (μ_{EN}) strengths and resonance energies (R_N) as well as theoretically calculated heat of formation values for the three complexes were also in the similar order (**Table 3**).

3.5. Computed Philicity Indices

The global electrophilicity index [28,29] ω measures the stabilization in energy when the molecular system acquires an additional electronic charge ΔN from the environment. It can be simply expressed in terms of electronic chemical potential (μ) and chemical hardness (η). These have been represented [24] by a very simple operational formulation in terms of the one-electron orbital energies of FMO, viz. the HOMO and LUMO given as:

$$\omega = \mu^2/2\eta \tag{7}$$

where

$$\mu \approx (\varepsilon_{\text{HOMO}} + \varepsilon_{\text{LUMO}})/2 \tag{8}$$

and

$$\eta \approx \varepsilon_{\text{LUMO}} - \varepsilon_{\text{HOMO}} \tag{9}$$

The global electrophilicity index includes the propensity of the electrophile to acquire an additional electronic charge as well as its resistance to exchange the electronic charge with the environment simultaneously. Thus, a good electrophile can be characterized by a high value of ω and a low value of η.

Again the recently introduced [23,30] global nucleophilicity index (N), has been based on the relationship $N = -IP$, where IP is the gas phase (intrinsic) ionization potential which can be directly extended to describe the local nucleophilicity. Within the simplest approximation to the nucleophilicity, the IP values can be approximated in terms of the HOMO energy in a molecule within a given molecular orbital (MO) scheme. The nucleophilicity index N for a given system, was therefore defined [23,40] as $N = \varepsilon_{\text{HOMO(Nu)}} - \varepsilon_{\text{HOMO(TCE)}}$ (in eV units) where $\varepsilon_{\text{HOMO(Nu)}}$ is the HOMO energy of the nucleophile and $\varepsilon_{\text{HOMO(TCE)}}$ corresponds to the HOMO energy of the tetracyanoethylene (TCE) taken as reference.

In **Table 5** we have presented the HOMO and LUMO energies of N1, K1, K2 and K3 along with their electronic chemical potential (μ), chemical hardness (η), global electrophilicity (ω) and the nucleophilicity index (N). The electronic chemical potential of N1 (−3.81 eV) is higher than that of all K's (−4.20, −3.82 and −4.19 eV) implying that the CT will take place from the nitrone (N1) to the ketone (K) in all the three cases with predominant HOMO$_{N1}$ - LUMO$_K$ interaction for the N1/K3 system presented in **Table 5** and **Figure 8**. According to the absolute scale of electrophilicity [41] based on the ω index, N1 and K1 - K3 all belong to the realm of strong electro-

Table 5. HOMO and LUMO energies, and the global reactivity indices for nitrone (N1) and three ketones (K1 - K3).

System	E_{opt} (a.u)	HOMO (eV)	LUMO (eV)	μ (eV)	η (eV)	ω (eV)	N (eV)
N1	−1091.51	−5.63	−1.98	−3.81	3.65	1.99	3.49
K1	−654.04	−6.31	−2.09	−4.20	4.22	2.09	2.81
K2	−768.56	−5.85	−1.78	−3.82	4.07	1.79	3.27
K3	−670.08	−6.22	−2.17	−4.19	4.05	2.17	2.90

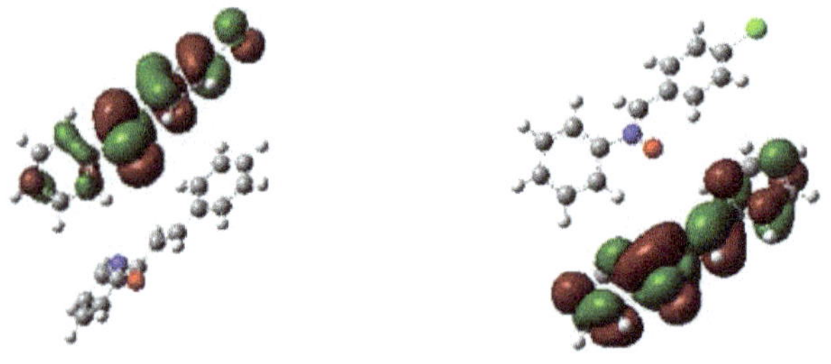

HOMO LUMO

Figure 8. Frontier molecular orbital picture of N1/K3 interacting system.

philes (values greater than 1.50 eV). Both K1 and K3 show greater values of ω than N1. These un-saturated ketones follow the electrophilicity order K3 > K1 > K2. Examining their nucleophilicity (N) indexes values, it can be noted that N1 is more nucleophilic (N = 3.49 eV) than the ketones. K1 (N = 2.81 eV) and K3 (N = 2.90 eV) have lower values of nucleophilicity index compared to K2 (N = 3.27 eV). Thus considering electrophilicity, K3 and K1 are both stronger electrophiles and N1 is the best nucleophile. Hence N1/K3 is expected to interact better as compared to the other N1/K, which was also observed experimentally (**Table 3**, 4th column). Experimentally, similar cycloadditions [42] were carried out using several nitrones which include our N1 and ketones including K1 and K2 of ours, but exactly their choice of reactant pairs do not coincide precisely with those of ours. However, they have reported the product ratios of different stereoisomers and also the total yields of the reactions carried out over a time span of about 20 hr. It is apparent from their work that the reaction gets faster from N1/K1 to N1/K2 producing better yields, thus supporting our conjecture that the sequence of weak binding constants between the corresponding reactant pairs follow: N1/K3 > N1/K1 > N1/K2. The philicity based computational results are in fair agreement with the experimental results in all respect.

4. Conclusions

From above, the following conclusions could be drawn.

- The nitrone favours charge transfer interaction with unsaturated ketones in the order K3 > K1 > K2 in terms of experimental binding constants (K), transition dipole strengths (μ_{EN}) and resonance energies (R_N) of the complexes and these are also in accordance with the electrophilicity (ω) and nucleophilicity (N) indexes of the ketones calculated at DFT/B3LYP/6-31G(d) level of theory on the optimized ground state geometries.
- In either case K2 has the least tendency to interact with the nitrone for entering into successful non-covalent charge transfer interaction.
- Overall, considering theory and experiment, K3 has strongest possibility to interact with this particular nitrone N1 in non-polar toluene medium.

5. Acknowledgements

Prof. Avijit Banerji of University College of Science, University of Calcutta (Department of Chemistry) has gifted the nitrone and ketones used in this study. The authors duly acknowledge him.

REFERENCES

[1] L. J. Prins, D. N. Reinhoudt and P. Timmerman, "Non-covalent Synthesis Using Hydrogen Bonding," *Angewandte Chemie International Edition*, Vol. 40, No. 13, 2001, pp. 2382-2426. doi:10.1002/1521-3773(20010702)40:13<2382::AID-ANIE2382>3.0.CO;2-G

[2] S. Bhattacharya, M. Banerjee and A. K. Mukherjee, "Study of the Formation Equilibria of Electron Donor-Acceptor Complexes between [60]-Fullerene and Methylbenzenes by Absorption Spectrometric Method," *Spectrochimica Acta Part A*, Vol. 57, No. 7, 2001, pp. 1463-1470. doi:10.1016/S1386-1425(00)00489-3

[3] S. Bhattacharya, S. Bhattacharya (Banerjee), K. Ghosh, S. Basu and Manas Banerjee, "Study of Electron Donor-Acceptor Complex Formation of *o*-ChloranilWith a Series of Phosphine Oxides and Tri-*n*-butyl Phosphate by the Absorption Spectrometric Method," *Journal of Solution Chemistry*, Vol. 35, No. 4, 2006, pp. 519-539. doi:10.1007/s10953-005-9013-x

[4] S. Bhattacharya, S. K. Nayak, S. K. Chattopadhyay, M. Banerjee and A. K. Mukherjee, "Absorption Spectroscopic Study of EDA Complexes of [70]Fullerene with a series of Methyl Benzenes," *Spectrochimica Acta Part A*, Vol. 57, No. 2, 2001, pp. 309-313. doi:10.1016/S1386-1425(00)00388-7

[5] Z. Zou, J. Ye and H. Arakawa, "Role of R in Bi2RNbO7 (R = Y, Rare Earth): Effect on Band Structure and Photocatalytic Properties," *Journal of Physical Chemistry B*, Vol. 106, No. 3, 2002, pp. 517-520. doi:10.1021/jp012982f

[6] A. Eychmuller and A. L. Rogach, "Chemistry and Photophysics of Thiol-Stabilized II-VI Semiconductor Nanocrystals," *Pure and Applied Chemistry*, Vol. 72, No. 1-2, 2000, pp. 179-188. doi:10.1351/pac200072010179

[7] T. H. Ghaddar, J. K. Whitesell and M. A. Fox, "Excimer Formation in a Naphthalene-Labeled Dendrimer," *Journal of Physical Chemistry B*, Vol. 105, No. 37, 2001, pp. 8729-8731. doi:10.1021/jp010933x

[8] A. S. Baranski and W. R. Fawcett, "Solvent Effects in Simple Electron Transfer Reactions," Winkler, K. Ed., *Journal of the Chemical Society, Faraday Transactions*, Vol. 92, 1996 pp. 3899-3904.

[9] M. Ricco, M. Bisbiglia, R. Derenzi and F. Bolzoni, "Observation of Superconductivity in TDAEC60," *Solid State Communications*, Vol. 101, No. 6, 1997, pp. 413-416.

[10] S. M. Andrade, S. M. B. Costa and R. Pansu, "Structural Changes in w/o TritonX-100/Cyclohexane-Hexanol/Water Microemulsions Probed by a Fluorescent Drug Piroxicam," *Journal of Colloid and Interface Science*, Vol. 226, No. 2, 2000, pp. 260-268. doi:10.1006/jcis.2000.6821

[11] A. Polozova and B. J. Litman, "Cholesterol-Dependent Recruitment of di22:6-PC by a G Protein-Coupled Receptor into Lateral domains," *Journal of Biophysics*, Vol. 79, No. 5, 2000, pp. 2632-2643. doi:10.1016/S0006-3495(00)76502-7

[12] C. Reichardt, "Solvatochromic Dyes as Solvent Polarity Indicators," *Chemical Reviews*, Vol. 94, No. 4, 1994, pp. 2319-2358. doi:10.1021/cr00032a005

[13] A. Padwa and W. H. Pearson, "Synthetic Applications of 1,3-Dipolar Cycloaddition Chemistry towards Heterocycles and Natural Products," John Wiley & Sons, New York, 2002. doi:10.1002/0471221902

[14] A. Banerji and P. Sengupta, "Recent Studies on 1,3-Dipolar Cycloadditions of Nitrones," *Journal of Indian Institute of Science*, Vol. 81, No. 3, 2001, pp. 313-323.

[15] A. K. Parhi and R.W. Franck, "A Weinreb Nitrile Oxide and Nitrone for Cycloaddition," *Organic Letters*, Vol. 6, No. 18, 2004, pp. 3063-3065. doi:10.1021/ol0489752

[16] Y. Zeng, B. T. Smith, J. Hershberger and J. Aube´, "Rearrangements of Bicyclic Nitrones to Lactams: Comparison of Photochemical and Modified Barton Conditions," *Journal of Organic Chemistry*, Vol. 68, No. 21, 2003, pp. 8065-8067. doi:10.1021/jo035004b

[17] K. B. Jensen, M. Roberson and K. A. Jorgensen, "Catalytic Enantioselective 1,3-Dipolar Cycloaddition Reactions of Cyclic Nitrones: A Simple Approach for the Formation of Optically Active Isoquinoline Derivatives," *Journal of Organic Chemistry*, Vol. 65, No. 26, 2000, pp. 9080-9084. doi:10.1021/jo001157c

[18] K. B. G. Torssell, "Nitrile Oxides, Nitrones and Nitronates in Organic Synthesis," VCH, Weinheim, 1988.

[19] E. Breuer, "Nitrones and Nitronic Acid Derivatives: Their Structures and Their Roles in Synthesis," In: S. Patai, Ed., *The Chemistry of Amino, Nitroso and Nitro Compounds and Their Derivatives*, John Wiley & Sons, New York, 1982.

[20] A. Banerji, D. Bandyopadhyay, T. Prangé and A. Neuman, "Unexpected Cycloadducts from 1,3-Dipolar Cycloaddition of 3,4-Dehydromorpholine *N*-Oxide to *N*-Cinnamoyl Piperidines—First Report of the Novel Formation of 2:1 Cycloadducts," *Tetrahedron Letters*, Vol. 46, No. 15, 2005, pp. 2619-2622. doi:10.1016/j.tetlet.2005.02.083

[21] R. Huisgen, "1,3-Dipolar Cycloaddition Chemistry," In: A. Padwa, Ed., Wiley, New York, 1984.

[22] K. V. Gothlf and K. A. Jorgensen, "Asymmetric 1,3-Dipolar Cycloaddition Reactions," *Chemical Review*, Vol. 98, No. 2, 1998, pp. 863-909. doi:10.1021/cr970324e

[23] L. R. Domingo, E. Chamorro and P. Pérez, "Understanding the Reactivity of Captodative Ethylenes in Polar Cycloaddition Reactions. A Theoretical Study," *Journal of Organic Chemistry*, Vol. 73, No. 12, 2008, pp. 4615-4624. doi:10.1021/jo800572a

[24] T. K. Das, S. Salampuria and M. Banerjee, "Computational DFT Study of the 1,3-Dipolar Cycloadditions of 1-Phenylethyl-*trans*-2-methyl Nitrone to Styrene and 1-Phenylethyl Nitrone to Allyl Alcohol," *Journal of Molecular Structure: THEOCHEM*, Vol. 959, No. 1-3, 2010, pp. 22-29. doi:10.1016/j.theochem.2010.08.001

[25] H. A. Benesi and J. H. Hildebrand, "A Spectrophotometric Investigation of the Interaction of Iodine with Aromatic Hydrocarbons," *Journal of American Chemical Society*, Vol. 71, No. 8, 1949, pp. 2703-2707. doi:10.1021/ja01176a030

[26] K. A. Connors, "Binding Constants: The Measurement of Molecular Complex Stabilility," John Wiley & Sons, New York, 1987.

[27] T. Chaudhuri, D. Goswami, M. Banerjee, S. Chattopadhyay and S. K. Nayak, "Supramolecular Selectivity of [60]-Fullerene among Equivalently Photoactive Porphyrins," *Journal of Luminescence*, Vol. 130, No. 10, 2010, pp. 1750-1755. doi:10.1016/j.jlumin.2010.04.004

[28] R. G. Parr, L. V. Szentpaly and S. Liu, "Electrophilicity Index," *Journal of American Chemical Society*, Vol. 121, No. 9, 1999, pp. 1922-1924. doi:10.1021/ja983494x

[29] A. Corsaro, V. Pistara, A. Rescifina, A. Piperno, M. Chiacchio and G. Romeo, "A DFT Rationalization for the Observed Regiochemistry in the Nitrile Oxide Cycloaddition with Anthracene and Acridine," *Tetrahedron*, Vol. 60, No. 31, 2004, pp. 6443-6451. doi:10.1016/j.tet.2004.06.052

[30] R. Contreras, J. Andres, V. S. Safont, P. Campodonico and J. G. Santos, "A Theoretical Study on the Relationship between Nucleophilicity and Ionization Potentials in Solution Phase," *Journal of Physical Chemistry A*, Vol. 107, No. 29, 2003, pp. 5588-5593. doi:10.1021/jp0302865

[31] J. P. Freeman, "Organic Synthesis, Collective Volume 1," R. L. Danheiser, (Ed.), 1941, p. 78. Volume 2, 1922, p. 1.

[32] A. Frisch, M. J. Frisch, F. R. Clemente and G. W. Trucks, "Gaussian 09 Cluster/LAN Parallel Version with Linda," Gaussian Inc., Wallingford, 2009.

[33] V. Gutmann, "The Donor-Aceeptor Approach to Molecular Interactions," Plenum Press, New York, 1978. doi:10.1007/978-1-4615-8825-2

[34] F. L. Riddle Jr. and F. M. Fowkes, "Spectral Shifts in Acid-Base Chemistry. 1. Van der Waals Contributions to Acceptor Numbers," *Journal of American Chemical Society*, Vol. 112, No. 9, 1990, pp. 3259-3264. doi:10.1021/ja00165a001

[35] M. J. Kamlet and R. W. Taft, "The Solvatochromic Comparison Method. I. The Beta-Scale of Solvent Hydrogen-Bond Acceptor (HBA) Basicities," *Journal of American Chemical Society*, Vol. 98, No. 2, 1976 pp. 377-383. doi:10.1021/ja00418a009

[36] T. Chaudhuri, S. Mula, S. Chattopadhyay and M. Banerjee, "Photophysical Properties of the 8-Phenyl Analogue of PM567: A Theoretical Rationalization," *Spectrochimica Acta Part A*, Vol. 75, No. 2, 2010, pp. 739-744. doi:10.1016/j.saa.2009.11.048

[37] T. Chaudhuri, P. Shukla, S. K. Nayak, S. Chattopadhyay and M. Banerjee, "Solvent Effect on Photophysical Properties and Mg^{2+} Binding of 1,3-Diphenyl-propane-1,3-dione," *Journal of Photochemistry and Photobiology A*: *Chemistry*, Vol. 215, No. 1, 2010, pp. 31-37. doi:10.1016/j.jphotochem.2010.07.017

[38] I. R. Gould, D. Noukakis, L. Gomez-Jahn, R. H. Young, J. Goodman and S. Farid, "Radiative and Nonradiative Electron Transfer in Contact Radical-Ion Pairs," *Chemical Physics*, Vol. 176, No. 2-3, 1993, pp. 439-456. doi:10.1016/0301-0104(93)80253-6

[39] G. Briegleb and J. Czekalla, "Intensity of Electron Transition Bands in Electron Donator-Acceptor Complexes," *Zeitschrift für physikalische Chemie* (*Frankfurt*), Vol. 24, 1960, pp. 37-54.

[40] P. Jaramillo, L. R. Domingo, E. Chamorro and P. Pérez, "A Further Exploration of a Nucleophilicity Index Based on the Gas-Phase Ionization Potentials," *Journal of Molecular Structure*: *THEOCHEM*, Vol. 865, No. 1-3, 2008, pp. 68-72. doi:10.1016/j.theochem.2008.06.022

[41] L. R. Domingo, M. J. Aurell, P. Pe´rez and R. Contreras, "Quantitative Characterization of the Global Electrophilicity Power of Common Diene/Dienophile Pairs in Diels-Alder Reactions," *Tetrahedron*, Vol. 58, No. 22, 2002, pp. 4417-4423. doi:10.1016/S0040-4020(02)00410-6

[42] N. Acharjee, A. Banerji and T. Prange, "DFT Study of 1,3-Dipolar Cycloadditions of C,N-Disubstituted Aldonitrones to Chalcones Evidenced by NMR and X-Ray Analysis," *Monatshefte für Chemie*, Vol. 141, No. 11, 2010, pp. 1213-1221. doi:10.1007/s00706-010-0393-2

9

Synthesis, Characterization and Third Order Non Linear Optical Properties of Metallo Organic Chromophores

Nallamuthu Ananthi[1], Umesh Balakrishnan[1], Sivan Velmathi[1*], Krishna Balakrishna Manjunath[2], Govindarao Umesh[2]
[1]Department of Chemistry, National Institute of Technology, Tiruchirappalli, India
[2]Department of Physics, National Institute of Technology, Surathkal, India
Email: *velmathis@nitt.edu

ABSTRACT

The organic imine and their metal complexes were synthesized and characterized by IR, UV and NMR. The third order non linear optical properties of the compounds were investigated. The measurements of second hyperpolarizabilites were performed using single beam Z-scan technique with 8 ns laser pulses. Ligand and its Copper, Zinc and Nickel complexes show good third order non linearity whereas Manganese complex did not show any activity.

Keywords: Third Order NLO Properties; Salen Ligand; Transition Metal Complexes; Z-Scan Technique

1. Introduction

The non linear optical properties of the π conjugated organic materials have been widely investigated due to their possible applications in a variety of optoelectronic and photonic applications [1-3]. In third order non linear optics, guidelines for the optimization of the second hyperpolarizability γ of any organic molecule have been steadily improving but the understanding is far less developed than for the first hyperpolarizability β [4,5]. So it is desirable to optimize the γ values. In particular the strong delocalization of π electrons or the presence of any hetero atoms in the organic backbone determines a very high molecular polarizability and thus remarkable third order optical nonlinearity. In general large hyperpolarizabilities are the result of an optimum combination of various factors such as π delocalization length, donor-acceptor groups, dimensionality, conformation and orientation for a given molecular structure [6,7].

Organic materials with high nonlinear optical (NLO) susceptibilities are considered to be potential alternatives to the inorganic materials for applications in optoelectronics [8-10]. Reports of devices, such as intra cavity frequency doublers based on the second-order NLO effect in an organic crystal, have already started to appear in the literature. As a result, there is an increasing interest in designing new organic materials with desired linear and nonlinear optical properties.

Crystal structure, spectroscopic studies, and nonlinear optical (NLO) properties of Schiff base metal complexes containing an N, S donor ligand have been studied by Zhao-Ming Xue *et al.* [11]. Third-order nonlinear optical properties of nickel(II) and copper(II) complexes with salen ligands, functionalized with electron donor/acceptor groups (DA-salen), have been investigated in solution by the Z-scan technique using an Nd:YAG laser (1064 nm).[12] $[Ag(L)_2](NO_3)\cdot(MeOH)\cdot(EtOH)$ and $[HgI_2(L)]$ {L = 1,2-bis [(ferrocen-l-yl methylene) amino] ethane} have been prepared, structurally characterized and third-order nonlinear optical properties have been studied. The results indicate that the two complexes exhibit very strong NLO absorption and strong self-focusing effects [13]. The third-order susceptibilities χ^3 of octa substituted metallophthalocyanines have been measured by third-harmonic generation experiments. The effect has been associated to the role of two-photon allowed metal possibly (*d-d*) transitions in the cobalt derivative [14]. A novel zinc (II) 1,3,5 -triazine-based complex $[Zn(TIPT)Cl_2]\cdot 2CH_3OH$ (TIPT = 2,4,6-tri(2-ispropylidene-1-ly)hydrazono-1,3,5-triazine) was prepared and structurally characterized and the hyperpolarizability γ value was found to be 8.26×10^{-30} esu [15].

In this paper we worked out the third order non linear optical parameters of the synthesized salen imine and its metal (Cu, Zn, Ni, Mn) complexes using Z-scan measurement using 7ns laser pulses at 532 nm. Herein we report, the effects of the chromophore hydroxyl group on this imine and also the role of different metals on the non linear optical property. The structures of the compounds studied are represented in **Scheme 1**.

2. Experimental Section

2,4-dihydroxy benzaldehyde, o-amino phenol were pur-

*Corresponding author.

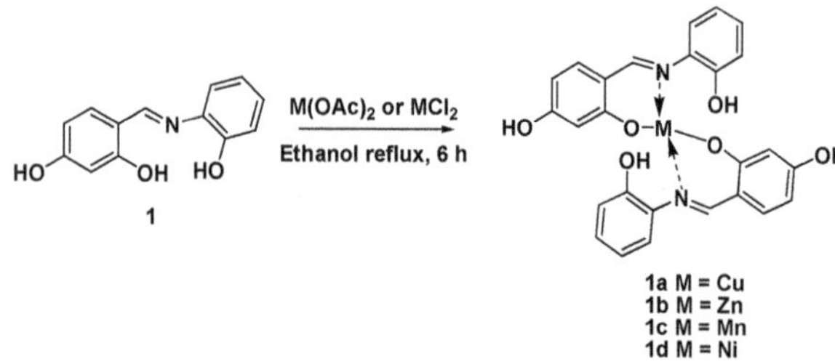

Scheme 1. Synthesis of Schiff base metal complexes.

chased from Aldrich chemicals and copper, zinc, manganese acetates and the nickel chloride were purchased from Loba chemie and used as received. Ethanol, Dimethy formamide (DMF) and other solvents were purified by the reported procedure.

2.1. Synthesis of Schiff Base 1

To the refluxing ethanolic solution of o-amino phenol (5 mmol, 0.546 g), 2,4-dihydroxy benzaldehyde (5 mmol, 0.4755 g) in ethanol was added by drops and then reflux in a water bath for three hours. Then the imine was separated by hexane wash in 80% yield.

Melting point (˚C): 78 - 80, IR data: ν (cm^{-1}) 3312 (-NH), 3390, 1056 (-OH), 1741(-$COOCH_3$), 1644 (C=N), ^{1}H NMR data ($CDCl_3$, TMS, δ, ppm): 6.4 (d, 1H, -CH), 6.7 (m, 2H, -CH), 7.12 (m, 2H, -CH), 7.54 (s, 1H, -CH), 7.78 (d, 1H, -CH), 8.87 (s, 1H, -HC=N), 9.85 (s, 3H, -OH). ^{13}C NMR ($CDCl_3$, TMS, δ, ppm): 103.7, 108.6, 110.9, 117.2, 122.5, 123.6, 128.7, 132.4, 141.2, 151.5, 159.8, 162.3, 162.5.

2.2. Synthesis of Metal Complexes 1a, 1b, 1c, 1d from Schiff Base 1

To the stirred solution of the imine (1 mmol), the metal starting compounds Cu $(CH_3COO)_2 \cdot H_2O$, Zn $(CH_3COO)_2 \cdot 2H_2O$, Mn $(CH_3COO)_2 \cdot 2H_2O$) and $NiCl_2 \cdot 6H_2O$ (1 mmol) was added in an ethanolic solution. Then the reaction mixture was reflux for six hours. Then the metal complexes were separated by washing with dichloromethane to remove the unreacted imine. Yield: 75%.

1a: Melting point (˚C): >360. IR data: ν (cm^{-1}): 3430, 1038 (-OH), 1622 (C=N).

1b: Melting point (˚C): >360. IR data: ν (cm^{-1}): 3433, 1035 (-OH), 1620 (C=N).

1c: Melting point (˚C): >360. IR data: ν (cm^{-1}): 3437, 1039 (-OH), 1616 (C=N).

1d: Melting point (˚C): >360. IR data: ν (cm^{-1}): 3436, 1037 (-OH), 1624 (C=N).

2.3. Transmission Measurements

The transmission coefficient of the samples synthesized was measured by Z-scan technique using Q-switched laser pulses at a wavelength of 532 nm. Z-scan technique introduced by Sheik-Bahae *et al.* [16] is an extensively utilized experimental tool for studying optical nonlinearities in a wide class of materials. The technique relies on the fact that the light intensity varies along the axis of a convex lens and is maximum at the focus. By moving the sample through the focus, the intensity dependent absorption is measured as a change of the transmission through the sample. This technique is an increasingly popular method for the measurement of the non linear absorption coefficient (β) and the non linear refractive index (η_2) of the samples and has the advantages that it immediately indicates the sign and type of non linearity (refraction or absorption). A frequency doubled Q-Switched Nd-YAG Laser (Spectra Physics GCR-170), which produces 8 ns laser pulses at 532 nm was used as the light source. The laser beam was focused by plano-convex lens of 25 cm focal length. For the measurements, liquid samples of the concentration 1×10^{-3} mol/liter were prepared in research grade DMF. The solutions were then taken in a quartz cuvette of thickness 1 mm and the cuvette was mounted on a motorized linear translation stage for translating the sample across the focal spot. The beam waist at the focus was estimated to be 18.9 μm and the corresponding Raleigh length was 2.11 mm. A circular aperture, of size 5 mm, is mounted in front of the photo detector placed at a distance of about 15 cm from the focal spot of the laser beam. The light intensity transmitted by the sample is measured as a function of the sample position along the Z-axis (the beam axis) thereby obtaining the NLO refractive index (η_2) "closed aperture" Z-scan data. The measurements were repeated after removing the aperture in order to obtain the NLO absorption index (β) "open aperture" Z-scan data. The closed aperture data was then divided by the open aperture data to generate the normalized transmission versus translation distance data, using which the nonlinear optical susceptibility was calculated following the analysis method of Sheik-Bahae *et al.* [16].

3. Results and Discussion

3.1. Synthesis and Characterization of Ligand and Metal Complexes

The Schiff base **1** was synthesized by the condensation of 2,4-dihydroxy benzaldehyde with o-amino phenol. The compound was well characterized by UV-vis, FTIR, ^{1}H and ^{13}C NMR spectroscopy methods before studying their NLO properties.From the electronic absorption spectra of the ligand, it was observed that there is no absorption band in the visible region. The bands in the region 207 - 230 nm could be assigned to excitation of the π electrons of the aromatic system. The observed bands in the range 316 - 340 nm can be attributed to the n-π^*

transition. The band in the region around 260 - 284 nm is due to the electronic transition between the π orbital localized on the azomethine group (C=N). The band in the region 208 - 225 nm could be assigned to excitation of the π electrons of the aromatic system. This UV-vis spectrum also indicates that there is negligible one photon absorption at a wavelength of 532 nm wavelength. Thus, in all the experiments, the nonlinear absorption behavior can be attributed to the two photon absorption (TPA).

In the electronic spectra of metal complexes, Nickel and copper complexes shows bands at 275 nm which are assigned to intra ligand electronic transitions (π-π^*). In zinc complex red shift has taken place compared to copper and nickel complexes, the band appears at 290 nm. This is due to the diamagnetic behavior of the Zn(II) ions whereas the Ni(II) and Cu(II) are paramagnetic. An intense peak around 420 nm are assigned to ligand to metal charge transfer. The bands with higher energy in the UV region are of intra-ligand π-π^* type or charge transfer transitions involving energy levels which are higher in energy than the ligand's Lowest Unoccupied Molecular Orbital (LUMO).

In the FTIR spectroscopy we get information about the presence of functional groups in the compound such as C=C, CH=N etc. by the position of peaks which arise due to stretching vibration of the bonds in the groups. The presence of C=N in the molecule is confirmed by the vibration observed between 1690 - 1640 cm^{-1} and aromatic C=C between 1600 - 1473 cm^{-1}. In the IR spectra of metal complexes, C=N bond stretching was slightly shifted to the lower region (1616 - 1624 cm^{-1}) confirming the coordination of C=N bond of the ligand to the metal. All the other function groups also show their stretching in the respective regions in the IR spectra.

NMR spectroscopy is one of the principal techniques which give us the structural information about molecules. NMR spectra were obtained using deuterated dimethyl sulphoxide (DMSO-d_6) as the solvent. Compound **1** gave a singlet at around δ 8.8 ppm corresponding to the CH=N proton indicating the formation of imine and the aromatic protons resonate in the δ 6.4 - 7.7 region. In addition, the O-H proton of phenol is shown as a singlet at around δ 9.8 ppm. In the case of ^{13}C NMR spectroscopy CH=N carbon came as a sharp peak at δ 145 - 150 ppm. In the case of the metal complexes the above given IR data shows the shift towards the lower frequency of the azomethine group in the metal complex compared to the free imine.

The spectroscopic lines also decreased in the case of the metal complex compared to that of the free imine because of the increased symmetry in the complexes. So the spectral data clearly shows the formation of the imine and their corresponding complexes.

3.2. Linear Optical, Thermal Stability and Solubility Studies of Complexes

From the electronic absorption spectra (**Figure 1**), it can be observed that there is no absorption band in the visible region.

After the formation of the metal complex, a new band appear around 390 nm - 420 nm which is due to the ligand to metal charge transfer (LMCT) of the resultant complex. The new peak is mainly attributed to the formation of co-ordination complex of Schiff base with Cu^{2+}, Zn^{2+}, Mn^{2+} and Ni^{2+} ion. The intensity of LMCT band was maximum for Nickel, followed by Zinc, Copper and Manganese in order.

The metal complexes synthesized were thermally stable above 360 degrees. They are completely soluble in polar solvents like methanol, ethanol, DMSO, DMF etc making the processability of the materials into thin films or devices easy.

3.3. Non Linear Optical Studies

The third order NLO studies were carried out for all the four compounds synthesized. All compounds showed the expected non linearity except the metal complex **1c**. The non linear absorption and refraction are expressed by the equations

$$\alpha(I) = \alpha + \beta I$$

$$n(I) = n + n_2 I$$

where α is the linear absorption coefficient, n is the linear refractive index, β is the nonlinear absorption coefficient and n_2 is the non linear refractive index. According to the theory of two photon absorption process, the change in intensity of the laser beam in the unit propagation length is expressed as $dI/dz + \alpha I + \alpha\beta I^2 = 0$, where α is the attenuation coefficient caused by linear absorption and scattering, β is the attenuation coefficient caused by non linear absorption. The non-linear two photon absorption coefficient value (β) is obtained from the Z-scan curves. **Figure 2(a)** represents normalized pure non linear re-

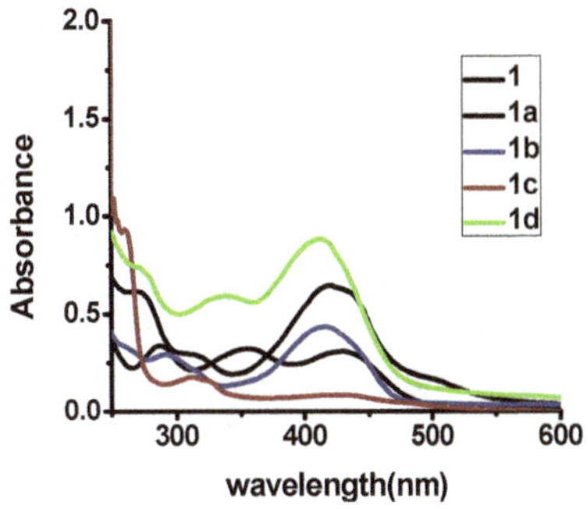

Figure 1. UV spectra of compounds (1, 1a, 1b, 1c & 1d) synthesized.

fraction curve which provides the non linear refractive index obtained by Z-scan measurements for **1a** in the concentration of 1 × 10^{-3} moles per litre. The measurement of transmittance without aperture enables the separation of non-linear refraction from the non linear absorption by dividing the closed aperture data by open one. The open aperture curve indicates the occurrence of non linear absorption. The graph in **Figure 2(a)** shows a peak followed by a valley which is a signature of negative optical nonlinearity for the samples synthesized.

This graph also represents pure refractive nonlinearity of the samples. From the graph, it is seen that the difference between the peak and the valley of the transmission curve is $\Delta T_{PV} = 1.5$. Similar curves are obtained for **1a**, **1b**, **1c** and **1d** in concentration of 1 × 10^{-3} M as shown in **Figures 2(b)-(e)**.

Large optical non-linearity in materials is commonly associated with resonant transitions which may be single or multi-photon nature. But for 532 nm outside resonant absorption, the non-linear two photon absorption coefficient value (β) is obtained by fitting our measured transmittance values to the expression

$$T(z) = \text{In}\left[1+q_0(z)\right]/q_0(z)$$

where $q_0(z) = \beta I_0 L_{eff}/\left(1+ z^2/z_0^2\right)$.

When there is no linear absorption, the linear absorption coefficient $\alpha \to 0$, $L_{eff} = L$, the thickness of the nonlinear medium. The calculated values of β for **1a** at 1 × 10^{-3} M per liter concentration is 0.2911 cm/GW and of non linear refractive index η_2 is −0.4028 × 10^{-11} esu.

Accordingly the real and imaginary parts, Re$\chi^{(3)}$ and Im $\chi^{(3)}$ are obtained by using Re $\chi^{(3)} = 2n_0^2\varepsilon_0 c n_2$ and Im $\chi^{(3)} = \left(n_0^2\varepsilon_0 c\lambda/2\pi\right)\beta$ and found to be −0.4315 × 10^{-13} esu and 0.0448 × 10^{-13} esu respectively (**Table 1**). Same procedure was repeated to calculate the nonlinear parameters of **1a, 1b** and **1d** in concentration 1 × 10^{-3} M. The values obtained are given in **Table 1** and are obtained by repeating Z-scan on each sample and the values are found to be consistent in all the trials. It should be pointed out that the Z-scan method is very sensitive as a way to locate the focus of the laser beam. Any uncertainty in the location may lead to the change in $\Delta T_{p\text{-}v}$ and the distortions appear on Z-scan signals. To compensate, the third-order nonlinearities are often associated with large error bars.

From the study it is clear that the ligand **1**, metal complex **1a**, **1b** and **1d** show good non linear optical properties. Among the metal complexes, manganese complex (**1c**) did not show any nonlinear optical properties. This is due to the d^5 configuration of Mn^{2+} ions. This half-filled orbital which is more stable configuration is responsible for the poor electron transfer between metal and the ligand system and results in poor nonlinear opti-

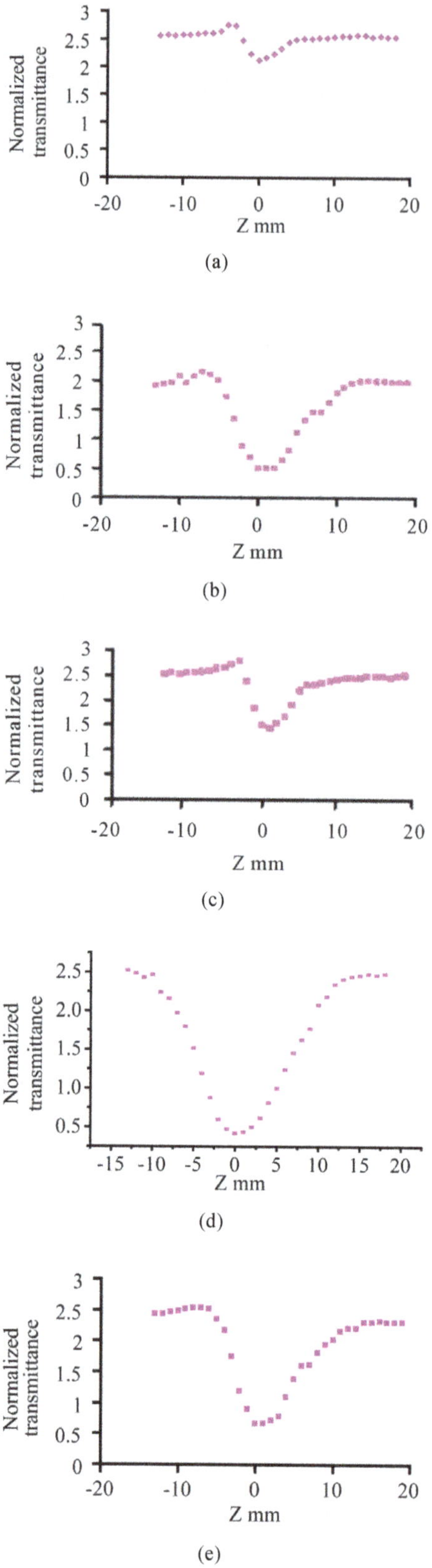

Figure 2. (a) Pure normalised Z-scan curve of 1; (b) Pure normalized Z scan curve of 1a; (2c) Pure normalized Z scan curve of 1b; (2d) Pure normalized Z-scan curve of 1c; (2e) Pure normalized Z-scan curve of 1d.

Table 1. Third order nonlinear optical parameters of Schiff base and their metal complexes.

Compound	η_2 ($\times10^{-11}$ esu)	β(cm/Gw)	Reχ^3 ($\times10^{-13}$ esu)	Imχ^3 ($\times10^{-13}$ esu)
1	−0.9591	2.2294	−1.0273	0.3426
1a	−0.4028	0.2911	−0.4315	0.0448
1b	−1.3374	3.4437	−1.4326	0.5293
1d	−0.9247	1.7910	−0.9905	0.2753

cal properties. In the case of the other metals Cu^{2+}, Zn^{2+} and Ni^{2+} the electron transfer between metal and ligand system are more feasible which results in better nonlinear optical properties.

Copper and nickel complexes were found to have better nonlinear optical properties compared to zinc complex. The transition metals especially first row transition metal ions are well known for their ability to form wide range of coordination complexes in which octahedral, tetrahedral, and square planar geometries predominate. Copper (II) is a typical transition metal ion to form complexes, but less typical in its reluctance to take up a regular octahedral (or) tetrahedral geometry. The magnitude of the splitting of the electronic energy levels in copper (II) complexes tend to be larger than other first row transition metals due to the presence of large Jahn-Teller distortion. Hence the electrons in the copper complexes can be easily transferred to the ligand system and vice versa. Nickel (II) complexes has d^8 electronic configuration, which is very susceptible for electronic transfer compared to zinc complex. Although zinc(II) ion is having much chemical behavior in common with copper (II) ion, due to the fully filled electronic configuration $[Ar]3d^{10}$, the electron transfer is not feasible compared to copper and nickel complexes.

4. Conclusion

The third-order nonlinear parameters of Schiff base and their metal complexes have been investigated by Z-scan technique. The molecules show negative nonlinearity. The third order nonlinearity arises due to the strong delocalization of the π-electrons and the electron transfer between ligand and metal. The four molecules studied in this investigation are of the donor-acceptor type. Among the samples (**1**, **1a-d**) examined for third-order NLO properties, manganese complex (**1c**) did not show NLO properties. Although the length of delocalization of electrons is the same for all metal complexes, the complexes differ in the electronic transfer taking place between metal and ligand. Hence they show different nonlinear response. Similarly the dimensionality and conformation of the metal complexes are also responsible for the difference in NLO properties.

REFERENCES

[1] D. R. Kanis, M. A. Ratner and T. J. Marks, "Design Synthesis and Properties of Molecule Based Assemblies with Large Second Order Optical Nonlinearities," *Chemical Reviews*, Vol. 94, No. 1, 1994, pp. 195-242. doi:10.1021/cr00025a007

[2] H. Ma, A. K. Jen and L. R. Dalton, "Polymer Based Optical Waveguide Materials Processing and Devices," *Advanced Materials*, Vol. 14, No. 19, 2002, pp. 1339-1365. doi:10.1002/1521-4095(20021002)14:19<1339::AID-ADMA1339>3.0.CO;2-O

[3] H. Kuhn and J. Robillard, "Nonlinear Optical Materials," CRC Press, Boca Raton, 1992.

[4] P. N. Prasad and D. J. Williams, "Introduction to Nonlinear Optical Effects in Organic Molecules and Polymers," Wiley, New York, 1991.

[5] D. S. Chemla and J. Zyss, "Nonlinear Optical Properties of Organic Molecules and Ctystals," Academic Press, Cambridge, 1987.

[6] D. J. Williams, "Nonlinear Optical Properties of Organic and Polymeric Materials," American Chemical Society, Washington DC, 1983.

[7] Ch. Bosshard, R. Spreitier, P. Gunter, R. R. Tykwinski, M. Schreiber and F. Diederich, "Structure-Property Relationship in Nonlinear Optical Tetraethynylethenes," *Advanced Materials*, Vol. 8, No. 3, 1996, pp. 231-234. doi:10.1002/adma.19960080309

[8] S. R. Marder, W. E. Torruellas, M. Blanch Desce, V. Riddi and G. I. Stegeman, "Large Molecular Third Order Optical Nonlinearities in Polarized Carotenoids," *Science*, Vol. 276, No. 5316, 1997, pp. 1233-1236. doi:10.1126/science.276.5316.1233

[9] J. L. Bredas, C. Adant, P. Tackx and A. Persoons, "A Third Order Nonlinear Optical Response in Organic Materials: Theoretical and Experimental Aspects," *Chemical Reviews*, Vol. 94, No. 1, 1994, pp. 243-278. doi:10.1021/cr00025a008

[10] P. Audebert, K. Kamada, K. Matsunaga and K. Ohta, "The Third Order NLO Properties of D-π—A Molecules with Changing a Primary Amino Group into Pyrrole," *Chemical Physics Letters*, Vol. 367, No. 1, 2003, pp. 62-71. doi:10.1016/S0009-2614(02)01575-0

[11] Z. M. Xue, Z. M. Tang, Y. W. Wu, J. Y. Tian, Y. P. Jiang, H. M. H. Fun and K. Usman, "A crystal Structure, Spectroscopic Studies, and Nonlinear Optical Properties of Schiff Base Metal Complexes Containing an N, S Donor Ligand," *Canadian Journal of Chemistry*, Vol. 82, No. 12,

2004, pp. 1700-1706. doi:10.1139/v04-142

[12] J. Tedim, S. Patricio, R. Bessada, R. Morais, C. Sousa, M. B. Marques and C. Freire, "Third Order Nonlinear Optical Properties of DA-Salen-Type Nickel(II) and Copper(II) Complexes," *European Journal of Inorganic Chemistry*, Vol. 17, No. 17, 2006, pp. 3425-3433. doi:10.1002/ejic.200600017

[13] H. Hou, G. Li, Y. Song, Y. Fan, Y. Zhu and L. Zhu, "Synthesis Crystal Structures and Third Order Nonlinear Optical Properties of Two Novel Ferrocenyl Schiff Base Complexes, "*European Journal of Inorganic Chemistry*, Vol. 12, No. 12, 2003, pp. 2325-2332. doi:10.1002/ejic.200200616

[14] M. A. Diaz Garcia, J. A. Duro and F. Fernandez Lazaro, "Third-Order Nonlinear Optical Susceptibilities of the Langmuir-Blodgett Films of Octa-Substituted Metallophthalocyanines," *Applied Physics Letters*, Vol. 69, No. 1, 1996, pp. 3-5. doi:10.1063/1.118150

[15] F. Yaoting, L. Gang, L. Zifeng, H. Hongwei and M. Hairong, "Synthesis, Structure and third Order Nonlinear Optical Properties of 1,3,5-Triazine-Based Zn(II) Three Dimensional Supramolecule," *Journal of Molecular Structure*, Vol. 693, No. 2, 2004, pp. 217-224. doi:10.1016/j.molstruc.2004.03.008

[16] M. Sheik-Bahae, A. A. Said, T. H. Wei, D. J. Hagan, E. W. Van Stryland, "Sensitive Measurement of Optical Nonlinearities Using a Single Beam," *IEEE Journal of Quantum Electronics*, Vol. 26, No. 4, 1990, pp. 760-769. doi:10.1109/3.53394

10

Attenuation in a Cylindrical Left Handed Material (LHM) Wave-Guide Structure

Hana Mohammed Mousa
Physics Department, Al-Azhar University, Gaza, Palestine
Email: H.mousa@alazhar-gaza.edu.ps

ABSTRACT

This paper tackles the wave attenuation along with a cylindrical waveguides composed of a left Handed material (LHM), surrounded by a superconducting or metal wall. I used the transcendental equations for both TE and TM waves. I found out that the waveguide supports backward TE and backward TM waves since both permittivity and magnetic permeability of LHM are negative. I also illustrated the dependence of the TE and TM wave attenuation on the wave frequency and the reduced temperature of the superconducting wall (T/T_c). Attenuation constant increases by increasing the wave frequency and it shows higher values at higher T/T_c . Lowest wave attenuation and the best confinement are achieved for the thickest TE waveguide. LHM-superconductor waveguide shows lower wave attenuation than LHM-metal waveguide.

Keywords: Left Handed Material; TM Waves; TE Waves; Attenuation Constant; Waveguides; Superconductor; Metal

1. Introduction

Recently, there has been a great interest in new type of electromagnetic materials called left-handed media [1]. Over the past, fifty years, Veselago was the first scientist to consider the left-handed meta-material (LHM). He defined it as media with simultaneously negative and almost real electric permittivity and magnetic permeability in some frequency range [2]. The electric and magnetic fields form a left-handed set of vectors with the wave vector [3]. These materials have exhibited unique properties, such as Snell law and Doppler shift. The negative refraction index has been recently observed by Shelby, Smith, and Shultz [4] in completely different systems. Negative refraction allows the fabrication of perfect lens [5] where an object can be reconstructed without any diffraction error and super-lens [6,7], and focusing by Plano-concave lens [8]. They can be made to be anisotropic and have indefinite index meta-materials which can be used to make hyper-lens [9]. This range of properties opens infinite possibilities to use meta materials in frequencies from micro-wave up to the visible wave. Circular and a planer waveguides have been widely applied in receivers of radio telescope [10,11]. Shabat and Mousa [12-15] discussed the propagation characteristics of nonlinear electromagnetic TE surface waves in a planer waveguide structure of a lateral anti-ferromagnetic/nonmagnetic super-lattices (LANS) film bounded by a nonlinear dielectric cover and a left-handed substrate. The backward and bi-stability behaviors have been noticed clearly. Huang, *et al.* [16] considered the wave propagation along with a cylindrical nanowire waveguide made of indefinite index meta-materials. They found out that the backward-wave modes can have very large effective index. These nanowires can be used as an ultra-compact optical buffer in integrated optical circuits. Yeap, *et al.* have discussed and developed a novel technique to compute the attenuation of waves propagating in circular waveguides with lossy and superconducting walls [17]. They have compared their results with by using the Stratton's method and the approximate perturbation method. The results from the three methods agree very well at a reasonable range of frequencies above the cutoff. The propagation of waves in superconductor media have drowned much attention and consideration [18]. The use of superconducting thin films in transmission line is advantageous for signal processing because films are low loss and they are with wide bandwidth [19]. The loss can be described in terms of the surface resistance of the superconductor and it is also related to the attenuation constant of the superconductor. It can be exerpted from the study of its propagation characteristics.

A superconductor like Yttrium barium copper oxide (YBCO), is a famous "high-temperature superconductor", achieves prominence because it is the first material to achieve superconductivity above (77 K), the boiling point of liquid nitrogen. It is associated with the formula

$YBa_2Cu_3O_{7-x}$ The superconducting properties of $YBa_2Cu_3O_{7-x}$ are sensitive to the value of x, its oxygen content with $0 \le x \le 0.65$. R.D Black *et al.* [20]. It describes a high-temperature superconducting-receiver system for use in nuclear magnetic resonance (NMR) microscopy. The implementation of thin-film YBCO receiver coils has improved the signal-to-noise ratio of nuclear magnetic resonance spectrometers by a factor of 3 compared to that achievable with conventional coils. This improvement enables the data acquisition time to be reduced by an order of magnitude. These coils are also potential applications in low-frequency magnetic resonance imaging (MRI). They are used in hospitals and clinics. The feasibility of applying high-temperature superconductor (HTS) technology to nonreciprocal microwave devices has been demonstrated in the form of isolators and circulators. They are used widely to achieve stability, reliability, and reproducibility in microwave circuit performance [21]. Pure $YBa_2Cu_3O_{7-x}$ ceramics normally display superconductivity and metallic conductivity below and above the critical temperature of the superconductor T_c respectively. Mohazzab, *et al.* described the synthesis of epoxy-modified YBCO ceramics and evaluates the semiconducting properties at temperature above T_c [22]. Of the new ceramic superconductors, only $YBa_2Cu_3O_{7-x}$ has been developed in thin-film form to the point of practical applications, and several devices are available. Intensive materials research have resulted in techniques, notably laser-ablation and radio-frequency sputtering.

In this analysis, a theoretical study of the propagation characteristics of TE and TM waves guided by an optical structure is presented. This structure consists of a left Handed material (LHM) cylinder with superconducting walls like $YBa_2Cu_3O_{7-x}$. ε_h and μ_h are electric permittivity and magnetic permeability of LHM respectively.

2. Constitutive Relations for TE and TM Waves

In this context, the structure geometry of the problem considered is shown in **Figure 1**. I considered wave propagation on a cylindrical waveguide. The axis of the waveguide is along with the z direction.

The longitudinal electric and magnetic fields E_z and H_z respectively, propagating in the waveguide can be derived by aid of Helmholtz's wave equation [23].

Decomposing Helmholtz's wave equation into a radial and a longitudinal part in cylindrical coordinates for the electric field $\boldsymbol{E}$ yields:

$$\nabla^2_{rad}\boldsymbol{E} + \frac{\partial^2 \boldsymbol{E}}{\partial z^2} + \mu_h \varepsilon_h k^2 \boldsymbol{E} = 0$$
$$\nabla^2_{rad} = \frac{1}{r}\frac{\partial}{\partial r}\left(r\frac{\partial}{\partial r}\right) + \frac{1}{r^2}\frac{\partial^2}{\partial \phi^2} \qquad (1)$$

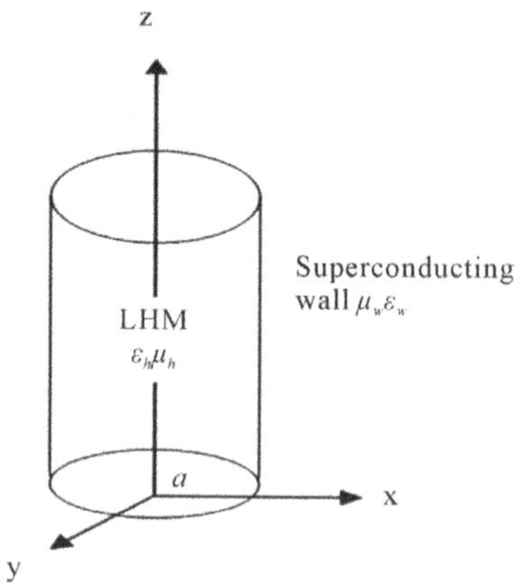

Figure 1. The proposed cylindrical waveguide composed of LHM and superconducting wall.

A plane wave solution for the electric field of the form

$$\boldsymbol{E}(r,\phi,z) = \boldsymbol{E}(r,\phi)\exp\left[i(\omega t - k_z z)\right]$$

is substituted into Equation (1) as:

$$\frac{1}{r}\frac{\partial}{\partial r}\left(r\frac{\partial \boldsymbol{E}}{\partial r}\right) + \frac{1}{r^2}\frac{\partial^2 \boldsymbol{E}}{\partial \phi^2} + h^2\boldsymbol{E} = 0 \qquad (2a)$$

with $h^2 = \mu_h \varepsilon_h k_0^2 - k_z^2$,

where ω represents the wave angular frequency, k_0 is the wave number in free space

$k_0^2 = \frac{\omega^2}{c^2} = \varepsilon_o \mu_o \omega^2$, ε_o and μ_o are the dielectric permittivity and magnetic permeability of free space respectively. k_z is the propagation constant. Both a negative dielectric permittivity and permeability are written as [3]:

$$\varepsilon_h(\omega) = 1 - \frac{\omega_p^2}{\omega^2}, \quad \mu_h(\omega) = 1 - \frac{F\omega^2}{\omega^2 - \omega_0^2}, \qquad (2b)$$

with plasma frequency ω_p and resonance frequency ω_0, μ_h is the magnetic permeability of LHM and ε_h is the electric permittivity of LHM.

Equation (2a) can be split into two equations by a separation of variables with the form:

$$r^2\frac{\partial^2 U(r)}{\partial r^2} + r\frac{\partial U(r)}{\partial r} + \left(h^2 r^2 - n^2\right)U(r) = 0 \qquad (3a)$$

$$\frac{\partial^2 \Phi(\phi)}{\partial \phi^2} + n^2\Phi(\phi) = 0 \qquad (3b)$$

Equation (3b) describes a simple harmonic oscillator and Equation (3a) is one form of the Bessel equations, whose solutions are the Bessel functions [24]. The following set of field equations are:

$$H_z = C_n' J_n(hr)\sin n\phi \qquad (4a)$$

$$E_z = C_n J_n(hr)\cos n\phi \qquad (4b)$$

where C_n and C_n' denote the coefficients of the longitudinal fields, r is the radial distance , $J_n(hr)$ is called the Bessel function of the first kind and n is the order of the Bessel function.

The propagation constant k_z is a complex variable which constitutes a phase constant β_z and an attenuation constant α_z as:

$$k_z = \beta_z - i\alpha_z. \quad (4c)$$

By Maxwell's curl equations, the transverse field components can be written as:

$$E_\phi = -\frac{i}{h^2}\left(\frac{k_z}{r}\frac{\partial E_z}{\partial \phi} - k_0\mu_h\frac{\partial H_z}{\partial r}\right) \quad (5a)$$

$$H_\phi = -\frac{i}{h^2}\left(k_0\varepsilon_h\frac{\partial E_z}{\partial r} + \frac{k_z}{r}\frac{\partial H_z}{\partial \phi}\right). \quad (5b)$$

Substituting Equation (4a) and Equation (4b) into Equation (5a) and Equation (5b), yields

$$E_\phi = \frac{1}{h^2}\left(\frac{ink_z}{r}C_n J_n(hr)\sin n\phi + ik_0h\mu_hC_n' J_n'(hr)\sin n\phi\right) \quad (6a)$$

and

$$H_\phi = -\frac{1}{h^2}\left(\frac{ink_z}{r}C_n' J_n(hr)\cos n\phi + ik_0h\varepsilon_hC_n J_n'(hr)\cos n\phi\right) \quad (6b)$$

At the wall, the tangential electric and magnetic fields E_t and H_t respectively are related to a surface impedance Z_s by [24,25]:

$$\boldsymbol{E}_t = -Z_s\left(\hat{a}_r x\boldsymbol{H}_t\right). \quad (6c)$$

With

$$\boldsymbol{E}_t = \hat{a}_\phi E_\phi + \hat{a}_z E_z \text{ and } \boldsymbol{H}_t = \hat{a}_\phi H_\phi + \hat{a}_z H_z. \quad (6d)$$

By substituting Equation (6d) into Equation (6c) one obtains:

$$E_\phi = Z_s H_z \quad (7a)$$

and

$$E_z = -Z_s H_\phi, \quad (7b)$$

Z_s can be expressed in terms of electrical properties of the wall material (superconductor) as:

$$Z_s = \sqrt{\frac{\mu_w}{\varepsilon_w}}, \quad (8)$$

ε_w is the permittivity of the superconductor. It is complex with the form [26]:

$$\varepsilon_w = \left(1 - \frac{1}{k_0^2\lambda_L^2}\right) - i\frac{\sigma}{\omega\varepsilon_0}. \quad (9a)$$

where

$$\lambda_L^2 = \frac{\lambda_0^2}{\left(1-(T/T_c)^4\right)}, \quad \sigma = \sigma_0 (T/T_c)^4 \quad (9b)$$

λ_0 is the field penetration depth at temperature $T = 0$ K, σ is the conductivity of the superconductor and T_c is the critical temperature of the superconductor. T/T_c is called the reduced temperature of the superconductor. At the boundary of the wall ($r = a$), by substituting Equation (4a), Equation (6a) and Equation (8) into Equation (7a), and dividing by $J_n(ha)$, one obtains:

$$\left(\frac{1}{h^2}\frac{ink_z}{a}\right)C_n + \left(\frac{ik_0\mu_h}{h}\frac{J_n'(ha)}{J_n(ha)} - \sqrt{\frac{\mu_w}{\varepsilon_w}}\right)C_n' = 0. \quad (10)$$

By substituting Equation (4b), Equation (6b) and Equation (8) into Equation (7b), and dividing it by $J_n(ha)$, one gets:

$$\left(\frac{ik_0\varepsilon_h}{h}\frac{J_n'(ha)}{J_n(ha)} - \sqrt{\frac{\varepsilon_w}{\mu_w}}\right)C_n + \left(\frac{1}{h^2}\frac{ink_z}{a}\right)C_n' = 0 \quad (11)$$

Equation (10) and Equation (11) constitute a homogeneous system which admits a non trivial solution only in case its determinant is zero. Solving the determinants of the coefficients C_n and C_n' in Equation (10) and Equation (11) results in the following transcendental equation:

$$\left(ih^2\sqrt{\frac{\mu_w}{\varepsilon_w}} + k_0h\mu_h\frac{J_n'(ha)}{J_n(ha)}\right)\left(ih^2\sqrt{\frac{\varepsilon_w}{\mu_w}} + k_0h\varepsilon_h\frac{J_n'(ha)}{J_n(ha)}\right) = \left(\frac{nk_z}{a}\right)^2 \quad (12)$$

The roots of Equation (12) are the allowed values of the propagation constant k_z. Thus, it determine the characteristics modes of propagation. For each value of n there is infinity of roots, any one of which can be denoted by the subscript m. Any root of Equation (12) can then be designated by k_{znm}. In Equation (12), since TE modes are determined by roots of $\frac{J_n'(ha)}{J_n(ha)} = 0$ [23], the attenuation constant of TE modes (α_z) can be obtained from k_z by extracting the imaginary part of Equation (4c).

An alternate form of the Equation (12) is required for TM modes by substituting Equation (4a), Equation (6a) and Equation (8) into Equation (7a), and dividing it by $J_n'(ha)$, then the result is :

$$\left(\frac{J_n(ha)}{J_n'(ha)}\frac{1}{h^2}\frac{ink_z}{a}\right)C_n + \left(\frac{ik_0\mu_h}{h} - \sqrt{\frac{\mu_w}{\varepsilon_w}}\frac{J_n(ha)}{J_n'(ha)}\right)C_n' = 0 \quad (13)$$

By substituting Equation (4b), Equation (6b) and

Equation (8) into Equation (7b), and dividing it by $J'_n(ha)$, the result is :

$$\left(\frac{ik_0\varepsilon_h}{h} - \sqrt{\frac{\varepsilon_w}{\mu_w}}\frac{J_n(ha)}{J'_n(ha)}\right)C_n + \left(\frac{ink_z}{h^2 a}\frac{J_n(ha)}{J'_n(ha)}\right)C'_n = 0 \quad (14)$$

In same way, the transcendental equation of TM modes is:

$$\left(ih^2\sqrt{\frac{\mu_w}{\varepsilon_w}}\frac{J_n(ha)}{J'_n(ha)} + k_0 h\mu_h\right)\left(ih^2\sqrt{\frac{\varepsilon_w}{\mu_w}}\frac{J_n(ha)}{J'_n(ha)} + k_0 h\varepsilon_h\right) = \left(\frac{nk_z}{a}\frac{J_n(ha)}{J'_n(ha)}\right)^2 \quad (15)$$

Since TM modes are determined by roots of $\frac{J_n(ha)}{J'_n(ha)} = 0$, the attenuation constant of TM modes (α_z) can be obtained [22].

3. Numerical Results and Discussion

In this paper, the numerical calculations for LHM cylinder with a superconducting wall like $YBa_2Cu_3O_{7-x}$, are taken with the following parameters: $\omega_p/2\pi = 10$ GHz, $\omega_0/2\pi = 4$ GHz and $F = 0.56$ [3], $\sigma_0 = 6.56\times10^6$ s/m and $\lambda_0 = 0.22$ μm, $\mu_w = 1$ [27]. The frequency range in which both ε_h and μ_h are negative is from 4 to 6 GHz. In this range, and at a definite thickness such as $a = 3$ mm, the solution for the phase constant β_z and attenuation constant α_z of TE waves is found by solving Equation (12). **Figure 2** displays the phase constant β_z of TE wave versus the wave frequency ($f = \omega/2\pi$) of the first band for different values of reduced temperature of the superconductor wall T/T_c; (0.7, 0.8, 0.9). Both the wave phase velocity $[v_p = \omega/\beta_z]$ and the group velocity $[v_g = \partial\omega/\partial\beta_z]$ dispersions are affected by the reduced temperature of the superconductor T/T_c, where v_p decreases to positive values, and v_g decreases to negative values as T/T_c increases; This means that the backward-TE waves are observed as effect of the LHM cylinder of increasing T/T_c. These backward TE waves have very large propagation lengths which is the same result as reported by Huang, *et al.* [16]. These waveguides can be used as phase shifters and filters in optics and telecommunications. **Figure 3(a)** displays the attenuation constant α_z of the second band of TE waves versus the wave frequency for increasing values of T/T_c. It illustrates that the attenuation constant increases by increasing the wave frequency as Yeap *et al.* reported [17]. Higher attenuation of waves is also observed at lower frequencies by increasing T/T_c, (*i.e.*, for curves ("1", "2" and "3"))where T/T_c increases to the values (0.5, 0.7 and 0.9) the higher attenuation of frequency value is respectively observed in (5.8, 5.7 and 5.6 GHz). By Equation (8), Equation (9a) and Equation (9b), increasing σ by increasing T/T_c will decreases ε_w. As a result, high Z_s and then high attenuation is achieved. The solution for the attenuation constant of TM waves is found by solving Equation (15). **Figure 3(b)**, describes the variation of the attenuation of the second band of TM guided waves for different values of reduced temperature of the wall. It displays that the attenuation is increased to negative values by increasing wave frequency and it indicates more attenuation at higher T/T_c. I believe that these negative values of attenuation could be a result of the effect of the negative values of ε_w. It affect the dispersion and decay constant of TM waves as noticed in [16, 28]. In the superconductor, ε_w values decreases from $(-0.101\times10^{10} - j0.19\times10^8)$ to $(-0.449\times10^9 - j0.129\times10^8)$ in the frequency range 4 to 6 GHz and at $T/T_c = 0.9$. As a comparison between the results of **Figures 3(a)** and **(b)**, the TE waveguide with $a = 3$ mm has lower attenuation than the TM wave-guide.

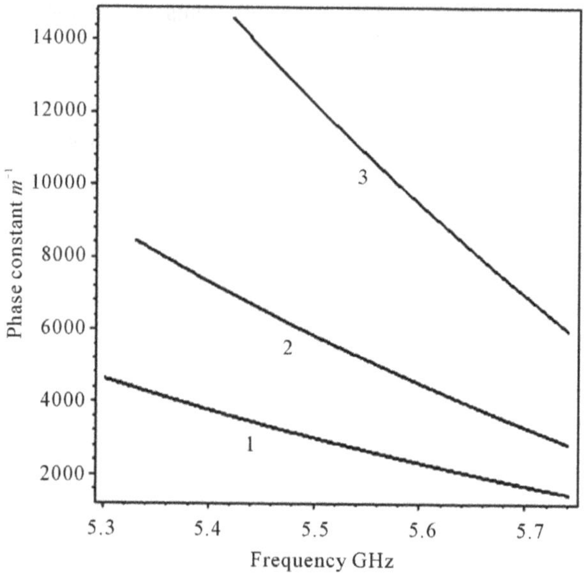

Figure 2. Dispersion curves of the first band TE guided waves of LHM-superconductor waveguide for (1) $T/T_c = 0.7$, (2) $T/T_c = 0.8$ and (3) $T/T_c = 0.9$. The curves are labeled with values of $\omega_p/2\pi = 10$ GHz, $\omega_0/2\pi = 4$ GHz $\sigma_0 = 6.56\times10^6$ s/m, $\lambda_0 = 0.22$ μm, $\mu_w = 1$, $a = 3$ mm and $\varepsilon_h \prec 0, \mu_h \prec 0$.

By increasing the band's order n to the values (1, 2, 3, 4, 5, 6, 7 and 8), roots of $[J'_n(ha)/J_n(ha) = 0]$ are increasing to the values (−1.84, −3, −4.2, −5.3, −6.4, −7.5 and −8.577) respectively. As a result, high attenuation of waves is realized at a higher band's order.

In **Figures 4(a)** and **(b)**, the wave frequency has been plotted against the attenuation constant for the first five

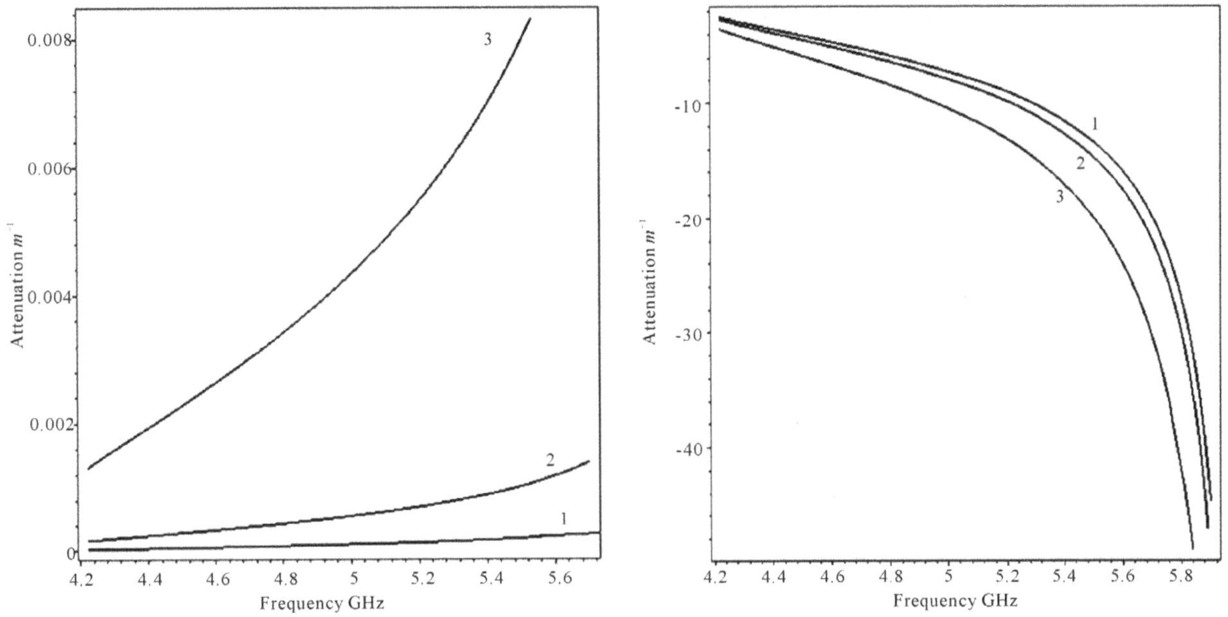

(a) (b)

Figure 3. Attenuation curves of the second band (a) TE and (b) TM guided waves of LHM-superconductor waveguide for (1) $T/T_c = 0.5$, (2) $T/T_c = 0.7$ and (3) $T/T_c = 0.9$. The curves are labeled with values of $\varepsilon_h \prec 0, \mu_h \prec 0$ and $a = 3$ mm.

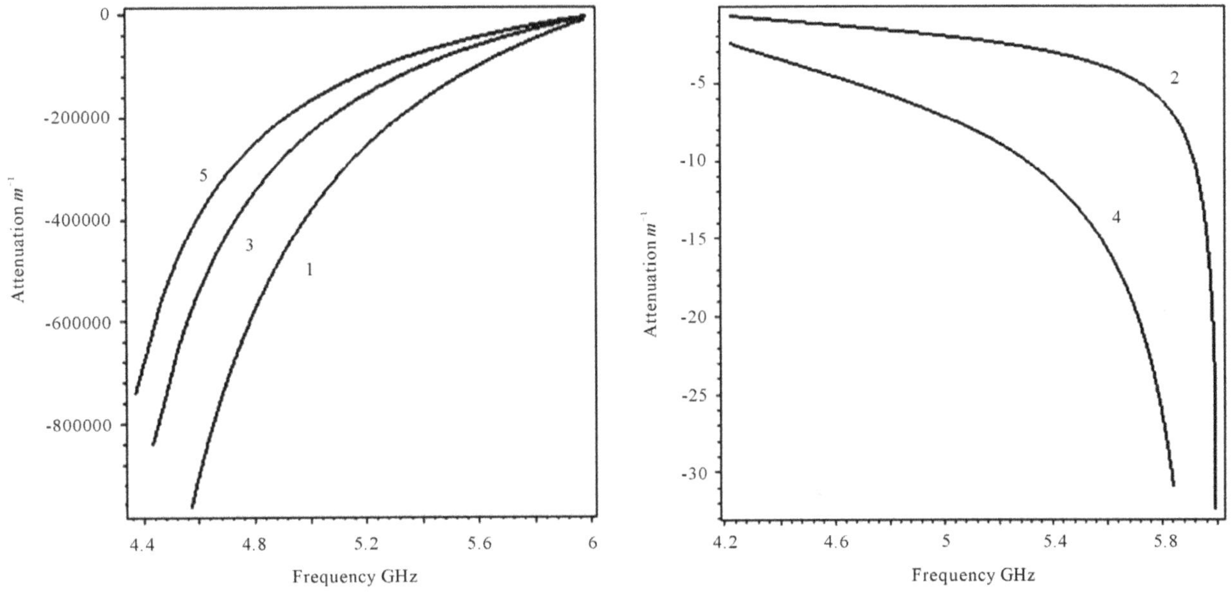

(a) (b)

Figure 4. (a) and (b): Attenuation curves of the first five bands TM guided waves of LHM-superconductor waveguide for $\varepsilon_h \prec 0, \mu_h \prec 0$, $T/T_c = 0.9$ and $a = 10$ mm.

TM bands. By increasing the band's order n to the values (1, 2, 3, 4 and 5), roots of $\left[J_n(ha)/J_n'(ha) = 0\right]$ are (−3.83, 30.56, −6.38, 49.3 and −8.7) respectively. For the odd band's order, (*i.e.*, (1, 3, 5)) as the frequency decreases further to cutoff, the attenuation rises to high negative values, signals propagation become almost impossible. The attenuation is decreased to high negative values as compared to attenuation of even band's order which is increased to small negative values by increasing frequency.

The effect of the waveguide thickness on attenuation of the second band of TE waves is noticed in **Figure 5(a)**. By decreasing the radius a to the values 10 mm, 7 mm, 3 mm, the attenuation increases to the values of (0.0015,

0.0028, 0.0084) respectively at frequency 5.6 GHz. The attenuation of the second band of TM waves is also noticed in **Figure 5(b)**. As the radius a decreases to the previous values, the magnitude of the TM attenuation rises to larger values nearly (−40).

This means that, in this waveguide, the lowest wave attenuation and the best confinement are achieved for the thickest TE waveguide. **Figure 6**, describes the attenuation curves when the superconductor wall $YBa_2Cu_3O_{7-x}$ is replaced by a metal like ferrite. Both ε_w and μ_w are replaced by ε_f and μ_f respectively. According to Lichtenecker's formula and in frequency range (4 to 5.8 GHz) [29], the estimated value of the dielectric permittivity of ferrite is $\varepsilon_f = 16.53 - j0.9$ and the magnetic permeability is $\mu_f = 1.31 - j0.48$. **Figure 6(a)** displays the attenuation constant of the second TE band versus the

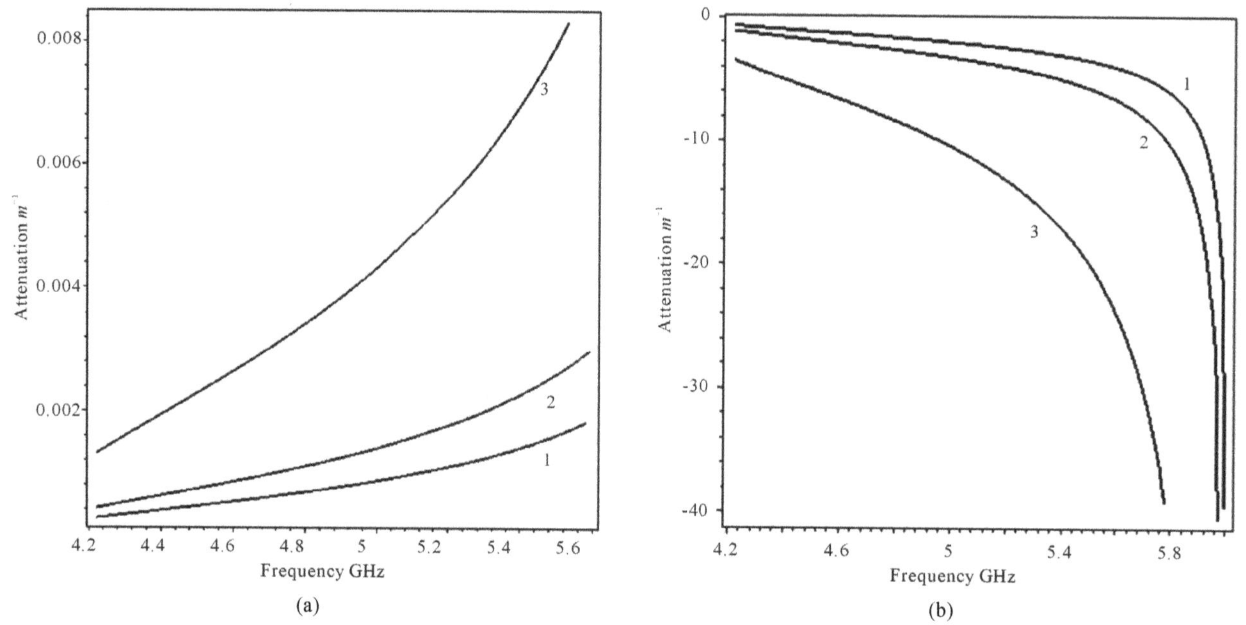

Figure 5. (a) Attenuation curves of the second (a) TE and (b) TM band of LHM-superconductor waveguide for (1) a = 10 mm, (2) a = 7 mm, and (3) a = 3 mm. The curves are labeled with values of $\varepsilon_h \prec 0, \mu_h \prec 0$, and $T/T_c = 0.9$.

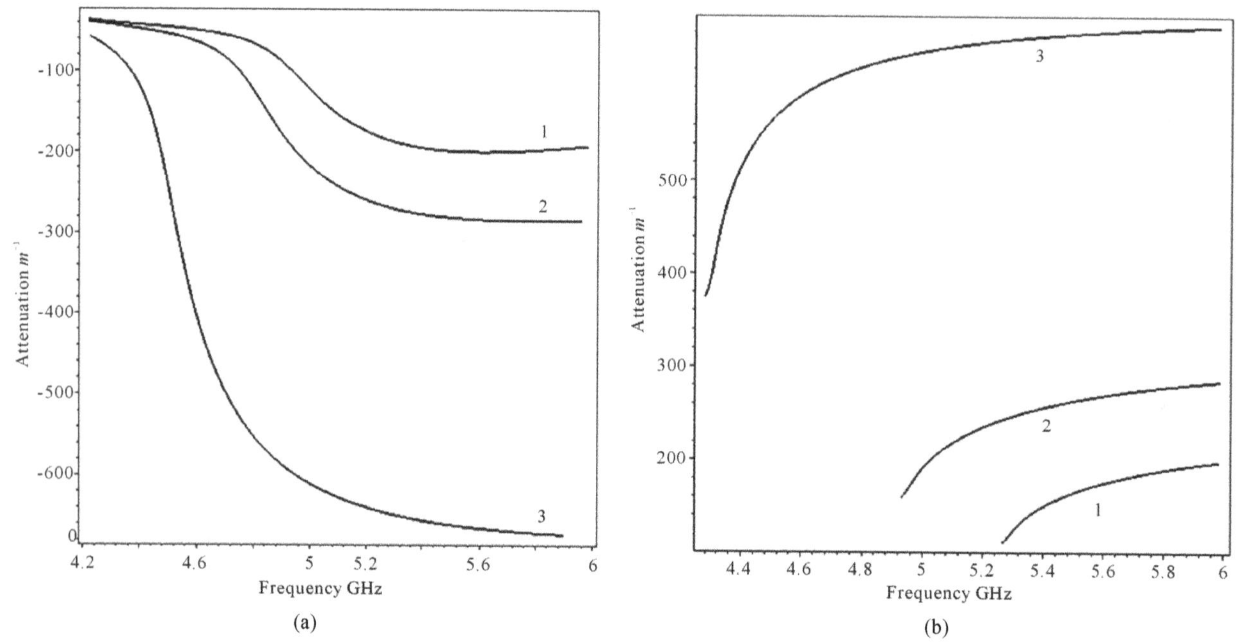

Figure 6. Attenuation curves of the second (a) TE band and (b) TM band of LHM-metal waveguide for (1) a = 10 mm, (2) a = 7 mm, and (3) a = 3 mm. The curves are labeled with values of $\varepsilon_h \prec 0, \mu_h \prec 0$ and $T/T_c = 0.9$.

wave frequency for different values of LHM-Ferrite radius. By decreasing the radius a to the values 10 mm, 7 mm, 3 mm, the attenuation increases to the values of (−200, −280, −660) respectively at frequency 5.6 GHz. For the same range of a, the TM second band's attenuation increases to the values of (160, 260, 660) respectively at frequency 5.6 GHz as displayed by **Figure 6(b)**.

As a comparison between **Figures 5** and **6**, the implementation of $YBa_2Cu_3O_{7-x}$ has reduced the TE and TM wave attenuation ratio by a factor of 10^5 and 10 respectively as compared to that achievable with LHM-Ferrite structure.

4. Conclusion

The attenuation characteristics of both TE and TM waves in a waveguide structure containing LHM-superconductor or LHM-metal are dicussed. I found out that, LHM stimulate the modes to be backward of large propagation lengths. The lowest wave attenuation and the best confinement are achieved for the thickest TE waveguide. I compared the loss of LHM-superconductor waveguide with that of LHM-metal waveguide. The LHM-superconductor waveguide is able to reduce the propagation losses, and to increase the mode's propagation lengths which are very promising results in designing some future microwave devices.

REFERENCES

[1] L. Hu and S. T. Chui, "Characteristics of Electromagnetic Wave Propagation in Uniaxially Anisotropic Left-Handed Materials," *Physical Review: B*, Vol. 66, No. 8, 2002, p. 085108. doi:10.1103/PhysRevB.66.085108

[2] V. G. Veselago, "The Electrodynamics of Substances with Simultaneously Negative Values of Permittivity and Permeability," *Soviet Physics Uspekhi*, Vol. 10, No. 4, 1967, p. 509. doi:10.1070/PU1968v010n04ABEH003699

[3] I. V. Shadrivov, A. A. Sukhorakov and Y. S. Kivshar, "Nonlinear Surface Waves in Left-Handed Material," *Physical Review: E*, Vol. 69, No. 1, 2004, p. 016617. doi:10.1103/PhysRevE.69.016617

[4] R. A. Shelby, D. R. Smith and S. Schultz, "Microwave Transmission through a Two-Dimensional, Isotropic, Left-Handed Meta-Material," *Applied Physics Letter*, Vol. 78, No. 4, 2001, pp. 489-491. doi:10.1063/1.1343489

[5] N. Garcia and M. Nieto, "Left-Handed Materials Do Not Make a Perfect Lens," *Physical Review Letter*, Vol. 88, No. 20, 2002, p. 207403. doi:10.1103/PhysRevLett.88.207403

[6] J. B. Pendry, "Negative Refraction Makes a Perfect Lens," *Physical Review Letter*, Vol. 85, No. 18, 2000, pp. 3966. doi:10.1103/PhysRevLett.85.3966

[7] W. T. Lu and S. Sridhar, "Flat Lens without Optical Axis: Theory of Imaging," *Optics Express*, Vol. 13, No. 26, 2005, pp. 10673-10680. doi:10.1364/OPEX.13.010673

[8] P. Vodo, W. T. Lu, Y. Huang and S. Sridhar, "Negative Refraction and Plano-Concave Lens Focusing in One-Dimensional Crystals," *Applied Physics Letter*, Vol. 89, No. 8, 2006, p. 084104. doi:10.1063/1.2338644

[9] I. I. Smolyaninov, Y. J. Hung and C. C. Davis, "Magnifying Superlens in the Visible Frequency Range," *Science*, Vol. 315, No. 5819, 2007, pp. 1699-1701. doi:10.1126/science.1138746

[10] J. R. Tucker and M. J. Feldman, "Quantum Detection at Millimeter Wavelengths," *Review of Modern Physics*, Vol. 57, No. 4, 1985, pp. 1055-1113. doi:10.1103/RevModPhys.57.1055

[11] M. J. Wengler, "Sub-Millimeter-Wave Detection with Superconducting Tunnel Diodes" *Proceedings of the IEEE*, Vol. 80, No. 11, 1992, pp. 1810-1826. doi:10.1109/5.175257

[12] H. M. Mousa, M. M. Shabat, H. Khalil and D. Jager, "Non-Linear Surface Waves along the Boundary of Magnetic Superlattices (LANS)," *Proceedings of SPIE*, Vol. 5445, 2003, pp. 274-278. doi:10.1117/12.560650

[13] H. M. Mousa and M. M. Shabat, "Non Linear TE Surface Waves on Magnetic (LANS) Superlattices," *International Journal of Modern Physics B*, Vol. 19, No. 29, 2005, pp. 4359-4369. doi:10.1142/S0217979205032796

[14] H. M. Mousa and M. M. Shabat, " Non linear TE Surface in a Left-Handed Material and Superlattices Wave-Guide structures," *International Journal of Modern Physics B*, Vol. 21, No. 6, 2007, pp. 895-906. doi:10.1142/S0217979207036746

[15] M. M. Shabat and H. M. Mousa, "The Propagation of Electromagnetic TE Surface Waves in Magnetic Superlattices (LANS) Film," *Proceedings of SPIE*, Vol. 6582, 2007. doi:10.1117/12.721062

[16] Y. J. Huang, W. T. Lu and S. Sridhar, "Nanowire Waveguide Made from Extremely Anisotropic Meta-Materials," *Physical Review: A*, Vol. 77, No. 6, 2008, p. 063836. doi:10.1103/PhysRevA.77.063836

[17] K. H. Yeap, C. Y. Tham, K. C. Yeong and H. J. Woo, "Wave Propagation in Lossy and Superconducting Circular Waveguides, "*Radioengeneering*, Vol. 19, No. 2, 2010, pp. 320-325.

[18] M. M.Shabat and D. Jager, "Magnetostatic Surface Wave in a Superconductor-Ferrite Structure," *5th International Workshop on Integrated Nonlinear Microwave and Millimeter wave Circuits*, Dusiburge, 1-2 October 1998.

[19] C. J. Wu, "Tunable Microwave Characteristics of a Superconducting Planar Transmission Line by Using a Nonlinear Dielectric Thin Film," *Journal of Applied Physics*, Vol. 87, No. 1, 2000, p. 493. doi:10.1063/1.371889

[20] R. D. Black, T. A. Early, P. B. Roemer, O. M. Mueller, A. Campero, L. G. Turner and G. A. Johnson, "A High-Temperature Superconducting Receiver for Nuclear Magnetic Resonance Microscopy," *Science*, Vol. 259, No. 5096, 1993, pp. 793-795. doi:10.1126/science.8430331

[21] E. Denlinger, R. Paglione, D. Kalokitis, E. Belohoubek, A. Pique, X. D. Wu, T. Venkatesan, A. Fathy, V. Pendrick, S. Green and S. Mathews, "Superconducting Nonreciprocal Devices for Microwave Systems," *IEEE Mi-*

crowave and Guided Wave Letters, Vol. 2, No. 11, 1992, pp. 449-451. doi:10.1109/75.165640

[22] G. Mohazzab and I. M. Low, "Electrical Properties of Epoxy-Modified YBCD Semiconducting Ceramics," *Journal of material Science Letters*, Vol. 16, No. 1, 1987, pp. 88-90. doi:10.1023/A:1018517305021

[23] J. A. Stratton, "Electromagnetic Theory," McGraw-Hill, Boston, 1941, pp. 527-542.

[24] D. K. Cheng, "Field and Wave Electromagnetics," 2nd Edition, Addison Wesley, Inc., Boston, 1989, pp. 547-557.

[25] C. Y. Tham, A. Mccowen and M. S. Towers, "Modelling of PCD Transients with Boundary Elements Method of Moments in the Frequency Domain," *Engineering Analysis with Boundary Elements*, Vol. 27, No. 4, 2003, pp. 315-323. doi:10.1016/S0955-7997(02)00119-4

[26] M. Tsutsumi, T. Fukusako and S. Yoshida, "Propagation Characteristics of the Magneto-Static Surface Wave in the YBCO-YIG Film-layered Structure," *IEEE Transactions on Microwave Theory and Techniques*, Vol. 44, No. 8, 1996, p. 1410. doi:10.1109/22.536023

[27] D. Mihalache, R. G. Nazmitdinov and V. K. Fedyanin, "Nonlinear Optical Waves in Layered Structure," *Soviet Journal of Nuclear Physics*, Vol. 20, 1989, pp. 86-107.

[28] H. M. Mousa and M. M. Shabat, "TM Plasmons in a Cylindrical Superlattices (LANS) Waveguide Structure," *Journal of Nano- and Electronic Physics*, Vol. 3, No. 3, 2011, pp. 15-17.

[29] H. Ebara, T. Inoue and O. Hashimoto, "Measurement Method of Complex Permittivity and Permeability for a Powdered Material using a Waveguide in Microwave Band," *Science and Technology Advanced Material*, Vol. 7, No. 1, 2006, pp. 77-83. doi:10.1016/j.stam.2005.11.019

11

Engineered Transitions in Photonic Cavities

Ali W. Elshaari[1], Stefan F. Preble[2]
[1]Electrical and Electronic Engineering Department, University of Benghazi, Benghazi, Libya
[2]Microsystems Engineering Department, Rochester Institute of Technology, Rochester, USA
Email: awe2048@rit.edu

ABSTRACT

We demonstrate for the first time, to best of our knowledge, that by engineering the states of a system of cavities it is possible to control photon transitions using non-adiabatic refractive index tuning. This is used to realize a novel photon transitions that are independent of the refractive index sign. In particular, we show through coupled mode theory and FDTD simulations that red shifts are possible in silicon resonators using the free-carrier plasma dispersion refractive index reduction.

Keywords: Integrated Optics; Resonators; Wavelength Shift

1. Introduction

The engineering and active control of resonant photonic structures has enabled unique optical functionalities, from isolators and delay elements [1-3] to 100% efficient adiabatic wavelength conversion by trapping light while tuning the state of a resonator [1-7]. However, these functionalities have been limited by the mechanism used to realize the refractive index change. This is particularly the case for adiabatic wavelength conversion on the Silicon photonics platform where the free-carrier plasma dispersion effect (PDE) is used to reduce the refractive index of a cavity [8-10]. This always results in a wavelength shift towards the blue, limiting the application of the effect [7,11]. In contrast, it was shown in [4] that when a resonator is *non-adiabatically* perturbed it is possible to transition photons to other resonant modes—even towards the red, albeit with a low efficiency. The reason for this is that the final state of the system couples to a continuum of output modes—with a dominant excitation of the adiabatic shift as seen in **Figure 1** [7]. Here we show that by carefully designing the states of a system of cavities nearly all of the light can be non-adiabatically transitioned to just one state—even towards the red. This opens the possibility of using the PDE for both blue and red shifts of light, which will enable rapidly reconfigurable wavelength converters for use in future on-chip wavelength-division-multiplexed systems. This is in contrast to non-linear wavelength converters relying on Raman scattering (RS) or four wave mixing (FWM) where high powers are required and wavelength changes are fixed by the wavelengths used in the system and can only be reconfigured by slowly tuning an external laser [12-14].

2. Designing System States

The proposed cavity system is shown in **Figure 2**. It consists of one input and output cavity each with large Free Spectral Range (FSR) and a transition cavity having closely spaced states (small FSR). By initially aligning one resonance of the input cavity and one resonance of the transition cavity it is ensured that the system only has one allowed state, which we call the input state, as shown as a double solid line in the figure (the double lines indicate that there is mode splitting). We note that although there are many internal degrees of freedom for the transition cavity, they are not allowed because they will all be off-resonance with respect to the input cavity. This is ensured by the phase matching and the ortho-normality of modes. So that only specific transition probabilities don't vanish. The probability of a transition between modes a and b is dictated by the overlap between the super-modes over the volume of the two rings in both transverse and longitudinal directions (γ is the coupling constant, $\varphi_{a,b}$ are the initial and final modes).

$$P_{ab} = \int \varphi_a \gamma_{ab} \varphi_{b\,\text{volume}}$$

Furthermore, light cannot couple to the output of the system because the output cavities resonance is detuned with respect to the input state (lower energy in **Figure 2** [-Eg/2]). As will be shown, this red-shifted output cavity determines the final state of the system, and consequently the new wavelength of the light.

To transition the light efficiently to the red-shifted output cavity we induce a non-adiabatic perturbation of

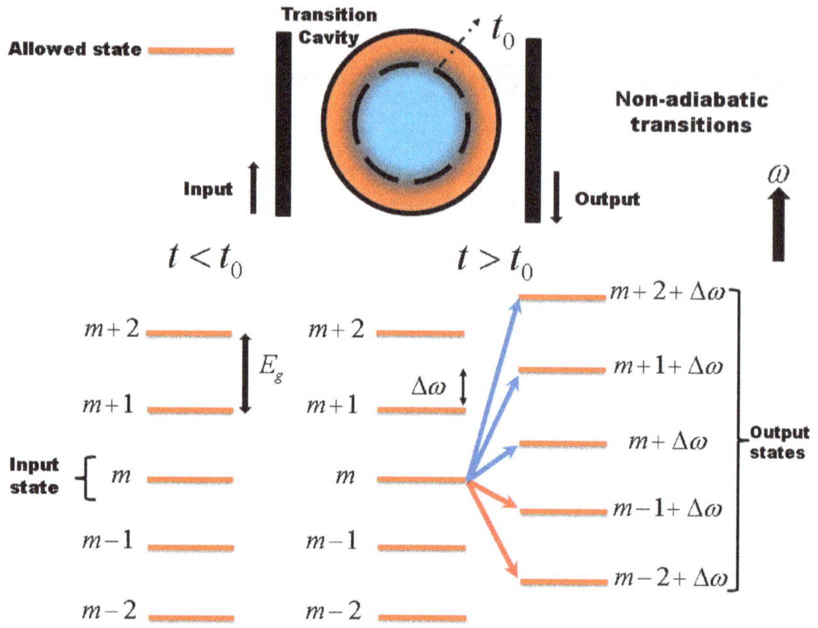

Figure 1. Light initially excites one state of the resonator ($t < t_0$). When the resonator is switched at a fast rate, multiple output states are excited ($t > t_0$). $\Delta\omega$ is the relative shift of all of the states due to the refractive index change.

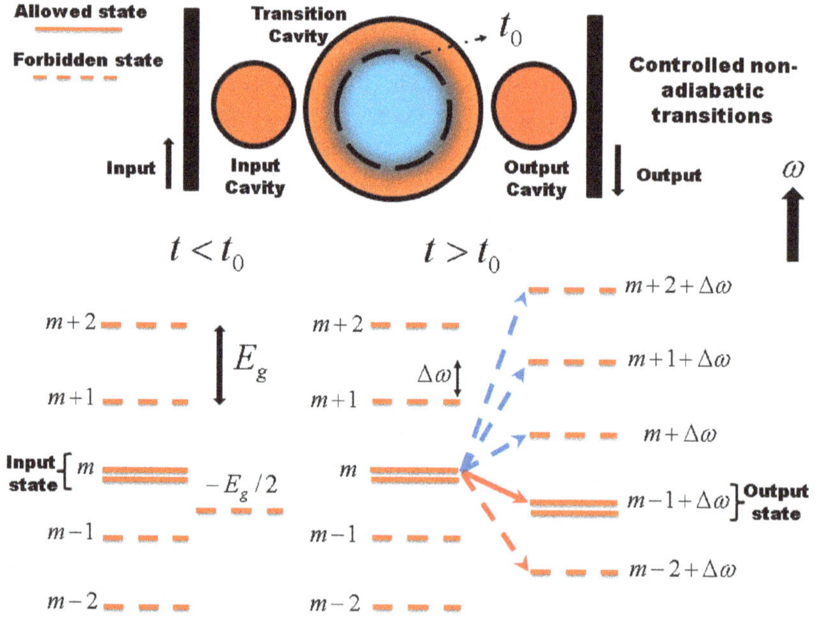

Figure 2. Photonic transitions in system consisting of threecoupled-cavities. Initially ($t < t_0$) the system only has one allowed input state—formed by the alignment of the input and transition cavity resonance (m). The light cannot escape to the output because the output cavity is in a forbidden state (-Eg/2). By non-adiabatically reducing the refractive index of the transition cavity ($t > t_0$) the states are shifted by Dw. Now the output cavity is in resonance with the (m^{-1}) mode of the transition cavity. This is the only allowed state of the system. The remaining transitions lie inside forbidden states in the energy diagram.

the transition resonator through a refractive index *reducetion*. This will blue-shift all of the states of the transition resonator, including the initially excited input state. However, now the output cavity will be on-resonance with the (m^{-1}) mode of the transition cavity—forming a newly, and singly, allowed state for the system. Since this state is at a lower energy than the original input state the wavelength of the light is efficiently red-shifted.

In order to demonstrate this concept we use time domain coupled mode theory to simulate the different mode dynamics [15-17]. The simulations takes into account waveguide and carrier losses in the system [9] and has been successfully applied to analysis of a multitude of dynamically controlled cavity systems [2,11,17-19]. In the model we consider realistic parameters—a typical value of waveguide loss of 3 dB/cm is assumed, and ring resonators are used as the cavities, with radii of $R_{input/output}$ = 10 μm and $R_{transition}$ = 200 μm (FSRs of 8.8 nm and 0.437 nm, respectively). The resonator states are seen in **Figure 3** where we see that the input ring and the transition ring have the same resonance. The output ring is placed at the midpoint of the gap between the (m) and (m^{-1}) mode of the transition resonator. It should be noted that for clarity the individual ring resonances were calculated separately in order to visualize the position of different resonances with respect to each other, while in the actual system they will be coupled—resulting in mode splitting.

3. Red Transitions with Refractive Index Reduction

The controlled transition of a pulse of light towards lower energies (red) is demonstrated in **Figure 4** by injecting a carrier density of $4E16/cm^3$ $(\Delta n/n)\% = 1\%$ in the transition resonator. The carriers are injected in a time of 100 fs, which is significantly shorter than the inter-mode coupling time $(t = 1/FSR_w)$ in order to ensure non-adiabatic transitions in the transition ring [4]. We note here that this fast switching speed in not a limitation of the system and can be relaxed by using resonators with a smaller FSR or by using resonant modulation as discussed below in Section 3. In **Figure 4** we see that most of the light is red shifted to the new (m^{-1}) mode of the transition cavity and there is negligible excitation of the other states. An important point to be emphasized

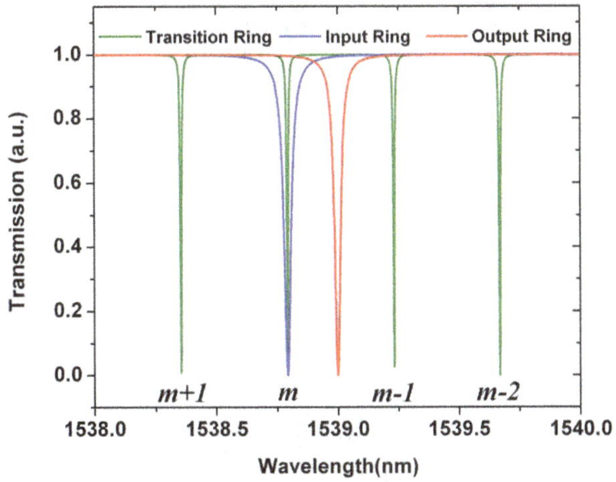

Figure 3. Transmission of different rings in the initial state of the system. The input resonator and the transition resonator have the same resonance condition, while the output resonator is purposely shifted to a lower energy—one-half FSR away.

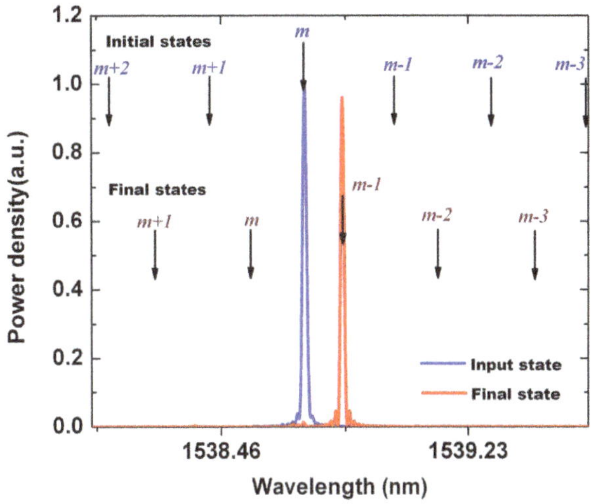

Figure 4. States before the switching (blue) and after switching (red) inside the transition ring. The conversion efficiency is 96%.

here is the fact that this is a true mode coupling in the transition ring, not a filtering effect—the spectrum shown in **Figure 4** is measured inside of the transition ring not at the output port. Lastly, we should note that the position of the converted light is not exactly at the half-FSR point but is slightly blue-shifted by 0.06 nm because of mode splitting of the coupled transition-output cavity state. Here we have tuned the resonators coupling constants in order to maximize coupling to only the blue-shifted split mode. In order to verify the results from coupled mode theory we numerically simulated the dynamic process by solving Maxwell's equations using the finite difference time-domain (FDTD) method [20]. The system has the same configuration as the one described earlier but with scaling differences in order to speed up the computation process (Radii of $R_{transition}$ = 24 μm and $R_{input/output}$ = 4 μm. These correspond to FSRs of ~4 nm and ~26 nm, respectively). We see in **Figure 5** that with an index shift of $(\Delta n/n)\% = 9\%$ the light is red-shifted by 2 nm. The double peak in the figure results from the interference of modes in the ring resonator which was evident in our previous experimental work [15].

The behavior is qualitatively the same as the results obtained with coupled mode theory. However, the conversion efficiency is slightly lower (88%) since the non-adiabatic transition is not as efficient with the large FSR used in the FDTD simulation. This is not a fundamental limitation provided the FSR of the transition ring is small enough. It all depends on the strength of the wavelength shift [7]. Provided a large enough refractive index, small rings with large FSR can be used.

4. Efficiency and State Design

We have shown here that the placement of the cavity

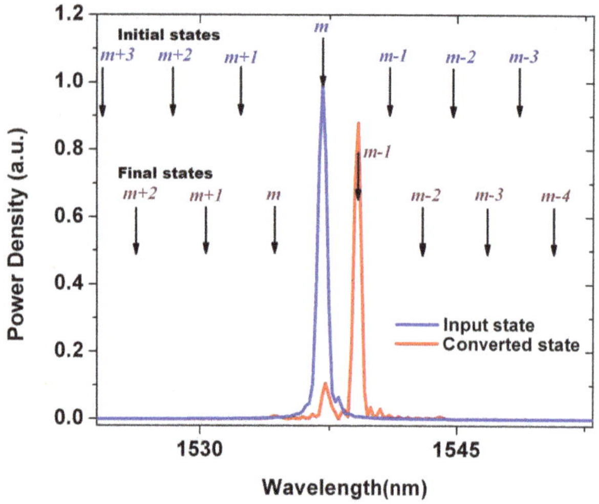

Figure 5. Finite-difference-time-domain verification of red-shifting results.

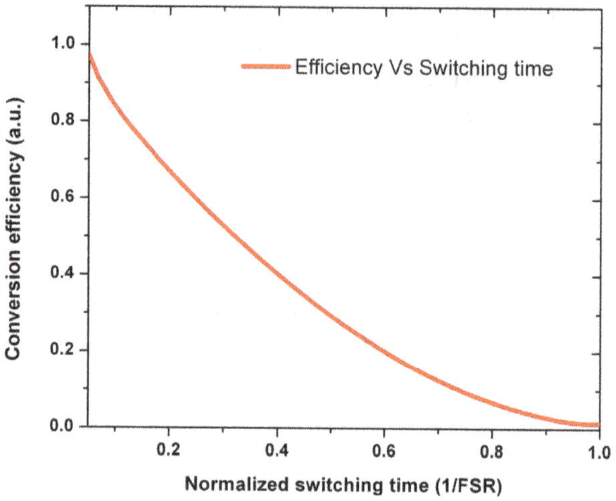

Figure 6. The efficiency decreases as switching speed is slowed due to the enhancement of the adiabatic shift in the resonator.

states will determine the new modes of the system. In addition, the maximum conversion efficiency is determined by the state design. For example, we found that this is achieved when the output cavities state is a half-FSR away from the initial transition cavities state (*i.e.* halfway between m and m^{-1}). This can be understood as follows. When this state is closer to the input state there is unintended adiabatic coupling to original (m) state. In addition, it would require a larger index change to achieve red-shifting since the initial (m^{-1}) mode would be even farther away from the output state. On the other hand, placing the out- put state closer to the initial (m^{-1}) state will also reduce the efficiency because the input and output state are initially further apart, but the refractive index change would be smaller—resulting in a weaker non-adiabatic transition. This could be overcome, however, by using resonant transitions where the resonator is switched at a rate corresponding to the difference in the state [2] spacing. We should add that without resonant transitions it is important to switch the transition cavity on the order of less than ~30% of 1/FSR (where FSR has units of frequency) in order to maximize the non-adiabatic transition process, as simulated in **Figure 6**.

5. Concluding Remarks

In conclusion, we demonstrated that by engineering the states of a system of cavities, and using non-adiabatic transitions, it is possible to obtain wavelength changes that are independent of the refractive index change sign. This new phenomena will enable more robust reconfigurable wavelength conversion systems where refractive index reductions can be used to both blue and red shift the frequency of light. In addition, the scheme proposed here for designing the states of a system of cavities could lead to novel dynamically controlled cavity systems for optical signal processing.

6. Acknowledgements

We would like to thank Dr. Edwin Hach for helpful discussions. The authors would also like to thank Dr. Gernot Pomrenke, of the Air Force Office of Scientific Research for his support and we thank Joseph Lobozzo II for the Lobozzo Optics Laboratory.

REFERENCES

[1] Z. Yu and S. Fan, "Complete Optical Isolation Created by Indirect Interband Photonic Transitions," *Nature Photos*, Vol. 3, 2009, pp. 91-95.

[2] J. N. Winn, S. Fan, J. D. Joannopoulos and E. P. Ippen, "Interband Transitions in Photonic Crystals," *Physical Review B*, Vol. 59, No. 3, 1999, pp. 1551-1554. doi:10.1103/PhysRevB.59.1551

[3] A. Khorshidahmad and A. G. Kirk, "Wavelength Conversion by Dynamically Reconfiguring a Nested Photonic Crystal Cavity," *Optics Express*, Vol. 18, No. 8, 2010, pp. 7732-7742. doi:10.1364/OE.18.007732

[4] P. Dong, S. Preble, J. Robinson, S. Manipatruni and M. Lipson, "Inducing Photonic Transitions between Discrete Modes in a Silicon Optical Microcavity," *Physical Review Letters*, Vol. 100, No. 3, 2008, pp. 1-4.

[5] T. J. Johnson, M. Borselli and O. Painter, "Self-Induced Optical Modulation of the Transmission through a High-Q Silicon Microdisk Resonator," *Optics Express*, Vol. 14, No. 2, 2006, pp. 817-831. doi:10.1364/OPEX.14.000817

[6] M. Notomi and S. Mitsugi, "Wavelength Conversion via Dynamic Refractive Index Tuning of a Cavity," *Physical Review A*, Vol. 73, No. 5, 2006, Article ID: 051803. doi:10.1103/PhysRevA.73.051803

[7] S. F. Preble, Q. Xu and M. Lipson, "Changing the Colour of Light in a Silicon Resonator," *Nature Photos*, Vol. 1,

No. 5, 2007, pp. 293-296. doi:10.1038/nphoton.2007.72

[8] W. M. Green, M. J. Rooks, L. Sekaric and Y. A Vlasov, "Ultra-Compact, Low RF Power, 10 Gb/s Silicon Mach-Zehnder Modulator," *Optics Express*, Vol. 15, No. 25, 2007, pp. 17106-17113. doi:10.1364/OE.15.017106

[9] R. A. Soref and B. R. Bennett, "Kramers-Kronig Analysis of E-O Switching in Silicon," *SPIE Integrated Optical Circuit Engineering*, Vol. 704, 1986, pp. 32-37.

[10] A. Liu, L. Liao, D. Rubin, H. Nguyen, B. Ciftcioglu, Y. Chetrit, N. Izhaky and M. Paniccia, "High-Speed Optical Modulation Based on Carrier Depletion in a Silicon Waveguide," *Optics Express*, Vol. 15, No. 2, 2007, pp. 660-668. doi:10.1364/OE.15.000660

[11] G. A. B. Daniel and D. Maywar, "Dynamic Mode Theory of Optical Resonators Undergoing Refractive-Index Changes," *Journal of the Optical Society of America B*, 2011, in Press.

[12] G. P. A. I. D. Rukhlenko, M. Premaratne and C. Dissanayake, "Continuous-Wave Raman Amplification in Silicon Waveguides: Beyond the Undepleted Pump Approximation," *Optics Express*, Vol. 34, No. 4, 2009, 536 Pages.

[13] H. Fukuda, K. Yamada, T. Shoji, M. Takahashi, T. Tsuchizawa, T. Watanabe, J.-I. Takahashi and S.-I. Itabashi, "Four-Wave Mixing in Silicon Wire Waveguides," *Optics Express*, Vol. 13, No. 12, 2005, pp. 4629-4637. doi:10.1364/OPEX.13.004629

[14] M. A. Foster, A. C. Turner, J. E. Sharping, B. S. Schmidt, M. Lipson and A. L. Gaeta, "Broad-Band Optical Parametric Gain on a Silicon Photonic Chip," *Nature*, Vol. 441, No. 7096, 2006, pp. 960-963. doi:10.1038/nature04932

[15] A. W. Elshaari, A. Aboketaf and S. F. Preble, "Controlled Storage of Light in Silicon Cavities," *Optics Express*, Vol. 18, No. 3, 2010, pp. 3014-3022. doi:10.1364/OE.18.003014

[16] C. Manolatou, M. J. Khan, S. Fan, P. R. Villeneuve, H. A. Haus and J. D. Joannopoulos, "Coupling of Modes Analysis of Resonant Channel Add-Drop Filters," *Quantum Electron*, Vol. 35, No. 9, 1999, pp. 1322-1331. doi:10.1109/3.784592

[17] S. Fan, M. F. Yanik, Z. Wang, S. Sandhu and M. L. Povinelli, "Advances in Theory of Photonic Crystals," *Journal of Lightwave Technology*, Vol. 24, No. 12, 2004, pp. 4493-4501. doi:10.1109/JLT.2006.886061

[18] Q. F. Xu, P. Dong and M. Lipson, "Breaking the Delay-Bandwidth Limit in a Photonic Structure," *Nature Physics*, Vol. 3, No. 6, 2007, pp. 406-410. doi:10.1038/nphys600

[19] C. R. Otey, M. L. Povinelli, S. Fan and S. Member, "Completely Capturing Light Pulses in a Few Dynamically Tuned Microcavities," *IEEE Journal of Lightwave Technology*, Vol. 26, No. 23, 2006, pp. 3784-3793. doi:10.1109/JLT.2008.2005511

[20] A. Taflove and S. Hagness, "Computational Electrodynamics: The Finite-Difference Time-Domain Method," Artech House Publishers, London, 2000.

12

The Design and Manufacturing of Diffraction Optical Elements to Form a Dot-Composed Etalon Image within the Optical Systems

Sergey Borisovich Odinokov, Hike Rafaelovich Sagatelyan
Bauman Moscow State Technical University (BMSTU), Moscow, Russia
Email: h_sagatelyan@mail.ru

ABSTRACT

The possibilities of manufacturing of diffraction optical elements (DOE), using the "Caroline 15 PE" plasma-etching machine were considered. It is established that at thickness of chromic mask of 100 nm the plasma-chemical etching (PCE) method reaches depth of surface micro-profile to 1.4 μm on optical glass. It allows increasing the diffraction efficiency of DOE to 0.3 - 0.35 on the second order of diffraction.

Keywords: Optical Glass; Plasma-Chemical Etching; Selectivity; Diffraction Gratings; Micro-Optics

1. Introduction

The development of optical subsystems, having minimal weight and size characteristics, but ensuring the reference-class location of several points in the plane of the image on a CCD-matrix (**Figure 1**), is an actual problem for a number of optical devices [1]. The most effective way to solve this problem is to use the diffractive optical element (DOE), which works on the transmission of light mode and contains several diffraction gratings (**Figure 2**)—one for each of points, formed by it.

The principle of operation of to be designed DOE is shown on the **Figure 3**. The set of parallel laser beams is falling on the DOE with angle of light incidence α. To minimize the size characteristic of optical device, it is necessary to provide as big as possible value of angle of incidence α. As it can be seen out of **Figure 3**, the DOE has to work on not zero order of the spectrum. So, the diffraction gratings of DOE are transferring echeletts—phase diffraction gratins, which have an ability to concentrate the diffracted light in the spectrum of defined (not zero) order.

At the present time only the echeletts, working on reflection of light and fabricated by engraving the pattern with a triangle micro profile, which creates an additional difference of travel in the limits of each period of grating [2], on the surface of the metal by especial cutting tool, are known. As far as the DOE, which has to be created, must work in transition mode, being made of optical inorganic glass, the cutting out of desired profile on its working surface is impossible. This kind of micro-optics workpieces can be created on basis of implementing the method of plasma chemical etching (PCE) [3].

The plasma-chemical etching method has a number of destinations, which allow forming on the worked surface narrow (width lower than 1 micrometer) gaps with relatively large (1 - 2 micrometer) deepness. The shortcoming of plasma-chemical etching consists in the fact, that by this method, from a practical point of view, it is possible to build only the binary micro-relief.

However, in the case of tilted fall of light on the diffraction grating [4], the binary rectangular micro-relief of the phase grating turns into a saw-like one (**Figure 4**). This allows one to create an echelett, manufactured by the method of plasma-chemical etching.

The object of this work was to create a transmitting DOE working on the 2nd order of diffraction. To achieve this goal the following tasks were solved:

- Theoretical determination of the required depth of grooves of phase diffraction gratings of DOE;
- Fabrication of DOE of optical glass by the PCE method including: 1) choice of technological equipment; 2) carrying out of experimental research on PCE of optical glass; 3) selection of optimal mode of PCE operation; 4) optical verification of fabricated DOEs;
- Discussion of acquired results of experimental research to explain the main regularities and peculiarities of PCE of optical glass.

The novelty of our works consist in establishment of interconnection between the conditions of PCE on manufacturing the DOE with it's operational behavior.

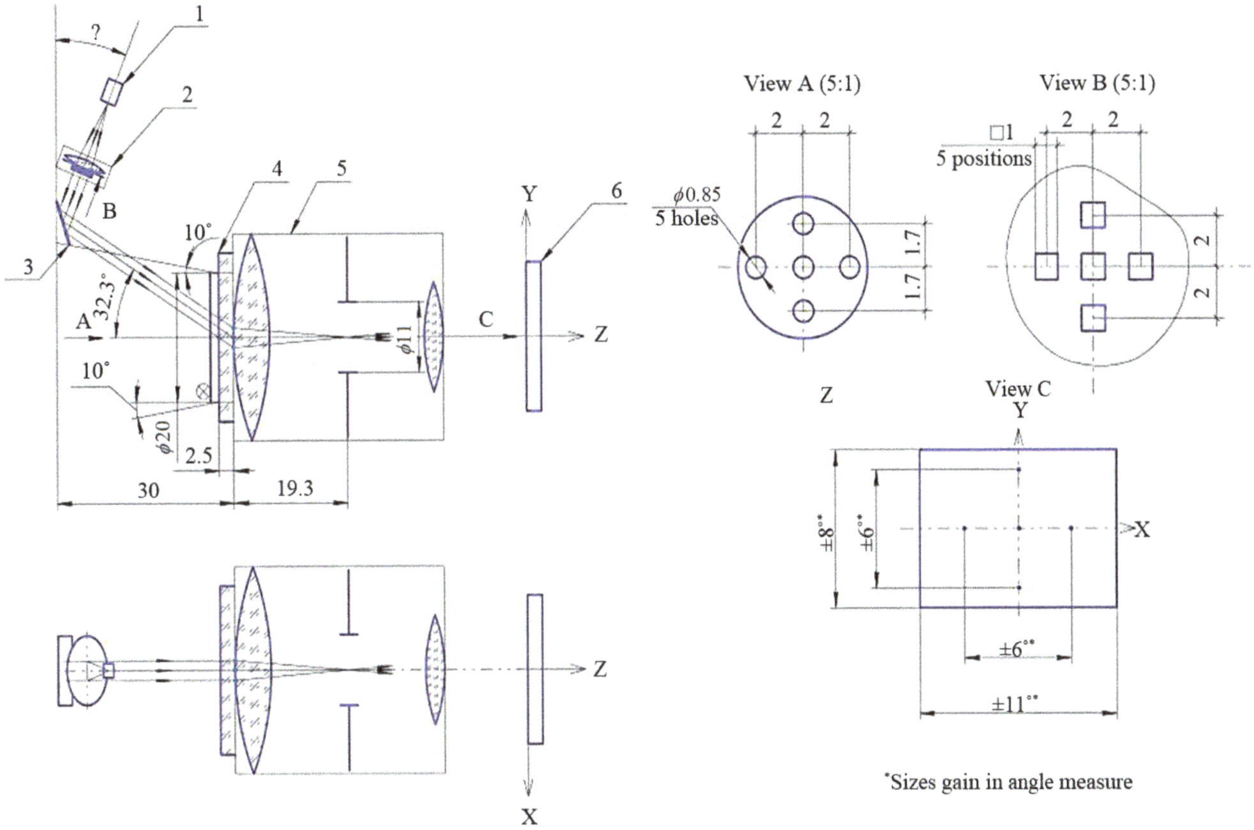

Figure 1. The optical scheme of device.

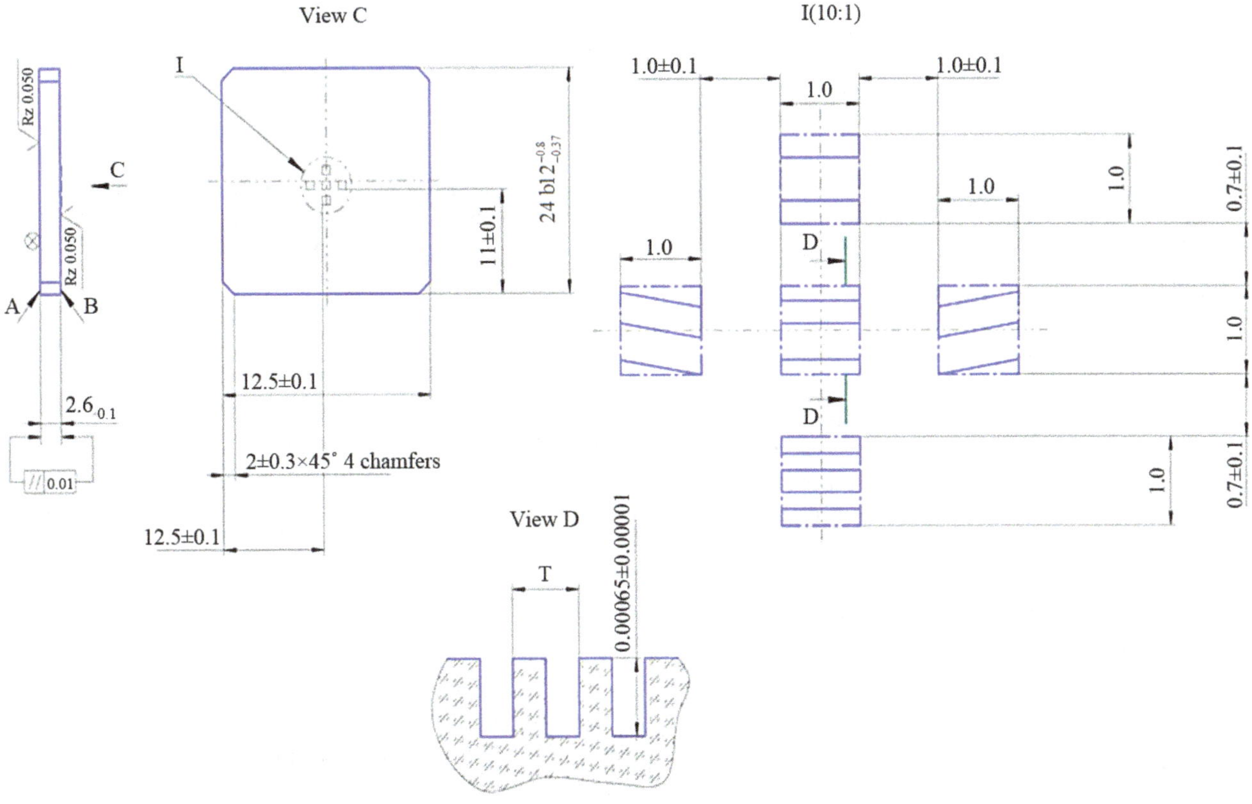

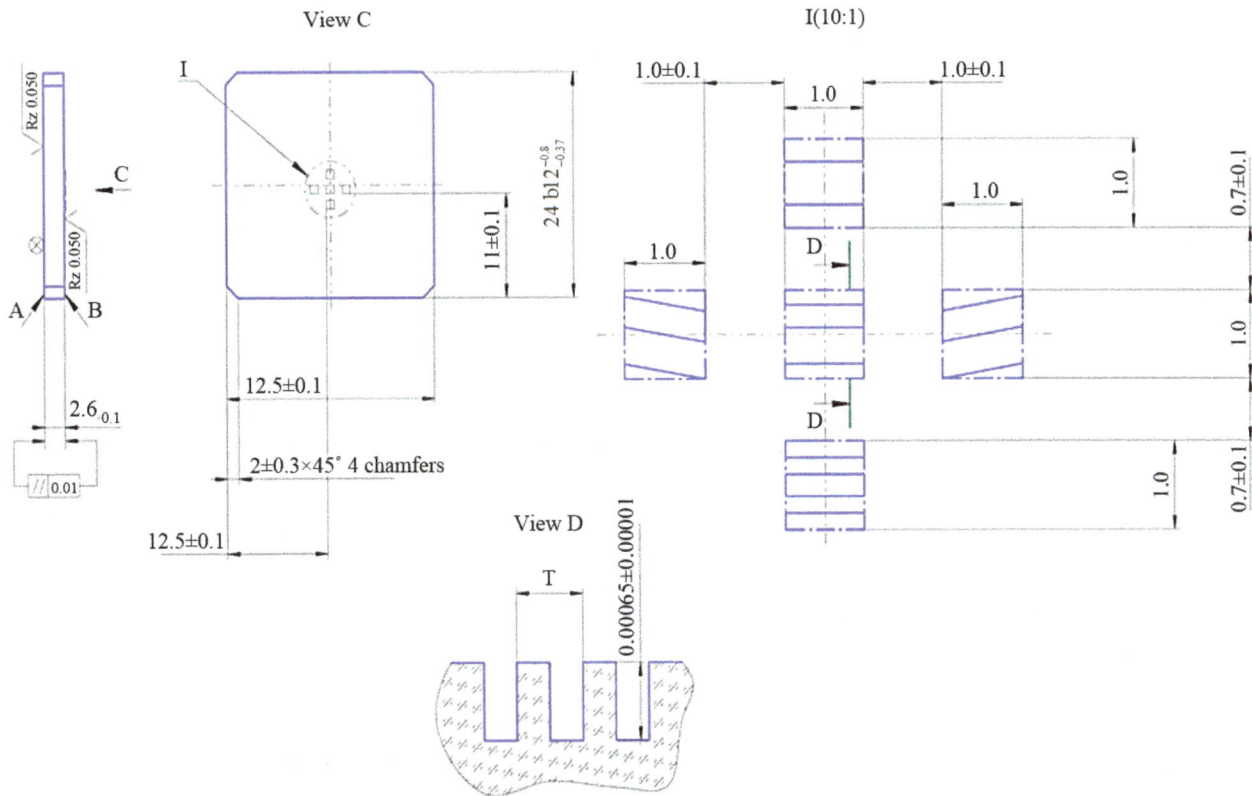

Figure 2. The design of DOE.

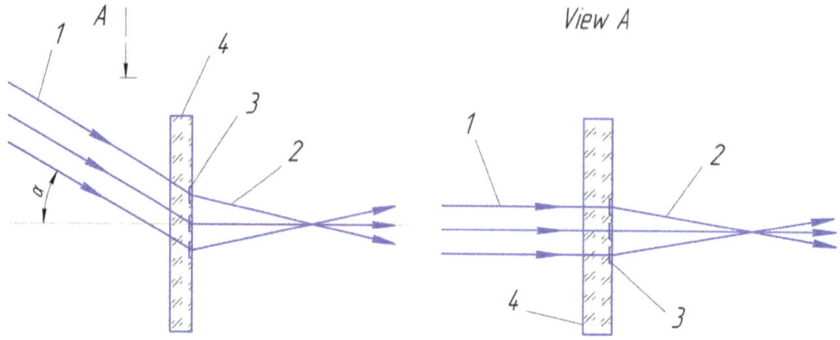

Figure 3. The scheme of DOE work: 1—Incident parallel beams; 2—Diffracted beams; 3—Diffraction gratings; 4—DOE.

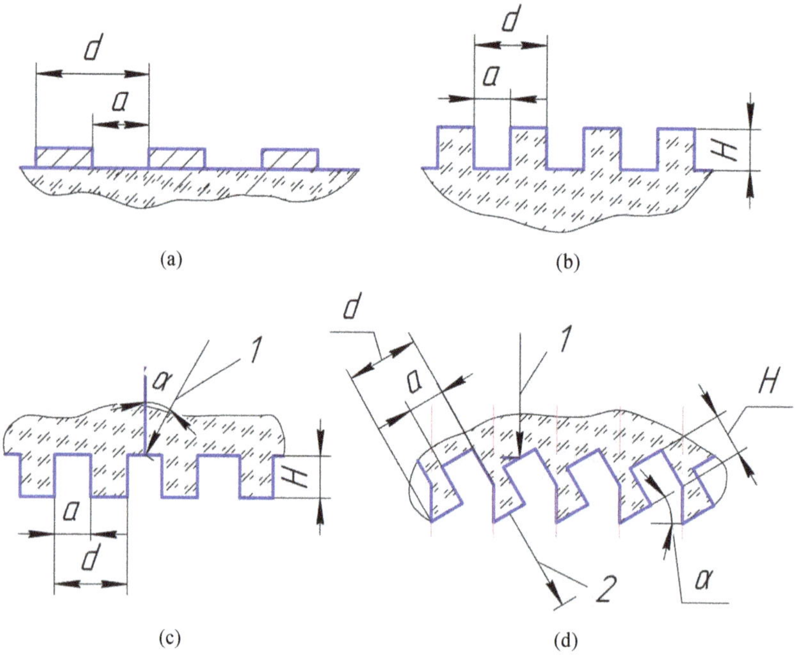

Figure 4. The scheme of manufacture of DOE, working as an transmitting echelett (1 and 2—the incident and the diffracted beams): (a) Creating of mask; (b) Plasma-chemical etching; (c) The incidence of beam on the phase grating; (d) The equivalent echelett.

2. Theoretical Justification

It is known [5], that for the concentration of energy in a given direction two conditions must be met, which, with regard to the case under consideration, can be formulated in the following way:

1) The direction of the zero order from a separate refracting element should coincide with the direction on the maximum of the required direction from the grating as a whole;

2) The direction on the zero order spectrum from the grating as a whole should coincide with the direction of the minimum while diffracting on the separate refracting element.

For the analysis of the first conditions we can write down the equation of the diffraction grating

$$d(\sin\alpha + \sin\beta) = n\lambda, \quad (1)$$

where d is the period of the diffraction grating; α—the incidence angle; β—the angle of required diffraction order; n—the number of order; λ—the wave length.

We shall consider the incidence of beam on the central diffraction grating of DOE (see **Figure 1**). Here we have $\beta = 0$ (the perpendicularity to the working plane of the DOE of the required direction on diffraction maximum). To work on the second order of diffraction ($n = 2$), expression (1) can be written in the form of

$$d\sin\alpha = 2\lambda. \quad (2)$$

The expression (2) specifies an unambiguous relationship between the required angle of incidence α and the period d of the diffraction grating. So, for $\alpha = 30°$ the period of the diffraction grating should be $d = 2.6$ μm.

Continuing to analyze the first condition for the energy concentration by an echelett, we can set the regularities of the formation of emission of zero order from a separate refracting element. It is obvious that in this case we have the irradiation, diffracted on the phase grating with a rectangular profile (**Figure 5(a)**).

The relative spectral intensity of radiation I_0 for the zero order in this case is described by expression [6]

$$I_0 = \left(1-\varepsilon^2\right)\cos^2\left[\pi\left(n-1\right)\frac{H}{\lambda}\right]+\varepsilon^2, \qquad (3)$$

where: $\varepsilon = \dfrac{d-2a}{d}$ is the asymmetry factor of rectangular phase grating; n is the refractive index of glass.

The plot of $I_0(H)$ dependence for $n = 1.5$; $\lambda = 0.65$ μm and $\varepsilon = 0$ is shown on **Figure 5**, *d*. It follows from this plot, that the maximum of relative spectral efficiency at the zero order of diffraction can be achieved at $H = 1.3$ μm. Thereby, the required H depth of grooves of the rectangular profile, which are formed by means of plasma-chemical etching (see **Figure 4(b)**), must be determined by the following formula, driven out from the expression (3):

$$H = \frac{\lambda}{n-1}, \qquad (4)$$

Let us consider the second condition for concentration of energy by the echelett, which in this case boils down to the fact, that toward the direction of ray 1 (see **Figure 4(d)**) there will be located the minimum of intensity of radiation, which is diffracted on the separate element. An appropriate model for the presented on the **Figure 4(d)** equivalent scheme is describing the echelett by a phase grating with a triangular profile (**Figure 5(b)**). Comparing **Figures 4(d)** and **5(c)**, we can assume that the deepness Δ of triangular relief of such a phase grating works out as

$$\Delta = d\sin\alpha + H\cos\alpha. \qquad (5)$$

The minimum of intensity of radiation for the zero order of diffraction on a separate element of profile determines from the following expression [6]:

$$I_0 = \left\{\frac{\sin\left[\pi\left(n-1\right)\dfrac{\Delta}{\lambda}\right]}{\pi\left(n-1\right)\dfrac{\Delta}{\lambda}}\right\}, \qquad (6)$$

where Δ is the deepness of relief (**Figure 5(c)**), calculated by formula (5).

The plot of $I_0(H)$ dependence for $n = 1.5$; $\lambda = 0.65$ μm; $\alpha = 30°$ and $d = 3$ μm is shown on **Figure 5(e)**. In this case, as well, as in previous one, we can see, that intensity of radiation of zero order of diffraction on the separate element of profile achieves its minimum at deepness of etching (see **Figure 4(b)**) $H = 1.3$ μm.

Coincidence of required deepness of etching $H = 1.3$ μm to maximize the intensity of radiation in the required direction of diffraction (see **Figure 5(b)**) as well as to minimize the intensity of radiation in the direction of zero order for the grating as a whole (see **Figure 5(e)**), indicates that the light transmitting echelett, designed to work on the second order of spectrum, can be produced by the method of plasma-chemical etching.

3. Experimental

The most appropriate way to create plasma, which is a source of positively charged ions and chemically active radicals, ensuring the removal of the processed material during the PCE of optical glass—an inorganic dielectric

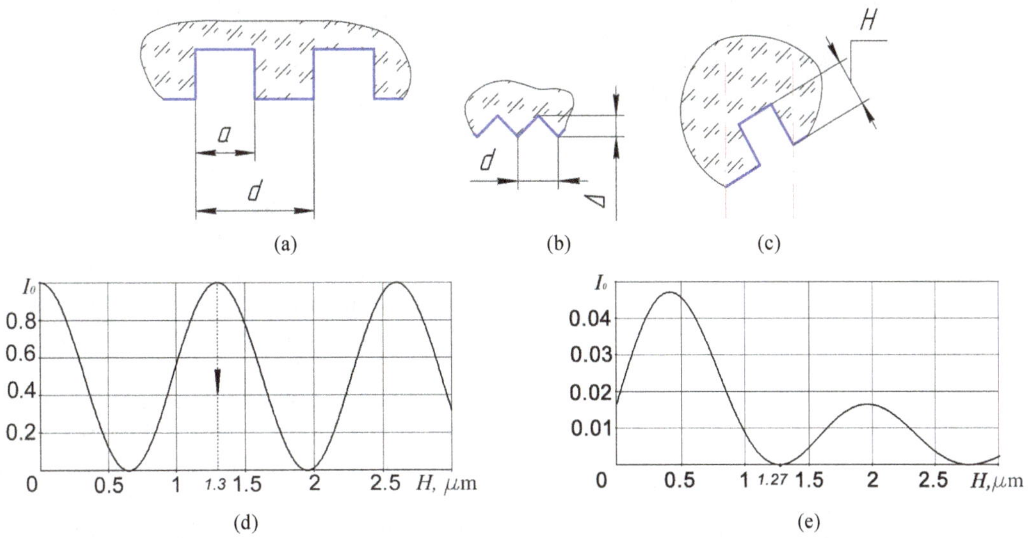

Figure 5. Phase diffraction gratings with a rectangular (a), a triangular (b) and an equivalent (c) profile as well as respective (d, e) relative spectral intensities of zero order.

material—is application of high-frequency inductive discharge. Such a discharge allows adjusting the power supply and, consequently, the temperature of plasma in a wide range, providing at the same time the sufficient number of free electrons to compensate the positive charge, which accumulates on the dielectric processed work-piece.

3.1. The Choice of Technological Equipment for Plasma-Chemical Etching (PCE) of Optical Glass

Experimental research of electro-physical characteristics of such facilities [7] allowed establishing a number of regularities of generation of currents (ion, electron) and voltages (in particular, the bias voltage) in the working chamber. However, these studies do not allow establishing any direct analytical relationships between the electro-physical parameters of the device and the output parameters of the PCE process, such, for example, as the deepness of micro-relief on the formed functional surface of DOE.

As the equipment to realize PCE process with regard to the manufacturing of DOE, the "Caroline 15 PE" device for plasma-chemical etching [8] was chosen, which is used in microelectronics to produce integrated circuits on the wafers having 200 mm diameter. The features of constructive arrangement of this device are presented on **Figure 6**.

While choosing of this device as an equipment to design the technology to produce DOE, made of optical glass (the main component of the composition of the furnace charge to melting such a glass is silicon dioxide), the information about a positive results of its using to produce parts, made of single-crystal quartz [9], was taken into consideration. Some results of experimental research of applying the "Caroline 15 PE" device to etching the single-crystal quartz at great depths—more than 200 μm —are presented in [9] without a detailed description of conditions and regimes of PCE operation. Particularly, the material of mask, the method of its gaining and the thickness, necessary to etch on depresses of order of tens of micrometers are not disclosed.

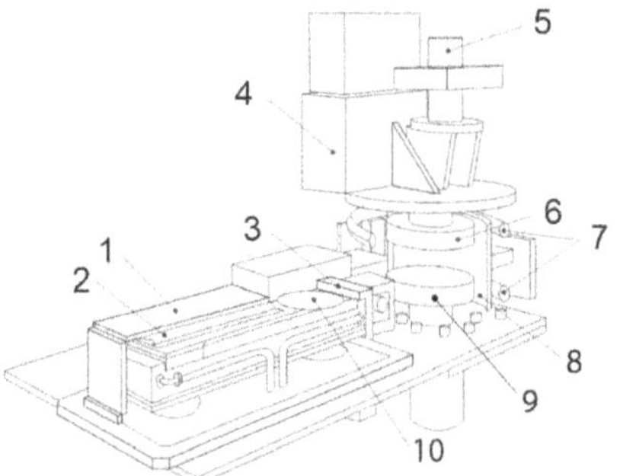

Figure 6. The "Caroline 15 PE" device for plasma-chemical etching: 1—sluice chamber; 2—transfer arm; 3—slit shutter; 4—impedance matcher; 5—laser control system; 6—antenna; 7—electromagnetic system; 8—housing of the working chamber; 9—wafer holding table; 10—carrier.

The results of our preliminary experimental studies on applying the "Caroline 15 PE" device to PCE of optical glass, using chrome and iron oxide as materials of the mask and blanks of reticles as crude product for DOE, are presented in [10].

3.2. Experimental Studies on Making Phase DOEs

Crude products (work-pieces) of DOE, having 35 mm in diameter, were gained by drilling out of blanks of reticles, used in integrated circuit technology. So, the material of phase DOE was the glass, used to make reticles. For the placement of DOE work-pieces having diameter 35 mm, on the wafer holding table of "Caroline 15 PE" device, which is intended to 200 mm silicon wafers, the special shielding carrier was designed.

The results of experimental studies of PHE operation for DOE manufacturing are shown below.

The influence of etching time t on the attained deepness of grooves of diffraction gratings while providing the PCE operation for phase DOE with chrome mask was investigated.

The following etching regime was fixed on "Caroline 15 PE" device: etching gas consumption (chlorofluorocarbon CF_4—Freon)—C_{CF4} = 2.4 l/min; plasma gas (argon) flow C_{Ar} = 0.8 l/min; table cooling gas (helium) consumption—C_{He} = 1.2 l/min; superposed magnetization current in coils, which embrace the chamber with plasma—I = 2.0 A; the power feed to the antenna—P_a = 250 W; the power feed to the table—P_{cm} = 250 W; rotating velocity n of the rotor of turbo—molecular pump—100% of nominal.

During the work of "Caroline 15 PE" device plasma-chemical etching conditions, which are associated with the functioning of two high-frequency generators with their impedance matchers, loading respectively at the antenna and the table, are formed in the process itself. These conditions include: the actual powers at the antenna and table, in both cases the values of powers of incent and reflected waves are registered at that; the bias voltage at the table U; the pressure in the working chamber p.

The resulting graphs of influence of PCE process duration on the deepness of grooving of diffraction gratings, obtained on the basis of research, carried out on the Surfcoder SE 1700 α by Kosaka Laboratory Ltd./Japan surface roughness measuring instrument, are presented on **Figure 7**. The deepness of profile was determined on two fabrication stages: 1) after the realization of PCE

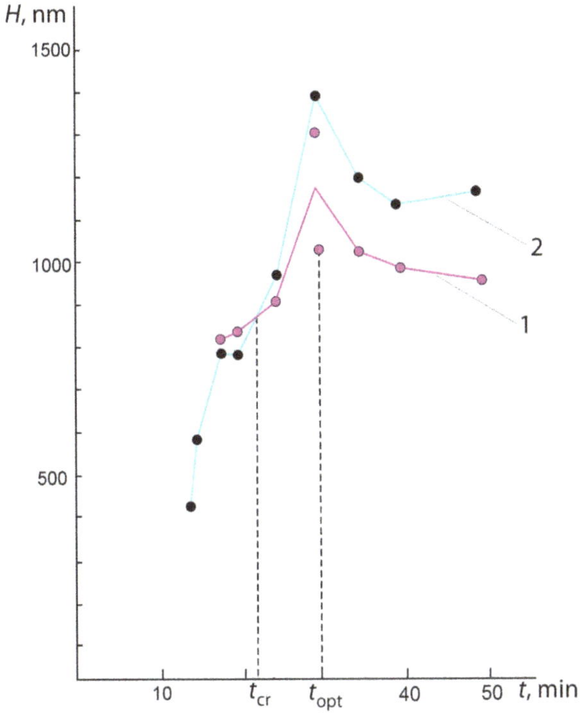

Figure 7. The dependences of diffraction grating's grooves deepness *H* on duration *t* of plasma-chemical etching (PCE) operation, gained after the PCE operation (curve 1) and after the operation of liquid acid etching, following the PCE operation (curve 2).

operation itself; 2) after the operation of acid etching, which was carried out to remove the remnants of chrome mask. On **Figure 7** the plot of dependence of diffractive grating's grooves, measured after the PCE operation itself corresponds to the broken curve 1, and the broken curve 2 corresponds to plot of dependence of diffraction grating's grooves depresses, obtained after the executing of acid etching operation, following the PCE operation on the same samples. Considering the curves 1 and 2 on the **Figure 7**, one can see the following regularities and peculiarities of the character of influence of the duration of PCE operation on the formed depth of diffraction grating's groove.

1) Both curves 1 and 2 on **Figure 2** have a maximum of grooves' depth on the PHE process duration t_{opt} = 30 min. In other words on $t < t_{opt}$ PCE process durations the deepness of grooves of diffractive gratings increases with the increase of PCE process duration, and on the PCE process durations $t > t_{opt}$ the deepness of grooves of diffractive gratings decreases with the increase of PCE process duration.

2) Curves 1 and 2 intersect on the PCE process duration t_{cr} = 22 min. On the $t < t_{cr}$ durations of PCE process, the deepness of grooves of diffractive gratings, measured after the PCE operation, exceeds the deepness of the same grooves, measured after the operation of acid etching, and on the $t > t_{cr}$ durations of PHE process the deepness of grooves of diffraction gratings, measured after the PHE operation, occur to be less than deepness of same grooving, measured after the operation of acid etching. In other words, in the latter case the acid etching operation leads to an increase in diffraction grating grooving's depth.

At $t = t_{cr}$ moment the deepness of grooves of diffraction gratings doesn't change after the execution of acid etching operation.

3) For the $t > t_{cr}$ PCE process durations the difference between the deepness of grooves after the PCE and after the acid etching increases with the raise of the PCE process's duration.

The analysis of the results of research of influence of duration of the PCE operation on the depth of relief of phase diffraction grating, presented on **Figure 7**, allows one to contend, that for the achievement of maximal glass etching deepness the single determining parameter is the selectivity of etching of glass related to material of mask. The operation of PCE of optical glass must be carried out on the most reasonable (optimal) mode. The optimum is the mode, which provides the maximum of the selectivity of etching. In this case, the ratio of etching speeds of the glass and the mask is understood as the selectivity of etching.

To find out the optimal mode of PCE, the series of experiments was carried out, using the iron oxide as the material of mask. The samples for these investigations were drilled out of ready to use reticles, used on integrated circuits' production. The mask for the PCE operation corresponded to not transparent areas of reticle.

The influence of relation of consumptions of the plasma generating (argon) and reactive (Freon) gases on a number of output parameters of PCE process was investigated. The result of mentioned series of experiments is presented on the **Figure 8** as a number of graphs. This dependences are acquired on fixing of other conditions of the process (the powers on antenna and the table, magnetization current, helium consumption, the speed of turbo-molecular pump) on the same levels, as at the previous experimental studies, results of which are presented on **Figure 7**.

Considering the graphs on **Figure 8**, one can see, that dependences of parameter of our interest—the selectivity *r* of etching—on consumptions of both argon and Freon, possess pronounced extremum nature, *i.e.* they've got maximums. We can assume that the optimal mode of PCE of glass on the "Caroline 15 PE" device is established—it corresponds to the consumption of argon C_{Ar} = 0.8 l/min at simultaneous consumption of Freon C_{CF4} = 2.4 l/min.

Practical application of plasma-chemical etching operation is conditioned on necessity to produce exactly phase diffraction gratings, which as opposed to ordinary

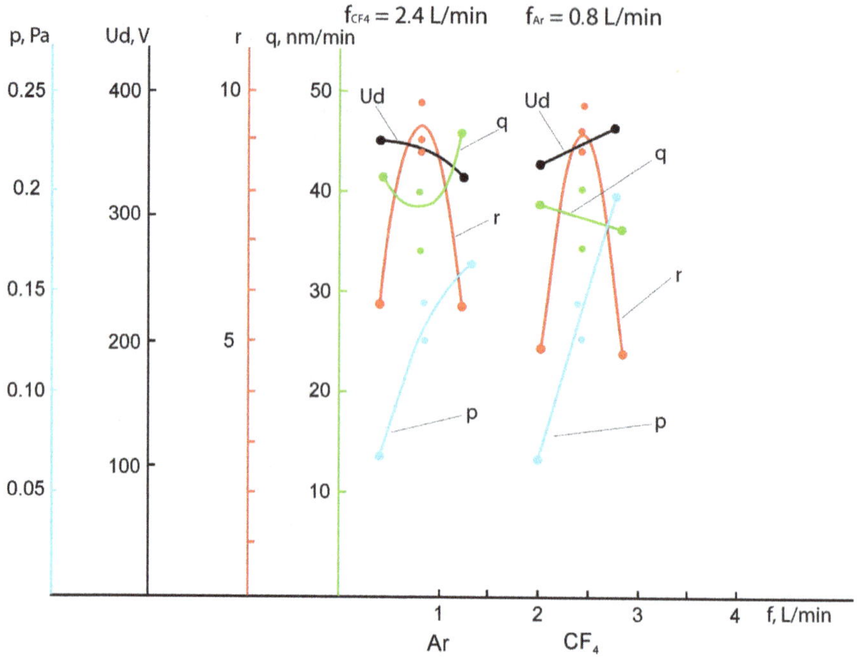

Figure 8. The dependences of productivity *q* and selectivity *r* of etching, as well as bias voltage U_d and vacuum pressure *p* in the chamber on consumptions *f* of argon (Ar) and Freon (CF_4).

(amplitude) ones provide heightened diffraction efficiency. The experimental study of diffraction efficiency of diffraction gratings of DOE with different deepness of grooves, fabricated with use of PCE method, was carried out on the specially designed optical stand.

The intensities of beams of 0. 1st, 2nd, 3d and −1st ("minus first") orders were measured with the help of special measuring instrument on tilted incidence of light. For amplitude diffraction gratings the diffraction efficiency decreases to the greatest extent on transition to the second and higher orders of diffraction. So, the most relevant is the question of determining the optimal deepness of diffraction gratings, providing the maximum of energy in the second order of diffraction.

Acquired dependencies of diffraction efficiency as the portion of the light's intensity of beam of the 2nd order of diffraction in the summarized intensity of light, transferred the DOE, on the deepness *d* of grooves of diffraction gratings with different periods are presented on the **Figure 9**, considering which one can see, that the partial share of second order of diffraction can reach and even exceed 30%.

Empirically established dependencies, which are presented as graphs on **Figure 9**, also indicate, that to the heighten the diffraction efficiency of phase diffraction gratings the deepness of grooves must be increased as much as possible.

4. Discussion

The heightening of diffraction efficiency of phase diffraction gratings in the 2nd order of diffraction with increasing of deepness of grooving of diffraction grating (see **Figure 9**) can be explained by decreasing of relative spectral intensity of radiation for the zero order of diffraction [11]. Generally speaking, in accordance with the theory of light diffraction on phase diffraction gratings with ideal rectangular profile of micro-relief, the relative spectral intensity of zero order of diffraction should be changed in waves, achieving the maximum (for our conditions) at grooving depths 700 nm, 1400 nm and etc. In practice, however, the obtained form of micro-relief of diffraction grating differs of ideal rectangular, so far as rounding occurs on the edges of ledges and the walls of micro-relief aren't upright, but are tilted on angle up to 30° to vertical line.

Thereby, the nominally rectangular shape of micro-relief of phase diffraction gratings in practice is close to the triangle, and for such a form of micro-relief, in accordance with the theory of diffraction [6], the relative spectral intensity of zero order of diffraction with increasing of depth of grooving of diffraction grating is changing wavy, but nevertheless tends to zero. Therefore, to raise the diffraction efficiency of phase diffraction gratings, the depth of grooving must be increased as possible.

As has already been noted above, maximally achieving by PCE deepness of diffraction grating's grooving determines mainly by the selectivity of etching of optical glass related to material of mask.

In connection with this, the detected maximums in $r = f(C_{Ar})$ and $r = f(C_{CF4})$ dependencies (see **Figure 8**) are

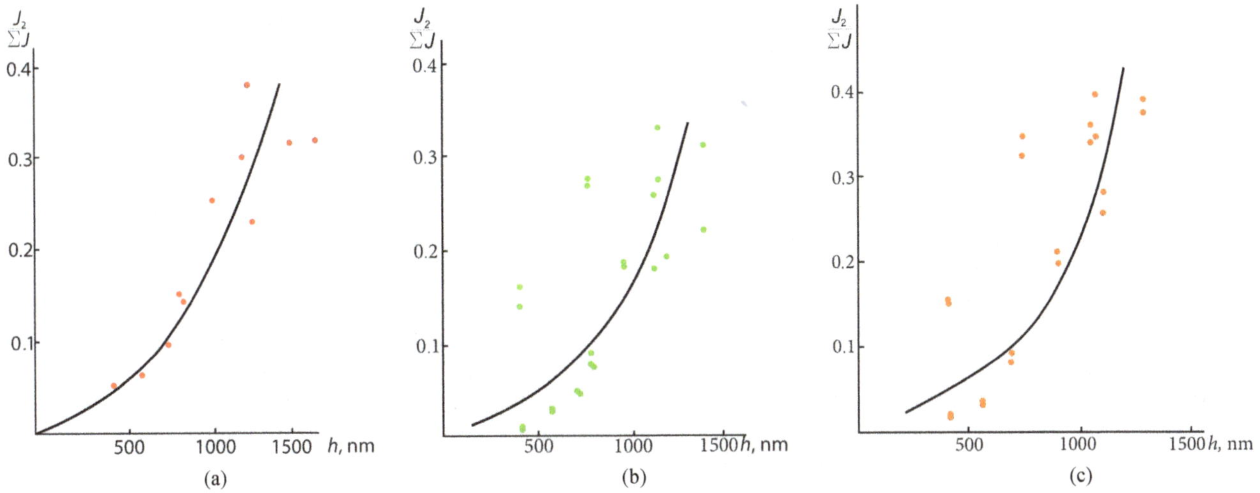

Figure 9. The dependences of ratio of light's intensity in the beam of 2nd order of diffraction I_2 to sum of intensities of light in the beams of 0, 1st, 2nd, 3rd and −1st orders of diffraction on the deepness h of grooves of diffraction gratings with the period $d = 2$ µm (a); $d = 2.5$ µm (b) and $d = 3$ µm (c).

gaining an exceptional importance. The graphs of other dependencies, presented at the same drawing, help to explain these maximums.

Among these graphs the most easily explainable are graphs of razing of vacuum pressure p with increasing of consumptions of argon and Freon. Explaining the characters of other dependencies is complicated and that is connected with the fundamental differences in the physical essence of the process, concomitant with the presence of atoms of argon and molecules of Freon in plasma. Whereas atoms of argon in high frequency plasma are dissociating on positively charged ions and electrons, the molecules of Freon in the same plasma are acquiring only the not-coupled electrons, becoming radicals (though formation of free electrons and complicated positively and negatively charged ions is possible as well).

Naturally, the rising of argon consumption leads to increasing of density of electrons in plasma, which causes decreasing of bias voltage U_d on the table, as well as to increasing of ion current through the work-piece of DOE under treatment, which results increasing of productivity q of glass etching. Existence of a minimum in the dependence of productivity of etching on argon consumption is explained by decreasing of the length of the free run of argon ions in the accelerating electric field with increasing of density of plasma: on heighted concentration of argon the physical destruction of glass with ions of argon is prevailed, and this is accompanied also by intensification of destruction of mask material and by decreasing of selectivity r of etching; on lowered concentration of argon it's ions are being successful to gain in the accelerating electric field an energy, which is enough to intensify the physical destruction of worked surface of glass (increasing of etching productivity q) as well as destruction of mask, which also brings to decreasing of etching selectivity r.

The disclosed regularities and features of the relationship of achievable etching depth of optical glass with the duration of PCE process could be explained as follows.

1) At the $t = t_{opt}$ moment (see **Figure 7**) an almost complete removal of chromium mask occurs—on areas, protected by the mask, only the separate, not connected with each other parts of mask can remain in form of isles. If the direction of the vector of speed of all the ions and radicals of plasma would be exceptionally on vertical line (perpendicular to treated surface), then, beginning of this moment, the changing of deepness of grooves should not occur. However, as far as for the realized conditions of PCE, other directions of vectors of speeds of ions and radicals (active particles) are as well typical, so the peaks of micro-profile of phase diffraction grating are worked out more intensively. The bottom of the micro-profile of diffraction grating is reach by not all of the particles, which could be there in case of the vertical flux, but the top points of micro-profile are reached by all of the particles, flying on them in the vertical flow, and by some of the particles, which are flying in tilted direction. Because of the more intense working out of peaks of micro-profile, the deepness of grooves decreases with increasing of etching time when $t > t_{opt}$.

Thereby, we can assume that the maximum of selectivity of etching of glass r is observing on the minimization of the etching rates q. On this, to reach the required depth of grooving it is necessary to increase the duration of PCE process, taking into account the fact that the dependence of achievable depth of etching on etching duration has a complicated character (see **Figure 7**).

2) The PCE process is characterized not only by removal of treating materials (glass and chromium), but as well—by the growth of some film on the processed sur-

faces (this effect in the practice of plasma-chemical processes is named as "the drop-off"). In accordance with the existing concepts, the growing film is a polymer—the fluoroplastic (because of it, the mentioned stray process carries as well a more exact name—"the polymerization"), which is created of Freon. The removal of worked materials happens from under the growing film of the polymer. The film grows both on chromium and glass and the rate of growth of the film on the chromium can be higher or lower than the rate of growth of the film on the glass. Simultaneously, the rate of etching of glass is much higher than the rate of chromium etching. At the moment of $t = t_{cr}$ it occurs, that the thickness of the film on the glass is equal exactly to the sum of two thicknesses—the thickness of remaining chromium layer and the thickness of the film, which had been grown on the chromium. Because of this, after the acid etching the deepness of grooves of diffraction grating doesn't change. Until $t < t_{cr}$ the thickness of remaining layer of chromium in the sum with the thickness of the polymer film, grown on it, exceed the thickness of polymer film, grown on the glass. So, as a result of acid etching, the deepness of grooves of diffraction grating decreases. If $t > t_{cr}$, then the thickness of polymer film, which is grown on the glass, exceeds the total thickness of remaining layer of chromium and the film, which is grown on chromium. This leads to increasing of depth of the diffraction grating's grooving after the execution of acid etching operation.

3) With the increase of the duration of PCE operation the thickness of film, that is dropped-off on the treated surface of glass as a result of spurious reaction of Freon polymerization, increases monotonically, asymptotically approaching the some limit value, consistent to the equality of rates of growing end etching of this film. If PCE operation is excessively long, *i.e.* such, in which the protecting Chromium mask is already removed, and then during the operation of acid etching, which follows the PCE operation, the removal only of mentioned film occurs. Because of that, the difference between the deepness of diffraction grating's grooving before and after the operation of acid etching monotonically increases with increasing of exceeding duration of PCE operation.

On the basis of analysis and explanation of physical essence of the regularities and peculiarities of PCE process, detected by experimental methods, the following conclusions and recommendations to design the PCE operation in the manufacture of DOE can be formulated.

1) To achieve the estimated depth of grooves of phase diffraction gratings by a method of automatically get the size as well as by the method of trial processes and measurements the correlation between the PCE operation duration t and it's critical duration t_{cr} must be taken into account. If $t < t_{cr}$, then as a result of acid etching operation, which follows the PCE operation and has the purpose of removing of remains of mask, the deepness of grooves will decrease, and if $t > t_{cr}$, then it will increase.

2) It is established, that with increasing of grooving's depth from 0 until 1.5 μm the diffraction efficiency of phase diffraction grating, manufactured by PCE method, rises monotonically, so to increase the diffraction efficiency the maximum possible deepness of grooving in the specified range should be sought.

3) To achieve the maximal depth of grooving of phase diffraction grating it is necessary to observe the $t = t_{opt}$ condition, where t_{opt} is the optimal duration of PCE process. With greater or lesser duration of PCE operation the deepness of grooves of phase diffraction grating will be less, than maximally achievable.

4) The complete use of the capabilities, which are assignable by thickness and material of mask, is possible by providing the maximal etching selectivity r of glass relating to mask material by selecting the optimal regime of PCE process.

5. Acknowledgements

This work is executed on carrying out the Research on the State contract N P950 from 27.05.2010 in the framework of Federal target program "Scientific and scientific-pedagogical personnel of innovative Russia" on years 2009-2013.

REFERENCES

[1] F. Ananasso and I. Bennion, "Optical Technologies for Signal Processing in Satellite Repeaters," *IEEE Communications Magazine*, Vol. 28, No. 2, 1990, pp. 55-64. doi:10.1109/35.46686

[2] E. G. Loewen and E. Popov, "Diffraction Gratings Applications," Marcel Dekker, Inc., New York, Basel, 1997.

[3] S. Jensen, "Inductively Coupled Plasma Etching for Microsystems," MIC, Lyngby, Denmark, 2004.

[4] T. Zhang, M. Yonemura and Y. Kato, "Optical Design of an Array-Grating Compressor for High Power Laser Pulses," *Fusion Engineering and Design*, Vol. 44, No. 1-4, 1999. pp. 127-131.

[5] Z. Peng, D. A. Fattl, A. Faraou, M. Fiorentino, J. Li and R.G. Beausoleil, "Reflective Silicon Binary Diffraction Grating for Visible Wavelengths" *Optics Letters*, Vol. 36, No. 8, 2011, pp. 1515-1517. doi:10.1364/OL.36.001515

[6] W. Friedl and B. Hartenstein, "Energy Distribution of Diffraction Gratings as a Function of Groove Form," *Journal of the Optical Society of America*, Vol. 45, No. 5, 1955, pp. 398-399.

[7] M. Mao, Y. N. Wang and A. Bogaerts, "Numerical Study of the Plasma Chemistry in Inductively Coupled SF6 and SF6/Ar Plasma Used for Deep Silicon Etching Applications," *Journal of Physics D: Applied Physics*, Vol. 44, No. 43, 2011, Article ID: 435202.

doi:10.1088/0022-3727/44/43/435202

[8] www.esto-vacuum.ru/oborudovanie/caroline/caroline-rr

[9] D. A. Zeze, R. D. Forrest, J. P. Cary, D. C. Cox, I. D. Robertson, B. L. Weiss and S. R. P. Silva, "Reactive Ion Etching of Quartz and Pyrex for Microelectronic Applications," *Journal of Applied Physics*, Vol. 92, No. 7, 2002, pp. 3824-3829. doi:10.1063/1.1503167

[10] S. B. Odinokov and G. R. Sagatelyan, "Technology of Manufacturing of Diffraction and Hologram Optical Parts with Functional Microrelief of Surface by Method of Plasma-Chemical Etching," *Vestnik MGTU, Priborostroenie*, No. 2, 2010, pp. 92-104.

[11] K. Knop, "Rigorous Diffraction Theory for Transmission Phase Gratings with Deep Rectangular Grooves," *Journal of Optical Society America A*, Vol. 67, No. 9, 1978, pp. 1206-1210. doi:10.1364/JOSAA.9.001206

13

Photoluminescence of $EuGa_2Se_4:Nd^{3+}$

Arif Mirjalal Pashayev[1,2], Bahadir Guseyn Tagiyev[1,2], Said Abush Abushov[1], Ogtay Bahadir Tagiyev[1,3], Fatma Agaverdi Kazimova[1]
[1]Institute of Physics, Azerbaijan National Academy of Sciences, Baku, Azerbaijan
[2]National Aviation Academy of Azerbaijan, Baku, Azerbaijan
[3]Lomonosov Moscow State University Baku Campus, Baku, Russia
Email: oktay58@mail.ru

ABSTRACT

The photoluminescence (PL) in temperature interval 77 - 300 K is investigated in Eu Ga_2Se_4:Nd polycrystals. It is established that broad band PL with maximum at 561nm is caused by intracentral transitions $4f^65d$ - $4f^7(^8S_{7/2})$ of Eu^{2+} ions. The intracentral emission of Nd^{3+}, corresponding to both transitions from $^4F_{3/2}$ level and higher situated levels, is observed at interband excitation. The essential intensity of transitions from $^2H_{11/2}$ and $^4F_{9/2}$ levels is the interested peculiarity of luminescence spectra of Nd^{3+} in these crystals.

Keywords: Photoluminescence; Intracentral Transitions; Europium and Neodymium Ions

1. Introduction

The series of $MGa_2S_4(Se_4)$, $M_2Ga_2S_4(Se_4)$, $M_3Ga_2S_6(Se_6)$, $M_4Ga_2S_7(Se_7)$, $M_5Ga_2S_8(Se_8)$ compounds is obtained in M-Ga-S(Se) system (where M is Ba, Sr, Ca, Eu, Yb) [1-4]. The compounds synthesized in M-Ga-S(Se) system can be combined in one group with general formula II_n-III_2-VI_m (where n = 1, 2, 3, 4, 5; m = n + 3; II are bivalent cations of Eu, Yb, Sm, Ca, Ba, Sr; III are trivalent cations of Ga, Al, In; VI are S and Se chalcogens). Many II_n-III_2-VI_m compounds [5] have large number of stoichiometric voids: the one-fourth fraction of free places unoccupied by cations in crystal lattice. The stoichiometric voids situated periodically in lattice, don't have the defect properties.

The compounds in M-Ga-S(Se) system, M cations of which are 4f elements (lanthanides), can be the active medium of semiconductor lasers, luminescent lamps, colored display screens and other information systems [6-8]. These compounds have the forbidden band width is 4.4 eV and effectively transform the energy of electric field, X-ray and ultraviolet emissions, and also electron beams into visible light. The excitation spectrum of these compounds covers the spectral region from near ultraviolet one up to 500 nm.

$EuGa_2Se_4$ is the one of comparably little-studied compounds in M-Ga-S(Se) system. $EuGa_2Se_4$ compound crystallizes in pseudo-orthorhombic sublattice at simultaneous existence of twinning and superstructure [9]. The lattice parameters are a = 20.760 Å, b=20.404 Å, c = 12.200 Å [10]. Eu atoms in $EuGa_2Se_4$ structure are in partial positions 16(e), 8(a) and 8(b) in space group D_{2h}^{24}-Fddd having eight Se atoms as nearest neighbors. Crystal structure and some physical properties of this compound are described in [2]. The photoluminescence investigation of $EuGa_2Se_4$ activated by neodymium ions is of the interest for the photoluminescence mechanism revealing and the determination of trap level energy spectrum and also practical application.

2. Experiment

$EuGa_2Se_4$ compound is synthesized from binary compounds EuSe and Ga_2Se_3, taken in stoichiometric relations, by solid-phase reaction in graphitized ampoules evacuated up to 10^{-4} millimeter of mercury. The synthesis is carried out at 1300 K in one-temperature furnace during 4 hours. The annealing during 24 hours at 1000 K is carried out after the synthesis. Activation by neodymium ions is realized using neodymium fluorides doping during synthesis process. PL is investigated in temperature interval 77 - 300 K. The continuous laser diode InGaN (λ = 405 nm) is the excitation source. The registration of emission spectrum is carried out on Spectral Diffraction Luminescense device. The emission receiver is photoelectric multiplier-39A. The luminescence of the samples activated by neodymium is excited by laser on Rodamine 6G (range 550 - 620 nm). The luminescence registration is carried out by Diffraction Lattice Monochromator with Photoelectric Multiplier and Boxcar-Integtator BC I-280.

3. The Results and Discussions

The PL spectra of $EuGa_2Se_4$ crystals at different temperatures are shown on **Figure 1**. The one intensive broad emission band covering the wavelength region 500 - 620 nm with maximum at 561 nm and two relatively narrow bands at 709 and 745 nm are observed in PL spectra. It is known that the broad emission band is usually observed in crystals containing Eu^{2+}. The energy position of luminescence band, caused by Eu^{2+}, changes in the dependence on crystal structure and percentage of Eu^{2+} [11]. The observable PL in region 700 - 900 nm (maxima at wave lengths 709 nm and 745 nm) is caused by electron-hall recombination. The excitation and emission spectra of $EuGa_2Se_4$ crystal at temperature 300 K are shown on **Figure 2**. As it is seen from the figure, PL excitation spectrum is significantly broad band one and consists of overlapping bands. One of these bands at 374 nm is very clear one, but others at 404, 415, 454, 471 nm aren't allowed bands at temperature 300 K. The dependence of PL band intensity (λ_{max} = 561 nm) on temperature is shown on **Figure 3**. As it is seen from the figure the intensity in temperature region 92 - 120 K very weakly depends on temperature, in region from 120 up to 170 K slowly decreases and further temperature growth (170 - 300 K) leads to strong decrease of emission intensity, *i.e.* the strong temperature quenching of PL takes place. The activation energy of temperature quenching of PL is defined by high-temperature inclination (170 - 300 K). It is 0.09 eV. The temperature dependence of band half-width of PL with the maximum at 561 nm in $\Gamma(T)$ and $T^{1/2}$ coordinates is shown on **Figure 4**. It is seen that this dependence in temperature region 77 - 300 K is linear one and can be described by the use of configuration coordinate model and Boltzmann distribution. The fol-

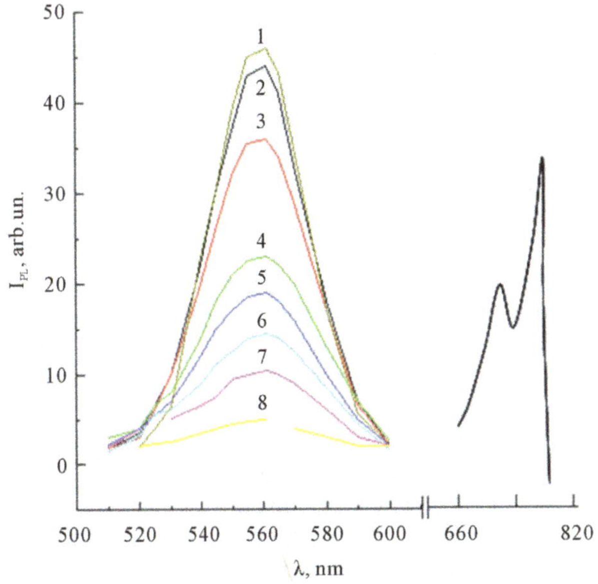

Figure 1. The PL spectra of $EuGa_2Se_4$ crystals at different temperatures (K): 1-92; 2-120; 3-144; 4-173; 5-184; 6-198; 7-215; 8-286.

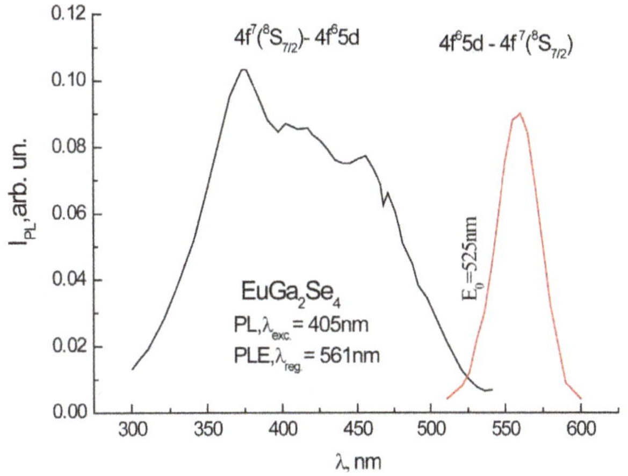

Figure 2. Excitation and emission spectra of $EuGa_2Se_4$ at 300 K.

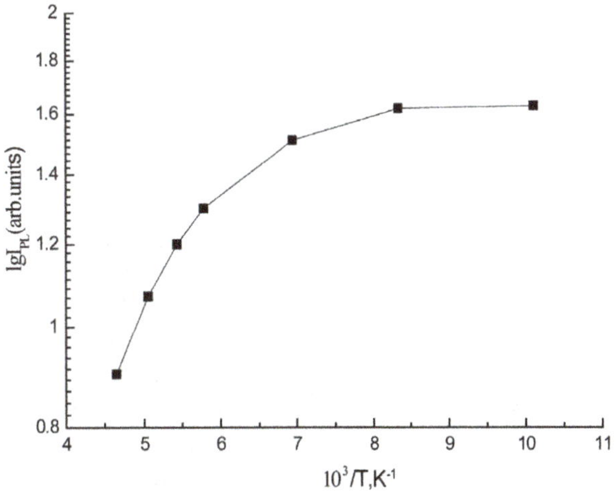

Figure 3. The temperature dependence of PL intensity of $EuGa_2Se_4$.

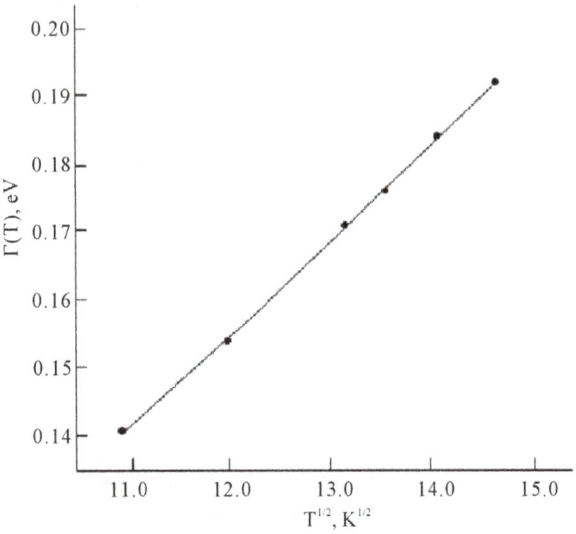

Figure 4. The dependence of PL band half-width of $EuGa_2Se_4$ with maximum at 561 nm on temperature.

lowing expressions, connecting Stokes shift ΔS, Huan-Rice factor s and phonon energy $\hbar\omega$ with the half-width temperature dependence $\Gamma(T)$, are obtained on the base of theoretical analysis of absorption and emission spectra in [12,13]:

$$\Delta S = (2S - 1)\hbar\omega \quad (1)$$

$$\Gamma(T) = 2.36\hbar\omega\sqrt{S}\sqrt{\coth\frac{\hbar\omega}{2KT}} \quad (2)$$

If $\hbar\omega < \kappa T$ then the expression under quadratic root in (2) can be expanded into series and limited by the first member:

$$\coth x = \frac{1}{x}, \text{ where } x = \frac{\hbar\omega}{kT} \quad (3)$$

Then expression (2) can be rewritten in the following form:

$$\Gamma(T) = 2.36\hbar\omega\sqrt{S}\sqrt{\frac{2kT}{\hbar\omega}} \quad (4)$$

or

$$\Gamma(T) = 2.36\sqrt{S}\sqrt{2kT\cdot\hbar\varpi} \quad (5)$$

The Equation (5) shows that half-width Γ linearly depends on $\sqrt{T}$. The values of Huan-Rice factor s and Stokes shift ΔS, which are equal to 8 ± 2 and 0.33 eV correspondingly, are found by experimental results. At calculation the phonon energy $\hbar\omega$ in $EuGa_2Se_4$ polycrystal is considered as equal to 23 meV according to data in [14]. The red shift D is obtained using formula $D = E_{ex} - E_{em}$ [12,13] (where E_{ex} is excitation state energy and E_{em} is emission energy of Eu^{2+} ion (4.19 eV and 2.21 eV correspondingly) $D = 1.98$ eV.

The constancy of energy position of broad band emission maximum at 561 nm (**Figure 1**) with temperature variation and temperature dependence of band half-width evidence about belonging of this emission band to Eu^{2+} ions, *i.e.* to intracentral transitions $4f^6 5d$ - $4f^7(^8S_{7/2})$ of Eu^{2+} ions.

The excitation spectra registered in the region of matrix self-absorption edge (**Figure 5**) evidence about fact, that the excitation on transition band-band isn't less effective, than one on intracentral transition $^4I_{9/2}$ - $^4G_{5/2}$. Note that the beginning of monotonously growing excitation spectrum part lies below than the region of matrix self-absorption edge, *i.e.* the transfer of excitation energy from small donor levels to Nd^{3+} ion ones is observed. The curve 2 is obtained under the conditions when the luminescence intensity linearly depends on pumping intensity; the curve 3 corresponds to nonlinear dependence of luminescence on pumping intensity (more detail about nonlinear dependence of luminescence intensity on pumping one can see below).

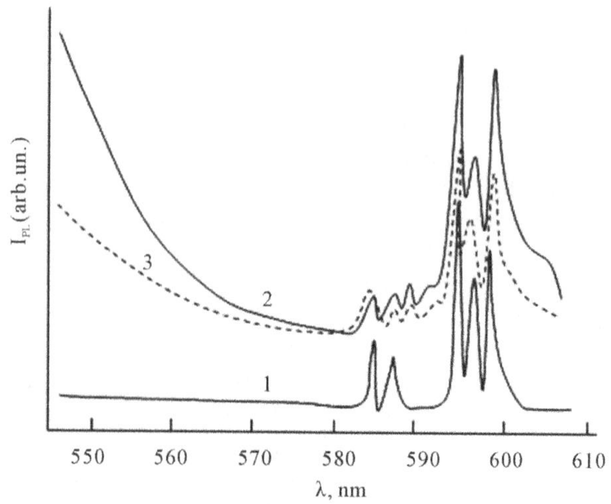

Figure 5. Excitation spectrum of $EuGa_2Se_4$:1 at % Nd^{3+} in 548 ÷ 605 nm wavelength range at 77 K (cur. 1) and 293 K (cur. 2, 3) 1 and 2 are linear parts of PL intensity dependence on pumping power 3 is nonlinear part of PL intensity dependence on pumping power.

It is interesting to note that the pumping at room temperature into matrix absorption band is more effective one for excitation of high-situated levels of Nd^{3+}, at 77 K the pumping is effective one for $^4F_{3/2}$ level. As a whole, the efficiency for excitation by means of matrix at 77 K is less than at room temperature.

Note that dependence of buildup effect of Nd^{3+} luminescence from $^4F_{3/2}$ level on excitation wavelength. The time passes of luminescence for $EuGa_2Se_4:Nd^{3+}$ at 77 K are given in the **Figure 6**. At 77 K the excitation on $^4G_{5/2}/\lambda_{exc} = 598$ nm level (curve 2) gives the clearly expressed buildup effect during ~15 μs. The transition to interband excitation, increase of pumping frequency $h\nu_{exc} > E_g$, $\lambda_{exc} = 545$ nm (curve 1) leads to disappearance of this effect. At room temperature the buildup isn't observed in both cases. The luminescence decay times for both cases of excitation are similar ones (~50 μs). At 77 K the luminescence decay times for $EuGa_2Se_4$ at small neodymium concentrations on transitions from $^4F_{3/2}$ and $^4F_{5/2}$ levels are equal to 50 μs and 5 μs correspondingly. Here we need to say several words about possible excitation mechanism Nd^{3+} in $EuGa_2Se_4$. In $EuGa_2Se_4$ at 293 K and interband excitation Eu^{2+} excites and it transfers the excitation on G-levels of neodymium. At temperature decrease Eu^{2+} band narrows, the overlapping with G-levels of Nd^{3+} disappears and that's why the levels causing the broad impurity emission bands which transfer the energy on $^4F_{3/2}$ and $^4F_{5/2}$ levels, mainly take part in energy transfer on Nd^{3+} at 77 K. As a result at 77 K $^4F_{3/2}$ level excites more effectively than G-levels as it is mentioned above. Moreover, the emission from $^4F_{3/2}$ takes place without buildup because of the energy is transferred on it directly from impurity levels. As a whole at

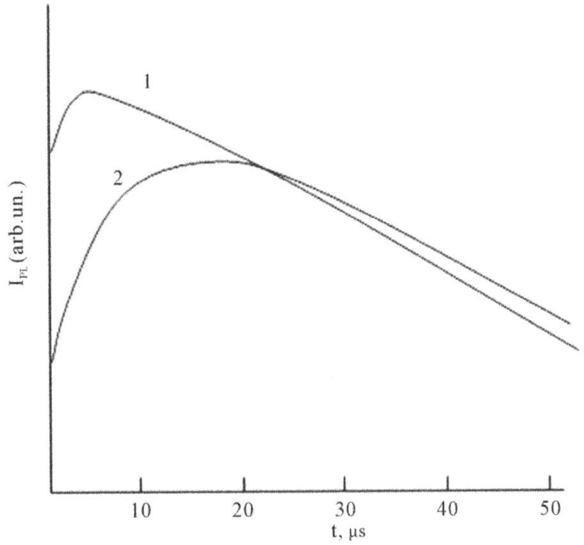

Figure 6. Time passes of PL from $^4F_{3/2}$ level for $EuGa_2Se_4$ <Nd> at 77 K; 1. $h\nu_{exc} > E_g$, λ_{exc} = 545 nm; 2. λ_{exc} = 598 nm.

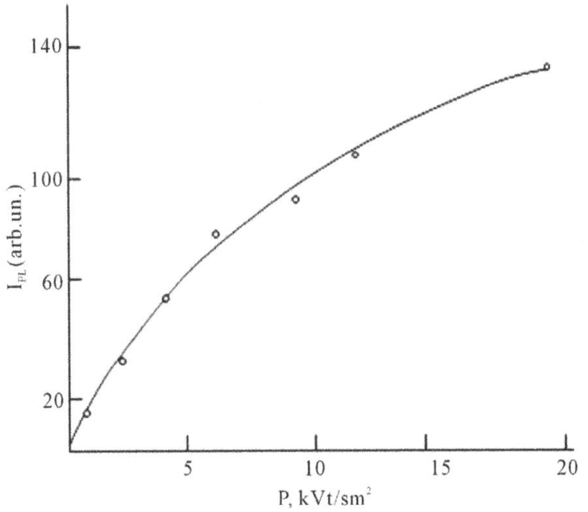

Figure 7. The dependence of PL intensity on pumping power for $EuGa_2Se_4:Nd^{3+}$, *T* = 293 K. t_{giv} =1 μs. λ_{reg} = 780 nm, λ_{exc} = 552 nm.

77 K the efficiency of energy transfer to Nd^{3+} ion by means of the interband transition goes down. The luminescence investigation results at transitions from $^4F_{3/2}$ level on $^4I_{9/2}$, $^4I_{11/2}$, $^4I_{13/2}$ evidence about the fact that in considered crystals not one type of luminescent centers exists as Stark component number exceeds the maximum possible one. This is also confirmed by the dependence of emission spectra on excitation wave length, on delay value of registration time moment and on sample history.

The effect of intensity quick saturation with pumping growth is the interest peculiarity of neodymium luminescence at interband excitation in the investigated crystals [15]. The dependence of PL intensity on pumping power is given on **Figure 7**. In this case the excitation wave length is 552 nm and emission registration wave length is λ_{reg} = 780 nm. It is seen that emission intensity nonlinearly depends on pumping power. The saturation is observed at room temperature and at liquid nitrogen temperature. This effect can be connected with pumping energy absorption by free carriers, growth of pumping reflection at increase of free carrier number, saturation of intermediate agents by means of witch the excitation is transferred from matrix to Nd^{3+} and etc. Note that luminescence intensity linearly depends on pumping power at excitation through absorption bands of Nd^{3+}.

Thus, in $EuGa_2Se_4$ crystals the broad band intensive luminescence is connected with intracentral transitions $4f^65d$ - $4f^7(^8S_{7/2})$ of Eu^{2+} ions. The carried investigations of triple compounds with neodymium show that nearest surrounding of Nd^{3+} has the near tetragonal symmetry in investigated crystals; the luminescence of Nd^{3+} effectively excites because of interband transitions; the intensive emission of Nd^{3+} from levels situated higher $^4F_{3/2}$ is observed, therefore $EuGa_2Se_4:Nd^{3+}$ can be used as narrow-band luminophors in 580 - 1090 nm range.

4. Acknowledgements

This work was supported by the Science Development Foundation under the President of the Republic of Azerbaijan, Grant № EİF-2011-1(3)-82/01/1.

REFERENCES

[1] M. R. Davolos, A. Garcia, C. F. Fousser and P. Hagenmuller, "Luminescence of Eu^{2+} in Strontium and Barium Thiogallates," *Journal of Solid State Chemistry*, Vol. 83, No. 2, 1989, pp. 316-323. doi:10.1016/0022-4596(89)90181-3

[2] P. C. Donahue and J. E. Hanlon, "The Synthesis and Photoluminescence of $M^{II}M_2^{III}(S,Se)_4$," *Journal of Electrochemical Society*, Vol. 121, No. 1, 1974, pp. 137-142.

[3] G. M. Niftiyev, B. G. Tagiyev, O. B. Tagiyev, V. A. Djamilov, T. A. Gulmaliyev, F. B. Askerov, B. M. Izzatov and Y. G. Talibov, "Optic and Luminescent Properties of $A^{II}B^{III}{}_2C^{VI}{}_4$ (A-Yb, Eu, Sm, Ca, Sr, Ba; B-Ga, In, C-S, Se)," *Inorganic Materials*, Vol. 28, No. 12, 1992, pp. 2269-2275.

[4] O. B. Tagiyev, Ch. M. Briskina, V. M. Zolin, G. M. Niftiyev, V. M. Markushev and G. K. Aslanov, "The Synthesis and Luminescent Properties Ternary Chalcogenides Compounds $EuGa_2S_4$ Doped with Lanthanides," *International Conference on Physics Chemistry and Technical Application of Chalckgenides*, Tbilisi, 1983, p. 50.

[5] N. A. Goryunova, "Complex Diamond-Type Semiconductors," Soviet Radio, Moscow, 1968, p. 265.

[6] S. Iida, T. Matsumoto, N. Mamedov, G. An, Y. Maruyama, O. Tagiev, A. Bayramov, R. Jabbarov and B. Tagiev, "Observation of laser oscillation from $CaGa_2S_4:Eu^{2+}$," *Japanese Journal Applied Physics*, Vol. 36, No. 7a, 1997, pp. 857-859. doi:10.1143/JJAP.36.L857

[7] G. P. Yablonskii, V. Z. Zubialevich, E. V. Lutsenko, A. M. Pashaev, B. G. Tagiev, O. B. Tagiev and S. A. Abushov, "Luminescence Propetties of Barium Thio-and Selenogallates Doped with Eu, Ce, and Eu+Ce," *Japanese Journal Applied Physics*, Vol. 50, No. 5, 2011, pp. 05FG02-05FG02-2.

[8] P. Benalloul, C. Barthou and J. Benoit, "II-III_2-S_4 Ternary Compounds: New Host Matrices for Full Color Thin Film Electroluminescence Displays," *Applied Physics Letters*, Vol. 63, No. 14, 1993, pp. 1954-1956. doi:10.1063/1.110612

[9] T. E. Peters and J. A. Baglio, "Luminescence and Structural Properties of Thiogallate Ce^{3+} and Eu^{2+} activated Phosphors," *Journal of Electrochemical Society. Solid-State Science and Technology*, Vol. 119, No. 2, 1972, pp. 230-236.

[10] P. G. Rustamov, O. M. Aliyev and T. H. Kurbanov, "Ternary Chalcogenides of Rare Earth Elements," Elm, Baku, 1981, p. 228.

[11] A. N. Georgobiani, S. A. Abushov, F. A. Kazymova, B. G. Tagiev, O. B. Tagiev, P. Benalloul and C. Barthou, "Luminescent Properties of $EuGa_2S_4$:Er^{3+}," *Inorganic Materials*, Vol. 42, No. 11, 2006, pp. 1188-1192. doi:10.1134/S0020168506110033

[12] P. Dorenbos, "f→d Transition Energies of Divalent Lanthanides in Inorganic Compounds," *Journal Physics: Condense Matter*, Vol. 15, No. 3, 2003, pp. 575-594.

[13] P. Dorenbos, "Energy of the First $4f^7 \rightarrow 4f^6 5d$ Transition of Eu^{2+} in Inorganic Compounds," *Journal of Luminescence*, Vol. 104, No. 4, 2003, pp. 239-260. doi:10.1016/S0022-2313(03)00078-4

[14] M. Jouanne, J.-F. Morhange, C. Barthou C. Charter, R. Jabbarov, J.-M. Trigerio, B. Tagiev and E. Gambarov, "Raman Investigation of Orthorhombic $M^{II}Ga_2(S,Se)_4$ Compounds," *Journal Physics: Condense Matter*, Vol. 14, No. 49, 2002, pp. 13693-13703. doi:10.1088/0953-8984/14/49/324

[15] G. K. Aslanov, Ch. M. Briskina, V. M. Zolin, V. M. Markushev, G. M. Niftiyev and O. B. Tagiyev, "Spectral and Luminiscent Investigation of Double Chalcogenides of Gallium and Rare-Earth Elements with Neodymium," *Inorganic Materials*, Vol. 22, No. 10, 1986, pp. 1630-1634.

Comparative Assessment on the Performance of Open-Loop and Closed-Loop IFOGs

Mohammad Reza Nasiri-Avanaki[1], Vahid Soleimani[2], Rohollah Mazrae-Khoshki[2]
[1]Electronics Department, Azad University of Karaj, Tehran, Iran
[2]Electronic Engineering Department, RAZI University, Kermanshah, Iran
Email: m_n_avanaki@yahoo.com

ABSTRACT

In this paper, we evaluated comprehensively the structure and operation of open-loop interferometric optical fiber gyroscopes (IFOG). To complete the previous works, a digital approach to derive the rotation angle in optical fiber gyroscopes is investigated theoretically. Results are simulated by the MATLAB software; therefore we could compare the results in simulated area with the values derived from theory. Also, feedback Erbium-doped fiber amplifier (EFDA) FOGs, called FE-FOG, is categorized in closed-loop IFOGs. The procedure of finding the Sagnac shift for open-loop and closed-loop IFOG have been studied and compared to one another. The signal processing in the open-loop IFOG was simulated using Matlab software and for the closed-loop IFOG by PSCAD. In the open-loop IFOG the analogue formulation of the IFOG in order to extract the phase shift is analyzed. A novel and promising method for derivation of Sagnac phase shift based on digital finite impulse response filtering is proposed. Based on our simulation results, the reliability and accuracy of the method is determined. In the closed-loop IFOG, the shift was derived through frequent use of Sagnac loop. The output signal is injected in the input again as feedback. The shift phase between clockwise and counterclockwise waves in each complete route, including primary and feedback route, is identified as Sagnac shift phase.

Keywords: Feedback Erbium-Doped Fiber Amplifier FOG (FE-FOG); Erbium-Doped Fiber Amplifier (EFDA); Digital Signal Processing; PSCAD; FIR Digital Filters; Interferometric Fiber Optic Gyro (IFOG); Sagnac Shift

1. Introduction

Angular rotation in vehicles is one of the parameters that need to be accurately measured in order to describe the position of an object. Traditionally, the angular momentum of a spinning rotor was used to determine the angular rate or displacement [1-3]. These devices are susceptible to damage from shock and vibration, exhibit cross-axis acceleration sensitivity and, for the lower cost versions; they have reliability problems [3].

Directions around the same closed optical path will experience path length difference that is proportional to the rotation rate of the setup [4,5]. Sagnac shift is the common feature among optical gyros. Sagnac is now extensively used in commercial inertial navigation systems for aircraft. Given the advantage of this effect, the actual path length difference due to the rotation is quite small, e.g. the gyro used in the aircraft navigation must detect rotation rates below 0.01 deg/hr [6]. FOG is used for measurement of the rotation based on Sagnact Effect. Two types of FOG are used: the interferometric FOG (I-FOG) in which a low coherence light source is used and the resonance FOG (R-FOG). In the R-FOG the differences of resonance frequency caused by sagnact effect is utilized for measuring the rotation. Therefore even a short length of optical fiber is sufficient for the measurement. R-FOG needs a coherent light source for which such effects like Kerr effect and reflective optic phenomenon must be considered. The FOG converts the Sagnact phase shift into a beam frequency between the clockwise and counterclockwise of laser modes. This implies sensing path length changes of about one part in 10^{16}, which corresponds to absolute length changes on the order of a nuclear dimension. Requirement of the precision of rotation measurement for spacecraft navigation lies between 0.01°/hr to 0.001°/hr.

The principle of operation of a typical fiber gyro is based on a phase modulation in both directions of an optical fiber loop, as if it acts as a delay line. The modulation frequency, $f_P = \frac{1}{2\tau}$, matches the half period of the transit time τ. Such a modulation scheme provides a sinusoidal response with a stable bias. However this response is nonlinear and the rotation rate proportional to the returning power is not perfectly stable.

Ring laser gyroscope (RLG) consists of a ring laser having two counter-propagating modes over the same path in order to detect rotation. It works based on Sagnac shift effect and requires high vacuum and precision mirror technology, which makes this technique very expensive [4]. The physical principal of RLG operation is analogous to the Doppler Effect, but it involves determination of the phase shift between two counter-propagating light beams in an evacuated mirrored cavity [6]. To address this drawback, interferometric fiber optic gyro (IFOG) is employed whose function resembles RLG, notwithstanding the fact that, in IFOG, the same effect is obtained in a fiber coil with the elimination of the high voltage and high vacuum, which results in a low-cost inertial rotation sensor [4,6]. In addition, IFOGs are abstracted in miniature devices; all-solid state with a limited number of components, therefore lower cost.

In terms of light source, RLG requires an external narrow band gas laser with its active gain medium which is an integral part of the sensing cavity, whereas IFOG works with an external broadband light source [7]. Path length measurement in RLG is performed by measuring the difference of resonant frequencies between two cavities [5]. Despite, IFOG rotation rate sensing is achieved through a direct measurement (open loop) or nulling (close loop) of the optical phase difference due to the rotation-induced Sagnac phase shift [4,6,7].

According to the fact that high precision gyro is not always required, e.g. in land automotive vehicle an IFOG with the accuracy between 10 to 200 deg/hr is sufficient [8], and also regarding to the least requirements with the most flexibility in the design of IFOG, the improvement of this sensor becomes more sensible.

2. Principle of Operation

The principle of operation of an IFOG is based on the phenomenon that a circuital system has different optical path lengths in the two propagation directions when it rotates [9]. The difference between counter propagating beams provides the amount of the rotation. The main structure includes a laser diode, usually a super luminescent diode (SLD) or Rare-earth doped fiber with improved wavelength stability, a coupler to split light into clockwise and counterclockwise directions, a length of optical fiber wound on a coil as the rotation sensing loop, and a photo detector to convert optical information into electrical signals for further processing. Light source temperature is controlled by using a thermo electric cooler (TEC) and the detector is a PINFET module with a high sensitivity and hybrid trans-impedance amplifier. The type of the fiber optic and the way of winding specify the particular category that IFOG belongs to, which suits it to the specific application of the sensor [10]. A complete IFOG has identical optical paths in the clockwise and counter clock wise. In this IFOG known as "minimum configuration", a polarizer and second coupler are employed [5,11].

Optical signal is received by the detector after passing through the fiber coil and couplers and converted into the electric current. An electrical amplifier is connected to the detector to change the current to the appropriate voltage. The obtained voltage is, then, amplified and fed into the signal processing circuit to extract the angular rotation rate [5,12].

EF-FOG work by frequent utilize of sagnact loop, in this manner output signal re-inputted as feedback. It performs like R-FOG but without high length coherence light source and never use the resonance effect. FE-FOG in comparison with Open-loop I-FOG has high sensibility and wide dynamic range. In this paper we compare the results of open loop and closed loop I-FOG.

A light source with a coherence length much shorter than the coil length allows only the wave pairs that have circulated the same number of times in the loop to interfere with each other and to produce an output related to the rotation-induced nonreciprocal phase shift. The resultant optical response is essentially the sum of all the optical responses of a series of conventional Sagnac Interferometric fiber-optic gyroscopes with their effective loop length in multiples of a single coil's length. The multiple-trip interference and the associated intensity summation produces a response resembling a resonance phenomenon, with the strength and sharpness of the resonance increasing with the number of interfering light waves. The proposed gyro has different from a resonant fiber-optic gyro (R-FOG) because it uses a low coherent light source. The low coherent source minimizes not only errors from Rayleigh back scattering, but also bias errors caused by the optical Kerr effect. This could not possibly be done in an R-FOG because of high coherent light source.

If the modulation frequency of phase modulator in sagnact loop consists of whole loop time delay, the output signal in IFOG will be in pulse shape; therefore, when rotation occurs in the system, the location of output peak pulse shifts by sagnact effect. Precision of measurement depends on sharpness of the output pulses. Sharpness of output pulse is determined by the phase modulation depth and EFDA gain.

Miniature fiber optic gyros have also been manufactured with all solid-state optical devices for precise measurements of mechanical rotation based on the sagnac principle.

Open-loop IFOG is a simple configuration of the IFOG. Closed-loop IFOG performs similar to the resonance FOG (R-FOG), both based on Sagnac effect, although used for different tasks.

3. Rotation Equation

When the optical ring is rotated with a tangential velocity v, the beam rotating with the ring will have an optical path longer than the counter-rotating beam by a distance ΔL given by [1]:

$$\Delta L = \frac{4\pi Rv}{c} \tag{1}$$

where, R is the ring radius and c is the speed of light in the vacuum. For a monochromatic light of wavelength λ, this change in optical path length results in the Sagnac phase difference, which is given by Equation (2).

$$\Delta\phi = \frac{2\pi}{\lambda}\Delta L = \frac{8\pi^2 Rv}{\lambda c} \tag{2}$$

The phase difference between two beams, after passing the ring with the area A and rotating with an angular velocity Ω, generates a phase difference which is given by [1]:

$$\Delta\phi = \frac{8\pi A\Omega}{\lambda c} \tag{3}$$

It is important to note that, the resultant phase shift is independent of the medium and of the exact shape of the loop. This compliant property of the IFOG is an advantage when it is required to design it to fit the volume constrains in the specific applications [1]. However, the resultant phase shift can be increased by additional loops (turns of fiber) *i.e.* if it is wound N turns of the fiber coil. The resultant phase shift becomes [13]:

$$\Delta\phi = \frac{8\pi A\Omega N}{\lambda c} \tag{4}$$

Alternatively, we can express the resultant phase shift in terms of coil diameter and fiber length by noting that:

$$A = \frac{\pi D^2}{4} \tag{5}$$

And,

$$L = N\pi D \tag{6}$$

So, the Sagnac phase shift can be rewritten as:

$$\Delta\phi = \frac{2\pi LD}{\lambda c}\Omega \tag{7}$$

As one example in a typical IFOG (200 m coil length, 10 cm-diameter coil) for measurement of the earth angular rotation (Ω = 15°/h = 0.73 μr/s), the sensor detects a phase difference of $\Delta\Phi$ = 36 μr, corresponding to an optical path difference of the order of 10 - 12 m [2,13].

Eventually the output signal for such configuration will be [2]:

$$I(t) = I_0\left(1+\cos(\Delta\phi)\right) \tag{8}$$

in which, I_0 is the current obtained when the ring is at the rest.

4. Open-Loop Configuration with Phase Modulation

As mentioned above, the major problem of the basic configuration is the output nonlinearity for small phase shift, $\Phi_s \approx 0$ which hinders high sensitivity measurements of the small rotation angles [14]. This limitation is overcome by transforming the base band cosine-dependence into a sinusoidal function [1,3]. Although for translating the output signal from base band to a carrier at angular frequency ω_{mod}, different solutions have been proposed, but today optical phase modulation technique is commonly used. The typical setup of a practical IFOG in all-fiber technology has a phase modulator which is inserted in the fiber coil close to a coupler output so that the phase delays are cumulated by the counter propagating waves. In all-fiber IFOG, phase modulator is constructed by winding and cementing a few fiber turns on a short, hollow piezoceramic tube (PZT). By applying a modulating voltage to the PZT, a radial elastic stress and a consequent optical path length variation due to the elasto-optic effect are generated [14].

5. The Open-Loop IFOG Modulation Equation

In this section, the modulation equations of extracting the phase shift are comprehensively described. In the configuration described in the previous section, the clockwise (CW) and counterclockwise (CCW) propagating waves experience a phase delay $\phi(t)$ and $\phi(t-\tau)$, respectively [15], where $\tau = L/v$ is the radiation transition time in a fiber with an overall length L. By applying a phase modulation at angular frequency ω_{mod}, the modulation equation will be [16]:

$$\phi_{\text{mod}}(t) = \phi_{\text{mod},0}\cos(\omega_{\text{mod}}t) = \phi_{\text{mod},0}\cos(2\pi f_{\text{mod}}t) \tag{9}$$

Here, $\phi_{\text{mod},0}$ is the amplitude of modulation signal, and f_{mod} is the modulation frequency.

Therefore, CW and CCW waves are modulated as phase shift by $\phi_{\text{mod}}(t)$ and $\phi_{\text{mod}}(t-\tau)$, respectively. So, the modulation phase shift difference is as follows:

$$\Delta\phi_{\text{mod}}(t) = \phi_{\text{mod}}(t) - \phi_{\text{mod}}(t-\tau) \tag{10}$$

Finally, the total phase shift in the output signal will be:

$$\Delta\phi_{total} = \phi_{ccw} - \phi_{cw} = \Delta\phi + \phi_{\text{mod}}(t) - \phi_{\text{mod}}(t-\tau) \tag{11}$$

Now, the output signal will become:

$$I(t) = I_0\left(1+\cos\left(\Delta\phi + \Delta\phi_{\text{mod}}(t)\right)\right) = I_0\left(1+\cos\left(\Delta\phi_{total}\right)\right) \tag{12}$$

Considering the fact that the SLD switched by $f_p = 1/2\tau$, with the help of Equations (9), (11) the phase shift becomes [16]:

$$\Delta\phi_{\text{mod}}(t) = \Delta\phi + \phi_{\text{mod},0}\cos(2\pi f_{\text{mod}} t) - \phi_{\text{mod},0}\cos(2\pi f_{\text{mod}}(t-\tau)) \quad (13)$$

$$\Delta\phi_{\text{mod}}(t) = \Delta\phi + 2\phi_{\text{mod},0}\sin(\pi f_{\text{mod}}\tau)\sin\left(2\pi f_{\text{mod}}\left(t-\frac{\tau}{2}\right)\right)$$
$$= \Delta\phi + 2\phi_{\text{mod},0}\sin\left(\frac{\pi}{2}\frac{f_{\text{mod}}}{f_p}\right)\sin\left(2\pi f_{\text{mod}} t - \frac{\pi}{2}\frac{f_{\text{mod}}}{f_p}\right) \quad (14)$$

Substituting the amplitude with Equation (15) as follow [1]:

$$\phi_{MI} = 2\phi_{\text{mod},0}\sin\left(\frac{\pi}{2}\frac{f_{\text{mod}}}{f_p}\right) \quad (15)$$

The output current of photodiode becomes:

$$I(t) = I_0\left[1+\cos\left(\Delta\phi + \phi_{MI}\sin\left(2\pi f_{\text{mod}} t - \frac{\pi}{2}\frac{f_{\text{mod}}}{f_p}\right)\right)\right] \quad (16)$$

Or can be rewritten as:

$$I(t) = I_0\left[1+\cos(\Delta\phi)\cos\left(\phi_{MI}\sin\left(2\pi f_{\text{mod}} t - \frac{\pi}{2}\frac{f_{\text{mod}}}{f_p}\right)\right) - \sin(\Delta\phi)\sin\left(\phi_{MI}\sin\left(2\pi f_{\text{mod}} t - \frac{\pi}{2}\frac{f_{\text{mod}}}{f_p}\right)\right)\right] \quad (17)$$

Using the below Bessel equations:

$$\cos(x\sin s) = J_0(x) + 2\sum_{m=1}^{\infty} J_{2m}(x)\cos(2ms) \quad (18)$$

$$\sin(x\sin s) = 2\sum_{m=1}^{\infty} J_{2m-1}(x)\sin((2m-1)s) \quad (19)$$

Equation (17) can be rewritten in a way that one can easily extract the relevant harmonics [16]:

$$I(t) = I_0\left[1+\cos(\Delta\phi)\cdot\left\langle J_0(\phi_{MI}) + 2\sum_{m=1}^{\infty} J_{2m}(\phi_{MI})\cos\left(2m\left(2\pi f_{\text{mod}} t - \frac{\pi}{2}\frac{f_{\text{mod}}}{f_p}\right)\right)\right\rangle - \sin(\Delta\phi)\cdot\left\langle 2\sum_{m=1}^{\infty} J_{2m-1}(\phi_{MI})\cos\left((2m-1)\left(2\pi f_{\text{mod}} t - \frac{\pi}{2}\frac{f_{\text{mod}}}{f_p}\right)\right)\right\rangle\right] \quad (20)$$

To have an optimum sensitivity, output power, photon noise, and signal to noise, ϕ_{MI} is selected to be as [16]

$$\phi_{MI}(opt) = 1.85 \text{ rad} \quad (21)$$

To simplify Equation (20), we consider $f_{\text{mod}} = f_p$ [1,16]. Using this assumption and Equation (15), we will have:

$$\phi_{MI} = 2\phi_{\text{mod},0} \rightarrow \phi_{\text{mod},0}(opt) = 0.925 \quad (22)$$

The amplitude of the output signal at f_{mod}, which gives the first harmonic of the output current, *i.e.* Equation (20) will be [2]:

$$I(f_{\text{mod}}) = -2C_1 I_0 \sin(\Delta\phi) J_1\left(2\phi_{\text{mod},0}\sin\left(\frac{\pi}{2}\frac{f_{\text{mod}}}{f_p}\right)\right) \quad (23)$$

$$I(f_p = f_{\text{mod}}) = -2C_1 I_0 \sin(\Delta\phi) J_1(2\phi_{\text{mod},0}) \quad (24)$$

Table 1 shows the four harmonics generated in the photodiode output current [2]:

Now, we can calculate $\Delta\phi$ by H_1 with the below equation [2,16]:

$$\Delta\phi = -\arcsin\left(\frac{H_1}{2CI_0 J_1(\phi_{MIi})}\right) \quad (25)$$

$\Delta\phi$ Could be also obtained with another approach using H_1, H_2, J_1 and J_2 [16]:

$$\Delta\phi = -\arctan\left(\frac{H_1 J_2(\phi_{MI})}{H_2 J_1(\phi_{MI})}\right) \quad (26)$$

A comfort and reliable method is proposed here is to use H_1, H_2, H_3 and H_4 to calculate $\Delta\phi$ with the help of the solution of Equation (27) and **Table 1** [2,16]:

$$\frac{2m}{x}J_m(x) = J_{m-1}(x) + J_{m+1}(x) \quad (27)$$

$$\frac{H_2}{H_4} = \frac{J_2(\phi_{MI})}{J_4(\phi_{MI})} \quad (28)$$

$$H_1 + H_3 = \frac{4H_2}{\phi_{MI}}\tan(\Delta\phi) \quad (29)$$

Table 1. Harmonics of the photodiode output current. $C_i = C$, $i = 0, \cdots, 4$.

frequency	Amplitude
DC	$H_0 = C_0 I_0(1+\cos(\Delta\phi)J_0(\phi_{MI}))$
f_{mod}	$H_1 = -2C_1 I_0 \sin(\Delta\phi) J_1(\phi_{MI})$
$2f_{\text{mod}}$	$H_2 = 2C_2 I_0 \cos(\Delta\phi) J_2(\phi_{MI})$
$3f_{\text{mod}}$	$H_3 = -2C_3 I_0 \sin(\Delta\phi) J_3(\phi_{MI})$
$4f_{\text{mod}}$	$H_4 = 2C_4 I_0 \cos(\Delta\phi) J_4(\phi_{MI})$

$$H_2 + H_4 = \frac{6H_3}{\phi_{MI}} \frac{1}{\tan(\Delta\phi)} \tag{30}$$

Owing to this, the Sagnac shift is obtained as Equation (31) which is the best suited for hardware implementation:

$$\Delta\phi = \arctan\sqrt{\frac{3H_3(H_1+H_3)}{2H_2(H_2+H_4)}} \tag{31}$$

6. Evaluation of Digital Filtering in an Open-Loop IFOG

The simulation results of the abovementioned procedure show that, this approach can perfectly be employed in the signal processing part of IFOGs. To achieve the required harmonics, four digital filters are designed. These filters were designed by SPtool in Matlab 7a. Since the phase linearity of the filters was important, finite impulse response (FIR) filters were taken into account to have the four required band pass filters. Regarding to the available data the beneath quantities are listed as input of the simulation:

Input rotation rate = 10 deg/s (AHRS range);
Coil length = 500 m;
Coil diameter = 10 cm;
Sampling frequency = 10 MHz;
Time domain = 2 ms;
Wave length = 1310 nm;
Refractive index of optical fiber core = 1.43;
Coil length transient time = (500/3e8) × 1.43.

The coil frequency is obtained as $1/2\tau = 1/2L/(c/1.43) = 0.208$ MHz. As mentioned before, this frequency is the PZT vibration and SLD switch frequency, as well.

Figure 1 shows the designed band pass filters with the same bandwidth (0.208 MHz) but different central frequencies. In the filters in **Figure 1**, F_p and F_s are, pass band frequency and stop band frequency, respectively. Also, R_p and R_s are ripples in pass band and stop band.

The filters are described as follows:

1) Band pass filter: Gain = 1, central frequency = 0.208 MHz;
2) Band pass filter: Gain = 1, central frequency = 0.416 MHz;
3) Band pass filter: Gain = 1, central frequency = 0.624 MHz;
4) Band pass filter: Gain = 1, central frequency = 0.832 MHz.

To extract the four harmonics from the IFOG output current, the filters are implemented by considering the direct form structure. **Figure 2** shows the amplitude of the filtered signals, H_1 to H_4. In the next step, rotation rate Ω, is calculated using H_1 to H_4 with the help of Equations (7) and (31).

The comparison of calculated Ω with the input rotation rate, *i.e.* 10 deg/s showed one percent difference. This discrepancy caused from utilizing unsharp filters. As it has been experienced, we will reach to the better accuracy if higher order filters are considered.

7. Closed-Loop EF-FOG Configuration and Operation

In this structure the weak feedback signal is amplified by the Er-Doped Fiber Amplifier (EDFA) to prevent the laser scattering. Fiber amplifier is assumed to be linear, which means there are no gain effects. The output signal is the sum of interferometric CW and CCW beams in the total route. The modulator is located in sagnac loop. If the phase modulator frequency is selected properly, the output signal will be a series of short pulses [17]. Hence modulation frequency of phase modulator and route delay must be approved by the equation $\omega_m\tau = 2n\pi$ where ω_m is angular frequency of the phase modulation, and τ is the time delay in whole round trip loop consisting of Sagnac loop and amplified fiber optic loop. If the sharpness of output pulses could not be adjusted by phase modulation, by changing EFDA gain it can be compensated. If this equation $\omega_m\tau = 2n\pi + \phi_0$ is not realized, for example due to detuning, there will be a phase shift ϕ_0 that can be demodulated as an error signal for phase modulated (PM) technology. The operation of the FE-FOG is sensitive to the EDFA gain [18]. In theory the first intereferometric output signal without feedback is equal to ordinary IFOG as:

$$\omega_m\tau = 2n\pi + P_1(t) = K_1\left\{1 + v\cos\left[\phi_s + \phi_e\cos(\omega_m t)\right]\right\} \tag{32}$$

where v is the interferometric coefficient and K_1 is a parameter that expresses the loss resulting from the Sagnac loop and the couplers of the Sagnac interferometer, ω_m is the modulation frequency, ϕ_e is the effective phase-modulation depth, which is expressed as

$$\phi_e = 2\phi_m \sin(\pi f_m \tau_s) \tag{33}$$

τ_s is the wave-propagation time through the Sagnac loop (please note the difference with the total time round-trip delay t) and is expressed as $\tau_s = nL/c$ (where L is the length of Sagnac loop), and ϕ_m is the phase-modulation depth. ϕ_s is the Sagnac phase shift induced by rotational movement, which is expressed as Equation (7) in the previous section.

By considering the effect of optical feedback, the second-time interference signal experienced after the feedback has been derived is

$$P_2(t) = AK_1^2K_2\left\{1 + v\cos\left[\phi_s + \phi_e\cos(\omega_m t - \omega_m\tau)\right]\right\} \times\left\{1 + \cos\left[\phi_s + \phi_e\cos(\omega_m t)\right]\right\}$$

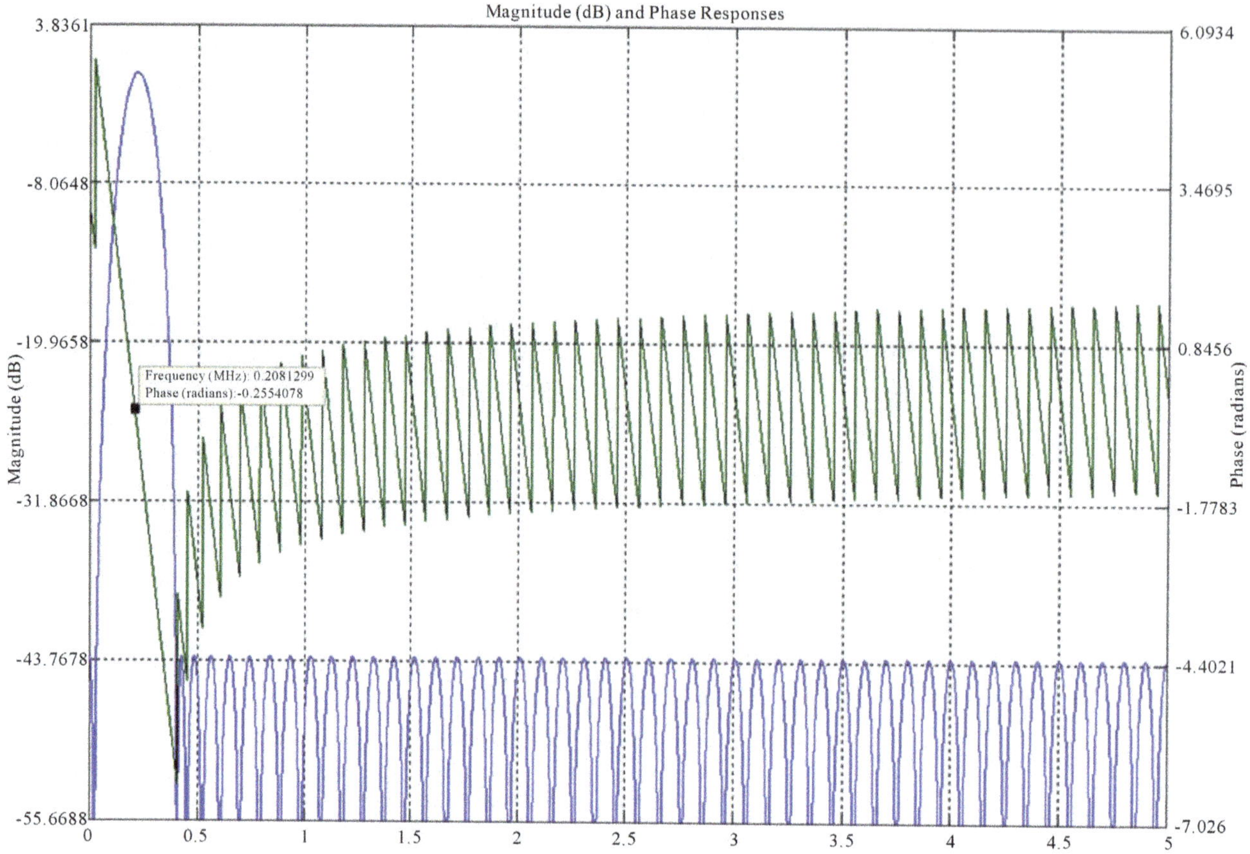
Magnitude (dB) and Phase Responses
Magnitude (dB)
Phase (radians)
3.8361
-8.0648
-19.9658
-31.8668
-43.7678
-55.6688
6.0934
3.4695
0.8456
-1.7783
-4.4021
-7.026
0
0.5
1
1.5
2
2.5
3
3.5
4
4.5
5
Frequency (MHz): 0.2081299
Phase (radians):-0.2554078

(a)

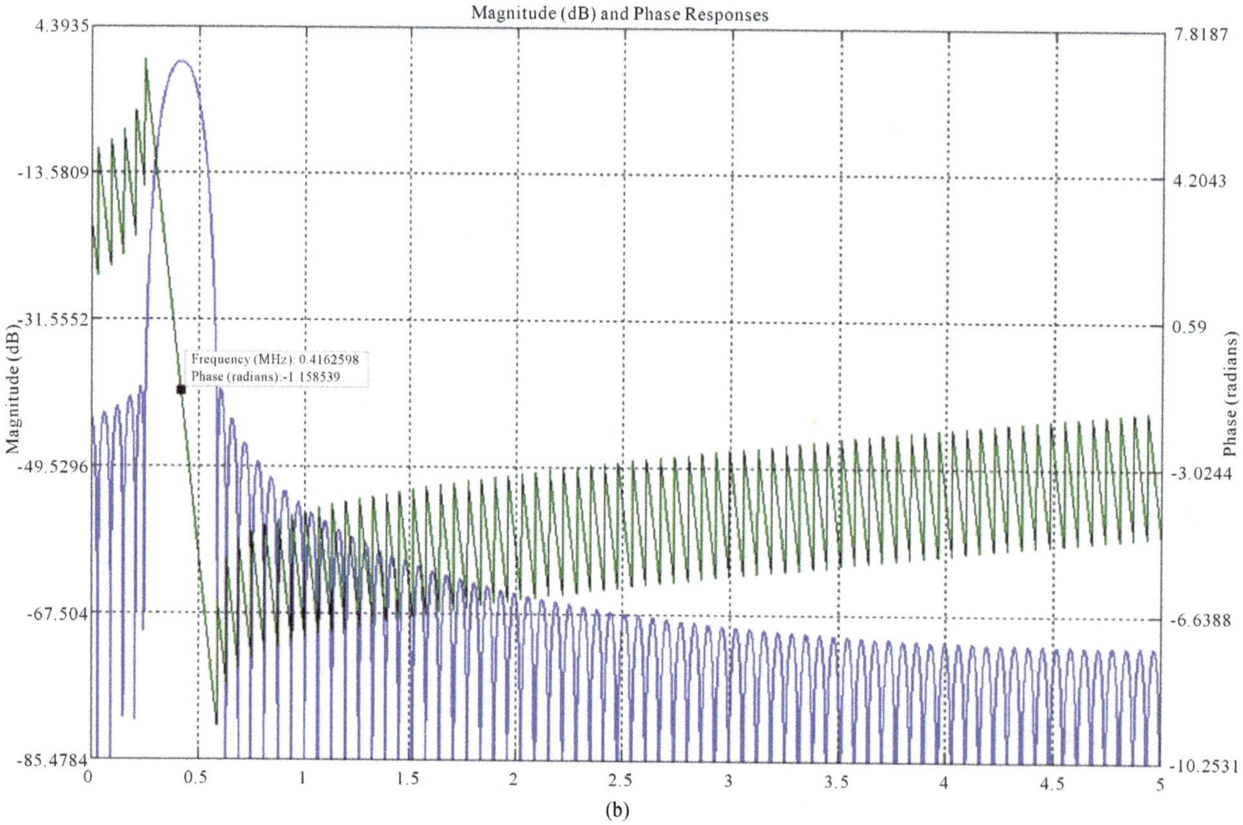
Magnitude (dB) and Phase Responses
Magnitude (dB)
Phase (radians)
4.3935
-13.5809
-31.5552
-49.5296
-67.504
-85.4784
7.8187
4.2043
0.59
-3.0244
-6.6388
-10.2531
0
0.5
1
1.5
2
2.5
3
3.5
4
4.5
5
Frequency (MHz): 0.4162598
Phase (radians):-1.158539

(b)

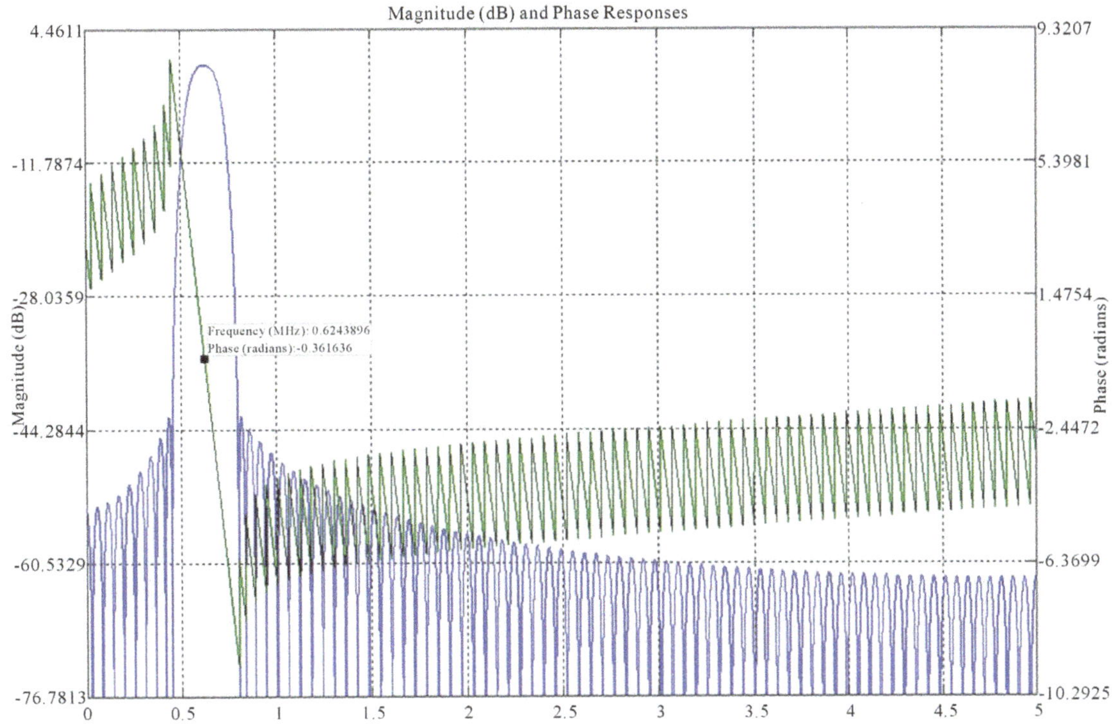

(c)

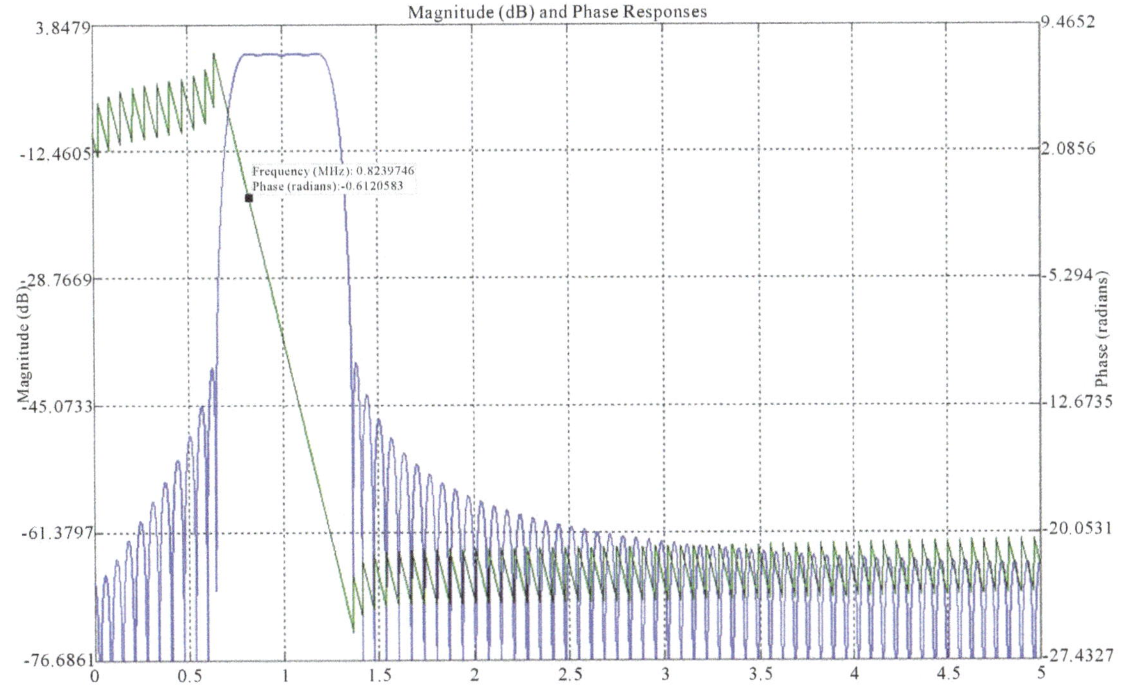

(d)

Figure 1. Band pass filters for extracting four harmonics from the photocurrent IFOG signal. FIR filter design is taken into account as these filters have phase linearity. (a) Band pass filter (harmonic 1) Kaiser window FIR with order = 100, F_{p1} = 0.202 MHz, F_{p2} = 0.21 MHz, F_{s1} = 0.334MHz, F_{s2} = 0.399 MHz, R_p = 0.3917, R_s = 39.45, Phase Delay = −0.255 rad; (b) Band pass filter (harmonic 2) Kaiser window FIR with order = 153, F_{p1} = 0.4 MHz, F_{p2} = 0.43 MHz F_{s1} = 0.249 MHz, F_{s2} = 0.587 MHz, R_p = 0.3276, R_s = 40.97, Phase Delay = −1.159 rad; (c) Band pass filter (harmonic 3) Kaiser window FIR with order = 162, F_{p1} = 0.597 MHz, F_{p2} = 0.653 MHz, F_{s1} = 0.457 MHz, F_{s2} = 0.802 MHz, R_p = 0.33, R_s = 40.29, Phase Delay = −0.361 rad; (d) Band pass filter (harmonic 4) Kaiser window FIR with order = 148, F_{p1} = 0.802MHz, F_{p2} = 0.849MHz, F_{s1} = 0.650 MHz, F_{s2} = 1.71MHz, R_p = 0.32, R_s = 40.25, Phase Delay = −0.612 rad.

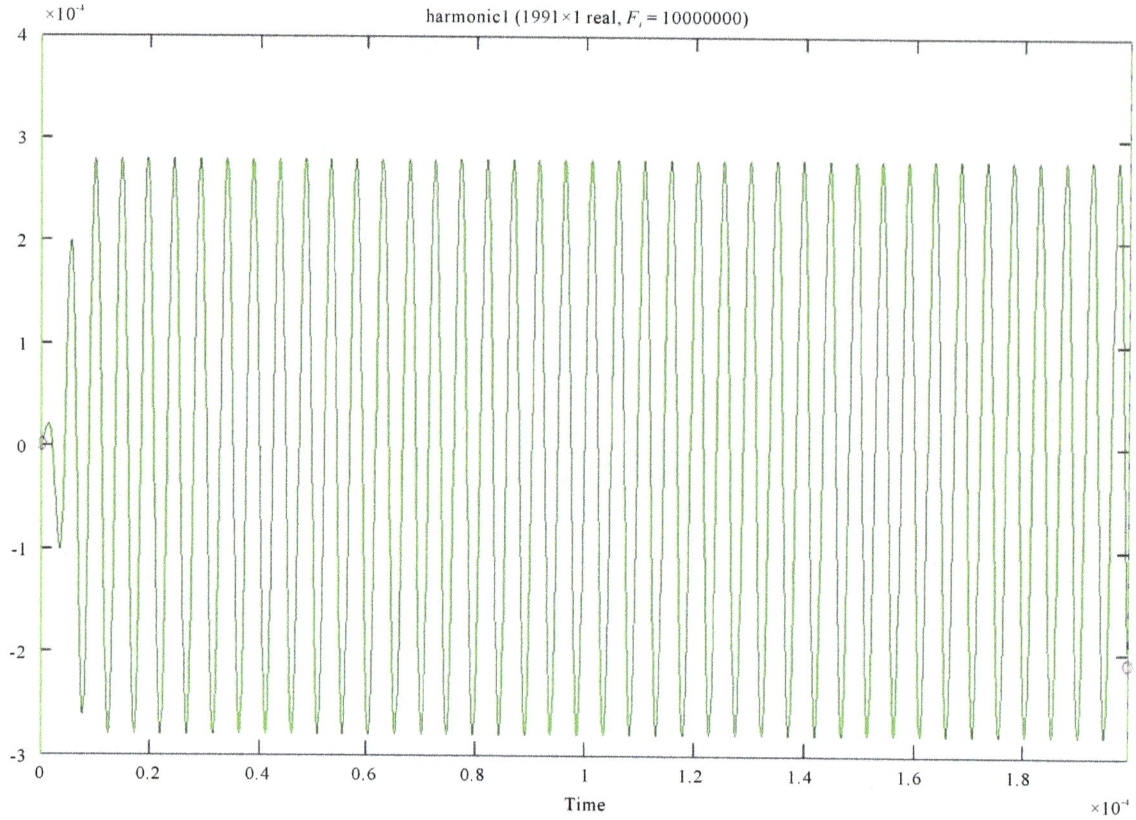
harmonic1 (1991×1 real, F_s = 10000000)
×10⁻⁴
4
3
2
1
0
-1
-2
-3
0
0.2
0.4
0.6
0.8
1
1.2
1.4
1.6
1.8
Time
×10⁻⁴

(a)

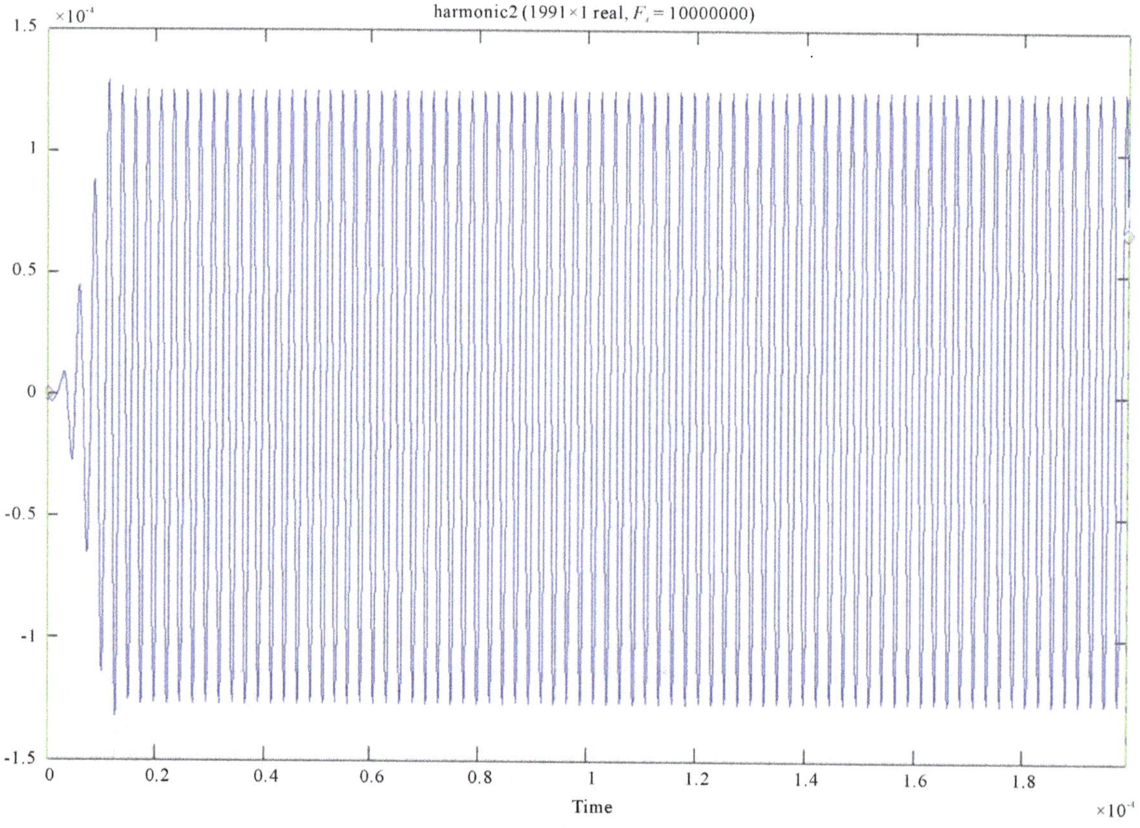
harmonic2 (1991×1 real, F_s = 10000000)
×10⁻⁴
1.5
1
0.5
0
-0.5
-1
-1.5
0
0.2
0.4
0.6
0.8
1
1.2
1.4
1.6
1.8
Time
×10⁻⁴

(b)

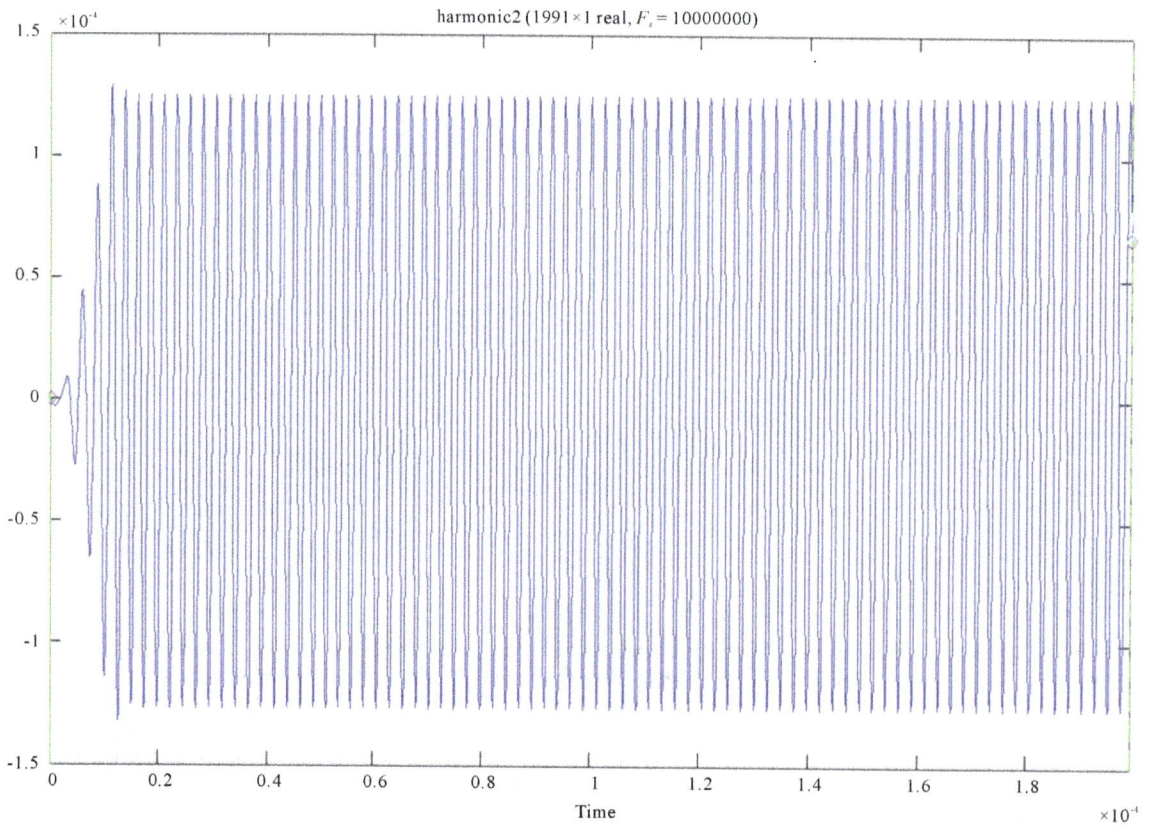

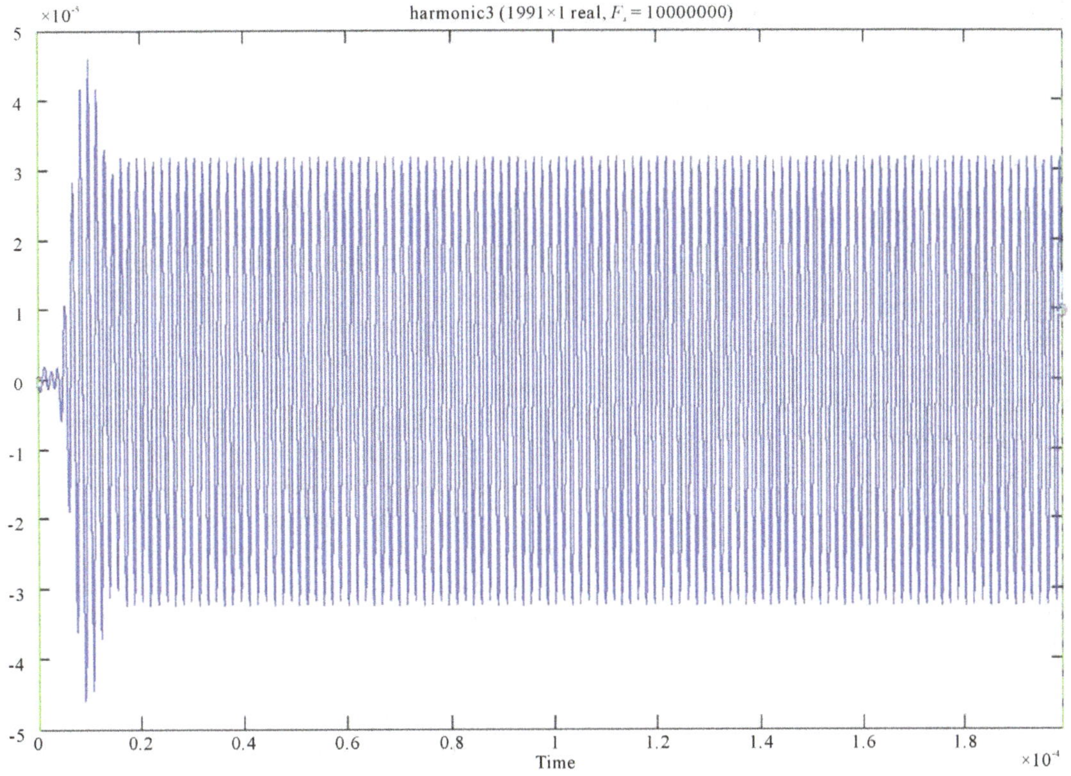

(c)

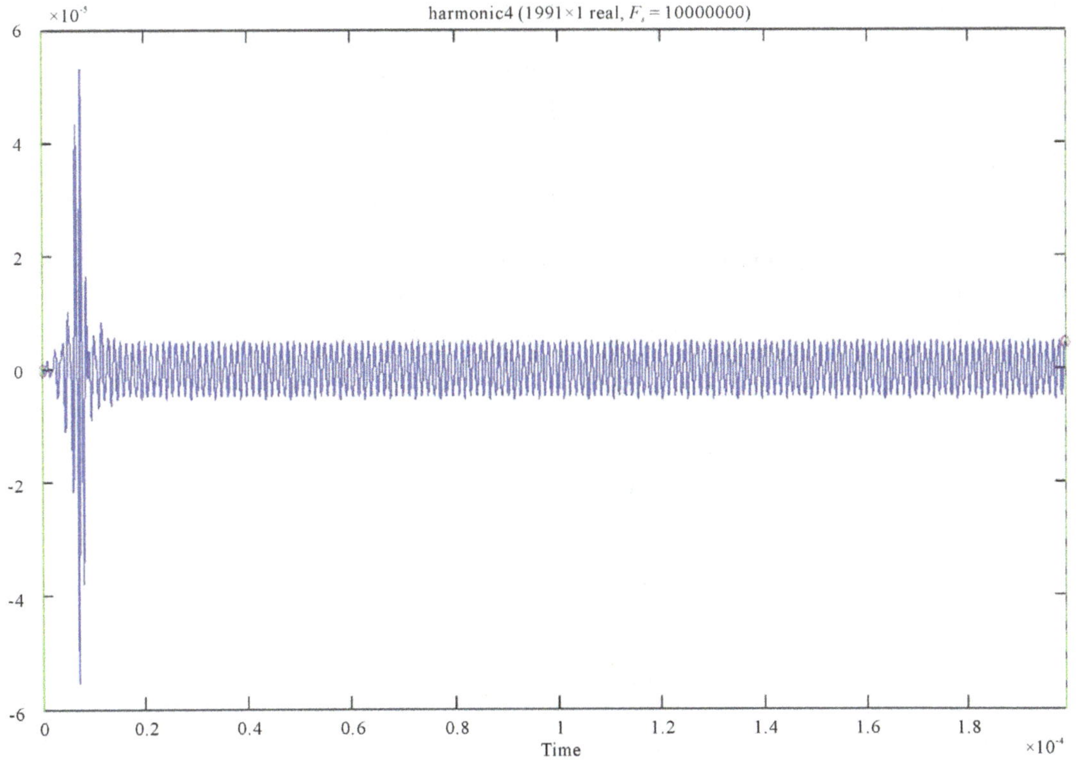

(d)

Figure 2. Signal harmonic extracted from FIR filtering. (a) Harmonic 1 filtered signal, Amplitude = 0.28 mV; (b) Harmonic 2 filtered signal, Amplitude = 0.125 mV; (c) Harmonic 3 filtered signal, Amplitude = 0.03 mV; (d) Harmonic 4 filtered signal, Amplitude = 0.008 mV.

and the third-time interference signal that occurs after feedback has been twice derived is

$$P_3(t) = A^2\left(K_1^3K_2^2\right)^2\left\{1+v\cos\left[\phi_s+\phi_e\cos\left(\omega_m t-2\omega_m\tau\right)\right]\right\} \times\left\{1+v\cos\left[\phi_s+\phi_e\cos\left(\omega_m t-\omega_m\tau\right)\right]\right\} \times\left\{1+v\cos\left[\phi_s+\phi_e\cos\left(\omega_m t\right)\right]\right\} \quad (35)$$

where A is the gain of the fiber amplifier and K_2 is a parameter that depends on the coupling ratio of the feedback couplers and the transmission loss in the feedback loop. For simplicity, we assume that $K' = K_1 \cdot K_2$. The total output at the photodetector is the summation of the number of above-mentioned interferences and is expressed as

$$P_{total}(t) = P_1(t) + P_2(t) + P_3(t) + \cdots \quad (36)$$

If $\omega_m\tau = 2n\pi$, the total photodetector output can be realized by proper adjustment of the modulation frequency to match the round-trip time, and we can have

$$P_{total}(t) = \frac{K\left\{1+v\cos\left[\phi_s+\phi_e\cos\left(\omega_m t\right)\right]\right\}}{1-AK'\left\{1+\left[\phi_s+\phi_e\cos\left(\omega_m t\right)\right]\right\}} \quad (37)$$

where K is the photodetector coefficient, because the total output is a series of short pulses, we can determine the peak value through the following equation:

$$\begin{aligned} &P'_{total}(t) = 0 \\ &Kv\sin\left[\phi_s+\phi_e\cos\left(\omega_m t\right)\right] = 0 \\ &\left[\phi_s+\phi_e\cos\left(\omega_m t\right)\right] = 2n\pi, n = 0,1,2,3, \end{aligned} \quad (38)$$

Above Equation is a condition required to find the peak position of the output pulse that is valid for both cases of rotation and nonrotation, please note that only those equations of even p correspond to the peak position. From this equation and condition we can see that the output pulse shifts if rotation occurs.

For Open-Loop Method for the FE-FOG In Equation (33) if ϕ_e is selected to be between $0 < \phi_e < 2\pi$ and $\phi_s = 0$ there is no rotation, Equation (33) is satisfied only when $n = 0$. In this case, the peak positions of the output pulse are determined through the following equation:

$$\omega_m t_0 = \frac{2i+1}{2}\pi, i = 0,1,2,\cdots \quad (39)$$

Here t_0 represents the peak positions corresponding to the nonrotation case and i denotes the peak number of the output pulse in the time axis. We see that the peak positions are not affected by the phase-modulation depth ϕ_e when there is no rotation. On the other hand, when rotation occurs ($\phi_s \neq 0$) the peak positions are affected by the Sagnac phase shift and can be determined by

$$\omega_m t_r = \arccos\left(\frac{\phi_s}{\phi_e}\right) + 2i\pi, i = 0,1,2, \quad (40)$$

where t_r denotes the peak positions when rotation occurs and i has the same meaning as in Equation (39) Comparing Equations (39) and (40), we see that the peak positions shift if rotation occurs, and the shift of the peaks is just equal to $\Delta t = t_r - t_0$. Therefore, the peak positions are in fenced by only the Sagnac phase shift if rotation occurs and if we fix the phase-modulation depth ϕ_e. We can thus determine the rotation rate by the detection of the peak shift Δt of the output pulse.

In this paper the wave-length of optical source is $\lambda_0 = 1.5\ \mu\text{m}$, the modulation frequency of the phase modulator, $f_m = 0.209790\ \text{MHz}$ ($\omega_m = 2\pi \cdot f_m$), and the radius of the Sagnac loop is R = 0.05 m. The round-trip length is 500 m, which is selected to match the modulation frequency for the pulse output, of which the Sagnac-loop length is 460 m and the length of the EDFA (including the pigtail of couplers) is 40 m. The interferometric coefficient is v = 0.96, the photodetection coefficient is K = 0.2, the parameter K = 0.06, the effective phase-modulation depth is $\phi_e = 0.6\ \text{rad}$, and the Sagnac phase shift is $\phi_s = 0$ for the nonrotation case.

Figure 3 demonstrates the principle of the rotation measurement for the open-loop method when the output pulse is shifted by rotation. The plots are normalized and $\phi_e = 0.18\ \text{rad}$. We see that the peak position shifts if rotation occurs, phase shift 0°/h and 3.6°/h. Further calculations show that the shift is increased as the rotation rate increases. The very sharp peak of the output pulse can result in a high-resolution rotation measurement. From Equations (7) and (38)-(40) (in Equation (39), i = 0, 2, 4), we can derive the rotation rate as:

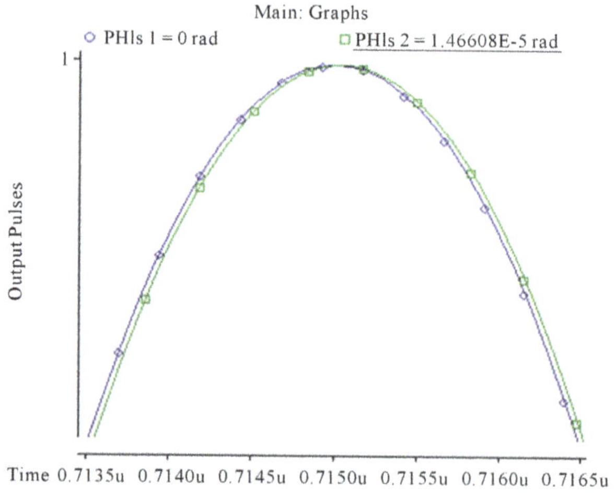

Figure 3. Rotation measurement for open-loop IFOG.

$$\Omega = -\frac{\lambda_0 c}{4\pi RL}\phi_e \cos\left(\omega_m t_0 + \omega_m \Delta t\right) = \frac{\lambda_0 c}{4\pi RL}\phi_e \sin\left(\omega_m \Delta t\right) \quad (41)$$

where Δt is the time shift of the output peak as a result of the rotation, In Closed-Loop Method for the FE-FOG, we selected ϕ_e with a condition of $2\pi < \phi_e < 4\pi$. In such a case the output pulse is shifted by only the rotation. In this section, we hope to show that the output pulse is affected by both rotation and phase modulation, so that we can use the phase modulation to compensate the Sagnac phase shift. Such a technical idea is consistent with the closed-loop method.

In Equation (38), if we assume that $\phi_s = 0$ (nonrotation) and we select $2\pi < \phi_e < 4\pi$, e.g., $\phi_e = 2.2\pi$, then Equation (38), can be satisfied by $n = 1$, $n = 0$, and $n = -1$, which means that the output pulse has three kinds of peaks. The peaks corresponding to $n = 0$ have the same properties as do those of the open-loop method and will not be affected by the phase modulation; however, the peaks corresponding to $n = 1$ or $n = -1$ will be in fenced by both the Sagnac phase shift and the phase-modulation depth. This performance can be used for the rotation measurement.

Figure 4 shows that the output pulses related to $\phi_e = 0.2\pi$ is simple, but, the output pulses related to $\phi_e = 2.2\pi$ are complicated and consist of three kinds of peaks designated A, B, and C. (Note: if we select ϕ_e to be greater than 4π, even more output peaks will appear, because in Equation (38) n also includes integers over. This case can also be analyzed similarly). The peak positions for A, B, and C are determined by $n = 1, 0, -1$, respectively, in Equation (38). In this paper, we look at the peak positions corresponding to peaks A (peaks C also have similar properties). In one time period, peaks A also consist of two additional peaks, marked with circles and squares, whose positions are determined by

$$\begin{aligned} \omega_m t_A &= \arccos\left(2\pi/\phi_e\right) + 2n\pi \\ \omega_m t_A &= 2\pi - \arccos(2\pi/\phi_e) + 2n\pi \left(n = 0, 1, 2, \cdots\right) \end{aligned} \quad (42)$$

$N = 0, 1, 2, \cdots$ respectively. If the Sagnac phase shift induced by rotation is canceled by the phase modulation of the phase modulator, the peak positions corresponding to peaks A will not shift in spite of rotation's occurring. To realize this phase compensation we can feed back an electrical signal at the phase modulator. The rotation rate can thus be evaluated through the value of the phase modulation. Further calculations also show that, although in the closed-loop case (when both rotation and feedback control of ϕ_e occur) peaks B and C are shifted, they cannot cross with peaks A. This is because, for in any case, the peak positions of peaks A, B, and C are determined from Equation (38) with $n = 1, 0, -1$, respectively, and they always have different values. The above result

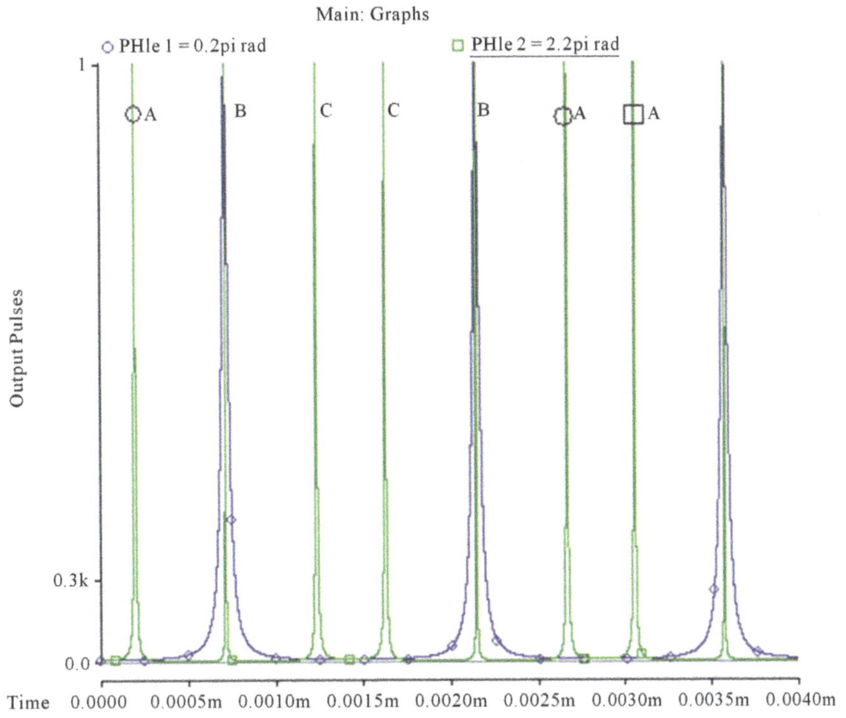

Figure 4. show the comparison of the output pulses for different phase modulations when one value of ϕ_e, $0 < \phi_e < 2\pi$ and $2\pi < \phi_e < 4\pi$. The plots are normalized.

tells us that in the closed-loop case the dynamic range is large.

A more detailed mathematical description is presented in the following paragraphs. We first consider the case of nonrotation ($\phi_s = 0$, for which the peak positions corresponding to peaks A can be determined by

$$\phi_{e0}\cos(\omega_m t_A) = 2\pi \tag{43}$$

where ϕ_{e0} is the phase-modulation depth corresponding to the nonrotation case. On the other hand, when rotation occurs and if the electrical-feedback signal is placed on the phase modulator, the peak positions corresponding to peaks A will not change, and we then have

$$\phi_s = 2\pi - (\phi_{e0} + \delta\phi_e)\cos(\omega_m t_A), \tag{44}$$

and from Equations (43) and (44) we can further derive

$$\phi_s = -\delta\phi_e \cos(\omega_m t_A) = -2\pi\frac{\delta\phi_e}{\phi_{e0}}, \tag{45}$$

We can evaluate the rotation rate by detecting $\delta\phi_e$.

Figure 5 shows the variation of the phase-modulation depth $\delta\phi_e$ as a function of the rotation rate for the closed loop. **Figure 5(a)** shows the output pulse when the rotation doesn't exist $\Omega = 0$ and $\phi_e = \phi_{e0} = 2.2\pi$ and **Figure 5(b)** shows the output pulse when rotate without electric feedback signal in phase modulator Ω = 3.65°/h and $\phi_e = \phi_{e0} = 2.2\pi$ therefore when rotation occur output pulses shifted. We can see the peak that shifted by rotation could back to the first position that $\delta\phi_e$ shows the variation of phase modulation induced by electric

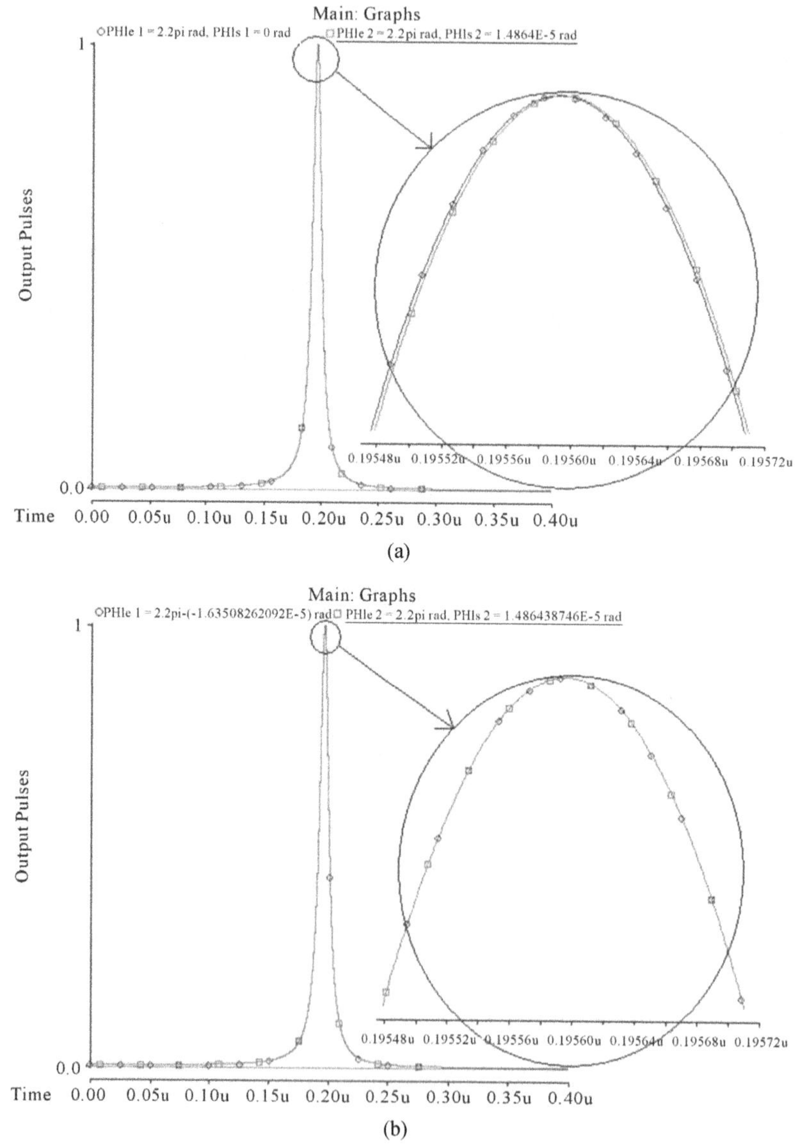

Figure 5. (a) Output pulse when the rotation doesn't exist $\Omega = 0$ and $\phi_e = \phi_{e0} = 2.2\pi$; (b) output pulse when rotate without electric feedback signal in phase modulator $\Omega = 3.65°$/h and $\phi_e = \phi_{e0} = 2.2\pi$.

feedback, that means the rotation rate of $\Omega = 3.65°/h$ needs the phase compensation $\delta\phi_e = -1.63508262092E-5$.

8. Conclusions

In this paper, a comprehensive formulation of open loop interferometric fiber optic gyroscope (IFOG) was studied. Through the signal processing explanation, an accurate approach was proposed to calculate the phase shift to make the hardware implementation simpler. The simulation results of the digital signal processing using FIR filters showed that the chosen computational method to extract Sagnac shift from IFOG signal current is promising. With pointing out the similarities and differences between calculated Ω with the input rotation rate of 10 deg/s, and the output of the digital filtering approach, just a slight difference of less than 1% was observed. This distinction appeared because of using unsharp filters. Regarding to the experiences, we will have more accurate IFOG if higher order of filters are utilized.

Also the functionality of the closed-loop fiber-optic gyroscope (I-FOG), based on multiple utilizations of the Sagnac loop and amplified optical feedback, has been proposed and theoretically investigated. The new gyroscope is termed as feedback Er-doped fiber-optic amplifier (FEDFA). A low-coherence light source is used in this FOG [19]. Amplification of a weak feedback gyroscope signal is performed by the incorporated EDFA. The final gyroscope output is pulsed if the modulation frequency of the phase modulator matches the round-trip time [20]. Sagnac phase shift can induce a shift of the output pulse, which is used for the rotation measurement.

REFERENCES

[1] M. Nasiri-Avanaki, "Full Progress of Digital Signal Processing in Open Loop-IFOG," *IEEE Optical Fiber Communication & Optoelectronic Exposition & Conference*, Shanghai, October 2006, pp. 1-10.

[2] M. Nasiri-Avanaki, "Digital Filtering to Obtain Sagnac Shift in Open Loop-IFOG," *North American Radio Science Conference*, Ottawa, July 2007, pp. 23-29.

[3] S. Merlo, M. N. Orgia and S. Donati, "Fiber Gyroscope Principles," Handbook of Fibre Optic Sensing Technology, John Wiley & Sons, Ltd., Hoboken, 2000, pp. 1-23.

[4] H. C. Lefevre, "Fundamentals of the Interferometric Fiber-Optic Gyroscope," *Optical Review*, Vol. 4, No. 1a, 1997, pp. 20-27. doi:10.1007/BF02935984

[5] M. Nasiri-Avanaki, "Open Loop-Interferometric Fiber Optic Gyroscope: Analysis and Methods to Implement," M.Sc. Thesis, Semnan University, Semnan, 2006.

[6] S. Oho, H. Kajioka and T. Sasayama, "Optical Fiber Gyroscope for Automotive Navigation," *IEEE Transactions on Vehicular Technology*, Vol. 44, No. 3, 1995, pp. 698-705. doi:10.1109/25.406639

[7] Mohammad R. N. Avanaki, "Full Progress of Digital Signal Processing in Open Loop-IFOG," *Proceeding of IEEE Optical Fibre Communication & Optoelectronic Conference*, Shanghai, 24-27 October 2006, pp. 1-10. doi:10.1109/AOE.2006.307367

[8] S. M. Bennett, S. Emge and R. B. Dyott, "Fiber Optic Gyroscopes for Vehicle Use," *IEEE Conference of Intelligent Transportation System*, Boston, 9-12 November 1997, pp. 1053-1057.

[9] B. Kelley, G. A. Sanders, C. E. Laskoskie and L. K. Strandjord, "Novel Fiber-Optic Gyroscopes for KEW Applications," *American Institute of Aeronautics and Astronautics: Aerospace Design Conference*, Irvine, February 1992, p. 1118.

[10] Mohammad R. N. Avanaki, "Principle to Optical Fiber," *UPS Journal*, Vol. 11, No. 3, 2005, pp. 23-26.

[11] S. Emge, S. M. Bennett, R. B. Dyott and D. Allen, "Reduced Minimum Configuration Fiber Optic Gyro for Land Navigation Application," *IEEE Aerospace and Electronic Systems Magazine*, Vol. 12, No. 4, 1997, pp. 18-21. doi:10.1109/62.575996

[12] Mohammad R. N. Avanaki and A. Toloei, "Evaluation to Digital Filtering to Obtain Sagnac Shift in Open Loop-IFOG," *Aircraft Engineering and Aerospace Technology Journal*, Vol. 81, No. 5, 2009, pp. 391-397.

[13] R. Y. Liu and G. W. Adams, "Interferometric Fiber Optic Gyroscope: A Summary of Progress," *IEEE Position Location and Navigation Symposium*, Las Vegas, 20-23 March 1990, pp. 31-35.

[14] M. S. Perlmutter, "A Tactical Fiber Optic Gyro with All-Digital Signal Processing," *Proceeding of Position Location and Navigation Symposium*, Northrop, 1994, pp. 170-175.

[15] Y. Liu and G. W. Adams, "Interferometric Fiber-Optic Gyroscope: A Summary of Progress," Honeywell, Inc., Morristown, 1994.

[16] C. Seidel, "Optimierungsstrategien für Faseroptische Rotationssensoren: Einfluss der spektralen Eigenschaften der Lichtquelle," Doctrine Thesis, Institut für Theoretische Elektrotechnik und Systemoptimierung (ITE), Karlsruhe, 2004

[17] C.-X. Shi, T. Yuhara, H. Iizuka and H. Kajioka, "New Interferometric Fiber-Optic Gyroscope with Amplified Optical Feedback," *Applied Optics*, Vol. 35, No. 3, p. 381. doi:10.1364/AO.35.000381

[18] C.-Y. Liaw, Y. Zhou and Y.-L. Lam, „Characterization of an Open-Loop Interferometric Fiber-Optic Gyroscope with the Sagnac Coil Closed by an Erbium-Doped Fiber Amplifier," *Journal of Lightwave Technology*, Vol. 16, No. 12, p. 2385. doi:10.1109/50.736607

[19] A. Noureldin, M. Mintchev, D. Irvine-Halliday and H. Tabler, "Computer Modelling of Microelectronic Closed Loop Fiber Optic Gyroscope," *IEEE Canadian Conference on Electrical and Computer Engineering*, Edmonton, 9-12 May 1999, pp. 633-638.

[20] B. Seçmen, "Simulation on Interferometric Fiber Optic Gyroscope with Amplified Optical Feedback," Middle East Technical University (METU), Ankara, 2003.

15

Optical Image Compression Using a Real Fourier Plane

Abdulsalam G. Alkholidi
Faculty of Engineering, Electrical Engineering Department, Sanaa University, Sanaa, Yemen
Email: Abdulsalam.alkholidi@gmail.com

ABSTRACT

Hastening transmission by efficiently providing compression is our goal in this work. Image compression consists in reducing information size representing an image. Elimination of redundancies and non-pertinent information enables memory space minimization and thus fast data transmission. Optics can offer an alternative choice to overcome the limitation of numerical compression algorithms. In this paper, we propose real-time optical image compression using a real Fourier plane to save time required for compression by using the principles of coherent optics. Digital and optical simulation results are presented and analyzed. An optical compression decompression setup is demonstrated using two different SLMs (SEIKO and DisplayTech). The purpose of this method is to simplify our earlier method, improve the quality of reconstructed image, and avoid the disadvantages of numerical algorithms.

Keywords: FT; Optical Compression; a Real Fourier Plane; Hologram; SLM

1. Introduction

In recent years, there has been much interest in image compression algorithms which are proposed in this paper. Our algorithm is oriented towards large size images. The goal of this work is to minimize the time required for compression with high compression ratio (C_r). One of the current technological challenges in image compression is to achieve real-time imaging rates with a minimal sacrifice in signal-to-noise ratio (SNR) [1]. We propose a novel technique of image compression using a real Fourier-plane capable of real-time image compression. In the frame of the present work our objective is to simplify our earlier optical image compression algorithm using the JPEG and optical JPEG standards [2]. We intend to eliminate the most complex part in the synoptic diagrams of the set up implementation compression and decompression described in [2,3].

To digitally decompress large images, they should be first transmitted to a calculator to enable applying one or several digital compression methods. The real-life application of this technique; consists first in optically compressing these images. Starting from an image, with size ($N \times N$) pixels (**Figure 1(a)**), we obtain a compressed version of only ($c \times c$) pixels as pointed out by **Figure 1(f)**. This results in a decrease of the volume of information to be transmitted to the calculator. The latter may provide additional compression, by applying digital methods. In this paper, we provide a detailed technical description of the proposed architecture. This exposition is interested to serve as a reader-friendly starting point for those interested in learning about optical JPEG compression. Although many details are included in our presentation, some details were necessarily omitted to focus on the main subject that is compression. The reader should, therefore, refer to standards of optical JPEG compression [2,3] before attempting an implementation. Indeed, much work has been done toward developing compression methods, most techniques reported, such as presented in [4-8].

The remainder of this paper is structured as follows: Section 2 provides a brief overview of the optical JPEG standards. This is followed, in Section 3, by a detailed description of mathematical approach of the proposed algorithm. In Section 4, optical architecture of compression/decompression is presented. Section 5, an optical setup of image compression/decompression of proposed method is presented and the results are presented in Section 6. Finally, we conclude with some closing remarks in Section 7.

2. Optical JPEG Compression

By using the principle of coherent optics, optical JPEG (OJPEG) compression and decompression; process the whole image at once and do not divide it into blocks. The disadvantage of OJPEG is the fact that its optical implementation is complicated [2,3]. To relax this constraint,

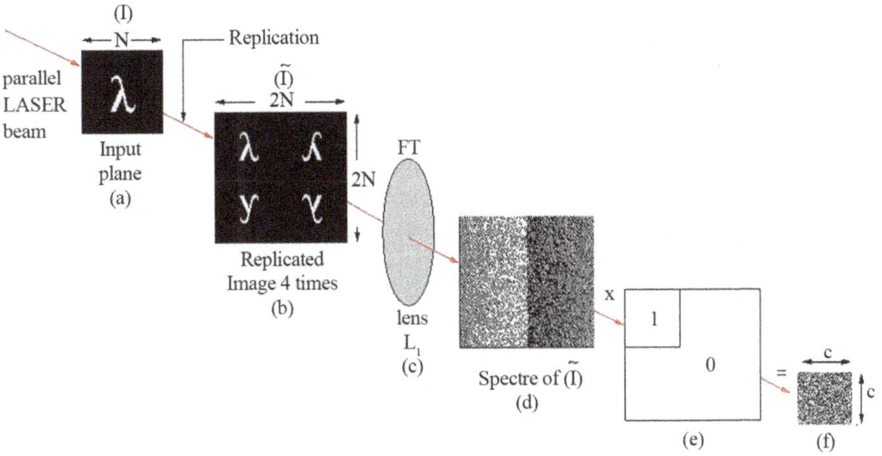

Figure 1. Synoptic diagram of the setup implementation of the optical image compression using a real Fourier plane, (a) original image I ($N \times N$) pixels, (b) replicated image I ($2N \times 2N$) pixels, (c) convergent lens to calculate FT, (d) real spectrum of replicated image, (e) information selection, (f) ($c \times c$) pixels is the size of compressed spectrum to be transmitted.

we propose a new optical image compression method using a real Fourier plane.

3. Mathematical Approach of Optical Image Compression Using a Real Fourier Plane

The main steps to compress an image in this work are as follows:

1) Replicate the original image ($N \times N$) pixels as showing in **Figure 2**. This operation comes down to making the image even. We then obtain a ($2N \times 2N$) pixels image with hermitian symmetry. Since the image is initially real, its hermetian version is merely even.

2) By applying Equation (1) to an image of ($2N \times 2N$) pixels, we obtain Equation (2).

Equation (1) gives the discrete version of the FT referred to as the 2 dimensional DFT of an ($N \times N$) pixels image.

$$F(p,q) = \frac{1}{N \times N}\sum_{m=0}^{N-1}\sum_{n=0}^{N-1} f(m,n) \times e^{-j2\pi\left(\left(\frac{mp}{N}\right)+\left(\frac{nq}{N}\right)\right)} \quad (1)$$

with

$$p \in 0 \text{ to } [N-1]$$

$$q \in 0 \text{ to } [N-1]$$

The term $f(m, n)$ stands for the discrete version of the two-dimensional signal $f(x, y)$.

For $N = 2^k$, k is an integer, fast algorithms FFT to compute DFT exist. For the obtained hermetian image, the DFT is expressed as follows:

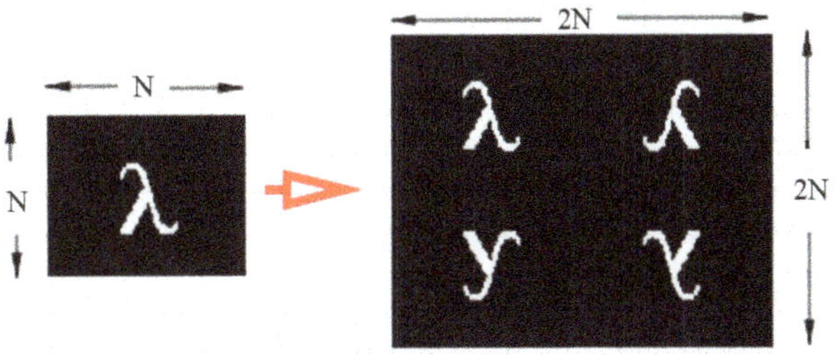

Figure 2. Characteristic replication of the input image.

$$\tilde{F}(p,q) = \frac{1}{4N \times N}\sum_{m=0}^{2N-1}\sum_{n=0}^{2N-1} \tilde{f}(m,n) \times e^{-j2\pi\left(\left(\frac{mp}{2N}\right)+\left(\frac{nq}{2N}\right)\right)} \quad (2)$$

with

$$p \in 0 \text{ to } [2N-1]$$

$$q \in 0 \text{ to } [2N-1]$$

3) The real spectrum obtained using Equation (2) after replication of the image **Figure 3(d)**, is then quantified.

We note that the input image with ($N \times N$) pixels presented in **Figure 3(a)** is real whereas its Fourier transform is complex. The main difference between the OJPEG compression presented in [2,3] and the proposed compression technique is that the spectrum obtained of replicated image ($2N \times 2N$) pixels showing in **Figure 2** is complex for the first one as presented in **Figures 3(a)** and **(c)**, because we presented a mathematical approach for this objective demonstrated in [3]. If we replicate an image to obtain ($2N \times 2N$) pixels, then its Fourier transform is real even. This presents the *core* of this work; that is,

$$F(u,v) = F^{*}(-u,-v) \quad (3)$$

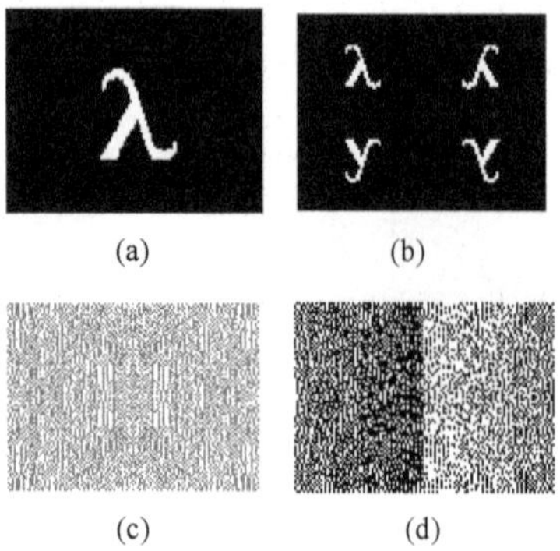

Figure 3. (a) Input image, symbol lambda I($N \times N$) pixels, (b) replicated image with I (($2N \times 2N$) pixels, (c) complex hermetian spectrum obtained from (a), (c) real even spectrum obtained from (b).

where "*" indicates the standard conjugate operation on complex number. From this, it follows that:

$$|F(u,v)| = |F(-u,-v)|, \tag{4}$$

Which says that the spectrum of the Fourier is symmetric [10,11]. The two spectra (c) and (d) are symmetric. But spectra (c) is symmetric complex and spectra (d) is symmetric real even.

4) After the real Fourier spectra step, we proceed to the quantification and selection. It consists firstly in mapping the spectra values on a set of selected values, (quantization), and secondly in eliminating redundant information in this spectra plane. These operations are performed together, thanks to a single threshold:

The quantification: this operation consists of dividing the values of the spectra by another matrix containing quantization values. The optical implementation of this operation is performed in the spectral domain by multiplying the spectrum coefficient by another computer generated hologram containing values of quantization. The filtering: to increase the compression ratio after quantization, a threshold operation is used to retain only the part of information that we want to transmit. This operation corresponds in general at a low-pass filter to preserve the low frequency information which is relevant for the spectrum obtained. High frequencies are omitted. The filter size is then reduced to (c), expressed as follows:

$$c = \sqrt{\frac{N \times N}{Q_s}} \tag{5}$$

- $c \times c$ is the size of compressed spectrum to be transmitted,
- Q_s is the quantification steps (in change the intensity of laser's energy).

Two compression parameters can be defined:

$$T_i = \frac{100 \times c \times c}{N \times N}\% = \frac{100}{Q_s} \tag{6}$$

$$C_{ro} = (100 - T_i)\% \tag{7}$$

- T_i is the transmitted information, percentage,
- C_{ro} is the optical compression rate.

The compression rate for digital image compression technique is defined as follows:

$$C_r = \frac{\text{Original Image Size}}{\text{Compressed Image Size}} \tag{8}$$

The main difference between two compression techniques is digital-JPEG reduces the number of bytes while optical image compression, using a real Fourier plane the number of pixels, is reduced as seen in Equations (7) and (8).

The optical compression ratio C_{ro} mentioned above Equation (7) consists in comparing the number of pixels remaining after compression with that of the initial image. As this ratio takes into account the remaining number of pixels, thus it is completely normal that the quality of the image is degraded when the compression ratio increases. The increase in the ratio means that there are fewer of pixels to be transmitted and thus less information on the image.

4. Optical Image Compression and Decompression Architectures Using a Real Fourier-Plane

4.1. Architecture of Optical Compression

We propose an optical architecture in a real Fourier plane by using a DFT. This optical compression technique can be processed numerically or optically. The synoptic diagram of the optical setup is demonstrated in **Figure 1**.

The synoptic diagram presented in **Figure 1** represents successive operations. The first one consists of replicating the input image **Figure 1(a)** in a specific way **Figure 1(b)**. The objectives of this replication operation are:

1) To obtain a real spectrum easy to compress in the frequency domain as illustrated in Equations (6) and (7).

2) To avoid the noise order of zero generated by the beam of LASER as demonstrated in **Figure 4** decompressed image.

The second stage performs the FT of this duplicated image by using a convergent lens "L_1" as shown in **Figure 1(c)**. The real spectrum **Figure 1(d)** is multiplied by the low-pass filter of **Figure 1(e)** to increase the compression ratio after quantization. In fact, this filter is used for the purpose of preserving the low frequency, and to select the transmitted data and to obtain the compressed optical part of information the input image. **Figure 1(f)**, this spectra, of size ($c \times c$) pixels (see Equation (5)), contains the data to be transmitted.

4.2. Architecture of Optical Decompression

On the receiver side, transmitted data should be decompressed to obtain the original image. The decompression consists in reversing the operations carried out during compression side. For this purpose, we propose the synoptic diagram of **Figure 5**. Information at the receiver is restricted to a ($c \times c$) pixels array **Figure 5(a)**, then we should first restore a ($2N \times 2N$) pixels spectrum array by filling zero padding the remaining zone by zeros as indicated in **Figure 5(b)**. We take the conjugate version of the image of **Figure 5(a)** and insert two duplicates of it in the image as illustrated in **Figure 5(b)**, where "c" is indicated. Two duplicates of the image of **Figure 5(a)**, itself are placed in the two zones where "c" is indicated in **Figure 5(b)**.

Each two duplicates in the diagonals of **Figure 5(b)** are point wise symmetrical with respect to the central point of the ($2N \times 2N$) pixels image. The objective of this reconstruction is to obtain the spectrum of the image ($2N \times 2N$) as shown in **Figure 5(c)**. Finally, an optical Inverse Fourier Transform (IFT) is obtained by a convergent lens "L_2" **Figure 5(d)**) which yields to the replicated image. Selecting one of the four quadrants gives the decompressed original image as showing in **Figure 5(e)**.

5. Implementations of Optical Image Compression/Decompression Using a Real Fourier Plane

5.1. Compression Implementation

After illustrating the formulation and compression/decompression architectures of optical image compression using a real Fourier plane, we turn to the optical implementation of the technique. For this purpose, we will present an all-optical setup which contains two stages of our method image compression/decompression. The diagram

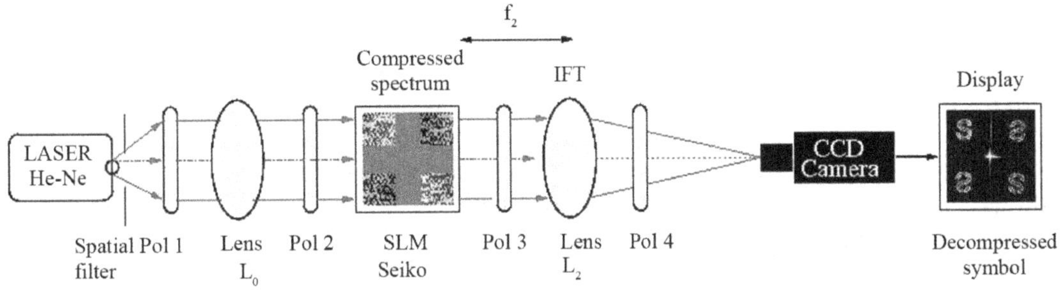

Figure 4. The optical setup decompression using SEIKO SLM.

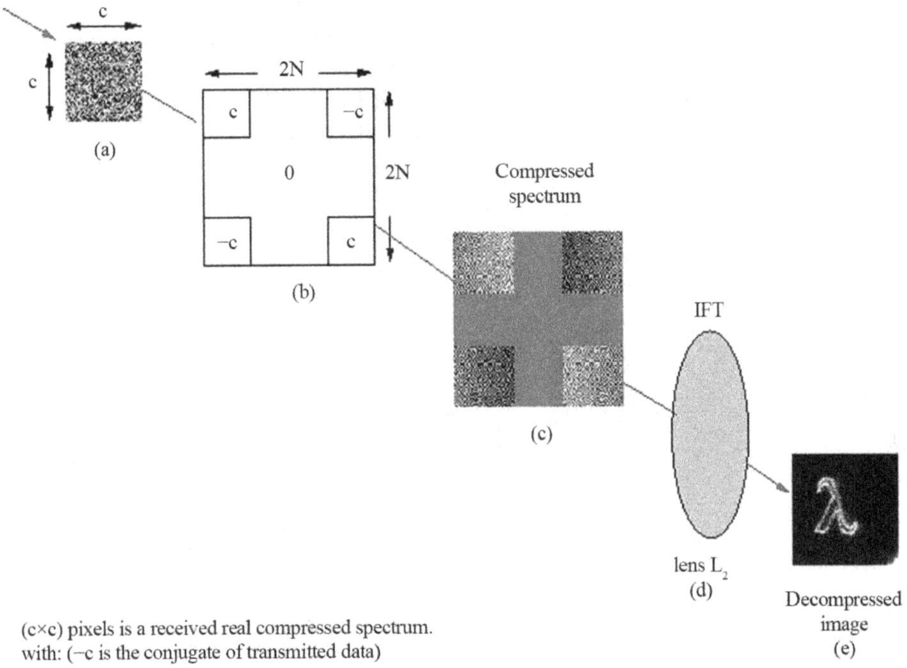

Figure 5. Synoptic diagram of the optical image decompression using a real Fourier plane, (a) received information, (b) restore a ($2N \times 2N$) pixels spectrum array by filling zero padding the remaining zone by zeros, (c) conjugate version of the image of Figure 5(a) and insert two duplicates of it in the image Figure 5(b) where "c" is indicated, (d) convergent lens to calculate IFT, (e) decompressed image.

of the optical compression setup of the proposed method is presented in **Figure 6**.

Technical characteristics of this montage of optical image compression using a real Fourier plane. The montage presented in **Figure 6** consists of the following dispositive:

A monochromatic light source LASER He-Ne source (λ = 633 nm, 25 μW), Spatial Filter, Three separators cubes, two polarizers, two mirrors, two lens L_0, L_1, the first one its focal is of 160 mm to be collimated the laser's beam and the second one to realize the Fourier transform. Compression implementation setup consists tow Spatial Light Modulators (SLM). A SLM generally consists of an addressing material and a modulating material. The optical property of the modulating material is changed by write-in information, and phase or amplitude of readout light is modulated in parallel, corresponding to the write-in information [12].

The first one SLM-1 of 640 × 480 pixels from Seiko with resolution of 42 μm is used to display the input plane of size ($2N \times 2N$) = (256 × 256) pixels as demonstrated in **Figure 6** (input plane). This VGA3 is a high resolution SLM based on Thin Film Transistor (TFT) twisted nematic (TN) display. It provides grey scale displaying capability at standard video frame rates. The second SLM-2 of 1280 × 780 pixels, from Display Tech to display the hologram necessary to realize the low pass filter (information selection).

The compression plane (output plane) consists of camera CCD of size 753 × 582 pixels with a resolution of 6 μm [13].

The first step for proposed optical montage is to calculate the focal distance (L_1) of the Fourier lens. For that, we are going to utilize the following equation:

$$f_1 = N \times d_e \times d_f / \lambda \qquad (9)$$

where λ is the operating beam wavelength, f_1 is the focal distance of L_1, N is the number of pixels of the various planes, d_i and d_o, respectively, represent the size of the pixel (resolution) of the input and output planes.

$$f_1 = \frac{N \times d_e \times d_f}{\lambda} = \frac{256 \times 42 \times 10^{-6} \times 13.2 \times 10^{-6}}{633 \times 10^{-6}} = 224.212 \text{ mm}$$

After having described the different components utilized in this optical setup presented in **Figure 6**, we are going to present the optical compression results. One of the disadvantages of this implementation is difficult to align the optical montage. The results of this multiplication, we obtain the spectra of compressed image ($c \times c$) pixels. That is registered using a CCD camera. The obtained result considers a numerical version of optical image compression using a real Fourier plane. The photo of the setup is given in **Figure 7**.

Figure 8 shows the image of size ($c \times c$) pixels that considers the result of optical compression implementation using a real Fourier plane, contains the data to be transmitted.

5.2. Decompression Implementation

After illustrating the optical compression implementation stage, we turn to the optical decompression setup. For this purpose, we will present an all-optical setup consisting of different components for implementing the image decompression using a real Fourier plane using two different SLMs. In this implementation, we are going to implement two spectra:

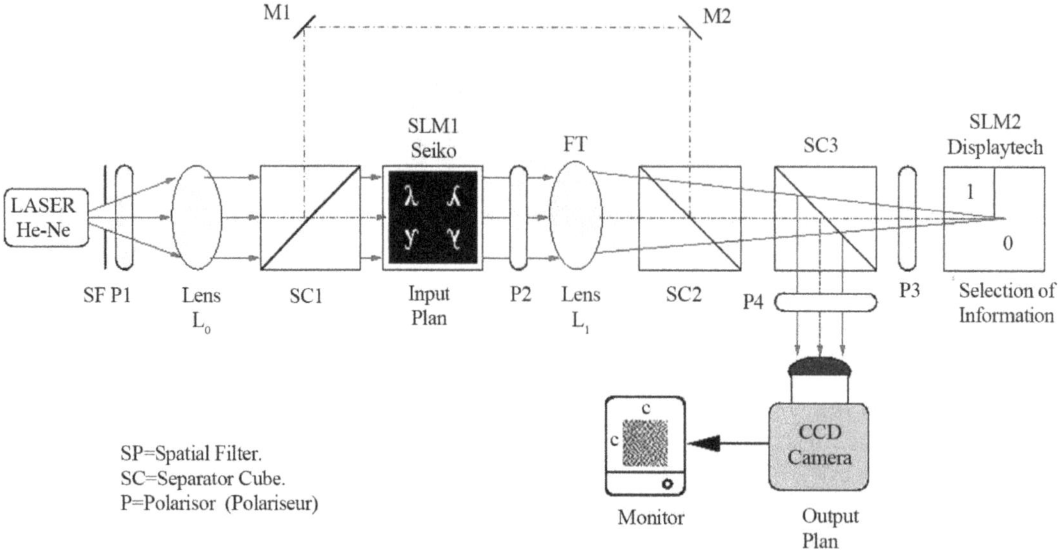

Figure 6. The optical setup used to optical image compression using a real Fourier plane.

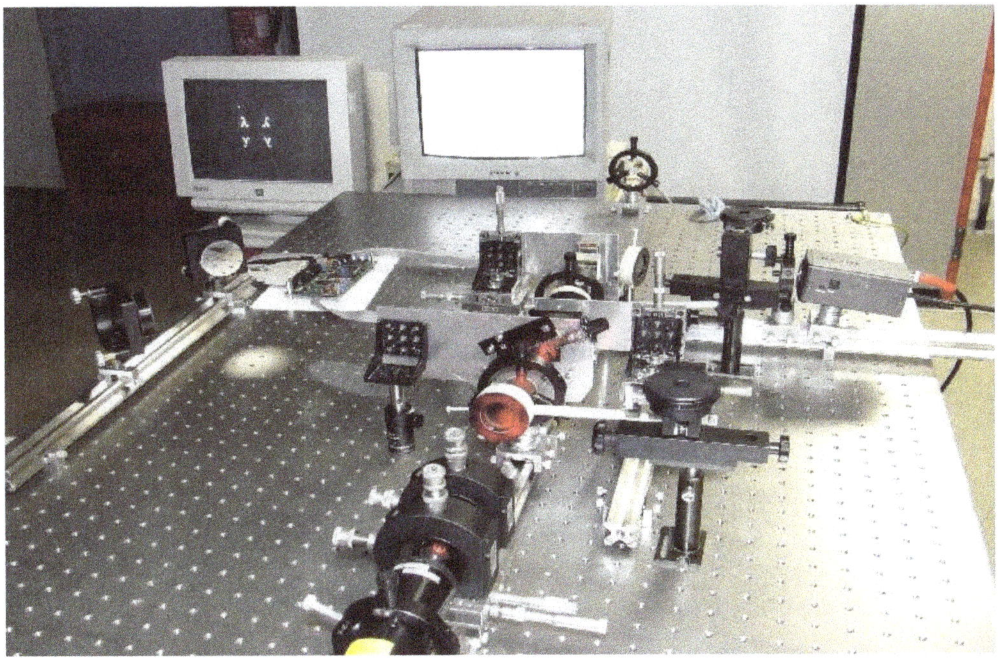

Figure 7. A photo of the setup used for the optical compression stage.

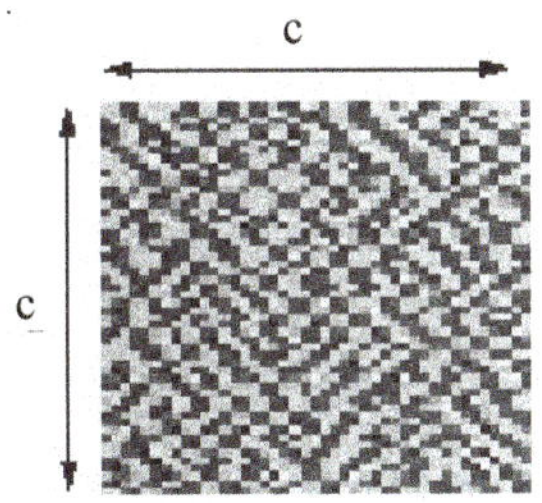

Figure 8. The result of optical compression implementation using a real Fourier plane.

- Compressed spectrum in grey level as demonstrated in **Figure 4**, where, we used SLM SEIKO.
- Binary compressed spectrum as demonstrated in **Figure 9**, where, we used DisplayTech SLM.

We used two SLM with different technical characteristics to compare the decompressed images. In fact, firstly we used the modulator SEIKO permits to display an image with several levels of grey, but with a weak-level of quality of reconstructed image. Secondly, we used DesplayTech binary modulator that displays revenue so important, but it does not permits to the display of the specters with several levels of grey. For this purpose, we propose the synoptic diagram of **Figure 4**. Information at the receiver is restricted to ($c \times c$) pixels. Two polarizers are placed before and after the SLM in order to operate in phase. A mask is used to isolate only the additive share of the SLM. The modulator is used to display the hologram that includes ($c \times c$) pixels representing the compressed spectrum of the image obtained digitally (**Figures 10(a)** and **(b)**).

6. Results of Optical Implementation

To validate the principle of proposed compression/decompression method of a real Fourier plane, we have considered several images as shown in **Figure 11**.

In column (a), we see the different symbols and an image used to validate our optical implementation of compression/decompression stages. Four images are considered including three binary letters images, namely "λ", "S", and "Ψ" and, a gray level image, namely the famous "Lena". Each input image extends over 128 × 128 pixels. In column (b), we have replicated each input image four times. Indeed, each image is mirrored horizontally, vertically and obliquely. In column (c), the calculated Optical Fourier Transform (OFT) of the different replicas is shown. We have optically implemented the IFT of these various spectra. These are used as references. The results are given in column (d). In column (e), we see the different compressed spectra with our proposed method. Finally, in column (f) the optically reconstructed versions of decompressed images are shown. The results, shown in column "f", validate the principle of our implementation method of the optical image compression/decompression using a real Fourier plane.

Figure 12 shows the optical implementation of the compression/decompression of proposed method. The input plane of this setup consists of a binary modulator DisplayTech to display the binary compressed spectra as demonstrated in **Figure 12(e)**, (a) original image; (b) replicated image; (c) replicated image spectra; (d) optical output plane of the replicated image spectra; (e) compressed spectra; (f) decompressed image. Finally, we

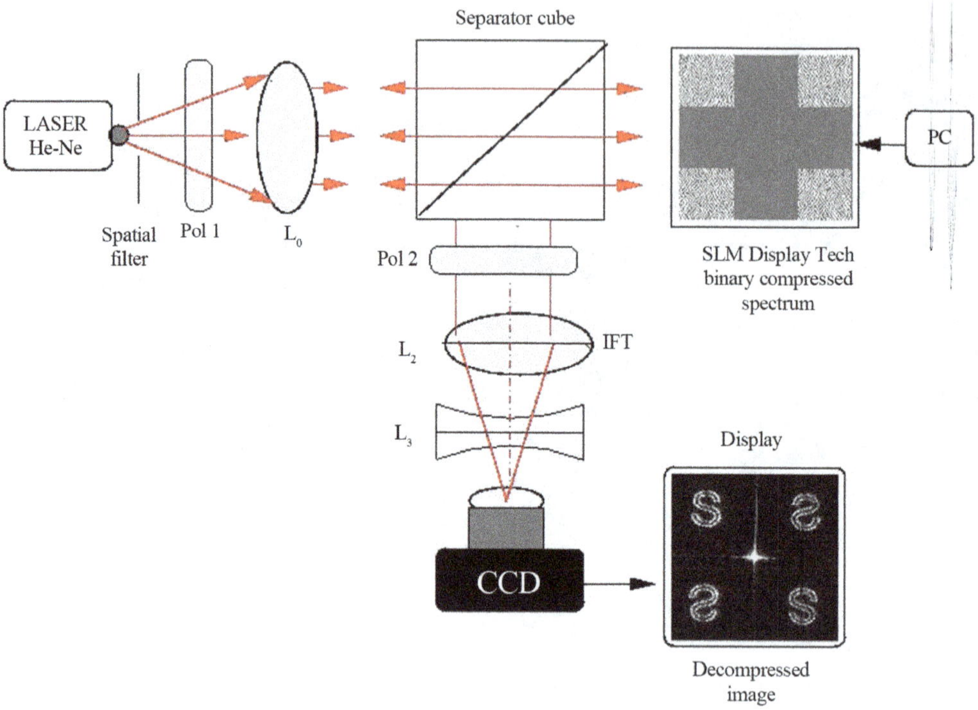

Figure 9. The optical setup decompression using DisplyTech SLM.

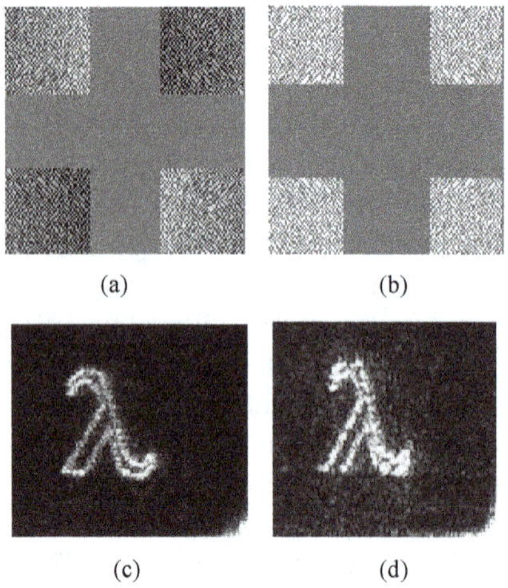

Figure 10. An examples of compressed spectra calculated numerically by applying our method, compressed spectra in grey level, (b) binary compressed spectra (c) decompressed symbol (λ) optically for (a), (d) decompressed binary symbol of (λ) for (b).

note that all optical setups had been realized in department of optoelectronic, ISEN-Brest, France.

7. Conclusions

In this paper, we have presented a novel method of optical image compression/decompression using a real Fourier plane. This optical compression technique processes the image in whole image at once and does not divide it into blocks, thanks to coherent optics. We successively simplified our earlier optical image compression method OJPEG. The proposed method has shown to be a good compression alternative for all types of images. In the practical part, we proposed two optical architectures: one for compression and another one for decompression. Indeed, we meet several difficulties with set up alignments of laser beam for assuring a multiplication between input plan (displayed over an SLM SEIKO) with low-pass filter whose select the pertinent information displayed over another SLM DisplayTech). In fact, decompressing an image is easy to implement optically where only one SLM is required as demonstrated in **Figures 4** and **9**. According to the above considerations, we successfully proof this novel optical compression/decompression method, reduced the information volume representing an image in pixels. Proposed method has several advantages compared to other algorithms. We can list the main features of proposed method according to obtained results:

- Minimize the time required for compression/decompression by using the principle of coherent optics.
- Gain on space memory.
- Improve the quality of reconstructed image numerically and optically.
- Avoid the problem that characterized digital JPEG as (mosaic effect).
- Simplify our earlier compression method optical JPEG.
- Apply this method in combination between numerical and optical (optical compression and numerical decompression), we can go ahead for all optics.

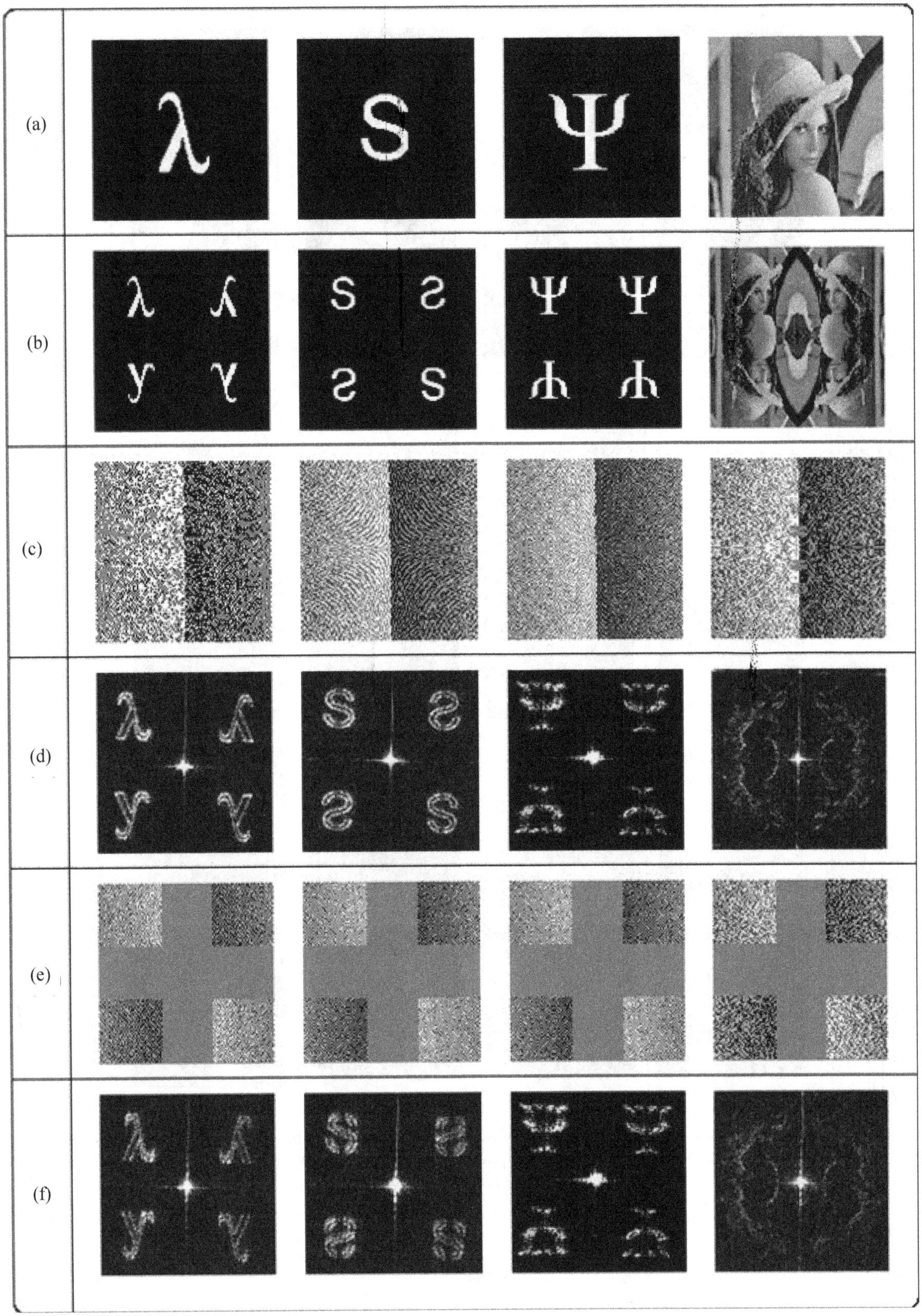

Figure 11. Results of optical compression/decompressionsetup of the real Fourier plane using SEIKO modulator: (a) original image I; (b) replicated image I; (c) replicated spectrum of the image; (d) optical output plane; (e) compressed spectra; (f) decompressed image.

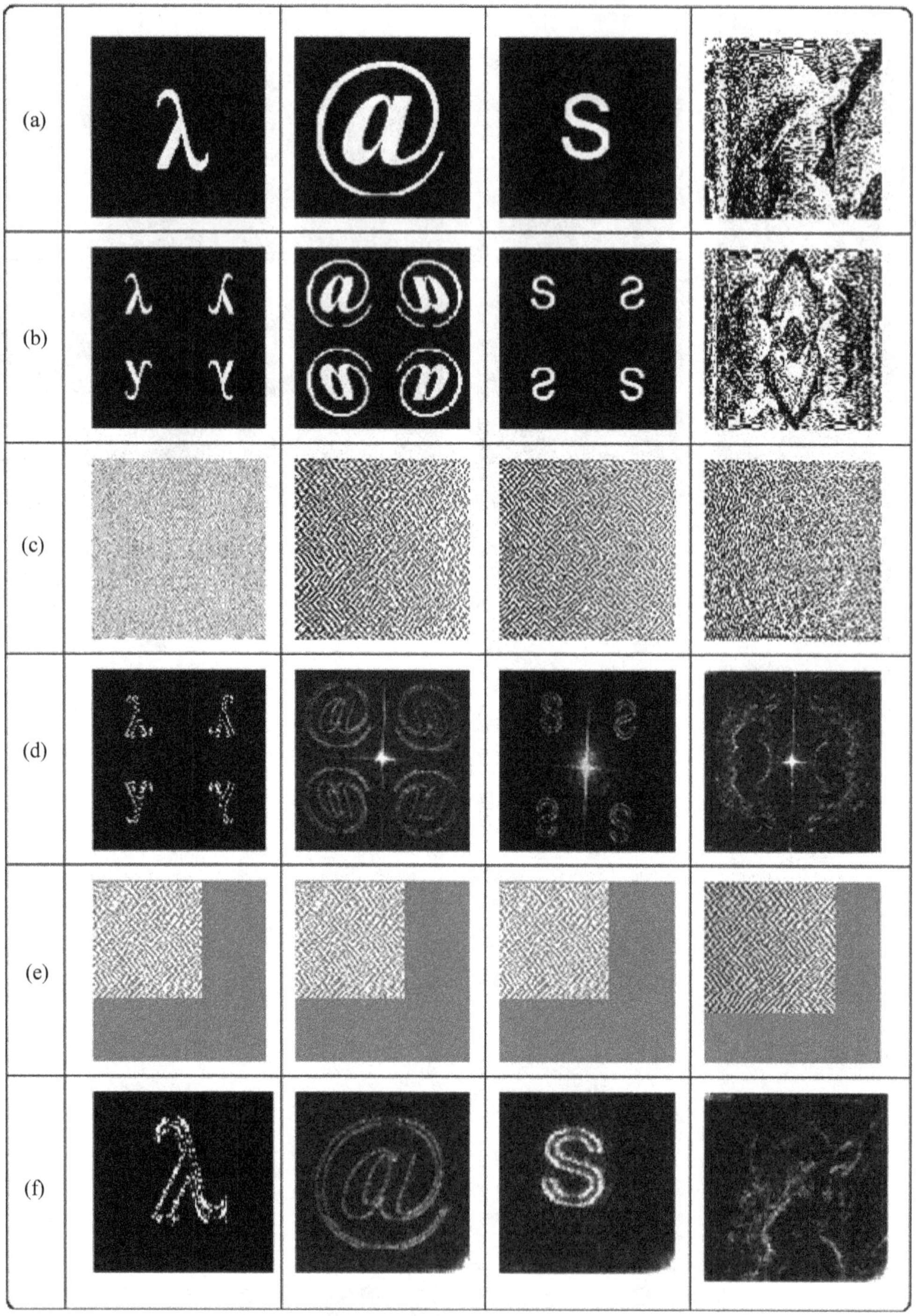

Figure 12. Validation results of optical compression/decompression implementation using a Display Tech modulator to display the binary compressed spectra, (a) input image I; (b) replicated image I; (c) replicated image spectra; (d) implementation of replicated image spectra; (e) compressed spectra; (f) decompressed image.

Finally the numerical and optical results obtained are generally considered to be more robust as appear in Section 6.

8. Acknowledgements

The author wants to thank Prof. Habib Hama (Moncton University), and Prof. M. Sujr (Sanaa University) for reading this paper and for their nice comments.

REFERENCES

[1] A. L. Oldenburg, J. J. Reynolds, D. L. Marks and S. A. Boppart, "Fast-Fourier-Domain Delay Line for *in Vivo*

Optical Coherence Tomography with a Polygonal Scanner," *Applied Optics*, Vol. 42, No. 22, 2003, pp. 4606-4611.

[2] A. Alkholidi, A. Alfalou and H. Hamam, "A New Approach for Optical Colored Image Compression Using the JPEG Standards," *Signal Processing*, Vol. 87, 2007, pp. 569-583.

[3] A. Alkholidi, "Analyse and Implementation of Compression/Decompression Processors Based on JPEG and JPEG 2000 Standards," Ph.D. Thesis, Université de Bretagne Ocedentale (UBO), Brest, 2007.

[4] A. Alkholidi, A. Cottour, A. Al falou, H. Hamam and G. Keryer, "Real Time Optical 2-D Wavelet Transform Based of the JPEG2000 Standards," *European Physical Journal (EPJ), Applied Physics*, Vol. 44, 2008, pp. 261-272.

[5] A. Alfalou, *et al.*, "Assessing the Performance of a Method of Simultaneous Compression and Encryption of Multiple Images and Its Resistance against Various Attacks," *Optics Express*, Vol. 21, No. 7, 2013, pp. 8025-8043. doi:10.1364/OE.21.008025

[6] A. E. Shortt, T. J. Naughton and B. Javidi, "Compression of Optically Encrypted Digital Holograms Using Artificial Neural Networks," *Journal of Display Technology*, Vol. 2, No. 4, 2006, pp. 401-410.

[7] C.-H. Chuang and Y.-L. Chen, "Steganographic Optical Image Encryption System Based on Reversible Data Hiding and Double Random Phase Encoding," *Optical Engineering*, Vol. 52, No. 2, 2013, Article ID: 028201.

[8] X. Q. Zhou, *et al.*, "Spatial-Frequency-Compression Scheme for Diffuse Tomography with Dataset," *Applied Optics*, Vol. 52, No. 9, 2013, pp. 1779-1792.

[9] H. Wang, S. S. Han and M. Kolobov, "Quantum Limits in Compressed Sensing of Optical Images," *Quantum Electronics and Laser Science Conference*, San Jose, 6 May 2012.

[10] R. C. Gonzalez and R. E. Woods, "Digital Image Compression," Prentice Hall, 2009.

[11] W. Goodman, "Introduction to Fourier Optics," MacGraw-Hill, New York, 1968.

[12] Y. Suzuki, "Spatial Light Modulators for Phase-Only Modulation," *IEEE*, 1999, pp. 1312-1313.

[13] G. Ntogari, *et al.*, "A Numerical Study of Optical Switches and Modulators Based on Ferroelectronic Liquid Crystals," *Journal of Optics A: Pure and Applied Optics*, Vol. 7, 2004, pp. 82-87.

16

Comparison between Analytical Solution and Experimental Setup of a Short Long Ytterbium Doped Fiber Laser

M. R. A. Moghaddam[1*], S. W. Harun[1,2], H. Ahmad[1]
[1]Photonics Research Center, University of Malaya, Kuala Lumpur, Malaysia
[2]Department of Electrical Engineering, University of Malaya, Kuala Lumpur, Malaysia
Email: *mramoghaddam@siswa.um.edu.my

ABSTRACT

In this research, Amplified Spontaneous Emission (ASE) spectrum characteristics for a highly Yb^{3+} doped glass fiber with different pump powers and pump wavelengths are scrutinized. ASE spectral profile and wavelength shift corresponding to different optical fiber lengths are measured. Highly-doped Yb^{3+} fiber lasers in a linear-cavity are both experimentally and analytically investigated. Rate equations are solved using quasi-numerical models. Numerical results are reported for wide range of operating conditions to enable design optimization. The model takes into account the scattering loss and the distributed laser loss power density in strongly pump condition. The effects of various parameters such as pump power, pump wavelength, signal wavelength and fiber length on the output power and laser threshold are studied. Theoretical results are shown to be in good agreement with the experimental data.

Keywords: Ytterbium Doped Fiber; ASE; Rate Equation; Runge-Kutta

1. Introduction

Fiber lasers have many advantages such as high conversion efficiency, immunity from thermal lensing effect, simplicity of optical construction and excellent beam quality. Ytterbium doping is attractive for high-power fiber lasers because of its high efficiency and strong pump absorption. Ytterbium-doped silica fibers exhibit very broad absorption and emission bands, from 800 nm to 1064 nm for absorption and 970 nm to 1200 nm for emission [1,2]. The simplicity of the level structure provides freedom from unwanted processes such as excited state absorption, multi-phonon nonradiative decay, and concentration quenching [3,4].

Much work on Ytterbium Doped Fiber lasers (YDFLs) focuses on increasing the efficiency of the Laser and some authors have studied Fiber lasers theoretically. In our earlier work a laser action was demonstrated with near 90% slope efficiency using a fabricated YDF as a gain medium [5].

In recent years, the output powers from doped fiber lasers and amplifiers have been scaled and one approach for further power scaling these light sources is to increase the concentration to reduce the required fiber length and avoid nonlinear effects.

However, pumping doped silica fiber with high concentrations can result in excess loss at the pump and signal wavelengths owing to photodarkening, which can significantly reduce the overall conversion efficiency and degrade the long-term performance [6].

In this article, we theoretically and experimentally analyze a YDF which is useful for low noise systems and very short application lengths. Rate equations are solved using semi-numerical models [7,8]. This model takes into account the scattering loss and the distributed laser loss power density in strongly pump condition [9]. The effects of various parameters such as pump power, pump wavelength, signal wavelength and fiber length on the output power and laser threshold are studied.

2. Experimental Setup

The experimental setup used for the lasing experiments is illustrated in **Figure 1**. In the work, different lengths (L) of YDF are pumped through a WDM by pigtailed diode lasers.

The YDF used is a heavily-doped Ytterbium silica fiber with a core absorption coefficient (α_0) of 1200 dB/m at 976 nm .This fiber has a core/cladding diameter of 4/125 μm, a numerical aperture of 0.2 and a cut-off wavelength of 1010 nm. The photodarkening effect in

*Corresponding author.

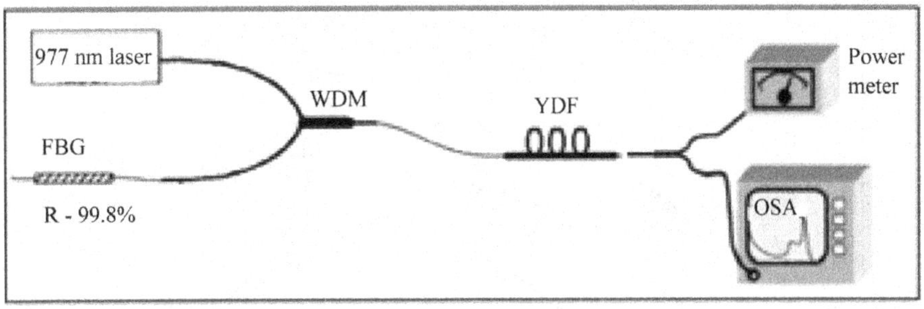

Figure 1. Experimental setup of laser cavity.

this fiber has been reduced by technique of direct nanoparticle deposition [10].

The output characteristics of the laser are measured using a power meter and an optical spectrum analyzer (OSA).

The reflectivity of output coupler is varied from 4% to 95%. For a lossless fiber laser, the output power is almost constant over reflectivity range since the higher reflectivity increases the power inside the cavity and simultaneously decreases a relative output power. For a lossy system, for example in a resonator with an imperfect back reflector or in an unmatched system with different numerical apertures, the higher output reflectivity increases a propagation length of the signal due to more round trips and it results in a higher loss. Therefore, the output power increases with the transmission of the output coupler.

In this work, the resonator cavity is formed between a Fiber Bragg Grating (FBG) and a perpendicularly cleaved YDF end. The experiments are carried out for two different FBGs having a peak reflectivity of about 95% with the peak wavelengths of 1064.94 nm, and 1028.02 nm. We compared the laser output characteristics for these two signal wavelengths. The forward and backward amplified spontaneous emission (ASE) spectrum at different pump wavelengths and power levels are also compared.

2.1. Comparison of ASE Spectrum

The ASE spectra of a forward 975 nm pumping scheme for five different lengths between 10 to 400 cm are shown in **Figure 2**. This figure shows that for a given pump power (200 mW), the wavelength corresponding to peak of ASE varies from 1055.42 to 1074 nm by increasing the YDF length from 10 to 400 cm. In other words there is an overall shifting of the ASE spectrum toward the higher wavelength side with increasing the length of the fiber. Meanwhile, the FWHM of the observed spectra decreases from 48 nm to 33 nm as the YDF length is increased from 10 to 400 cm as shown in **Figure 3**.

In fact, for longer fibers the ASE would be reabsorbed in the fiber as it propagates beyond the point where the pump is attenuated. Since the absorption cross-section is higher on the lower wavelength side, the re-absorption is more prominent in this region.

From residual transmitted pump intensity, it is found that the pump power is completely absorbed even in a short piece of this fiber. Inset figure verifies that the shape of the ASE spectra don't show any considerable change with the pump power.

For a given fiber length, the experimental results show that the ASE (fluorescence) bandwidth can be further increased as the pumping wavelength decreases. From the result obtained for 100 mW, 975 nm forward pumping, an ASE bandwidth of 32 nm is obtained for 4 m long YDF. While under the same condition, the ASE bandwidth is measured to be 55 nm at a shorter pumping wavelength (967 nm). However the ASE bandwidth decreases from 55 nm to 28 nm as the pump power is increased from 100 to 600 mW.

A comparison of the ASE peak position for two different pumping wavelengths in a forward pumping scheme having a piece of 4 m long YDF is shown in **Figure 4**. From the figure, we can deduce that moving toward longer pump wavelength results in shifting ASE peak toward longer wavelength.

As shown in **Figure 5**, the backward ASE does not suffer the ASE re-absorption. Therefore it shows a wider bandwidth in the figure inset.

2.2. Laser Setup

In a lossless fiber laser, the output power does not change with the fiber length. However, in reality, an increase in the fiber length makes a larger absorption and a larger signal loss. **Figure 6** shows the 1064 nm laser output power versus 975 nm launched pump power. The output power is compared in this figure for different fiber lengths between 0.3 to 2 m.

Using a numerical aperture (NA) of 0.2 for YDF used and a NA of 0.14 for FBG in the setup, the maximum output power is measured to be 110 mW for 2 m of YDF at maximum pump power. As also shown in the figure lasing starts at a different threshold power depending on the YDF length.

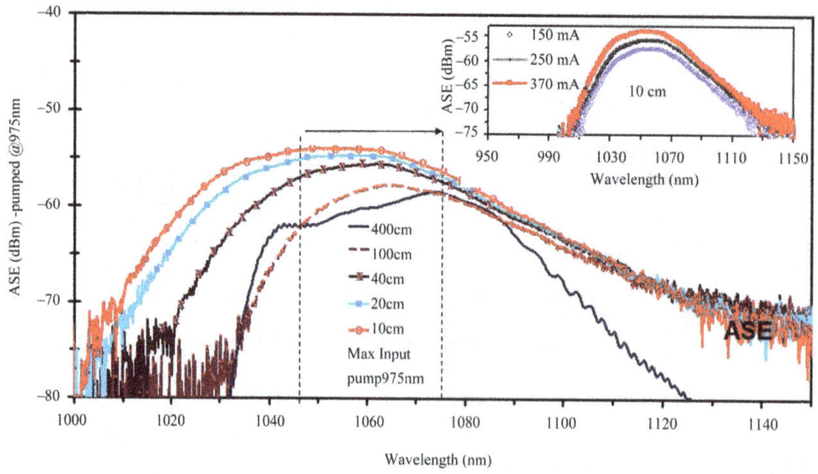

Figure 2. ASE spectrum shift for different fiber lengths and input powers.

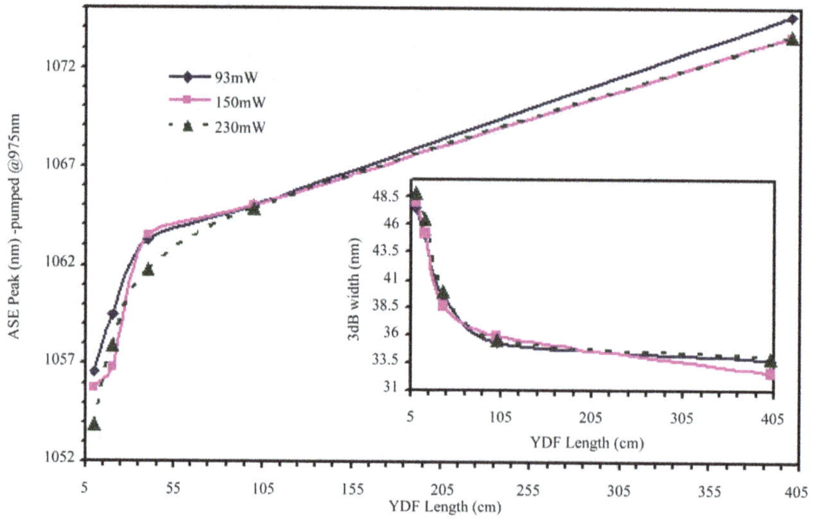

Figure 3. The variation ASE peak and bandwidth under 975 nm forward pumping for different fiber lengths.

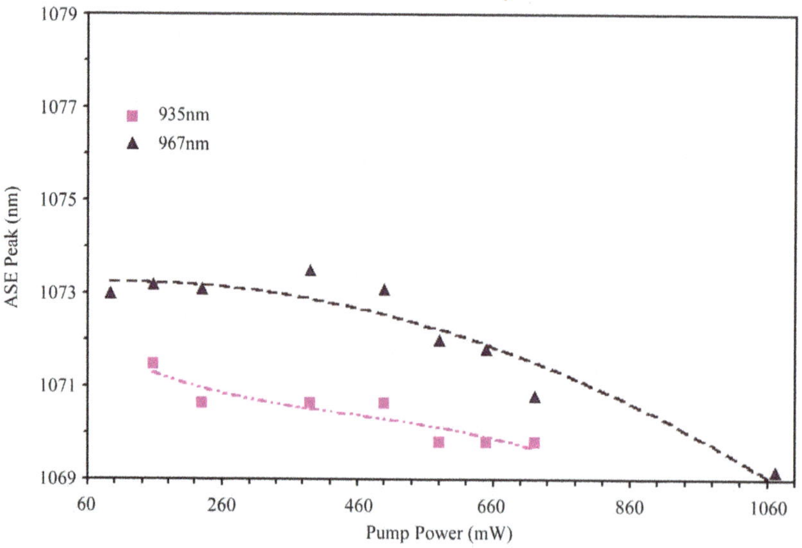

Figure 4. The variation of ASE peak position with pumping wavelength.

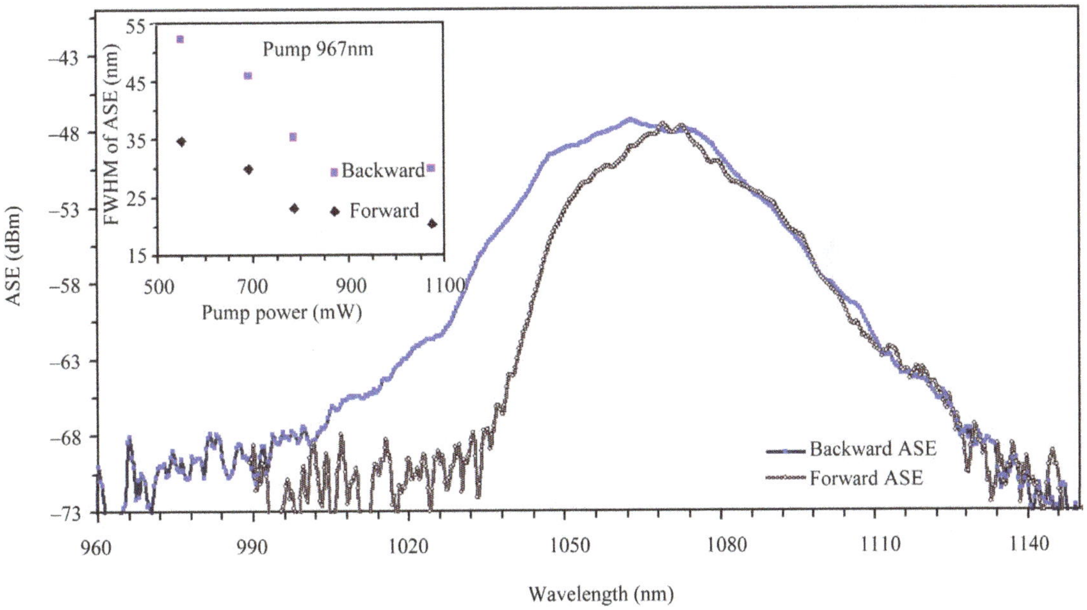

Figure 5. Comparison of ASE in forward and backward pumping schemes.

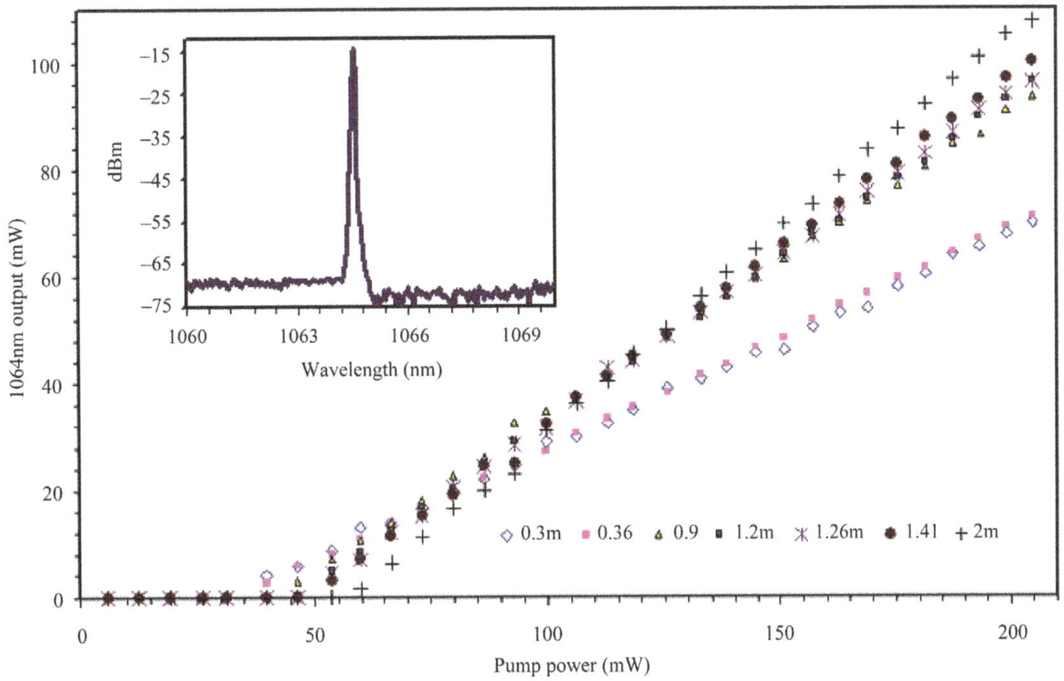

Figure 6. Output power as a function of the 975 nm pump power for 1064 nm wavelength. Inset shows typical laser spectrum.

On the other hand, for shorter operating wavelength the output power is optimized at a shorter YDF length near 30 cm. **Figure 7** shows the the output power of the laser versus pump power when the operating wavelength (λ) is 1028.02 nm. This figure indicates that a long YDF does not have the higher slope efficiency and shows that the laser threshold decreases as the length of YDF decreases. It is also found that the bandwidth of laser line (0.08 nm) remains unchanged with the pump power.

3. Rate Equations

The spectroscopy of the Yb ion is simple as compared to other rare-earth ions. Nonetheless, when Ytterbium is doped in the amorphous silica glass, the emission and absorption spectrum broaden beyond that which would be found in a crystalline structure due to the sub-bands of the two manifolds, $^2F_{5/2}$ and $^2F_{7/2}$ [11,12]. **Figure 8** shows the absorption and emission cross sections for YDF used in this work [13]. The three levels

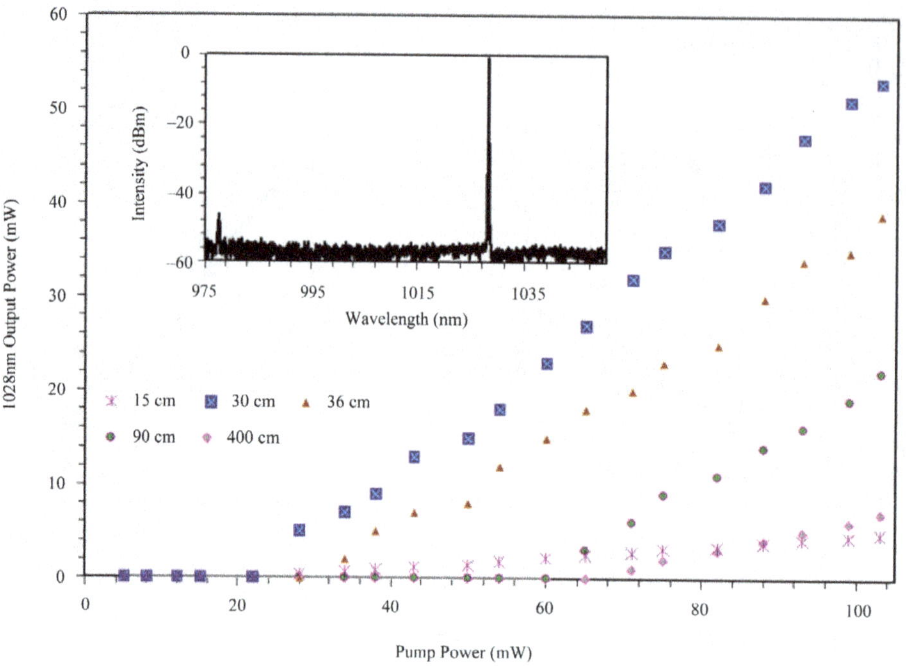

Figure 7. Output power as a function of the pump power for operating at 1028 nm.

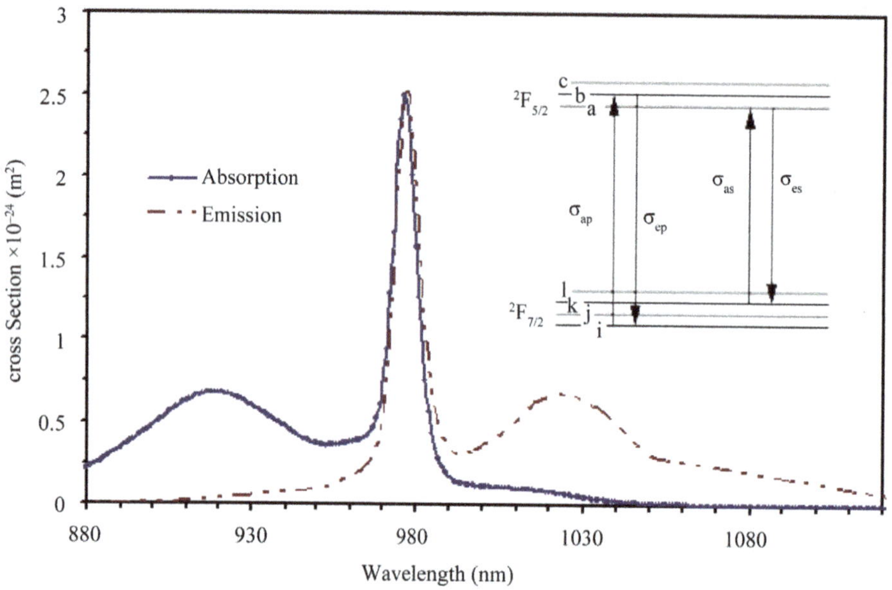

Figure 8. Absorption (solid line) and emission (dashed line) cross-sections used for numerical calculations. Inset-The energy levels structure of Yb^{3+} ions in silica.

in the upper manifold ($^2F_{5/2}$) are denoted by a, b, and c in the inset of the figure.

The analytical solutions of the rate equations in the presence of optical scattering loss and strongly pumped condition has been reported in [14]. The set of coupled equations and boundary conditions to be solved are given as:

$$\frac{N_2(z)}{N}=\frac{\dfrac{\Gamma_p\sigma_{ap}\left[P_p^+(z)+P_p^-(z)\right]}{h\nu_pA}+\dfrac{\Gamma_s\sigma_{as}\left[P_s^+(z)+P_s^-(z)\right]}{h\nu_pA}}{\dfrac{\Gamma_p\left(\sigma_{ap}+\sigma_{ep}\right)\left[P_p^+(z)+P_p^-(z)\right]}{h\nu_pA}+\dfrac{1}{\tau}+\dfrac{\Gamma_s\left(\sigma_{as}+\sigma_{es}\right)\left[P_s^+(z)+P_s^-(z)\right]}{h\nu_pA}} \quad (1)$$

$$\pm\frac{dP_p^{\pm}(z)}{dz} = -\Gamma_p\left[\sigma_{ap}N-\left(\sigma_{ap}+\sigma_{ep}\right)N_2(z)\right]P_p^{\pm}(z)-\alpha_p P_p^{\pm}(z) \quad (2)$$

$$\pm\frac{dP_s^{\pm}(z)}{dz} = \Gamma_s\left[\left(\sigma_{as}+\sigma_{es}\right)N_2(z)-\sigma_{as}N\right]P_s^{\pm}(z)-\alpha_s P_s^{\pm}(z) \quad (3)$$

where τ ascertains the spontaneous lifetime, and $\alpha_P(\alpha_S)$ denotes scattering loss coefficient of pump (oscillating) light which is independent of z. Moreover $\sigma_{ap}(\sigma_{as})$ and $\sigma_{ep}(\sigma_{es})$ are absorption and emission cross sections of the pump (laser) light, respectively (see **Figure 8**).

Equation (1) describes the distribution of the upper level population N_2 at position z along the YDF, while the evolution of the co propagating $\left(P_P^{+}(z)\right)$ and counter propagating $\left(P_P^{-}(z)\right)$ pump powers and laser powers (are given by Equations (2) and (3). It is worth to point out that the concentration density of Yb^{3+} ions $N = N_1(z) + N_2(z)$ are set to a constant value and the strongly pumped condition is used based on $N_2(z) \ll N$ and $\sigma_{as} \ll \sigma_{es}$.

In the above equations, h, c, v_p and v_S denote Planck's constant, the speed of light in vacuum, the pump and oscillation frequencies respectively. Some of the important parameters used in the numerical calculations and the procedure to obtain these parameters have been reported in [15]. In the modeling the active fiber with a length of L and a core area of $A = 1.28 \times 10^{-11}$ m^2 has an average constant concentration per unit volume (N) of 1.02×10^{26} ions/m^3. The excited state lifetime, the background loss at 1000 nm, Γ_P and Γ_S (power filling factors) are also set at 0.88 ms, 40.99 dB/km, 0.7 and 0.9 respectively. Where Γ_S represents the contribution of laser power in the core and Γ_P illustrates the fraction of the pump power actually coupled to the core.

According to the **Figure 9**, the pump power is injected at $z = L$ or 0 at a known level and the laser beam is generated and propagated along the fiber. At the ends of the YDF, a portion of the power is ultimately reflected. Hence the rate equations are to be solved subject to the boundary conditions [7]:

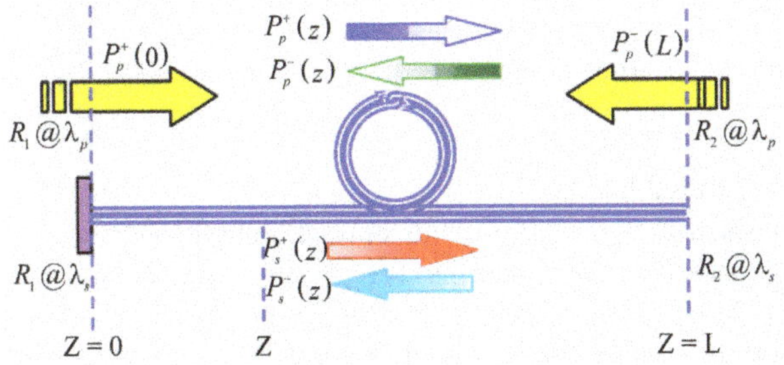

Figure 9. Schematic illustration of end-pumped YDFL.

$$P_s^{+}(0) = R_1 P_s^{-}(0) \quad (4)$$

$$P_s^{-}(L) = R_2 P_s^{+}(L) \quad (5)$$

where $R_2(\lambda)$ and $R_1(\lambda)$ are the reflectivities of λ and $P_S^{\pm}(L)$ and $P_S^{\pm}(0)$ are the oscillating powers at $z=L$ and 0 respectively. In numerical methods, the initial values of $P_S^{\pm}(0)$ or $P_S^{\pm}(L)$ would be determined using shooting methods [8].

In this work, Equations (1)-(3) are solved without neglecting the scattering loss and approximating in the distributed loss laser power density [7]. The calculations are run numerically in fourth-order Runge-Kutta method by using Matlab.

4. Simulation Results and Discussion

Figure 10(a) compares the numerical and experimental results of threshold power as a function of the fiber length for operating wavelengths of 1028 nm and 1064 nm. The numerical and experimental output powers as a function of the fiber length are also compared for operating wavelengths of 1028 nm and 1064 nm as shown in **Figure 10(b)**. In the figures, forward pumping power is fixed at 93 mW.

As demonstrated in **Figure 11** for a 2 m long YDF and pumping at 974 nm, the minimum of the threshold power is obtained at operating wavelength of 1080 nm.

For a given length of the resonator (e.g. 2 m long YDF), by choice of an appropriate pumping and operating wavelength, the threshold power can be reduced by more than six times as shown in **Figure 12**.

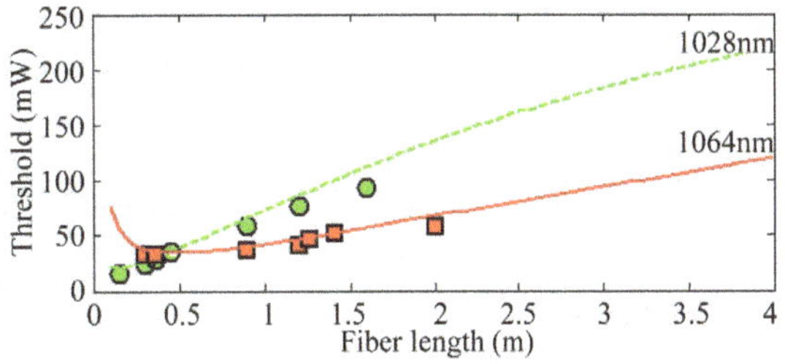

(a)

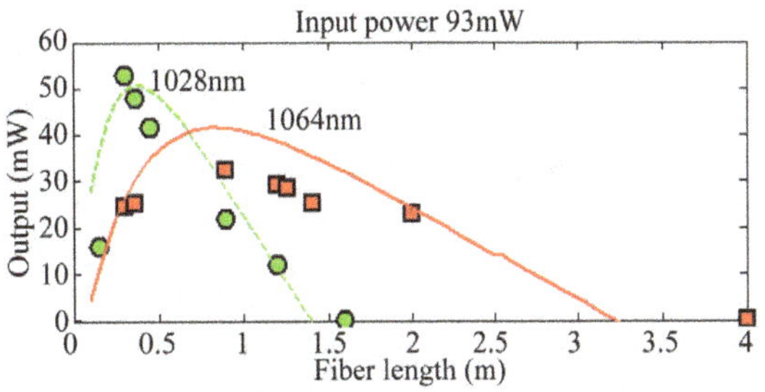

(b)

Figure 10. (a) Threshold power and (b) output power against fiber length at 93 mW of input pump power.

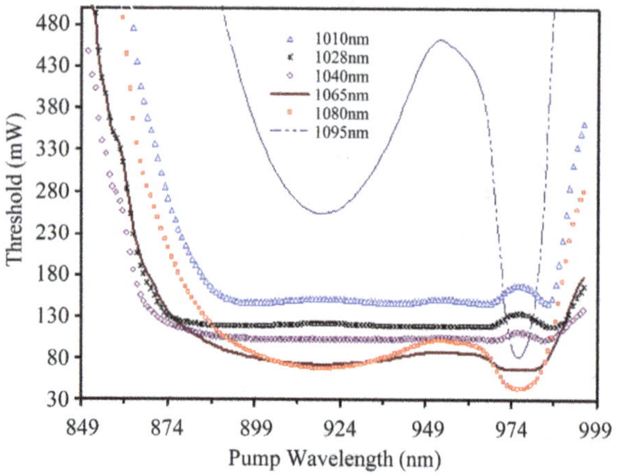

Figure 11. Laser threshold as a function of pumping wavelength for various operating wavelengths. The YDF length is fixed at 2 m.

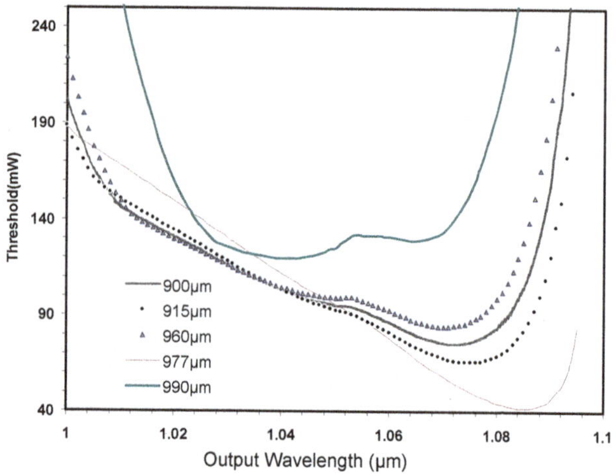

Figure 12. Laser threshold for 2 m long YDF as a function of operating wavelength.

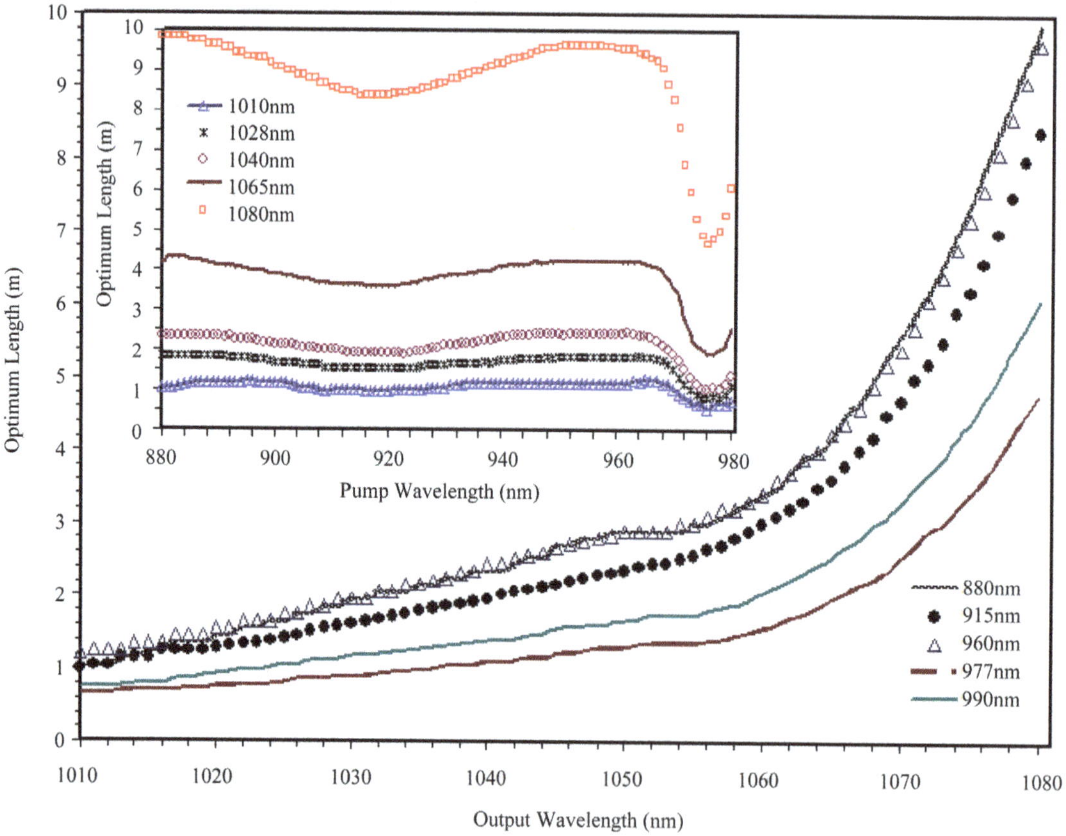

Figure 13. Optimum length as a function of operating wavelength for a fixed input power of 200 mW. Inset shows the dependency of optimum length on the pumping wavelength.

Figure 13 shows the optimum length of proposed YDF laser as a function of operating wavelength when the forward Pump power is fixed at 200 mW. It can be observed from the figure that, the optimum fiber length increases with increasing the pump power. It is clear that the efficiency declines when the YDF is longer or shorter than presented lengths.

As shown in the figure the optimal length shortens when operating wavelengths are shorter. As demonstrated in the figure inset, for pumping near 915 nm or 975 nm, the optimal length is further shortened while, for pumping near 960 nm, longer lengths of the fiber are required.

5. Conclusion

In this research, the ASE spectrum characteristics for a highly Yb^{3+} doped fiber at different pump power levels and pump wavelengths were analyzed. Output power, laser threshold and optimum length were also measured and calculated under wide range of operating conditions. It was shown that, the measured ASE bandwidth (FW-HM) would be approximately 32 nm under 975 nm pumping scheme while it can be reached up to 55 nm under 967 nm forward pumping scheme. Then, we proved that, by setting pumping and operating wavelength to certain proper values, one can reduce the threshold power by more than six times. The threshold power is minimum near pumping wavelength of 974 nm and at operating wavelength of 1080 nm. Finally the optimal YDF length shortens when operating wavelengths are shorter.

REFERENCES

[1] V. A. Akulov, D. M. Afanasiev, S. A. Babin, D. V. Churkin, S. I. Kablukov, M. A. Rybakov and A. A. Vlasov, "Frequency Tuning and Doubling in Yb-Doped Fiber Lasers," *Laser Physics*, Vol. 17, No. 2, 2007, pp. 124-129. doi:10.1134/S1054660X07020120

[2] S. Kablukov, E. Dontsova, V. Akulov, A. Vlasov and S. Babin, "Frequency Doubling of Yb-Doped Fiber Laser to 515 nm," *Laser Physics*, Vol. 20, No. 2. 2010, pp. 360-364. doi:10.1134/S1054660X10040043

[3] A. S. Kurkov, "Oscillation Spectral Range of Yb-Doped Fiber Lasers," *Laser Physics Letters*, Vol. 4, No. 2. 2007, pp. 93-102. doi:10.1002/lapl.200610094

[4] M. R. A. Moghaddam, S. Harun, S. Shahi, K. Lim and H. Ahmad, "Comparisons of Multi-Wavelength Oscillations Using Sagnac Loop Mirror and Mach-Zehnder Interferometer for Ytterbium Doped Fiber Lasers," *Laser Physics*, Vol. 20, No. 2, 2010, pp. 516-512. doi:10.1134/S1054660X10030138

[5] S. W. Harun, M. C. Paul, M. R. A. Moghaddam, S. Das, R. Sen, A. Dhar, M. Pal, S. K. Bhadra and H. Ahmad, "Diode-Pumped 1028 nm Ytterbium-Doped Fiber Laser with near 90% Slope Efficiency," *Laser Physics*, Vol. 20, No. 3. 2010, pp. 656-660. doi:10.1134/S1054660X10050051

[6] A. D. Guzman Chavez, A. V. Kir'yanov, Y. O. Barmenkov and N. N. Il'ichev, "Reversible Photo Darkening and Resonant Photobleaching of Ytterbium Doped Silica Fiber at In-Core 977 nm and 543 nm Irradiation," *Laser Physics Letters*, Vol. 4, No. 10. 2007, pp. 734-739. doi:10.1002/lapl.200710053

[7] L. Xiao, P. Yan, M. Gong, W. Wei and P. Ou, "An Approximate Analytic Solution of Strongly Pumped Yb-Doped Double-Clad Fiber Lasers without Neglecting the Scattering Loss," *Optics Communications*, Vol. 230, No. 4-6, 2004, pp. 401-410. doi:10.1016/j.optcom.2003.11.017

[8] Z. Lali-Dastjerdi, F. Kroushavi and M. H. Rahmani, "An Efficient Shooting Method for Fiber Amplifiers and Lasers," *Optics & Laser Technology*, Vol. 40, No. 8. 2008, pp. 1041-1046. doi:10.1016/j.optlastec.2008.02.006

[9] D. Xue, Q. Lou and J. Zhou, "Comparison of Yb-Doped Fiber Laser with One-End and Double-End Pumping Configuration," *Optics & Laser Technology*, Vol. 39, No. 4, 2007, pp. 871-874. doi:10.1016/j.optlastec.2005.12.005

[10] J. Koponen, M. Söderlund, H. J. Hoffman, D. A. V. Kliner, J. P. Koplow and M. Hotoleanu, "Photodarkening Rate in Yb-Doped Silica Fibers," *Applied Optics*, Vol. 47, No. 9. 2008, pp. 1247-1256. doi:10.1364/AO.47.001247

[11] G. Demirkhanyan, "Intensities of Inter-Stark Transitions in YAG-Yb^{3+} Crystals," *Laser Physics*, Vol. 16, No. 7. 2006, pp. 1054-1057. doi:10.1134/S1054660X0607005X

[12] M. Javadi-Dashcasan, F. Hajiesmaeilbaigi, H. Razzaghi, M. Mahdizadeh and M. Moghadam, "Optimizing the Yb: YAG Thin Disc Laser Design Parameters," *Optics Communications*, Vol. 281, No. 18. 2008, pp. 4753-4757. doi:10.1016/j.optcom.2008.05.055

[13] T. W. Huang and W. P. Lin, "Braodband Tunable 1060 nm Fiber Ring Laser Band on Hybrid Amplifier with 45 nm Tuning Range," *IEEE Conference on Cross Strait Quad-Regional Radio Science and Wireless Technology*, Harbin, 26-30 July 2011, pp. 270-273.

[14] I. Kelson and A. Hardy, "Optimization of Strongly Pumped Fiber Lasers," *Journal of Lightwave Technology*, Vol. 17, No. 5, 1999, pp. 891-897. doi:10.1109/50.762908

[15] P. Nandi and G. Jose, "Ytterbium-Doped P_2O_5-TeO_2 Glass for Laser Applications," *IEEE Journal of Quantum Electronics*, Vol. 42, No. 11. 2006, pp. 115-1121. doi:10.1109/JQE.2006.882557

17

Effect of Experimental Parameters on the Fabrication of Gold Nanoparticles via Laser Ablation

Hisham Imam[1], Khaled Elsayed[2], Mohamed A. Ahmed[2], Rania Ramdan[2]
[1]National Institute of Laser Enhanced Sciences, Cairo University, Giza, Egypt
[2]Materials Science Laboratory (1), Physics Department, Faculty of Science, Cairo University, Giza, Egypt
Email: hishamimam@niles.edu.eg

ABSTRACT

In this study we report the effect of laser parameters such as laser energy, laser wavelength as well as focusing condition of laser beam on the size and morphology of the gold nanoparticles (GNPs) prepared in deionised water by pulsed laser ablation. The optimum conditions at which gold nanoparticles obtained with controllable average size have been reported as these parameters affected the size, distribution and absorbance spectrum. Effect of energy was studied. The laser energy was divided into three regions (low, middle and high). A noteworthy change was observed at each region, as the average size changed from 5.9 nm at low energy to 14.4 nm at high energy and the gold nanoparticles reached a critical size of 8 nm at 100 mJ. The Effect of the wavelength on the particle size was examined at 1064 nm, 532 nm. It was found that, the optimum ablation laser wavelength was 1064 nm. Finally, significant results obtained when the effect of focusing conditions studied.

Keywords: Laser Ablation; Goldnanparticle

1. Introduction

In the last few years and due to unique physicochemical characteristics of gold nanoparticles and their wide usages in different fields, the number of publications on the preparation and characterization of gold nanoparticles has extensively increased [1,2]. For example, the recently recognized behavior of gold to act as a soft Lewis acid and large surface-to-volume ratio of nanogold as well as its inert property have widely enhanced the application of gold nanoparticles as a catalyst in the field of organic synthesis [1]. The ability of gold to produce heat after absorbing light provides a medicinal usage named as photothermal therapy [3]. All mentioned usages together with application of nanogold in gene and drug delivery increased studies on development of methods for gold nanoparticles production [4,5]. Although preparation of nanogold by physical procedures (such as laser ablation) provides gold nanoparticles with narrow range of particle size, it needs expensive equipments and has low yield [6]. Hazardous effects of organic solvents, reducing agents and toxic reagents applied for synthesis of gold nanoparticles on environment, encouraged researchers to developeco-friendly methods for preparation of gold nanoparticles [7,8].

Also due to their size-dependent physical and chemical properties [9-11], gold nanoparticles have optical, electrical, magnetic and mechanical properties which make them suitable for many applications such as drug and gene delivery [12,13], the ability to generate table immobilization of biomolecules [14] and in several targeting application [15]. There are several methods for preparing metal nanoparticles such as pulsed laser deposition [16], flame metal combustion [17], chemical reduction [18], photo-reduction [19], electrochemical reduction [20], solvothermal [21], electrolysis [22], green method [23], microwave induced [24], sonoelectrochemical [25], aerosol flow reactor [26], photochemical reduction [27], chemical fluid deposition [28], spray pyrolysis [29,30], and spark discharge [31]. Among them, the pulsed laser ablation in liquids PLAL has become an increasingly popular top-down approach [32] for producing nanoparticles. It's a relatively new method that was first introduced by Fojtik *et al.* in 1993 [33] as is a promising technique for the controlled fabrication of nanomaterials via rapid reactive quenching of ablated species at the interface between the plasma and liquid with high-quality nanoparticles free from chemical reagents. Therefore, PLAL process has received much attention as a novel NPs production technique. The most advanced method for producing nanostructured materials has been developed performing pulsed laser ablation of gold plate in liquids [34-38]. Advantages of this method include the relative simplicity of the procedure and the absence of chemical reagents in the final preparation, but the size distributions of the

GNPs prepared by this technique tend to be broadened due to the agglomeration and ejection of large fragments during laser ablation. To achieve the particlesize reduction, different surfactants can be used [39-41]. Another important advantages gold nanoparticles prepared by PLAL process were stable for a period of months. The main aim of the present study is to prepare pure gold nanoparticles in easy, fast and one step method via PLAL process in deionised water at various laser fluencies and wavelengths. The effect of the focusing conditions on the formation and fabrication of metal nanoparticles has not been approached by many researchers before. In this paper we discuss the focusing conditions that play an important role in the formation of small and narrowly dispersed gold nanoparticles and elucidate the mechanisms of the nanoparticles growth.

2. Experimental Setup

The formation of gold nanoparticles is fabricated via pulsed laser ablation of the corresponding gold metal plate (99.99%) of 3 mm thickness and its dimensions 1.5 × 1 cm^2 immersed into the liquid as shown in **Figure 1**. The gold plate was thoroughly washed with ethanol and deionised water to remove organic contamination and placed at the bottom of a glass vessel filled with 50 ml of an aqueous solution of deionized water. The gold metal plate was kept at 15 mm below the liquid surface. Pulsed Q-Switched laser Nd:YAG (Surelite II, Continuum) was used. The laser beam was operated at fundamental wavelength 1064 nm or second harmonic 532 nm and 10 Hz. The laser beam was focused by a plano-convex lens with focal length of 5 cm. The energy of the laser pulse was measured by power meter (OPHIR-NOVA). The ablation process was typically done for 5 minutes at room temperature. The laser beam was focused on the surface of the target and it was scanned over by using X-Y stage to avoid the craters on the surface of target. Morphology and size distribution of nanoparticles were characterized

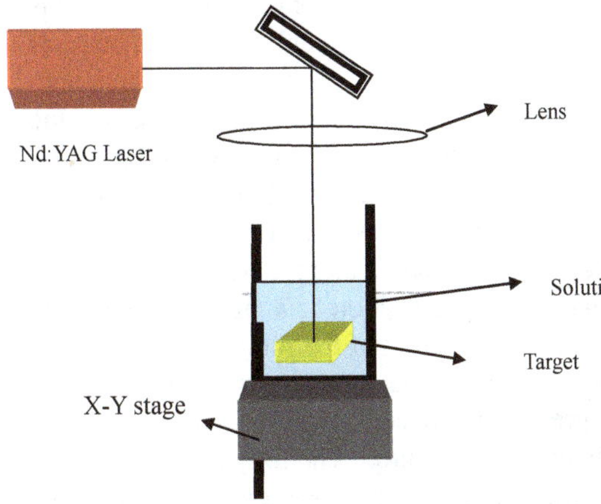

Figure 1. Schematic diagram of experimental setup.

by HTEM (JEOL 1200). The solution containing GNPs was inserted in the bath of ultrasonicator for half of an hour and a drop of solution containing gold nanoparticles onto carbon coated copper grids and kept to completely dry at room temperature. The diameter of more than 100 nanoparticles in sight on a given micro graph was directly measured and the distribution of the particle diameters was obtained. The optical absorption spectrum of the solution was measured with UV-Vis spectrophotometer (PG instrument Ltd., T80+).

3. Results and Discussion

3.1. Effect of Energy of Laser Beam

Laser energy was adjusted by inserting different glass sheets in the path of the laser beam before it reached the focus of the lens. In this section, the effect of laser fluence was studied on the obtained gold nanoparticles, their distribution and their optical spectrum. The energy of laser ranged between (10 - 250 mJ), which can be divided into three regions, low energy (10 mJ), middle region (20 - 100 mJ) and high energy (150 - 250 mJ). And there are three different laser ablation mechanisms in these three regions.

3.1.1. At Low Energy of Laser Beam (10 mJ)

Under the action of the laser at low energy the target is heated, but due to the strong confinement of the liquid at the surface, the vaporization rate is strongly restricted and no plume forms. In the absence of a vapor plume, the hot target is remained in contact with water promoting the oxidation of the nanoparticle oxides [42]. The reaction is initiated with the oxidation of the molten target surface by oxygen splitting of water molecules at the hot target so the hydroxide nanoparticles exist on the target surface. The gold hydroxide material on the surface then desorbs from the hot target surface and diffuses into the water and it isn't aggregated due to their negative charge so the average size of produced gold nanoparticles is small and they are produced by thermal vaporization as shown in **Figure 2**.

The average size is 5.9 nm and as it's shown the particles are small and have spherical shape and the yield is small and this is due to the negative charge on gold nanoparticles which prevents them from aggregation and their yield is small due to presence of water at the surface of the target which restricts their growing as explained above.

3.1.2. Intermediate Energy of Laser Beam (Plume Mixing Zone) (20 - 100 mJ)

In this stage the plume develops more slowly and is limited to a size much smaller than in a gas atmosphere [43]. The large pressure in the confined vapor plume

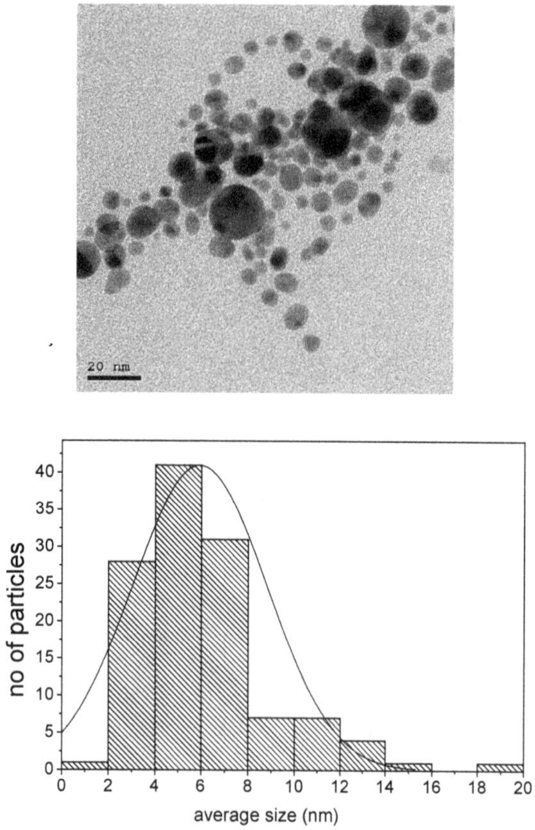

Figure 2. The TEM micrograph and histogram of gold nanoparticles prepared in 10 mJ.

results in an expansion beyond the equilibrium point, the internal plume pressure equals the hydrostatic pressure of the liquid, increasing the difference between the pressure inside and outside the plume decrease the expansion until it is stopped, the hydrostatic pressure then collapse, the plume back into the target. In this region, the nanoparticles fall into two distinct size distributions those are attributed to:

1) Target surface vaporization;

2) Explosive ejection of molten droplets directly from the target. And this leads to abroad size distribution as shown in **Figures 3(a)-(e)**.

The surface vaporization was discussed above, the explosive ejection occurs when the temperature approaches the thermodynamics critical temperature, thermal fluctuation is amplified and the rate of homogenous bubble nucleation rises dramatically and the target makes rapid transition from superheated liquid to a mixture of vapor and equilibrium liquid droplet. At this fluence, the momentum of a plume allows it to expand further out into the liquid, increasing the plasma life time and this results in an increase of screening of laser light from surface of bulk Gold target [44,45].

The data in the histograms show that, the average size of gold nanoparticles prepared in pure water at 20, 30, 40, 50, 100 mJ are 14.3, 13.2, 12.9, 12.2 and 8.08 respectively. As it is shown the average size decreases with the increasing of laser energy and from TEM micrographs it is found that, as laser energy increases, the particles become homogenous and have spherical shape and also the concentration increases until it reaches to their critical size at 100 mJ, and then the particles appear as diffused particles. The decrease in the average size of gold nanoparticles can be attributed to large energy which excited the gold nanoparticles in a solution, the photon energy is readily converted to the internal modes of the nanoparticles as during a single laser pulse, one gold nanoparticle is considered to absorb consecutively more than one thousand photon and it's temperature rises significantly so that the nanoparticle starts to fragment. After the single laser pulse the diffused into the solution and the temperature of nanoparticles return to room temperature before the next one arrive. The heating and cooling of nanoparticles occur in every laser pulse.

3.1.3. At High Energy of Laser Beam (Plasma Etching) (150 - 250 mJ)

At the high energy, laser energy is absorbed in the liquid to the target resulting material removal by reactive sputtering rather than direct laser ablation, as the intensity of laser in the presence of ablated material in the water, as this happen the amount of the light reaching to the target goes to zero, plasma formation in the water creates a cavitations bubble that expand and then collapse, driving highly energetic species into target [46]. In this region the average size of gold nanoparticles begins to increase again as shown in **Figures 4(a)-(c)**.

From the above figure, the average size of gold nanoparticles prepared in pure water at 150, 200 and 250 mJ are 11, 12.8 and 14.4 nm respectively it is observed that when laser energy exceed certain value (above 100 mJ) the average size and concentration begin to increase and the shape of prepared GNPs begins to take the spherical shape again after their shape was diffused when it prepared at 100 mJ this can be explained as when the GNPs reach their critical size, small fragments such as gold atoms and small aggregates resulting from photo fragmentation are dispersed in solution, therefore the nanoparticles present in the solution grow by attracting these small fragments. The fragmentation rate must increase with increasing the laser energy because the internal energy of irradiated nanoparticles increases. On the other hand, the aggregation rate increases with the increase of the concentration of the small fragments, after the laser is off, the fragmentation is terminated so that only the aggregation proceeds until the gold atoms and small fragments are consumed. The fragmentation rate of each nanoparticle decreases because it's absorption coefficient per atom decreases with it's diameter. In contrast

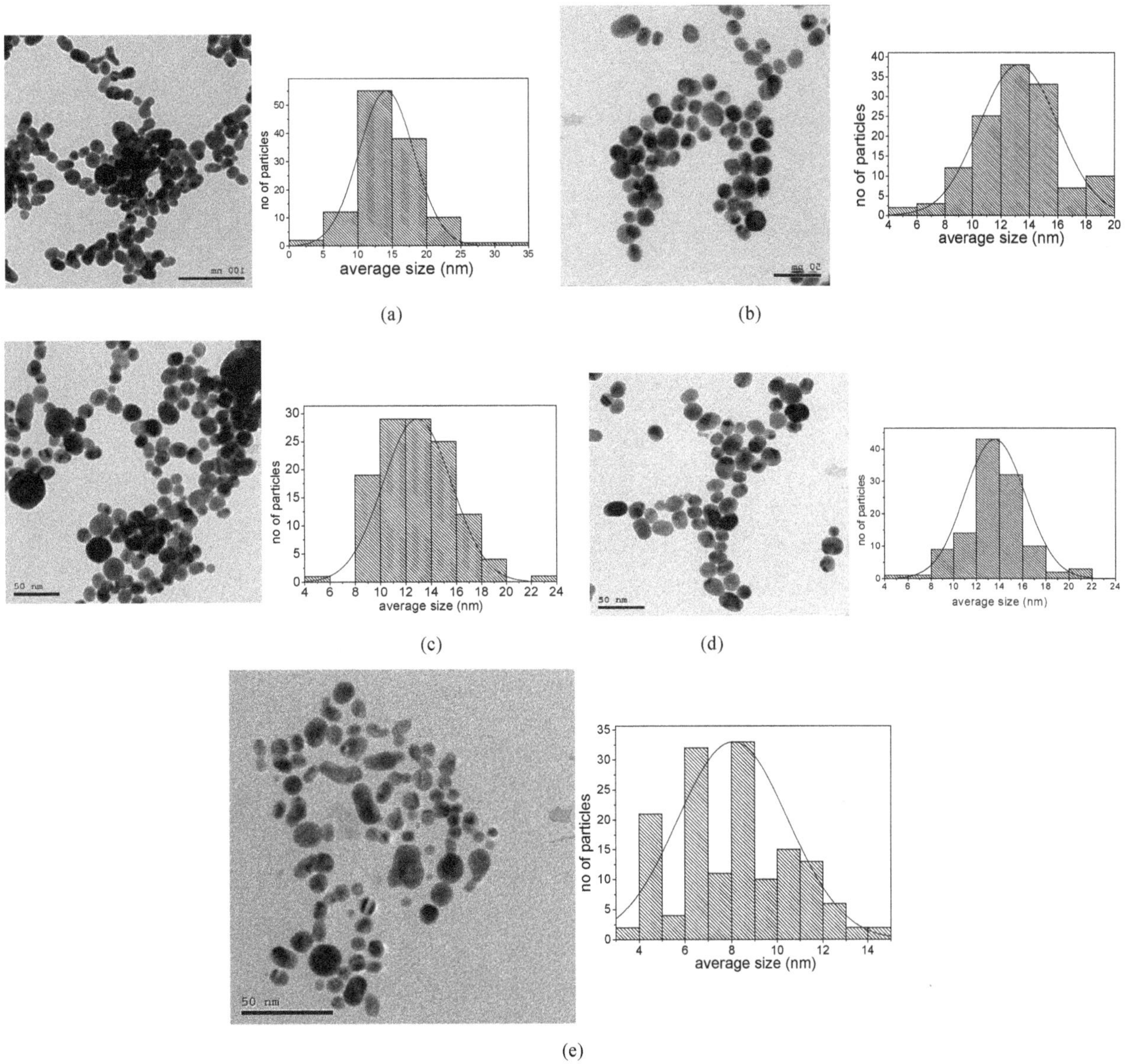

Figures 3. (a)-(e) TEM micrographs and their histograms of gold nanoparticles prepared in 20, 30, 40, 50, 100 mJ respectively.

the aggregation rate increases because the concentration of the small fragments increases, therefore, the minimum diameter is realized when the rate of fragmentation is equal to that of the aggregation [46]. From the above discussion, the average size decreases as the energy increases until it reach to the critical size then it begins to increases again as shown in the **Table 1** and **Figure 5**. The absorbance spectrum is studied which give indication on the concentration and size distribution of the gold nanoparticles prepared and this will ensure the results obtained from TEM as shown in **Figure 6**.

From the **Figure 6**, as the laser energy increases, the height of the Plasmon peak increases which indicate to the increase in the concentration of nanoparticles, till the nanoparticles reach it's critical size and the absorption coefficient of gold nanoparticles decreases and they couldn't absorb any more as they couldn't be fragment again, red shift in the Plasmon peak ensure the decrease in the average size of gold nanoparticles. As the laser energy increases, the Plasmon peak becomes narrower indicating that the particles have a homogenous distribution as the laser energy increases. The absorbance spectrum obtained at high laser energy is shown in **Figure 7**.

At high laser energy (150 - 250 mJ) as shown above, the absorbance shows an increase in the Plasmon peak with increasing the laser energy, indicating to the increase in the concentration of gold nanoparticles as the energy increase the peak becomes more narrow and there is a blue shift in Plasmon peak as the energy increase as the size of the gold nanoparticles increase. These results are matching as obtained from TEM.

So, the operating above or below the threshold will

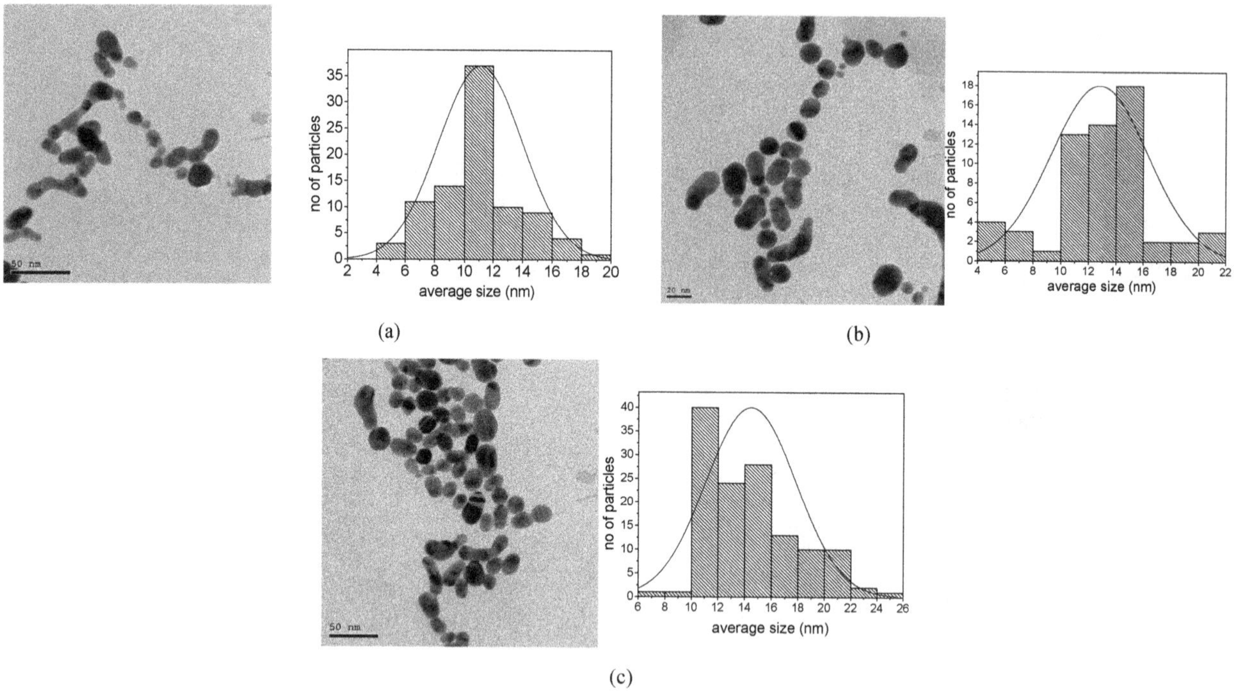

Figures 4. (a)-(c) TEM micrographs and histogram of gold nanoparticles prepared in pure water at 150, 200 and 250 mJ respectively.

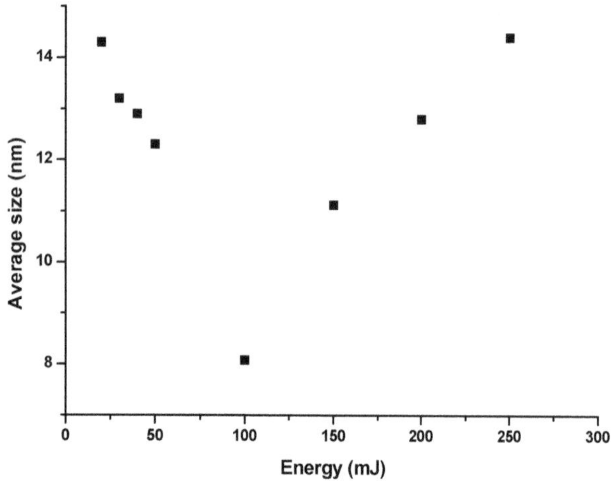

Figure 5. The relation between the energy and the average size.

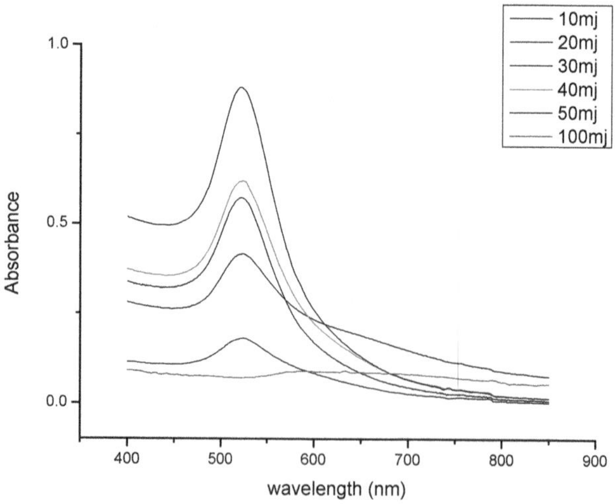

Figure 6. The absorbance spectrum of gold nanoparticles prepared in pure water at (10 - 100 mJ).

therefore allows the experimenter to choose between a low quality of small, narrowly distributed nanoparticles and much larger quantity but with a wider size distribution.

3.2. Effect of Wavelength

Effect of laser wavelength was examined at 1064 nm, 532 nm, the gold nanoparticles was fabricated using two laser beam at 1064 nm and 532 nm. From TEM as shown in **Figure 8** the particles distributions of 1064 are more homogenous and the average size of GNPs prepared at 532 nm is smaller than that prepared at 1064 nm.

The average size of GNPs prepared at 532 nm is 13.7 nm and 14.5 nm at 1064 mm. This can be contributed to two factors: The value of the absorption coefficient of bulk gold is higher at 532 nm, so the intensity of absorbance peak of 532 nm higher than 1064 nm. The photon energy of 532 nm is higher than that at 1064 nm, which lead to the fragmentation of GNPs prepared at 532 nm, the average size of GNPs prepared at 532 nm is smaller than that prepared at 1064 nm. in terms of the amount of ablated Au as determined by the area of surface Plasmon

Table 1. The relation between energy and average size of obtained gold nanoparticles.

Energy (mJ)	Average size(nm)
20	14.3
30	13.2
40	12.9
50	12.3
100	8.08
150	11.12
200	12.8
250	14.4

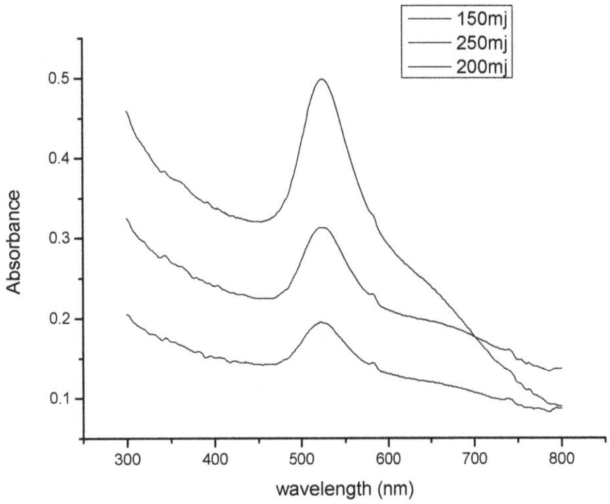

Figure 7. The absorbance spectrum of gold nanoparticles prepared in pure water at high laser fluence.

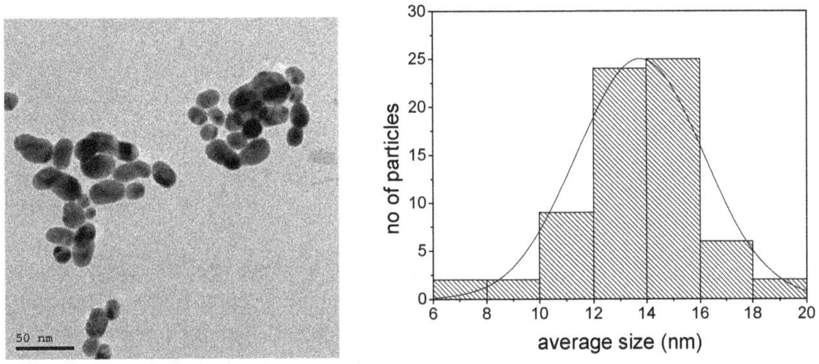

(a)

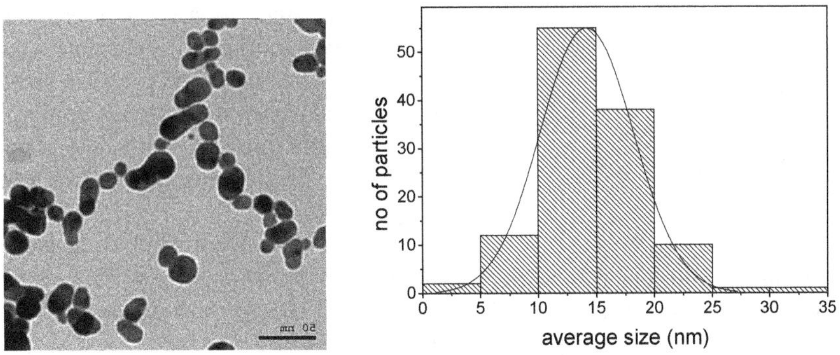

(b)

Figure 8. The TEM and histogram of GNPS prepared at 532 nm, 1064 nm respectively.

peak of the resulting gold nanoparticles [47]. Laser ablation efficiency increased when the laser wavelength increasing. It was found that, the optimum ablation laser wavelength was 1064 nm as shown in **Figure 9**. From the above figure, the intensity of surface Plasmon peak of GNPs prepared at 1064 nm is higher than that prepared at 532 nm, which indicated that the amount of GNPs prepared at 1064 nm higher than that prepared at 532 nm.

3.3. Effect of Geometry of Focusing Conditions

In this section two parameters (focal length and the target position) which affect on the shapes and size distribution of obtained GNPs which prepared in pure water are studied. The target position: there are three positions are studied (above focus, at focus and below focus) for a lens has 5 cm focal length. As shown in **Figure 10**, TEM for gold nanoparticles were prepared above focusing

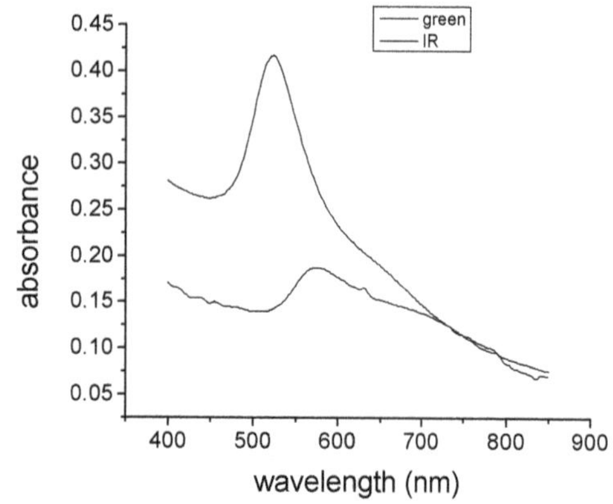

Figure 9. Shows the absorbance spectrum of gold nanoparticles prepared in deionised water at 1064 nm and 532 nm.

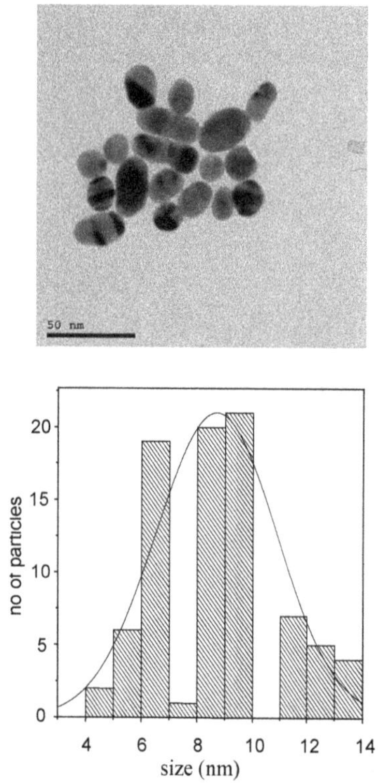

Figure 10. The TEM micrograph and histogram of GNPs prepared in pure water above focusing condition.

conditions.

From **Figure 10**, it's shown that the average size of the obtained particles is 8.6 nm, the yield is small and particles are diffused, this may be due to the low fluence and water breakdown doesn't take place [48], at this fluence the target heated and due to the the strong confinement of the liquid at the surface and the vaporization rate is strongly restricted and no plume forms [49]. **Figure 11** shows GNPs prepared at focusing condition.

As shown from the figure the average particle size is 8.9 nm, which is larger than gold nanoparticles prepared at defocusing condition and the particles have a spherical

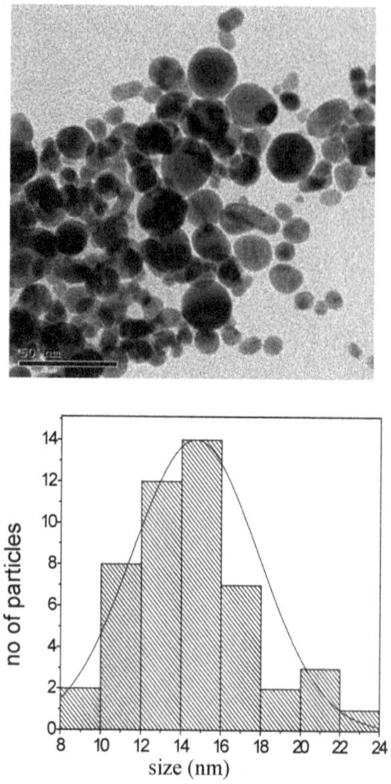

Figure 11. The TEM micrograph and histogram of GNPs prepared in pure water at focusing condition.

shapes and are also agglomerated, this can be explained as, when the ablation takes place at focus the plasma pressure, temperature and density increase more than in the defocusing condition. High temperature of plasma temperature excites, ionizes and dissociates the surrounding water layer, the nucleation starts directly. The high plasma density increases the molecular interaction and particles growth so as a result the particles size increase as observed. When laser is off plasma cooling and recombination take place, pressure and temperature will reduce but particle growth continue until certain value above which it's growth stopped.

When the gold nanoparticles prepared below focusing condition as shown in **Figure 12**, unexpected results obtain as the average size of the obtained particles decreased to 7.2 nm.

In this case the laser irradiation onto the target, most of the energy at the focal region is absorbed by the water, the energy reaches to the target reduces [43], also the liquid increasing the amount of screening of laser light [46] so the energy reached to the target is low and the same results obtained when lens of 10 cm focal length is used as shown in **Table 2**.

Another parameter of the focusing conditions that affect the shape and size distributions of the gold nanoparticles is the lens effect. Since each lens produces a different energy density distribution in space. In each case the sample has been placed at the focus. The minimum

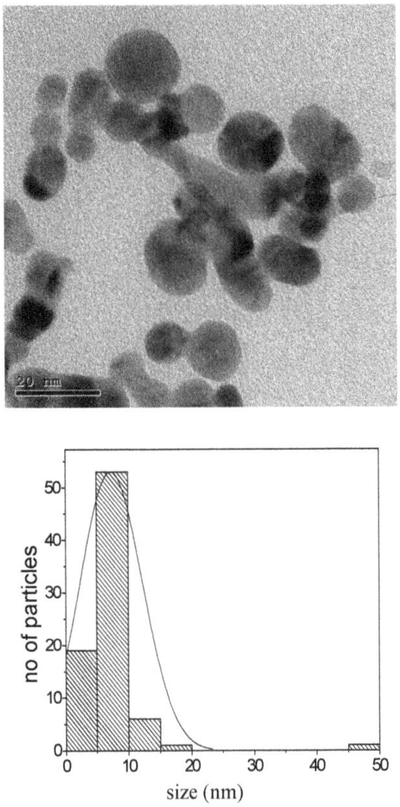

Figure 12. shows the TEM micrograph and histogram for GNPs prepared in pure water below focusing condition.

Table 2. The average size and the particle distribution of GNPs at different focal length and target positions.

	Target position	Average size (nm)	Particle distribution (nm)
f = 5 cm	Above focusing	8	8 - 14
	At focusing	9	15 - 23
	Below focusing	7	9 - 45
f = 10 cm	Above focusing	8	8 - 15
	At focusing	10	10 - 19
	Below focusing	8	6 - 16

diameter of laser beam spot at the focus changes with changing the lens. This in turns changes the laser fluence and hence producing gold nanoparticles with different shapes and sizes. Laser plasma is known to be highly dependent on the laser power density at the surface [40]. **Figures 13(a)-(c)** show TEM and histogram of GNPs prepared by using lenses with different focal length (5, 10, 20 cm), the average size of obtained GNPs are 8.9, 9.8, 17.8 nm respectively.

From the **Figures 14**, **15** and **Table 3**, the laser fluence decrease as the focal length increase as a result the average size increase and the yield of obtained particles also increase. At small focal length (f = 5 cm), very small and hot plasma produced due to high energy density (0.454 MW/m^2) that is localized at the focal point and as a result the average size of produced GNPs is small and have spherical shape and the yield is large and this may be due to the absorption of laser energy in liquid reaches to the target resulting material removing by reactive sputtering [45]. while GNPs prepared at focal length (f = 10 cm), the laser fluence is lower (0.113708 MW/m^2), so the particles obtained are enlarged and the distribution became abroad because the laser pulses producing collides above the sample, the next pulse is partially scattered by these collides particles, so the energy scattered and only too small portion of the energy reached to surface of the target. At focal length (f = 20 cm) the average size of GNPs is larger than that obtained at focal length 5, 10 cm as the laser fluence became smaller (0.028427 MW/m^2) due to the long (tail) along laser beam as the same collides effect and the energy of laser beam is scattered so very small amount of energy is reached to the target resulting in small yield, large average size and broadened distribution.

Figure 16 shows the absorbance spectrum of GNPs prepared at focal length 5, 10, 20 cm, the height of peak decrease with increasing the focal length and that contributed to the small yield obtain as focal length increase also there is a red shift in wavelength with increasing the focal length and that contributed to the increase in the

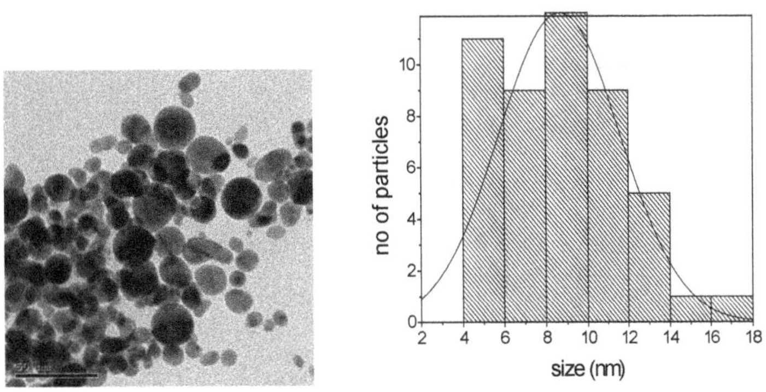

(a)

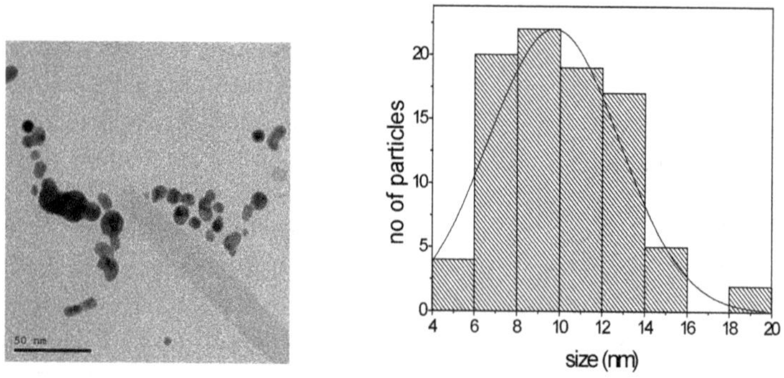

(b)

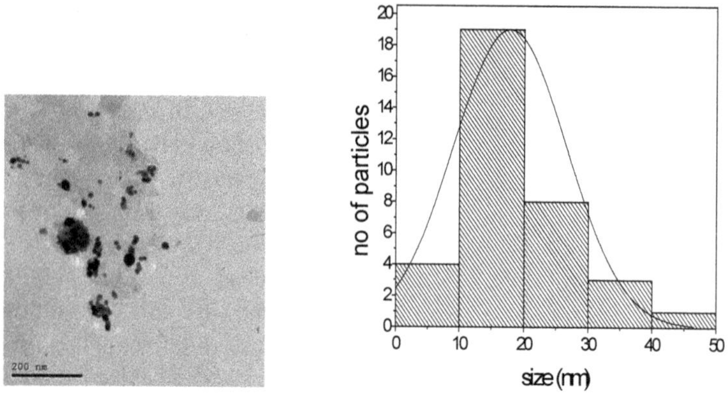

(c)

Figures 13. (a)-(c) TEM micrograph and histogram of GNPs prepared in pure water at focal length 5, 10, 20 cm.

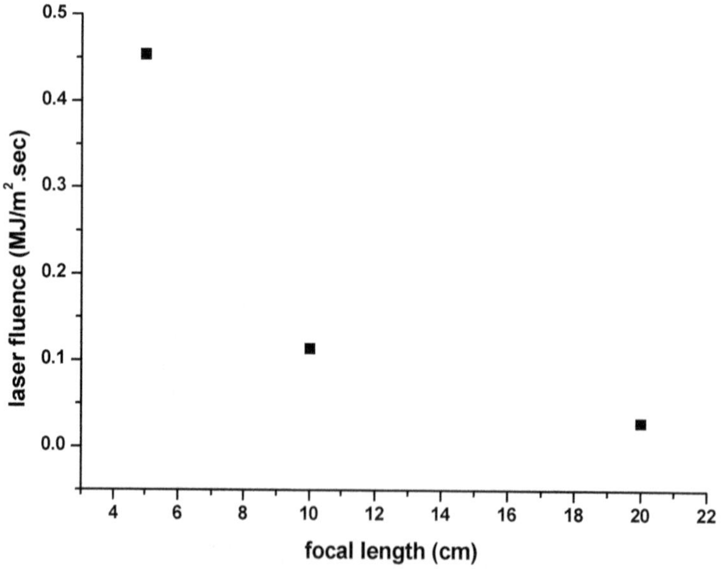

Figure 14. Show the dependence of the focal length on the average size and laser fluence respectively.

Table 3. Show relation between the focal length, laser fluence and the average size.

Average size (nm)	$d_{min} = 2.44\, f\lambda/D$ m	Irradiance (I = P/A) MW/m^2	Focal length (cm)
9	2.16×10^{-6}	0.454	5
10	4.32×10^{-5}	0.113708	10
18	8.6×10^{-5}	0.028427	20

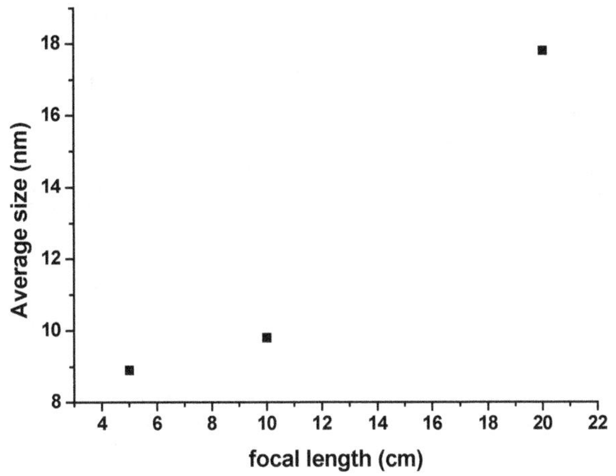

Figure 15. Show the dependence of the focal length on the average size and laser fluence respectively.

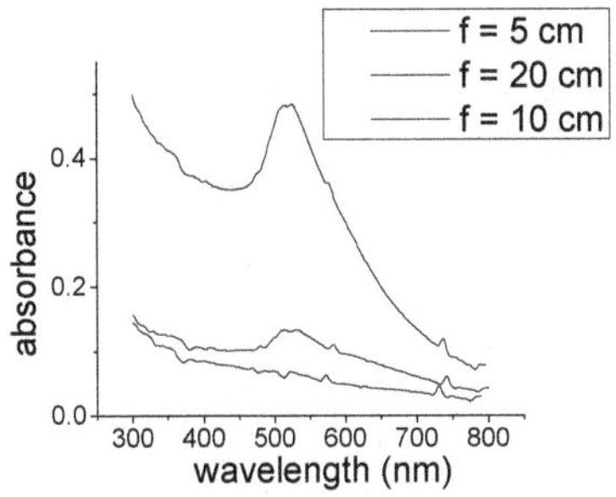

Figure 16. Absorbance spectrum of GNPs prepared at different focal length.

average size of GNPs as focal length increase. That matches with the TEM results.

4. Conclusion

Gold nanoparticles (GNPs) were successively prepared by laser ablation in water. The energy of laser beam affected on the prepared nanoparticles as the energy increases the size of nanoparticles decreases until they reached their critical size below which the particle not sensitive to the laser energy. As the energy increases above this value the nanoparticles begin to agglomerate again and the size increases. Also the wavelength affects the size and yield of the obtained nanoparticles. Smaller size distribution and yield was obtained at shorter wavelength of 532 nm rather than 1064 nm. The focal length and the position of the samples both have a notable affect on the shape and size distribution of the gold nanoparticles. The size distribution and shape can be controlled by optimizing the laser parameter such as energy of laser beam, wavelength or by optimizing the focusing condition of the laser beam.

REFERENCES

[1] F. K. Alanazi, A. A. Radwan and I. A. Alsarra, "Biopharmaceutical Applications of Nanogold," *Saudi Pharmaceutical Journal*, Vol. 18, No. 4, 2010, pp. 179-193.

[2] E. J. Yoo, T. Li, H. G. Park and Y. K. Chang, "Size-Dependent Flocculation Behavior of Colloidal Au Nanoparticles Modified with Various Biomolecules," *Ultramicroscopy*, Vol. 108, No. 1, 2008, pp. 1273-1277.

[3] M. Shakibaie, H. Forootanfar, K. Mollazadeh-Moghadam, Z. Bagherzadeh, N. Nafissi-Varcheh, A. R. Shahverdi and M. A. Faramarzi, "Green Synthesis of Gold Nanoparticles by the Marine Microalga Tetraselmis Suecica," *Biotechnology and Applied Biochemistry*, Vol. 57, No. 2, 2010, pp. 71-75. doi:10.1042/BA20100196

[4] C.-M. Shih, Y.-T. Shieh and Y.-K. Twu, "Preparation of gold Nanopowders and Nanoparticles Using Chitosan Suspensions," *Carbohydrate Polymers*, Vol. 78, No. 2, 2009, pp. 309-315. doi:10.1016/j.carbpol.2009.04.008

[5] K. Kalishwaralal, V. Deepak, S. R. K. Pandian and S. Gurunathan, "Biological Synthesis of Gold Nanocubes from Bacillus Licheniformis," *Bioresource Technology*, Vol. 100, No. 21, 2009, pp. 5356-5358.

[6] K. Kalishwaralal, V. Deepak, S. R. K. Pandian, M, Kottaisamy, S. BarathManiKanth, B. Kartikeyan and S. Gurunathan, "Biosynthesis of Silver and Gold Nanoparticles Using *Brevibacterium casei*," *Colloid and Surface*, *B*: *Biointerfaces*, Vol. 77, No. 2, 2010, pp. 257-262. doi:10.1016/j.colsurfb.2010.02.007

[7] I. Maliszewska, L. Aniszkiewicz and Z. Sadowski, "Biological Synthesis of Gold Nanostructures Using the Extract of *Trichoderma koningii*," *Acta Physica Polonic A*, Vol. 116, 2009, p. S-163.

[8] P. Mukherjee, S. Senapati, D. Mandal, A. Ahmad, M. I. Khan, R. Kumar and M. Sastry, "Extracellular Synthesis of Gold Nanoparticles by the Fungus *Fusarium oxysporum*," *Biochemistry* (*Chemical Biology*), Vol. 3, No. 5, 2002, pp. 461-463. doi:10.1002/1439-7633(20020503)3:5<461::AID-CBIC461>3.0.CO;2-X

[9] Y. Y. Fong, J. R. Gascooke, B. R. Visser, G. F. Metha and M. A. Buntine, "Laser-Based Formation and Properties of Gold Nanoparticles in Aqueous Solution: Formation Kinetics and Surfactant-Modified Particle Size Dis-

tributions," *The Journal of Physical Chemistry C*, Vol. 114, No. 38, 2010, pp. 15931-15940. doi:10.1021/jp9118315

[10] F. Mafuné, J. Y. Kohno, Y. Takeda and T. Kondow, "Growth of Gold Clusters into Nanoparticles in a Solution Following Laser-Induced Fragmentation," *The Journal of Physical Chemistry B*, Vol. 106, No. 34, 2002, pp. 8555-8561.

[11] F. Mafuné, J. Y. Kohno, Y. Takeda and T. Kondow, "Dissociation and Aggregation of Gold Nanoparticles under Laser Irradiation," *The Journal of Physical Chemistry B*, Vol. 105, No. 38, 2001, pp. 9050-9056. doi:10.1021/jp0111620

[12] D. Pissuwan, T. Niidome and M. B. Cortie, "The Forthcoming Applications of Gold Nanoparticles in Drug and Gene Delivery Systems," *Journal of Controlled Release*, Vol. 149, No. 1, 2011, pp. 65-71. doi:10.1016/j.jconrel.2009.12.006

[13] P. Ghosh, G. Han, M. De, C. K. Kim and V. M. Rotello, "Gold Nanoparticles in Delivery Applications," *Advanced Drug Delivery Reviews*, Vol. 60, No. 11, 2008, pp. 1307-1315. doi:10.1016/j.addr.2008.03.016

[14] D. T. Nguyen, D.-J. Kim and K.-S. Kim, "Controlled Synthesis and Biomolecular Probe Application of Gold Nanoparticles," *Micron*, Vol. 42, No. 3, 2011, pp. 207-227. doi:10.1016/j.micron.2010.09.008

[15] A. B. Etame, C. A. Smith, W. C. W. Chan and J. T. Rutka, "Design and Potential Application of PEGylated Gold Nanoparticles with Size-Dependent Permeation through Brain Microvasculature," *Nano Medicine: Nanotechnology, Biology and Medicine*, Vol. 7, No. 6, 2011, pp. 992-1000. doi:10.1016/j.nano.2011.04.004

[16] T. Donnelly, S. Krishnamurthy, K. Carney, N. McEvoy and J. G. Lunney, "Pulsed Laser Deposition of Nanoparticle Films of Au," *Applied Surface Science*, Vol. 254, No. 4, 2007, pp. 1303-1306. doi:10.1016/j.apsusc.2007.09.033

[17] S. Yang, Y.-H. Jang, C. H. Kim, C. Hwang, J. Lee, S. Chae, S. Jung and M. Choi, "A Flame Metal Combustion Method for Production of Nanoparticles," *Powder Technology*, Vol. 197, No. 3, 2010, pp. 170-176. doi:10.1016/j.powtec.2009.09.011

[18] C. Wu, X. Qiao, J. Chen, H. Wang, F. Tan and S. Li, "A Novel Chemical Route to Prepare ZnO Nanoparticles," *Materials Letters*, Vol. 60, No. 15, 2006, pp. 1828-1832. doi:10.1016/j.matlet.2005.12.046

[19] H. Jia, J. Zeng, W. Song, J. An and B. Zhao, "Preparation of Silver Nanoparticles by Photo-Reduction for Surface-Enhanced Raman Scattering," *Thin Solid Films*, Vol. 496, No. 2, 2006, pp. 281-287. doi:10.1016/j.tsf.2005.08.359

[20] P. Y. Lim, R. S. Liu, P. L. She, C. F. Hung and H. C. Shih, "Synthesis of Ag Nanospheres Particles in Ethylene Glycol by Electrochemical-Assisted Polyol Process," *Chemical Physics Letters*, Vol. 420, No. 4-6, 2006, pp. 304-308. doi:10.1016/j.cplett.2005.12.075

[21] M. J. Rosemary and T. Pradeep, "Solvothermal Synthesis of Silver Nanoparticles from Thiolates," *Journal of Colloid and Interface Science*, Vol. 268, No. 1, 2003, pp. 81-84. doi:10.1016/j.jcis.2003.08.009

[22] M. Szymańska-Chargot, A. Gruszecka, A. Smolira, J. Cytawa and L. Michalak, "Mass-Spectrometric Investigations of the Synthesis of Silver Nanoparticles via Electrolysis," *Vacuum*, Vol. 82, No. 10, 2008, pp. 1088-1093. doi:10.1016/j.vacuum.2008.01.022

[23] H. Huang and X. Yang, "Synthesis of Polysaccharide-Stabilized Gold and Silver Nanoparticles: A Green Method," *Carbohydrate Research*, Vol. 339, No. 15, 2004, pp. 2627-2631. doi:10.1016/j.carres.2004.08.005

[24] J. Gu, W. Fan, A. Shimojima and T. Okubo, "Microwave-Induced Synthesis of Highly Dispersed Gold Nanoparticles within the Pore Channels of Mesoporous Silica," *Journal of Solid State Chemistry*, Vol. 181, No. 4, 2008, pp. 957-963. doi:10.1016/j.jssc.2008.01.039

[25] Y.-C. Liu, L.-H. Lin and W.-H. Chiu, "Size-Controlled Synthesis of Gold Nanoparticles from Bulk Gold Substrates by Sonoelectrochemical Methods," *Journal Physical Chemistry B*, Vol. 108, No. 50, 2004, pp. 19237-19240. doi:10.1021/jp046866z

[26] H. Eerikainen and E. Kauppinen, "Preparation of Polymeric Nanoparticles Containing Corticosteroid by a Novel Aerosol Flow Reactor Method," *International Journal of Pharmaceutics*, Vol. 263, No. 1-2, 2003, pp. 69-83. doi:10.1016/S0378-5173(03)00370-3

[27] K. L. McGilvray, M. R. Decan, D. Wang and J. Scaiano, "Facile Photochemical Synthesis of Unprotected Aqueous Gold Nanoparticles," *Journal of the American Chemical Society*, Vol. 128, No. 50, 2006, pp. 15980-15981. doi:10.1021/ja066522h

[28] M. Duocastella, J. M. Fernandez-Pradas, J. Dominguez, P. Serra and J. L. Morenza, "Printing Biological Solutions through Laser-Induced Forward Transfer," *Applied Physics A*, Vol. 93, No. 4, 2008, pp. 941-945. doi:10.1007/s00339-008-4741-6

[29] Y. Itoh, M. Abdullah and K. Okuyama, "Direct Preparation of Nonagglomerated Indium Tin Oxide Nanoparticles using Various Spray Pyrolysis Methods," *Journal of Materials Research*, Vol. 19, No. 4, 2004, pp. 1077-1086. doi:10.1557/JMR.2004.0141

[30] S. Y. Yang and S. G. Kim, "Characterization of Silver and Silver/Nickel Composite Particles Prepared by Spray Pyrolysis," *Powder Technology*, Vol. 146, No. 3, 2004, pp. 185-192. doi:10.1016/j.powtec.2004.07.010

[31] N. S. Tabrizi, M. Ullmann, V. A. Vons, U. Lafont and A. Schmidt-Ott, "Generation of Nanoparticles by Spark Discharge," *Journal of Nanoparticle Research*, Vol. 11, No. 2, 2009, pp. 315-332. doi:10.1007/s11051-008-9407-y

[32] J.-P. Sylvestre, A. V. Kabashin, E. Sacher, M. Meunier and J. H. T. Luong, "Stabilization and Size Control of Gold Nanoparticles during Laser Ablation in Aqueous Cyclodextrins," *Journal of the American Chemical Society*, Vol. 126, No. 23, 2004, pp. 7176-7177. doi:10.1021/ja048678s

[33] A. Fojtik, A. Henglein and B. Bunsen-Ges, "Laser Ablation of Films and Suspended Particles in Solvent-Formation of Cluster and Colloid Solutions," *Chemical Physics*, Vol. 97,

No. 2, 1993, pp. 252-254.

[34] S. Machmudah, Wahyudiono, Y. K. M. Sasaki and M. Goto, "Nano-Structured Particles Production Using Pulsed Laser Ablation of Gold Plate in Supercritical CO_2," *Journal of Supercritical Fluids*, Vol. 60, 2011, pp. 63-68. doi:10.1016/j.supflu.2011.04.008

[35] N. V. Tarasenko, A. V. Butsen, E. A. Nevar and N. A. Savastenko, "Synthesis of Nanosized Particles during Laser Ablation of Gold in Water," *Applied Surface Science*, Vol. 252, No. 13, 2006, pp. 4439-4444. doi:10.1016/j.apsusc.2005.07.150

[36] H. J. Kim, I. C. Bang and J. Onoe, "Characteristic Stability of Bare Au-Water Nanofluids Fabricated by Pulsed Laser Ablation in Liquids," *Optics and Laser in Engineering*, Vol. 47, No. 5, 2009, pp. 535-538. doi:10.1016/j.optlaseng.2008.10.011

[37] S. I. Dolgaev, A. V. Simakin, V. V. Voronov, G. A. Shafeev and F. B. Verduraz, "Nanoparticles Produced by Laser Ablation of Solids in Liquid Environment," *Applied Surface Science*, Vol. 186, No. 1-4, 2002, pp. 546-551. doi:10.1016/S0169-4332(01)00634-1

[38] N. Haustrup and G. M. Oconnor, "Nanoparticle Generation during Laser Ablation and Laser-Induced Liquefaction," *Physics Procedia*, Vol. 12, Part B, 2011, pp. 53-64.

[39] T. Sakai, H. Enomoto, K. Torigoe, H. Sakai and M. Abe, "Surfactant- and Reducer-Free Synthesis of Gold Nanoparticles in Aqueous Solutions," *Colloids and Surfaces A: Physicochemical and Engineering Aspects*, Vol. 347, No. 1-3, 2008, pp. 18-26. doi:10.1016/j.colsurfa.2008.10.037

[40] P. Calandra, C. Giordano, A. Longo and V. TurcoLireri, "Physicochemical Investigation of Surfactant-Coated Gold Nanoparticles Synthesized in the Confined Space of Dry Reversed Micelles," *Materials Chemistry and Physics*, Vol. 98, No. 2-3, 2006, pp. 494-499. doi:10.1016/j.matchemphys.2005.09.068

[41] F. K. Liu, "Extremely Highly Efficient On-Line Concentration and Separation of Gold Nanoparticles Using the Reversed Electrode Polarity Stacking Mode and Surfactant-Modified Capillary Electrophoresis," *Analytica Chimica Acta*, Vol. 694, No. 1-2, 2011, pp. 167-173. doi:10.1016/j.aca.2011.03.056

[42] J. P. Sylvestre, S. Poulin, E. Sacher, M. Meunier and J. H. T. Luong, "Surface Chemistry of Gold Nanoparticles Produced by Laser Ablation in Aqueous Media," *Journal of Physical Chemistry B*, Vol. 108, No. 34, 2004, pp. 16864-16869. doi:10.1021/jp047134+

[43] B. Xu, R. G. Song, P. H. Tang, J. Wang, G. Z. Chai, Y. Z. Zhang and Z. Z. Ye, "Preparation of Ag Nanoparticles Colloid by Pulsed Laser Ablation in Distilled Water," *Key Engineering Materials*, Vol. 373-374, 2008, pp. 346-349.

[44] A. Sasoh, K. Watanabe, Y. Sano and N. Mukai, "Behavior of Bubbles Induced by the Interaction of a Laser Pulse with a Metal Plate in Water," *Applied Physics A: Material Science & Processing*, Vol. 80, No. 7, 2005, pp. 1497-1500.

[45] W. T. Nichols, T. Sasaki and Naoto Koshizaki, "Laser Ablation of a Platinum Target in Water. I. Ablation Mechanisms," *Journal of Applied Physics*, Vol. 100, No. 11, 2006, pp. 114911-114917. doi:10.1063/1.2390640

[46] W. T. Nichols, T. Sasaki and N. Koshizaki, "Laser Ablation of a Platinum Target in Water. III. Laser-Induced Reactions," *Journal of Applied Physics*, Vol. 100, No. 11, pp. 112006-114913. doi:10.1063/1.2390642

[47] P. Smejkal, J. Pfleger, B. Vlckova and O. Dammer, "Laser Ablation of Silver in Aqueous Ambient: Effect of Laser Pulse Wavelength and Energy on Efficiency of the Process," *Journal of Physics: Conferences Series*, Vol. 59, 2007, p. 185.

[48] A. Natha, S. S. Lahaa and A. Khare, "Effect of Focusing Conditions on Synthesis of Titanium Oxide Nanoparticles Via Laser Ablation in Titanium-Water Interface," *Applied Surface Science*, Vol. 257, No. 7, 2011, p. 3118.

[49] V. Bulatov, L. Xu and I. Schechter, "Spectroscopic Imaging of Laser-Induced Plasma," *Analytical Chemistry*, Vol. 68, No. 17, 1996, pp. 2966-2973. doi:10.1021/ac960277a

18

Electromagnetically Induced Transparency Using a Artificial Molecule in Circuit Quantum Electrodynamics

Hai-Chao Li, Guo-Qin Ge
School of Physics, Huazhong University of Science and Technology, Wuhan, China
Email: lhc2007@hust.edu.cn, gqge@hust.edu.cn

ABSTRACT

Electromagnetically induced transparency (EIT) having wide applications in quantum optics and nonlinear optics is explored ordinarily in various atomic systems. In this paper we present a theoretical study of EIT using supercon- ducting circuit with a V-type artificial molecule constructed by two Josephson charge qubits coupled each other through a large capacitor. In our theoretical model we make a steady state approximation and obtain the analytical expressions of the complex susceptibility for the artificial system via the density matrix formalism. The complex susceptibility has additional dependence on the qubit parameters and hence can be tuned to a certain extent.

Keywords: EIT; Artificial Molecule; Complex Susceptibility

1. Introduction

Electromagnetically induced transparency (EIT) [1,2] through quantum coherent effects has attracted considerable interest due to its extensive applications in quantum optics and atomic physics. The first experimental demonstration of EIT was based on a Λ-type atomic system [3]. EIT has also been observed experimentally in the V-type [4] and cascade-type [5] energy level configurations. It's of particular interest to indicate EIT how to appear via quantum interference in a V-type system because population trapping isn't involved. In contrast to the usual weak probe regime, EIT can be realized in the strong probe regime [6], where population inversion is not correlated with optical gain and the traditional correspondence between inversion and gain is not satisfied.

Circuit quantum electrodynamics(QED) [7,8], where transmission line resonator plays the role of cavity and superconducting qubit [9,10] behaves as artificial atom to replace the natural atom, has recently become a new testbed for quantum optics. Compared with the conventional cavity QED with atomic gases, superconducting circuits as artificial quantum systems in solid-state devices have significant advantages, such as offering long coherence time to implement the quantum gate operations [11], huge tunability and controllability by external electromagnetic fields [12]. As an on-chip realization of cavity QED, circuit QED has reproduced many quantum optical phenomena, including Kerr and cross-Kerr nonlinearities [12,13], the Mollow Triplet [14], Autler-Townes effect [15], EIT [16, 17]. Further- more, circuit QED can be used to realize ultrastrong coupling regime [18] previously inaccessible to atomic systems and explore novel optical phenomena emerging only in this regime.

Although have being extensively studied in traditional atomic systems, investigations of EIT phenomena in superconducting circuits based on mesoscopic Josephson junctions are still scarce. Recently experimental observation of EIT has been reported by using a single artificial atom coupled to a 1D transmission line [16] and EIT can be utilized as a sensitive probe of decoherence in superconducting circuits [19]. Besides, a nanomechanical resonator can provide additional auxiliary energy levels to a superconducting Cooper-pair box so that EIT can be realized in the system [20].

Motivated by these investigations, we propose a scheme to perform EIT employing V-type artificial molecule, which is constructed by two superconducting charge qubits coupled each other through a large capacitor. In our EIT scheme, a weak probe field with Rabi frequency Ω_1 and frequency ω_1 couples the $|1\rangle \leftrightarrow |3\rangle$ transition while a strong control field with Rabi frequency Ω_2 and frequency ω_2 couples the $|1\rangle \leftrightarrow |2\rangle$ transition, as shown in **Figure 1**.

This paper is organized as follows. We first describe the theoretical model and gain the energy spectrum of the V-type artificial system in Section 2. Then, we give steady-state analysis of EIT by utilizing the density matrix method and acquire the complex susceptibility for the superconducting system in Section 3 and our conclusions are given in Section 4.

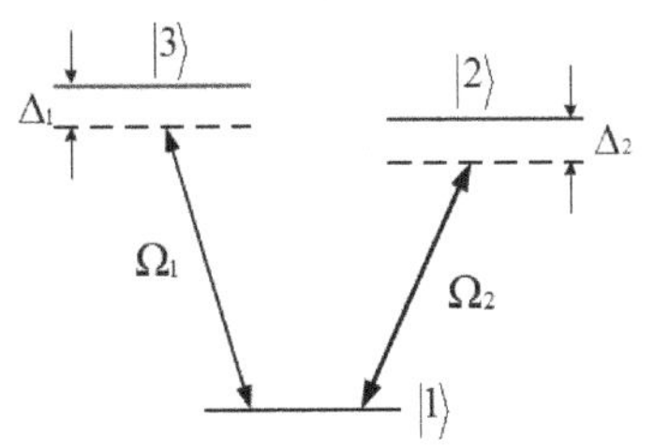

Figure 1. Schematic illustration of EIT for the artificial molecule.

2. The Model of Artificial Molecule

Let us consider two interacting superconducting charge qubits which are electrostatically coupled to each other by a large capacitor C_m. Each charge qubit has a superconducting quantum interference device (SQUID) ring geometry biased by an external flux and so the effective Josephson coupling energy can be varied from zero up to its maximum value. The Hamiltonian of coupled qubits reads

$$H = E_{c1}(n_1 - n_{g1})^2 - E_{J1}\cos\beta_1 + E_{c2}(n_2 - n_{g2})^2 - E_{J2}\cos\beta_2 + E_m(n_1 - n_{g1})(n_2 - n_{g2}) \quad (1)$$

The first four terms represent two independent qubits and the last term describes the interaction between the qubits due to the electrostatical coupling of the capacitor. E_{J1} and E_{J2} are the effective Josephson coupling energy for the corresponding SQUID; β_1 and β_2 are the phases of the SQUID; E_{c1} and E_{c2} are the effective Cooper-pair charging energies for the qubits; n_i and n_{gi} for I = 1,2 are the number operator of excess Cooper-pairs on the island and the normalized gate induced charge; E_m is the capacitive coupling energy between the charge qubits.

Working in the vicinity of one degeneracy point ($n_{gi} \in$ [0,1]), only two adjacent charge states $|0\rangle$ and $|1\rangle$ on the island are relevant while all other charge states, having a much higher energy, can be ignored [10]. In this case the Hamiltonian can be written as

$$H = \sum_{i=1,2}\left(-\frac{1}{2}B_{zi}\sigma_{zi} - \frac{1}{2}E_{Ji}\sigma_{xi}\right) + \frac{1}{4}E_m\sigma_{z1}\sigma_{z2} - \frac{1}{2}E_m\delta_2\sigma_{z1} - \frac{1}{2}E_m\delta_1\sigma_{z2} \quad (2)$$

where $B_{zi} = E_{ci}(1-2n_{gi})$ for I = 1,2 are the difference of the electrostatic energy between the states $|0\rangle$ and $|1\rangle$, σ_z and σ_x are the Pauli matrices and $\delta_i = 1/2 - n_{gi}$. Switching to the eigenbasis $|e\rangle$ and $|g\rangle$ of the qubits and exactly at the co-resonance point $\delta_i = 0$, the Hamiltonian takes the form

$$H = \frac{1}{2}E_{J1}\rho_{z1} + \frac{1}{2}E_{J2}\rho_{z2} + \frac{1}{4}E_m\rho_{x1}\rho_{x2} \quad (3)$$

To avoid confusion we introduce a second set of Pauli operator $\boldsymbol{\rho}$ acting on the eigenstates of qubits. Without loss of generality, we assume that the two superconducting charge qubits are identical (i.e., $E_{c1} = E_{c2} = E_c$, $E_{J1} = E_{J2} = E_J$). So the eigenvalues of coupled qubits are readily written as

$$\begin{aligned} E_4 &= \frac{1}{4}(E_m^2 + 16E_J^2)^{\frac{1}{2}} \\ E_3 &= \frac{1}{4}E_m \\ E_2 &= -\frac{1}{4}E_m \\ E_1 &= -\frac{1}{4}(E_m^2 + 16E_J^2)^{\frac{1}{2}} \end{aligned} \quad (4)$$

with the corresponding eigenstates being

$$\begin{aligned} |4\rangle &= \sin\alpha|gg\rangle + \cos\alpha|ee\rangle \\ |3\rangle &= \frac{1}{\sqrt{2}}(|ge\rangle + |eg\rangle) \\ |2\rangle &= \frac{1}{\sqrt{2}}(|ge\rangle - |eg\rangle) \\ |1\rangle &= \cos\alpha|gg\rangle - \sin\alpha|ee\rangle \end{aligned} \quad (5)$$

Here the parameter α satisfies the following relations

$$\sin 2\alpha = \frac{E_m}{\sqrt{E_m^2 + 16E_J^2}} \quad \cos 2\alpha = \frac{4E_J}{\sqrt{E_m^2 + 16E_J^2}} \quad (6)$$

It is worthwhile to note that arbitrary transitions can not be allowed in the above four states due to selection rules for superconducting qubits. By calculating the matrix elements of ρ_{x1} and ρ_{x2} between the eigenstates, we find that the transitions $|1\rangle \leftrightarrow |4\rangle$ and $|2\rangle \leftrightarrow |3\rangle$ are forbidden while the other transitions with nonzero matrix elements are allowed. choosing the three levels with lowest eigenenergies shown in **Figure 1**, we obtain the V-type artificial system.

3. Complex Susceptibility

EIT phenomenon of a closed three level system interacting with a weak probe field and a strong control field can be demonstrated by adopting the density matrix formalism. In the eigenbasis of the qubits, the interaction Hamiltonian between the three-level artificial molecule and two semiclassical fields is expressed as ($\hbar$ = 1)

$$H_{int} = -\frac{1}{2}\left(\Omega_1 e^{-i\omega_1 t}|e\rangle_{1\,1}\langle g| + \Omega_2 e^{-i\omega_2 t}|e\rangle_{2\,2}\langle g| + H.c.\right) \quad (7)$$

In the basis $\{|1\rangle, |2\rangle \text{ and } |3\rangle\}$ of the V-type artificial system and with the rotating-wave approximation, the interaction Hamiltonian is given by

$$H_{int} = -\frac{1}{2}\left(\Omega_1 e^{-i\omega_1 t}\frac{1}{\sqrt{2}}\cos\alpha|3\rangle\langle 1| + \Omega_2 e^{-i\omega_2 t}\frac{1}{\sqrt{2}}\cos\alpha|2\rangle\langle 1| + H.c.\right) \quad (8)$$

In the interaction picture, the Hamiltonian of the system reads

$$H_I = \Delta_1 |3\rangle\langle 3| + \Delta_2 |2\rangle\langle 2| - \frac{1}{2}\left(\xi_1 |3\rangle\langle 1| + \xi_2 |2\rangle\langle 1| + H.c.\right) \quad (9)$$

where $\Delta_1 = \omega_{31}\omega_1$ is the detuning of the probe field, $\Delta_2 = \omega_{21}\omega_2$ is the detuning of the control field,

$$\xi_1 = \Omega_1 \cos\alpha/\sqrt{2} \text{ and } \xi_2 = \Omega_2 \cos\alpha/\sqrt{2}.$$

We can select the frequencies of the fields so that the probe field ω_1 and the control field ω_2 are near resonant with the transitions $|1\rangle \leftrightarrow |3\rangle$ and $|1\rangle \leftrightarrow |2\rangle$, respectively. In this case, other transitions can be ignored in our discussion. The evolution of the system is governed by the set of density matrix equations of motion

$$\begin{aligned}
\dot{\rho}_{33} &= -\Gamma_3\rho_{33} + \frac{i}{2}\xi_1\rho_{13} - \frac{i}{2}\xi_1^*\rho_{31} \\
\dot{\rho}_{22} &= -\Gamma_2\rho_{22} + \frac{i}{2}\xi_2\rho_{12} - \frac{i}{2}\xi_2^*\rho_{21} \\
\dot{\rho}_{31} &= -(\gamma_{31} + i\Delta_1)\rho_{31} + \frac{i}{2}\xi_1(\rho_{11} - \rho_{33}) - \frac{i}{2}\xi_2\rho_{32} \\
\dot{\rho}_{21} &= -\gamma_{21}\rho_{21} + \frac{i}{2}\xi_2(\rho_{11} - \rho_{22}) - \frac{i}{2}\xi_1\rho_{23} \\
\dot{\rho}_{32} &= -(\gamma_{32} + i\Delta_1)\rho_{32} - \frac{i}{2}\xi_2^*\rho_{31} + \frac{i}{2}\xi_1\rho_{12} \\
1 &= \rho_{11} + \rho_{22} + \rho_{33}
\end{aligned} \quad (10)$$

Here we further assume that the control field frequency ω_c matchs the level spacing between the states $|2\rangle$ and $|1\rangle$, i.e. $\Delta_2 = 0$. In these equations we have introduced phenomenologically the relaxation rates $\Gamma_i (I = 1,2,3)$ for the levels as well as the total dephasing rates $\gamma_{ij} = (\Gamma_i+\Gamma_j)/2+\tau_\varphi$ including the relaxation and pure dephasing processes. Since we are interested in the dispersion and absorption properties of the V-type artificial system, only first-order perturbation expansion of the equations of matrix elements are necessary. For the system we set[21]

$$\rho_{33}^{(0)} = 0 \quad \rho_{22}^{(0)} = \frac{B}{\Gamma_2 + 2B} \quad \rho_{11}^{(0)} = \frac{\Gamma_2 + B}{\Gamma_2 + 2B} \quad (11)$$

where B is the rate of pumping by the control field

$$B = \frac{1}{2}\frac{|\Omega_2|^2}{\gamma_{21}} \quad (12)$$

Taking into account the steady-state solution (i.e., all derivatives are set equal to zero), we have the first order matrix element

$$\rho_{31}^{(1)} = \frac{i\xi_1}{2}\left[\rho_{11}^{(0)}(\gamma_{32} + i\Delta_1) - \frac{|\xi_2|^2(\rho_{11}^{(0)} - \rho_{22}^{(0)})}{4\gamma_{21}}\right] \times \left[(\gamma_{31} + i\Delta_1)(\gamma_{32} + i\Delta_1) + \frac{|\xi_2|^2}{4}\right]^{-1} \quad (13)$$

Combining the relation

$$\varepsilon_0\chi\Omega_1 = 2|\mu_{31}|^2\rho_{31} \quad (14)$$

with equation (12), we have the following expressions of the complex susceptibility $\chi = \chi' + i\chi''$:

$$\chi' = -\frac{\cos\alpha|\mu_{31}|^2\Delta_1}{\sqrt{2}\varepsilon_0 Z}[\rho_{11}^{(0)}(\frac{|\xi_2|^2}{4} - \Delta_1^2 - \gamma_{32}^2) + \frac{|\xi_2|^2}{4\gamma_{21}}(\rho_{11}^{(0)} - \rho_{22}^{(0)})(\gamma_{31} + \gamma_{32})] \quad (15)$$

$$\chi'' = \frac{\cos\alpha|\mu_{31}|^2}{\sqrt{2}\varepsilon_0 Z}\{\rho_{11}^{(0)}[\gamma_{32}(\gamma_{31}\gamma_{32} + \frac{|\xi_2|^2}{4}) + \gamma_{31}\Delta_1^2] - \frac{|\xi_2|^2}{4\gamma_{21}}(\rho_{11}^{(0)} - \rho_{22}^{(0)})(\frac{|\xi_2|^2}{4} + \gamma_{31}\gamma_{32} - \Delta_1^2)\} \quad (16)$$

where ε_0 is the vacuum permittivity, μ_{31} is the transition dipole moment and

$$Z = (\frac{|\xi_2|^2}{4} + \gamma_{31}\gamma_{32} - \Delta_1^3)^2(\gamma_{31}\Delta_1 + \gamma_{32}\Delta_1)^2 \quad (17)$$

It seems that the above expressions are similar to the susceptibilities of the conventional three level atomic systems, but here the complex susceptibility of the artificial molecule has additional dependence on the tunable Josephson coupling energy E_J and the capacitive coupling strength E_m through the parameter α and hence can be tuned to a certain extent.

Figures 2(a) and **(b)** plot both the real and imaginary parts of the susceptibility χ as a function of the probe detuning Δ_1 and the dimensionless ratio of the Josephson coupling energy E_J to the interaction energy E_m according to the equations (15) and (16). **Figure 2** shows that the absorption profile is even symmetric and the dispersion profile is odd about the zero probe detuning point $\Delta_1 = 0$. We can observe that the absorption is minimum at the zero point $\Delta_1 = 0$ and increases with the growth of the Josephson coupling energy E_J, but the absorption value does not become large after the E_J is increased to a cer- tain value, as can be seen from **Figure 2(b)**.

Moreover, absorption coefficient can be dominated greatly by the control field strength Ω_c, as depicted in **Figure 3**. From the drawing, we see that single absorption peak appears in the regime of weak control field and indicates strong absorption to probe field. As the control power is increased, the doublet spacing of absorption curve increases and absorption value between the two peaks gradually tends to zero EIT, i.e. EIT effect arises in the higher control intensities.

4. Conclusions

In conclusion, we have theoretically investigated the EIT

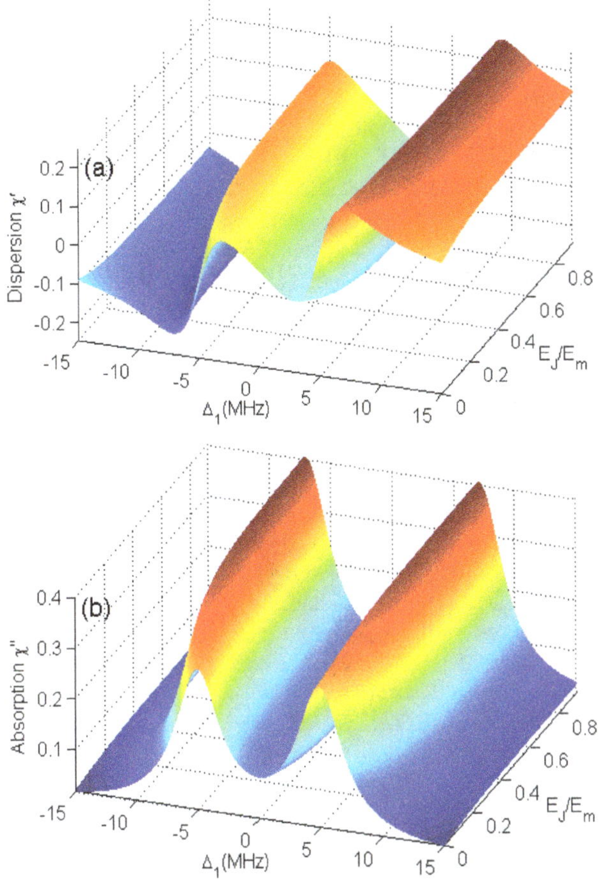

Figure 2. Real part χ' and imaginary part χ'' of the complex susceptibility versus the probe detuning Δ_1 and the dimensionless ratio E_J/E_m. Here parameters E_{Jmax} = 14.5GHz, E_m = 15.7GHz, $\gamma_{31} = \gamma_{21}$ = 2MHz, γ_{32} = 2.5MHz, Γ_2 = 1/0.7MHz, Ω_c = 20MHz.

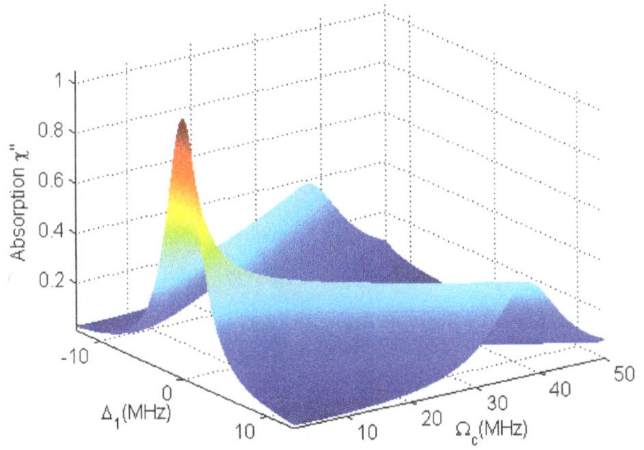

Figure 3. Absorption coefficient χ'' as a function of the probe detuning Δ_1 and the control field intensity Ω_c (from 2 to 50 MHz). Parameters E_J = 14GHz, E_m = 15.7GHz, $\gamma_{31} = \gamma_{21}$ = 2MHz, γ_{32} = 2.5MHz, Γ_2 = 1/0.7MHz.

effect in a V-type artificial system derived from two coupled superconducting charge qubits. Using the density matrix formalism, we obtain the analytical expressions of the complex susceptibility which have extra dependence on qubit parameters E_J and E_m. As a result, EIT can be tuned to a certain extent by changing the Josephson coupling energy E_J compared with the conventional EIT phenomenon in the atomic systems where atomic parameters are uaually fixed.

5. Acknowledgements

This work was supported in part by the National Natural Science Foundation of China under the Grant No. 11 274132 and the Nature Science Foundation of Hubei Province.

REFERENCES

[1] D. J. Fulton, S. Shepherd, R. R. Moseley, B. D. Sinclair, and M. H. Dunn, "Continuous-wave Electromagnetically Induced Transparency: a Comparison of V, Λ, and Cascade Systems," *Physical Review A*, Vol. 52, No. 3, 1995, pp. 2302-2311. doi: 10.1103/PhysRevA.52.2302

[2] M. Fleischhauer, A. Imamoglu and J. P. Marangos, "Electromagnetically Induced Transparency: Optics in Coherent Media," Reviews of Modern Physics, Vol. 77, No. 2, 2005, pp. 633-673. doi:10.1103 /RevModPhys.77.633

[3] K. J. Boller, A. Imamoglu and S. E. Harris, "Observation of Electromagnetically Induced Transparency," Physical Review Letters, Vol. 66, No. 20, 1991, pp. 2593-2596. doi: 10.1103/PhysRevLett.66.2593

[4] A. Lazoudis, T. Kirova, E. H. Ahmed, P. Qi, J. Huennekens and A. M. Lyyra, "Electromagnetically Induced Transparency in an Open V-type Molecular System," *Physical Review A*, Vol. 83, No. 6, 2011, pp. 063419. doi: 10.1103/PhysRevA.83.063419

[5] J. Gea-Banacloche, Y. Q. Li, S. Z. Jin and Min Xia, "Electromagnetically Induced Transparency in Ladder-type Inhomogeneously Broadened Media: Theory and Experiment," *Physical Review A*, Vol. 51, No. 1, 1995, pp. 576-584. doi: 10.1103/PhysRevA.51.576

[6] S. Wielandy and A. L. Gaeta, "Investigation of electromagnetically induced transparency in the strong probe regime," *Physical Review A*, Vol. 58, No. 3, 1998, pp. 2500-2505. doi: 10.1103/PhysRevA.58.2500

[7] J. Q. You and F. Nori, "Atomic Physics and Quantum Optics Using Superconducting Circuits," *Nature*, Vol. 474, No. 7353, 2011, pp. 589-597. doi: 10.1038/nature10122

[8] S. M. Girvin, M. H. Devoret and R. J. Schoelkopf, "Circuit QED and Engineering Charge-based Superconducting Qubits," *Physica Scripta*, Vol. 2009, No. T137, 2009, pp. 014012. doi:10.1088/0031-8949/2009/T137/014012

[9] J. Q. You and F. Nori, "Superconducting Circuits and Quantum Information," *Physics Today*, Vol. 58, No. 11, 2005, pp. 42-47. doi: 10.1063/1.2155757

[10] Y. A. Pashkin, O. Astafiev, T. Yamamoto, Y. Nakamura and J. S. Tsai, "Josephson Charge Qubits: a Brief Review," *Quantum Information Processing*, Vol. 8, 2009, pp. 55-80. doi: 10.1007/s11128-009-0101-5

[11] B. C. Sanders, "Quantum Optics in Superconducting Circuits," *AIP Conference Proceedings*, Vol. 1398, 2011, pp. 46-49. doi: 10.1063/1.3644209

[12] Y. Hu, G. Q. Ge, S. Chen, X. F. Yang and Y. L. Chen, "Cross-Kerr-effect Induced by Coupled Josephson Qubits in Circuit Quantum Electrodynamics," *Physical Review A*, Vol. 84, No. 1, 2011, p. 012329. doi: 10.1103/PhysRevA.84.012329

[13] S. Rebić, J. Twamley and G. J. Milburn, "Giant Kerr Nonlinearities in Circuit Quantum Electrodynamics," *Physical Review Letters*, Vol. 103, No. 15, 2009, p. 150503. doi: 10.1103/PhysRevLett.103.150503

[14] O. Astafiev *et al.*, "Resonance Fluorescence of a Single Artificial Atom," *Science*, Vol. 327, No. 5967, 2010, pp. 840-843. doi: 10.1126/science.1181918

[15] M. A. Sillanpää et al., "Autler-Townes Effect in a Superconducting Three-Level System," *Physical Review Letters*, Vol. 103, No. 19, 2009, p. 193601. doi: 10.1103/PhysRevLett.103.193601

[16] A. A. Abdumalikov, Jr., O. Astafiev, A. M. Zagoskin, Yu. A. Pashkin, Y. Nakamura and J. S. Tsai, "Electromagnetically Induced Transparency on a Single Artificial Atom," *Physical Review Letters*, Vol. 104, No. 19, 2010, p. 193601. doi: 10.1103/PhysRevLett.104.193601

[17] J. Joo, J. Bourassa, A. Blais and B. C. Sanders, "Electromagnetically Induced Transparency with Amplification in Superconducting Circuits," *Physical Review Letters*, Vol. 105, No. 7, 2010, p. 073601. doi: 10.1103/PhysRevLett.105.073601

[18] T. Niemczyk et al., "Circuit Quantum Electrodynamics in the Ultrastrong-coupling Regime," *Nature Physics*, Vol. 6, No. 10, 2010, pp. 772-776. doi: 10.1038/nphys1730

[19] K. V. R. M. Murali, Z. Dutton, W. D. Oliver, D. S. Crankshaw and T. P. Orlando, "Probing Decoherence with Electromagnetically Induced Transparency in Superconductive Quantum Circuits," *Physical Review Letters*, Vol. 93, No. 8, 2004, pp. 087003. doi: 10.1103/PhysRevLett.93.087003

[20] X. Z. Yuan, H. S. Goan, C. H. Lin, K. D. Zhu and Y. W. Jiang, "Nanomechanical-resonator-assisted Induced Transparency in a Cooper-pair Box System," *New Journal of Physics*, Vol. 10, 2008, p. 095016. doi: 10.1088/1367-2630/10/9/095016

[21] G. R. Welch, "Observation of V-Type Electromag- netically Induced Transparency in a Sodium Atomic Beam," *Foundations of Physics*, Vol.28, No. 4, 1998, pp. 621-638. doi: 10.1023/A:1018765706887

19

A Magnetically Levitated Precise Pointing Mechanism for Application to Optical Communication Terminals

Thomas Edward Donaldson Frame, Alexandre Pechev
Surrey Space Centre, University of Surrey, Guildford, UK
Email: T.E.Frame@surrey.ac.uk

ABSTRACT

Increasing data bandwidth requirements from spacecraft systems is beginning to pressure existing microwave communications systems. Free-Space optical communications allows for larger bandwidths for lower relative power consumption, smaller size and weight when compared to the microwave equivalent. However optical communication does have a formidable challenge that needs to be overcome before the advantages of the technology can be fully utilized. In order for the communication to be successful the transmitter and receiver terminals need to be pointed with a high accuracy (generally in the order of ≤10 μradians) for the duration of communication. In this paper we present a new concept for the precise pointing of optical communications terminals (termed the Precise Pointing Mechanism). In this new concept we combine the separate pointing mechanisms of a conventional optical terminal into a single mechanism, reducing the complexity and cost of the optical bench. This is achieved by electromagnetically actuating the whole telescope assembly in 6 degrees-of-freedom with an angular resolution of less than ±3 μradians within a 10 (Az. El.) field of view and linear resolution of ±2 μm. This paper presents the new pointing mechanism and discusses the modelling, simulation and experimental work undertaken using the bespoke engineering model developed.

Keywords: Optical Communications; Magnetic Levitation; Control; Pointing; Active Anti-Vibration

1. Introduction

Communication is an ever growing and advancing field that is driven by many factors. As space borne instrumentation and data gathering/processing systems advance, an inherent increase in the data rate and bandwidth of the communications system follows. Optical terminals can provide advantages such as larger data bandwidths, lower relative power consumption, and smaller size/weight over microwave systems due to the nature of the medium used [1]. Also as the RF communication frequency spectrum becomes more crowded an optical link can offer an attractive alternative without the need for introduction of complex channel filters and algorithms [2] along with a greater immunity and security. However due to the narrow divergence of the medium used in optical communications, a very accurate and responsive pointing system is required in order to maintain the link between the communication terminals. A pointing error greater than a few micro radians can dramatically reduce the power seen on the receiver. Therefore a stable and precise pointing mechanism is required for the duration of communications.

Generally optical terminals consist of a fixed Telescope acting as the antenna, and a series of mirror/optics assemblies (the Coarse Pointing Assembly (CPA), Fine Pointing Assembly (FPA) and the Point Ahead Assembly (PAA)).

The CPA traditionally consists of a gimbal mechanism using stepper motors and gear boxes to achieve movement in azimuth and elevation, such as in the ESA SOUT terminal [3]. An alternative is to use a form of hybrid stepper motor as in the ISLFE optical terminals [3]. The FPA consists of a pair of mirrors that work together to focus the beam onto the receiver unit. In the SOUT terminal the FPA uses a single mirror that is manipulated using a permanently excited DC motor [3]. In the ISLFE terminal however the FPA consists of a single mirror that is connected to Lorentz force actuators that manipulate it using capacitive sensors to control the angular position [4].

Another major requirement of the pointing system is to overcome spacecraft disturbances and maintain the optical link for the duration of communication. Magnetic levitation technology provides a solution to overcoming disturbances, and offers infinite resolution that is only constrained by the sensor resolution and noise. In this research the functionality of the separate pointing mechanisms are combined into a single unit that can reject disturbances shown in [5], [6] and fulfils the requirements of the separate mechanisms [7,8]. This single unit will magnetically levitate and actuate the entire telescope assembly to reduce the number of individual optical elements and reduce complexity of the optical bench.

Magnetic levitation has been considered and implemented in many areas and differing fields, and achieved high precision positioning, as demonstrated by [9] and [10]. However this technology has not been applied to a space-borne magnetically levitated optical telescope. The system developed by [9] levitates a 6-DOF triangular platform that demonstrates a 5 nm position resolution for 6 axes but has a very small travel range in the order of micro-meters. Used effectively this technology can provide the accuracy requirements in order to maintain an effective optical communications link [9,11,12]. Applications of this technology are terrestrial inter building links, space-borne deep space communications, or a GEO-LEO relay satellite in the micro-sat range. The technology presented in this paper could also be adapted to form part of the pointing mechanism of an imaging system where a form of active anti-vibration would be advantageous and improve images [13]. Another application due to recent ESA activity is a direct Earth to Mars link that would significantly improve communications [14].

The purpose of this paper is to introduce the Precise Pointing Mechanism developed here at the Surrey Space Centre. An engineering model that can completely magnetically levitate and actuate a 60 mm telescope antenna in all 6 degrees of freedom has been developed and used to demonstrate the use of magnetic levitation to provide two pointing resolutions (±10 mRadians and ±3 μRadians) for application to an optical communications scenario.

This mechanism is developed to be an enabling technology for use in terrestrial and free-space optical communications, or applications where active anti-vibration and precise pointing would be advantageous. The design is scalable dependent upon the gain requirements of the telescope.

This paper discusses an example application of the engineering model developed, gives an overview of the control strategy used and finally presents some results to demonstrate the resolution of the terminal.

2. Optical Communications Using the PPM

This section presents some applications that can be considered for the engineering model PPM presented in this paper. Using the PPM as the actuation mechanism for optical terminals, it is envisaged that the field of view of the PPM would require that the telescope assembly element be mounted on each spacecraft and Pre-set (As shown in **Figure 1**) to point in the direction of the next spacecraft to allow communications with the target satellite.

As can be seen from **Figure 1** the OCT is not aligned to the spacecraft axis it is aligned to point at the target spacecraft. This mounting should be configured so that it is aligned to compensate for the separation of the spacecraft at a pre-defined altitude. Assuming an orbit with an eccentricity of 0.001 (near circular) an assumption for height variations between the two spacecraft is made at ±10 km. If the link distance is in the region of 1000 km then 8% of the terminals azimuth and elevation range is used to compensate for altitude errors. Assuming that the typical attitude error is 1°, the ±5° azimuth and elevation range of the terminal of the engineering model in this research will be able to adequately compensate for altitude and attitude errors of the two spacecraft. If the link distance reduces to 200 km then the pointing requirement for the terminal changes to 77% of the ±5° limit of the engineering model.

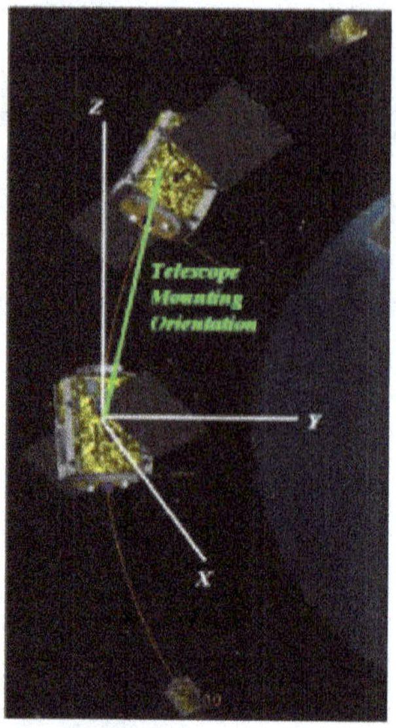

Figure 1. Pre-Set mounting configuration for current T1004-PPM applied to an OCT. Elevation is controlled by the PPM and the Azimuth range is controlled by the slip ring on which the PPM is mounted.

Considering a GEO satellite to LEO satellite link the usefulness of the PPM can be demonstrated. The GEO satellite has a single PPM terminal mounted on a slip ring on the satellite that allows communication with the LEO satellite. A LEO satellite that has the same orbit properties as the UK-DMC also has a single PPM mounted on a slip ring on it to facilitate communication with the GEO satellite.

The PPM on the GEO satellite is mounted on the Earth facing side of the spacecraft and has a Field of View (FOV) of ±5°. It has an increased azimuth range of 280° due to mounting on a slip ring. By modelling this scenario in AGI's Satellite Tool Kit (STK) the access between the two satellites can be determined.

The actuation that is required from the terminal is shown in **Figure 2**. This shows the range through which the PPM terminal has to actuate in order to achieve the

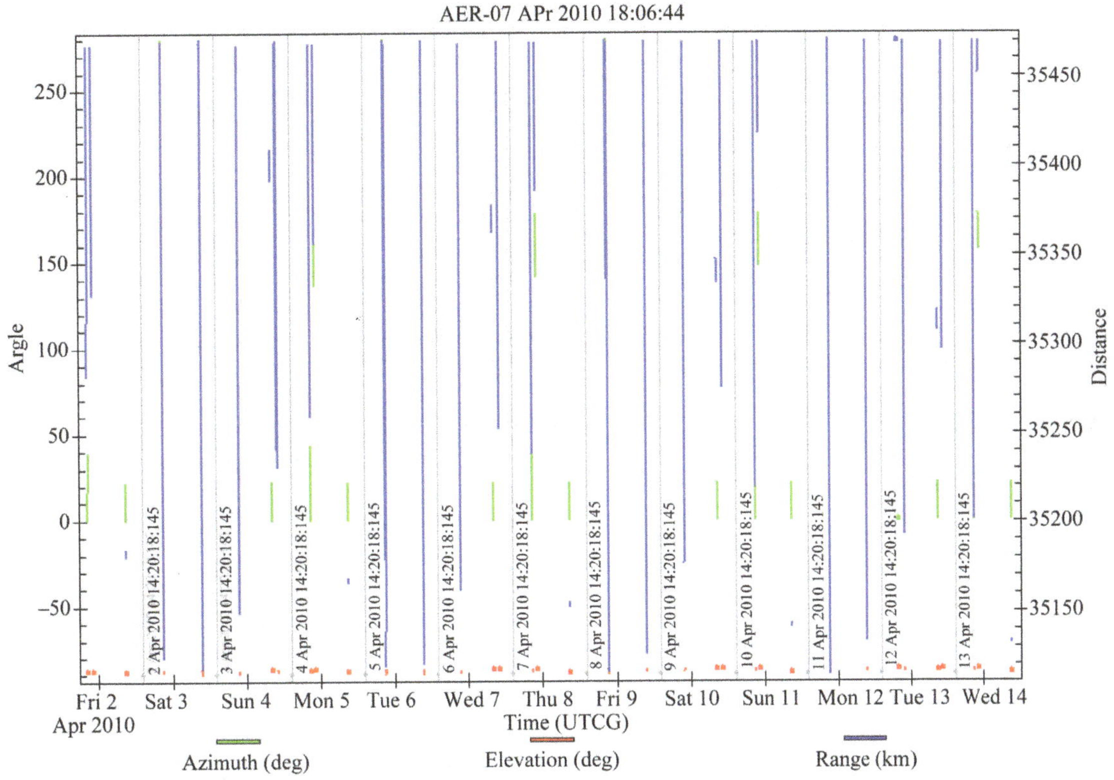

Figure 2. Azimuth and Elevation range for access between UK-DMC and a GEO relay sat. The azimuth range (green) is actuated by the slip ring on which the PPM is mounted. The Elevation (red) is actuated by the PPM and the change in link distance is shown in blue.

access between the GEO and LEO satellites.

The total access for the two week period is 80.82 minutes, at 1 Gbit/s this converts to a total of 4849 Gbit of data transferred. This limited Azimuth and Elevation range is a limitation of the engineering model and not the technology. Future work is considering increasing this range to ±10°.

3. The Precise Pointing Mechanism (PPM)

This research differs from existing Optical Communications Terminals (OCT) by magnetically levitating the entire telescope assembly of the OCT using a combination of reluctance force electromagnets and permanent magnets. The telescope element is then actuated in 6 degrees of freedom using the electromagnets to compensate for optical pointing errors, thus maximizing received power on an Avalanche Photo Diode (APD). For this evaluation model the pointing controller can compensate for pointing errors within a 10° FOV. By actuating the telescope of the OCT the complexity and number of optical components required in the optical bench is reduced. This has the effect of reducing the cost and complexity of the optical bench. A trade off is an increase in power consumption, but this is offset by a saving in mass.

The antenna of this optical communications system consists of a telescope providing a gain to overcome transmission losses and maximize the amount of optical power received from the transmitted source, and thus maximizing the signal to noise ratio and reducing the probability of a bit error rate. OCT applications use reflecting telescopes that utilise the property of paraboloidal mirrors to concentrate and magnify the optical signal. In traditional systems the telescope remains fixed while mirror assemblies are manipulated in order to provide the pointing. This research differs by manipulating the entire telescope assembly in order to fulfill the requirements of the separate mirror assemblies. The operation of the optical system of the PPM is shown diagrammatically in **Figure 3**.

The type of telescope used in the PPM engineering model will be a Cassegrain configuration. The main driving force for the selection of this type of telescope is the gain and speed achieved from a very small compact terminal (when compared to other types such as Newtonian). A typical Cassegrain configuration consists of two mirrors (a primary parabolic, and a secondary mirror), where the primary has a central aperture drilled that allows the focused light to be reflected through it from the secondary mirror. The focal length of this telescope assembly has been designed to give an F-number of 5 which combines

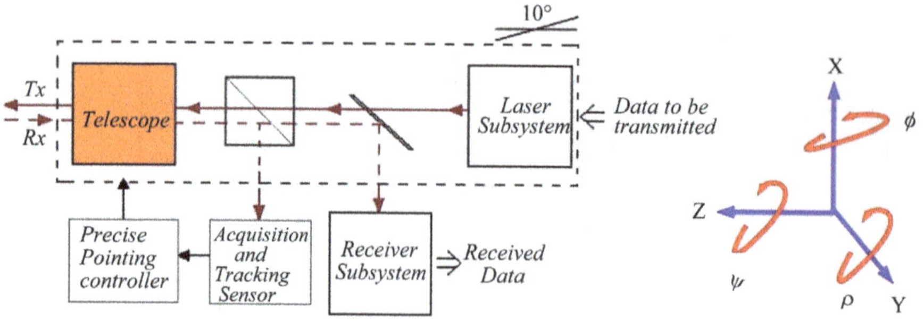

Figure 3. Configuration of OCT using the PPM as the actuation method for the telescope assembly. The optical axis of the telescope assembly is aligned with the Z axis shown.

speed and brightness of image.

3.1. System Configuration

The PPM developed in this research consists of eight reluctance force electromagnets and four pairs of permanent magnets that are used to levitate and actuate the telescope element of the OCT that consists of a catadioptric Schmidt Cassegrain spider configuration. A computer aided design (CAD) image of the evaluation model is shown in **Figure 4**. The model shown in **Figure 4** is only a receiver version of the PPM OCT, but we also have a version that includes a laser transmitter (with some additional optics and different sensor mounting).

The system uses two nested control loops and feedback networks to provide coarse pointing and fine tracking. Coarse pointing is achieved through the use of three low cost Eddy Current Probes (ECP) mounted underneath the telescope assembly (PRS-04 Fabricated by Sensonics). These provide positional data with accuracy around 2 μm which is not adequate to sustain an optical link (as will be discussed later). Fine tracking occurs once an optical signal is present on entrance of the telescope and thus focused onto the Position Sensing Device (PSD), the PDM-10 produced by On-Trak Photonics. This is a dual axis PSD and tracks the laser beacon to an accuracy of 0.1 μm (±6 μm seen on the active area of the PSD) which can sustain the proposed optical link without excessive degradation of the communication channel. These measured errors in pointing are then used by the controllers to actuate the telescope to reduce the overall pointing error.

The engineering model was fabricated by the workshops of the University of Surrey with some parts. Such as the laminated electromagnet cores being sub contracted. **Figure 5** shows the completed PPM setup in the SSC laboratories, mounted on a passive anti-vibration table. This is to reduce the influence of stray vibrations on the PPM during experimentation.

The engineering model was designed to demonstrate the levitation and actuation of the telescope of an OCT using low cost COTS, and to demonstrate the use of layered control loops combining lower resolution sensors and low cost higher resolution sensors, resulting in a range of pointing accuracies required to maintain an optical link.

Although the data rate of the OCT scenario is not evaluated in experimentation the divergence of the link scenario is facilitated using a small lens mounted on the aperture of the laser. This diverges the laser at an angle of approximately 30°. The entrance pupil to the telescope assembly is then 150 mm away from the laser source which allows the optics (which were designed for a 3000 km optical link) to focus the laser source back to a fine spot on the PSD. It was found during experimentation that the quality of the low cost laser introduced a small positional error caused by the elliptical beam shape at the exit of the laser.

A system diagram that describes the configuration of the PPM experiment is shown in **Figure 6**.

The PPM OCT is divided into a number of different elements that have been developed to achieve a pointing resolution to sustain a free-space optical communications link.

3.2. System Model

In order to simulate the PPM system a model was developed in Matlab that represents the dynamics of the ter minal and the hardware from which the engineering model is constructed. **Figure 6** shows the system diagram that contains models of the Telescope Antenna dynamics, Sensor Amplifiers, Actuator Coils and Current Amplifier models in order to generate an accurate system model for development of the control system. This model was implemented in the Simulink environment of Matlab and used to simulate performance of the PPM before implementation on the bespoke hardware. The model initially included linearised models of the electromagnetic cores in order to allow the development of the control system. Once this was completed and demonstrated in simulation the model for the electromagnetic actuators was replaced with its non-linear version to model the real life behavior of the actuators within the system. Experimentation was also undertaken to ascertain the resolution

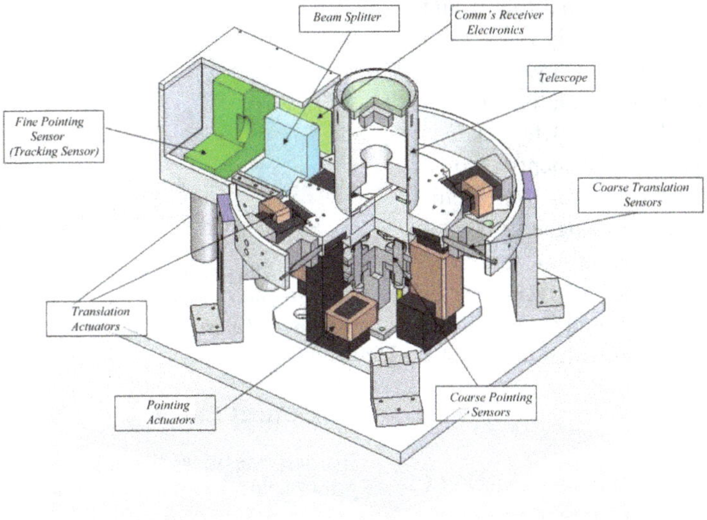

Figure 4. CAD image showing layout and elements of the PPM.

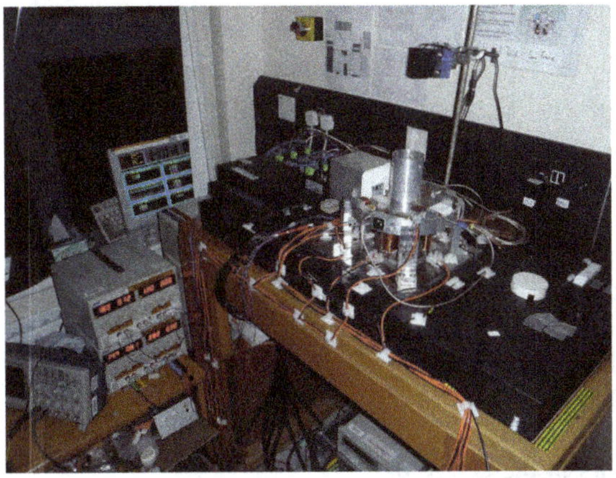

Figure 5. Implemented PPM mounted on anti vibration table.

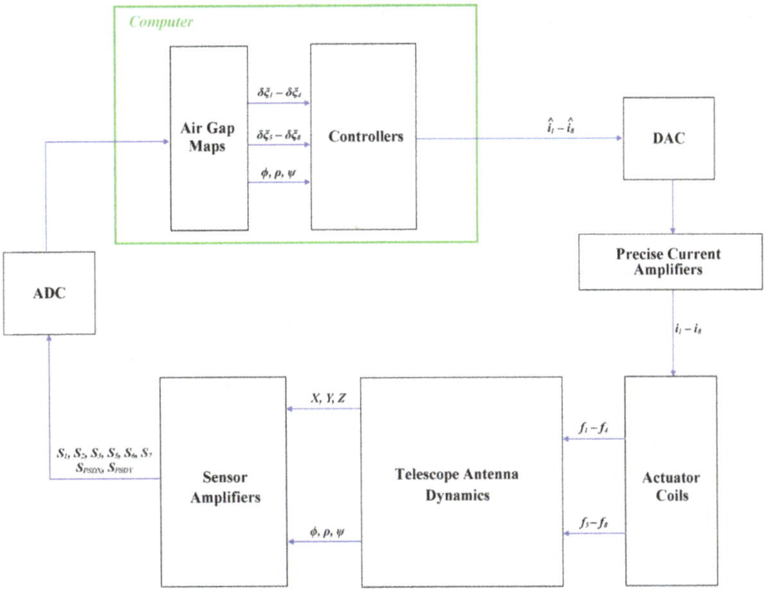

Figure 6. Overview of the setup of the PPM experiment that is shown in the photo of Figure 5.

of the ECP's and fine tracking sensor so that they could be modeled as accurately as possible. Finally noise from the bespoke precise current amplifiers that drive the current in each actuator was found experimentally for each actuator channel and then modeled for completeness.

In hardware all but the "Computer" elements (shown in the green box) are replaced with their hardware equivalents. National Instruments 6023E Analogue to Digital Converter (ADC) card and 6703 Digital to Analogue Converter (DAC) card are used as the interface components between the PPM hardware, and the host computer running XPC real time environment on which the control system is implemented.

Telescope Assembly Dynamics

The configuration of the Telescope Antenna of the PPM contains both translation and rotation components.

The behaviour of the PPM in translation along the X, Y, or Z axis is governed by Newton's Second Law as in Equation (1),

$$F_X = M_{PPM}\alpha_X \tag{1}$$

where F_X is the force acting along the X axis, M_{PPM} is the mass of the telescope assembly and α_X is the acceleration along the X axis. The accelerations (α_X, α_Y and α_Z) are found in the PPM by taking the second derivative of the terminal displacement along each respective axis, which is measured through the use of the coarse sensors (ECP's).

The Telescope Assembly (shown in **Figure 7**) is designed so that Centre of Mass (CM) is also the centre through which the Telescope Assembly rotates. This location is shown in **Figure 7**, and occurs in the centre of the 45° mirror that is used to divert the focused laser spot onto the fine tracking sensor. This reduces the complexity in the control and reduces the amount of power needed to rotate the telescope. If the CM of the Telescope Assembly was located at the entrance pupil more power would be needed in the actuators to maintain stability of the system.

The Inertia of the telescope assembly shown in **Figure 7** can be described as Equation (2).

$$I_{PPM} = \begin{bmatrix} 264 & -0.005 & -0.008 \\ -0.008 & 263 & 0.008 \\ 0.008 & -0.005 & 283 \end{bmatrix} \times 10^{-6}\ \text{kgm}^2 \tag{2}$$

These align with the X, Y and Z axis of the telescope assembly. Using Equation (2) in Euler's equation of motion to describe the rotational motion of the telescope assembly, we yield Equation (3).

$$T_X = I_{XX}\dot{\omega}_X + \omega_X \times I_{XX}\omega_X \tag{3}$$

where the torque about the X axis is given by T_X, (and similarly for the Y and Z axis, T_Y and T_Z respectively), ω_X is the angular acceleration about the X axis.

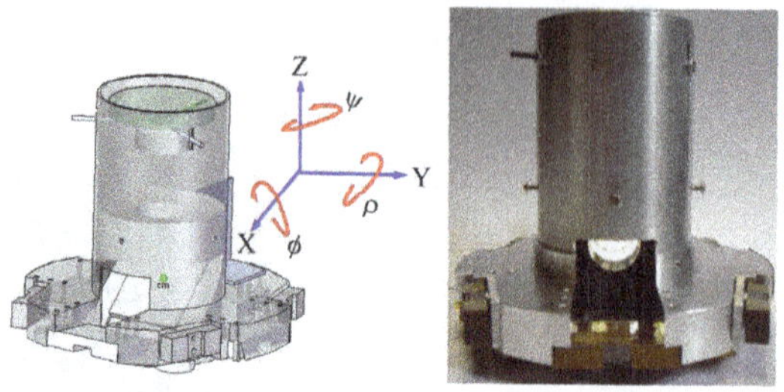

Figure 7. Illustration showing the location of the Centre of Mass of the Telescope Assembly (left), and a photo of the fabricated Telescope Assembly for use in the engineering model (right).

3.3. Optical Bench

This part considers the simulation of the antenna element of the OCT and considers some sources of error that will affect the performance of it, which will impact the fine pointing resolution of the PPM. An F5 catadioptric Cassegrain telescope has been developed as this offers a good compromise between image intensity at the focal plane and speed. A note to consider is that this Telescope Assembly would need to be used at least 15° off the sun-line in order to prevent damage to the receiver and optics. This requirement can be modified by developing baffles. The 60 mm diameter parabolic mirror and telescope aperture gives a gain of approximately 106 db generated by the telescope to offset some of the losses associated with free space optical communications.

The telescope assembly is tilted about φ and ρ (X and Y shown in **Figure 7** to correct for pointing errors, while the remaining four degrees of freedom remain fixed (X, Y, Z and rotation about Z, ψ).

Figure 8 shows the ray trace for the engineering model with a zero degree pointing error in the incoming rays. The laser source is modelled as a Laser Diode placed 150 mm away from the entrance to the telescope with a 30° divergence angle. This is representative of the engineering model configuration and the divergence angle of the laser is achieved through the use of a diffusing lens mounted at the exit of the laser source.

In **Figure 9** the incoming rays from the laser source are focused onto the centre of the Tracking Sensor and APD (via the beam splitter hard mounted onto the stator element of the terminal). The radius of the focused laser signal spot is approximately 2 microns RMS seen on both the centre of the Tracking Sensor and the APD. An image of the spot using relative intensity values can be seen in **Figure 9**. The relative illumination on the fine tracking sensor and the spot position can be seen in **Figures 10** and **11** respectively.

3.4. Magnetic Actuators

The PPM consists of two electromagnetic systems that provide levitation and control over the pointing and

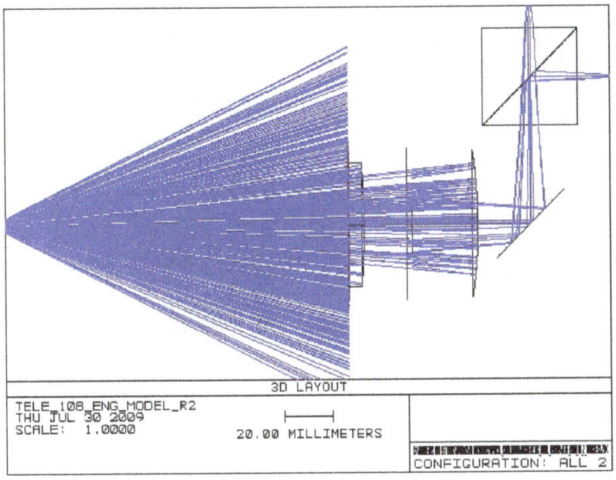

Figure 8. Ray trace image of terminal with 0° pointing error.

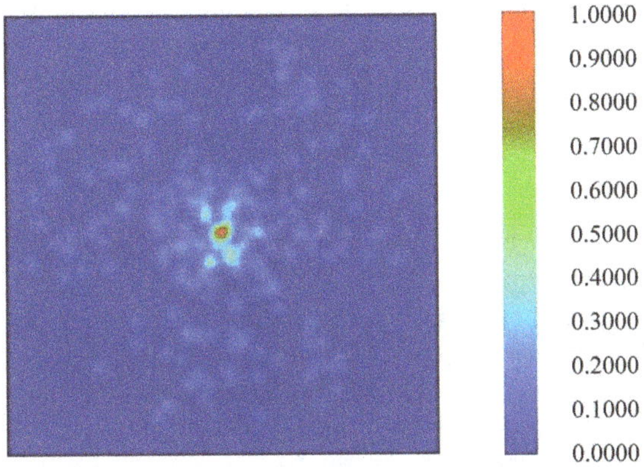

Figure 9. Relative intensity spot image on the tracking sensor with a 0° pointing error. This slide is 10 µm square at a resolution of 128 by 128 pixels.

translation of the telescope antenna in 6 Degrees of Freedom (X, Y, Z, φ, ρ and ψ). Two sets of electromagnetic cores were designed and manufactured by Photofabrication.

The magnetic systems are illustrated graphically in **Figure 12**.

The actuators highlighted in blue control translation along Z, rotation about X (φ) and rotation about Y (ρ). The actuators highlighted in green control translations along X and Y and rotation about Z (ψ).

Pointing in this context refers to φ, ρ and ψ, and translation refers to linear motion along axis X, Y and Z.

Permanent magnets (PM's) are used in a repulsive configuration that have been sized and selected in order to provide a levitation force along the Z axis (**Figure 14**) are shown in **Figure 13**. The repulsive force generated by these permanent magnet pairs serves two purposes. For Earth testing of the PPM this force offsets the effects of gravity on the mass of the telescope antenna. It also provides a force that opposes the reluctance force generated

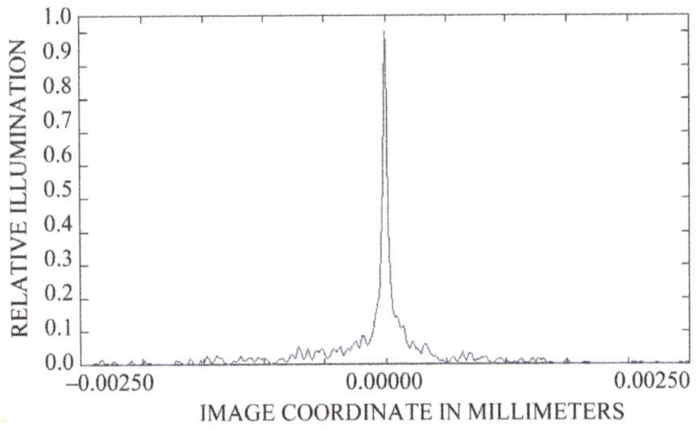

Figure 10. 2D relative intensity on the X axis of the tracking sensor with 0° pointing error.

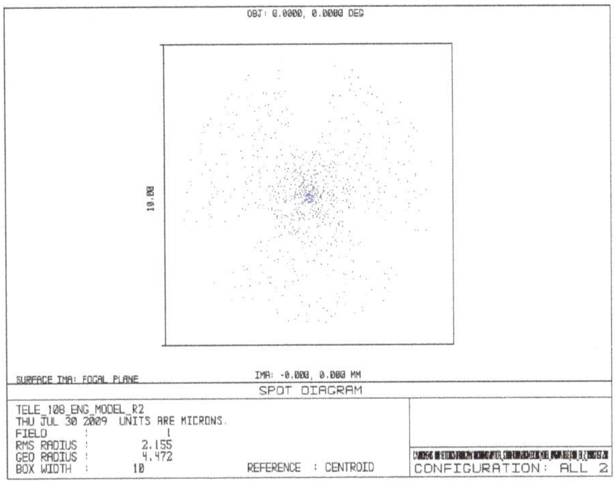

Figure 11. Spot size and position on the active area of the tracking sensor with 0° pointing error.

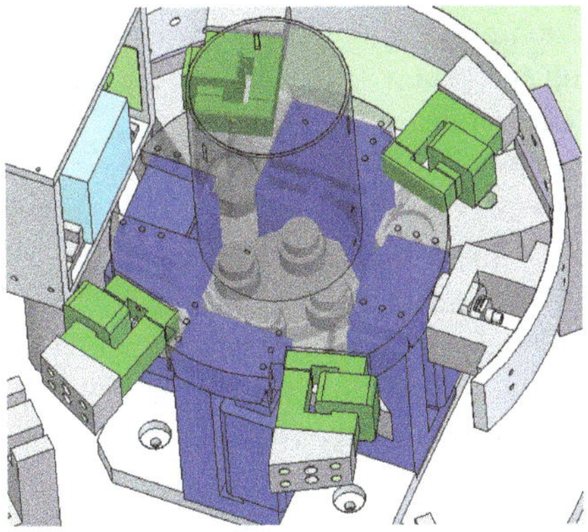

Figure 12. Illustration showing the magnetic systems that levitate and actuate the PPM.

by the four electromagnets, thus allowing control over the Z displacement. The forces produced by the pointing actuators for this engineering model total a net force of 12.04 N.

The force allows the effects of gravity on the telescope assembly to be offset, as well as providing a net force of approximately 0.04 N (in reference position) for the pointing actuators (f_1, f_2, f_3 and f_4) to overcome, allowing

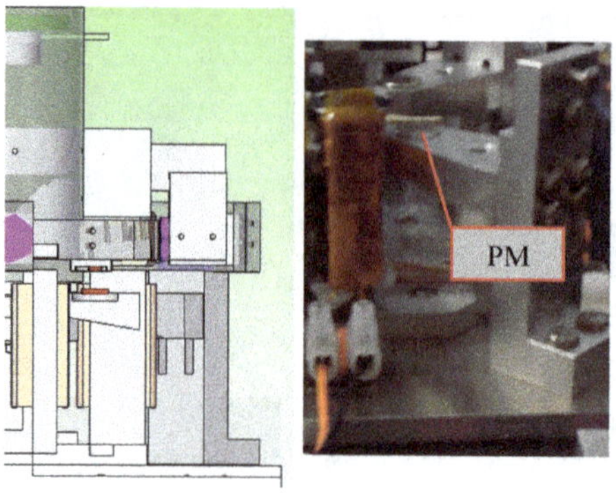

Figure 13. CAD image showing position of the permanent magnets highlighted in red (left) with a photo showing the position of the lower permanent magnet within the PPM. The top PM is embedded in the telescope assembly (right).

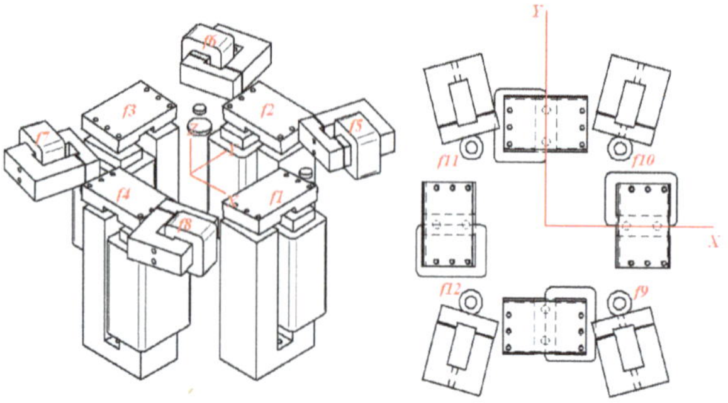

Figure 14. Layout showing placement of electromagnets and forces.

levitation to be achieved.

The repulsive forces produced by the permanent magnets is a function of the air gap between the pairs, therefore this is adjustable in order to allow these forces to be tuned on the engineering model.

The permanent magnets also produce undesired forces that induce a torque (T_Z) about the Z axis, and unwanted translation along X and Y.

The effects of these are reduced as much as possible through offsetting the positions of the permanent magnets. Through magnetic modelling using the Ansoft Maxwell 3D package the unwanted forces generated were found to be less than 10% of the force produced by the electromagnets when a full 5° (scenario in which worse case unwanted disturbances are experienced), rotation about φ and ρ was commanded, so there is adequate margin in the electromagnets to overcome these disturbances.

3.5. Control System

The control architecture of the PPM consists of six Single Input Single Output (SISO) decoupled controllers that each controls a single degree of freedom utilising a coarse sensor network that comprises Sensonics PRS04 and PRS02 Eddy Current Probes.

A further two controllers are then implemented using a fine tracking sensor that consists of the On-Trak PSM-10 position sensing device (PSD). This control strategy is summarised in **Figure 15**.

It is this combination that allows for the two resolutions achievable using the PPM (Coarse and Fine). The telescope assembly is actuated by manipulating the magnetic field in order to change the displacement of the telescope assembly.

In essence it obeys the relationship of Equation (4), where F_X is the force along the X axis, f_{bias} is the bias force applied to that actuator and $\delta\hat{f}$ is the demanded force from the control system to correct for disturbances on the telescope assembly.

$$F_X = f_{bias} + \delta\hat{f} \tag{4}$$

Using the relationship between the force produced by the actuators and the current required to generate that force (Equation (5)) together with the inverse of the terminal dynamics the change in each individual actuator f_n,

$$f_n = -\left(\frac{\mu_0 N^2 A_n}{4}\right)\left(\frac{i_n^2}{\xi_n^2}\right) \tag{5}$$

where i_n is the current in the nth coil and ξ_n is the air gap between the rotor and core parts (the rotor being attached to the telescope ring), μ_0 is the permeability of free-space, N is the number of turns of conductor on the electromagnets core and A_n is the area of a single actuator pole face. The force generated by the electromagnets is a reluctance force which is always attractive therefore the minus sign in Equation (5) represents this attraction.

The current demands from each controller are then mapped onto the required electromagnetic cores to generate these desired force changes to correct for disturbances and pointing errors influencing the telescope assembly. The controllers are implemented in the form of Lead-Lag compensator's in the form of Equation (6).

$$K = K_X \times \left[\frac{\sqrt{\beta_X}}{\beta_X} \times \frac{s + \sqrt{\beta_X}\,\omega_{cX}}{s + \sqrt{\beta_X}\,\dfrac{\omega_{cX}}{\beta_X}}\right] \times \left[\frac{s + \omega_{iX}}{s}\right] \tag{6}$$

where K_X is a gain term for the axis in question, β_X is the damping coefficient and ω_{cX} and ω_{iX} are the corner frequencies of the lead and lag terms respectively. The gain term K_X was derived from the dynamic model for the telescope assembly when the gain of the controllers causes the magnitude to be 1 at the cross-over frequency. The gains are then tuned once when implemented in simulation or on the hardware. When implementing the control strategy on the simulation model or hardware the controllers did not initially include an integration term (the lag term in the controller), which was added once the

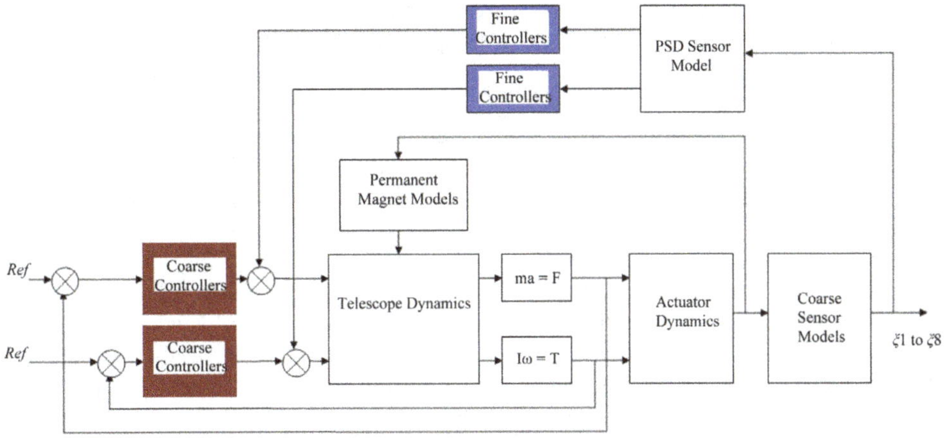

Figure 15. Overview of control strategy to levitate and track an optical signal. In red are the coarse controllers that levitate and actuate the telescope assembly using the eddy current probes, and in blue are the fine controllers that tilt the telescope about X and Y (φ and ρ).

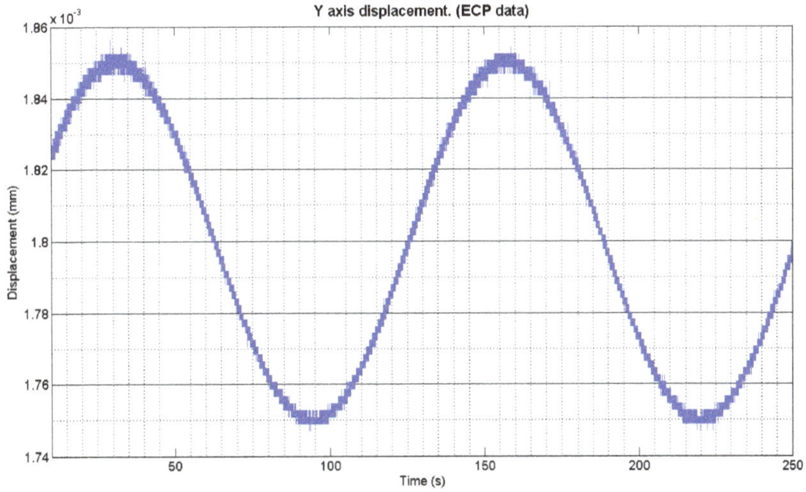

Figure 16. Response of the Y axis (S_6 ECP data) to a sine wave using the coarse control loops and sensor network.

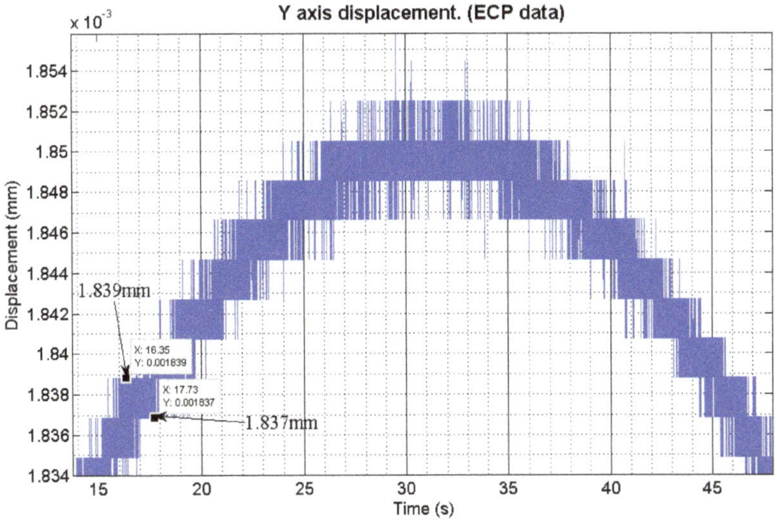

Figure 17. Zoomed in top section of Y axis response. The quantisation noise can clearly be seen and the positional resolution of the axis is ±2 µm.

system was stabilised. This made it easier to achieve levitation and actuation, although as expected the steady-state error was quite high for each axis.

It was at this point that the fine controllers were introduced to the system to provide an extra level of feedback from the fine tracking sensor that manipulates φ and ρ to compensate for pointing errors using the position error of the laser spot on the PSD.

They take the same form as shown in Equation (6), and did not include the lag term initially. A commanded sine wave was used to demonstrate the response of the terminal along the Y axis. The response of the telescope assembly, along with the quantisation noise can be seen in **Figures 16** and **17**. This also shows the step sizes achievable along this axis.

A commanded rotation for φ was driven with a similar sine wave (±8 milli-radians @ 0.05 rad/s). Rotation φ is driven by a commanded torque T_X about the X axis. **Figure 18** shows the rotation of the telescope assembly in response to the commanded position changes.

Figure 19 shows the position of the terminal once the Laser is passed through the telescope optics and incident on the active area of the PSD.

Due to the mechanical fixing of the PSD a positive change in angle φ corresponds to a negative peak on the active area of the fine tracking sensor (the PSD). This is illustrated in the time differences in positive peaks in **Figures 18** and **20**.

The positional resolution of the telescope assembly while under coarse control is within ±5 μradians for both φ and ρ, which is short of the resolution required to maintain an optical communications link.

However during simulation of the engineering model the resolution of the eddy current probes was expected to be lower than the 2 μm (expected to be around ±4 μm) that could be expected under typical operating conditions. This is shown in the histogram of **Figure 21**.

During implementation the performance of the ECP

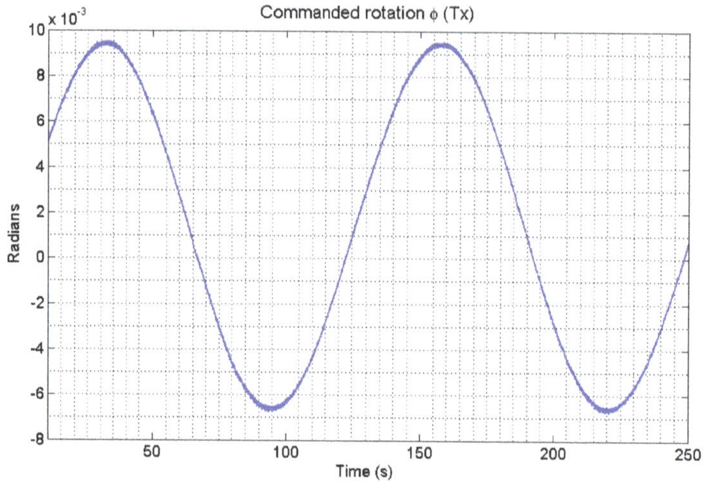

Figure 18. Response of the telescope assembly to a commanded rotation of φ.

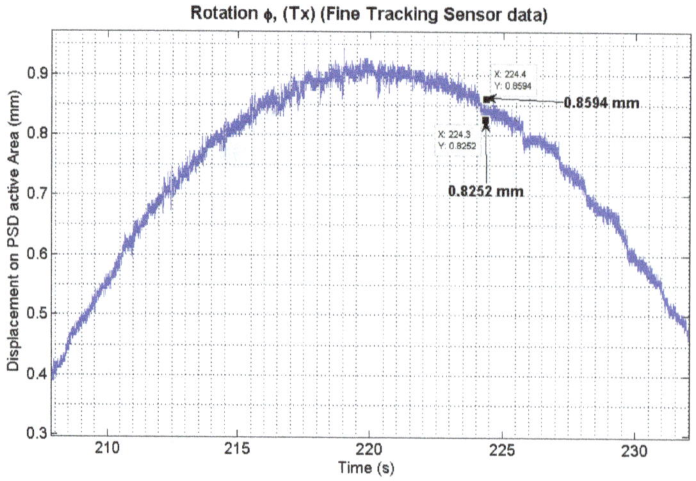

Figure 19. Rotation φ as seen on the respective axis of the fine tracking sensor (PSD). This corresponds to a resolution of approximately 0.18 mRadians under coarse control.

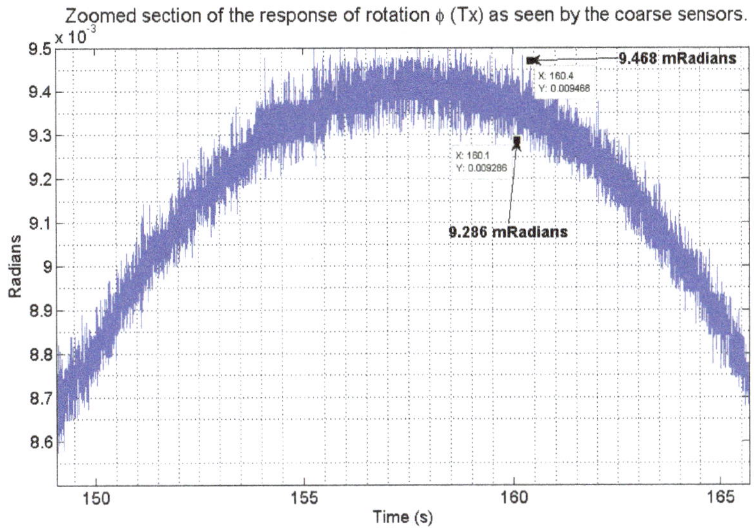

Figure 20. Zoomed view of the response of φ shown in Figure 18. This shows the positional resolution to be approximately 0.18 m Radians.

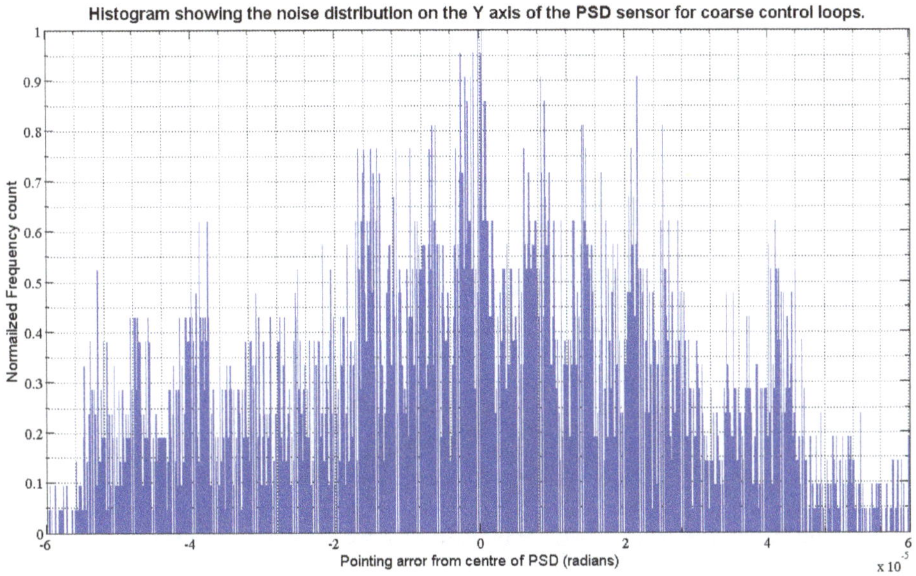

Figure 21. Pointing resolution of the simulated coarse control loops using the ECP sensors.

sensors was worse than modelled in simulation (In the order of ±6 μm. This was due to the metal used in fabrication of the telescope assembly and the small size of sensing area available on the telescope assembly. The performance was dramatically improved however with the addition of very thin steel sheets that were glued to the sensing surfaces of the telescope assembly.

This has the effect of dramatically increasing the conduction of the eddy currents which in turn improves the resolution and performance of the ECP's.

At this point if an optical signal is incident on the PSD tracking sensor then the fine control loops are activated and the terminal actuated to track the signal.

The resolution of the fine loop is plotted over the top of the coarse control loops to demonstrate the two resolutions achievable with the PPM of this research (**Figures 22** and **23**).

The histogram is taken over 400 seconds of steady state data. The experiment was repeated a number of times over a large range of pointing positions. The curve shown in black is the modelled pointing resolution requirements for a 3000 km optical link [15].

4. Conclusions

This research has seen the investigation, implementation and successful testing of a new concept for the precise pointing of, in this instance an optical antenna for communications. The realisation of the PPM allows for sub

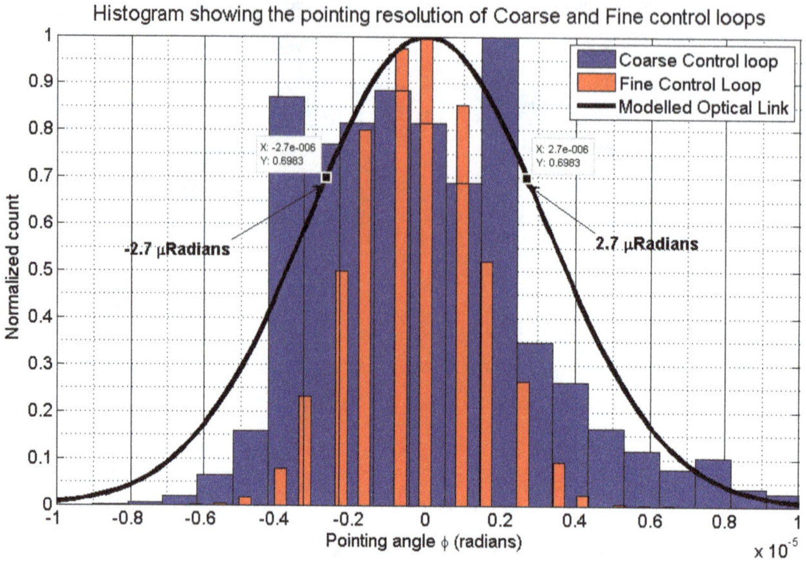

Figure 22. Coarse and fine pointing control loops plotted against the received optical power as a function of pointing angle error for φ.

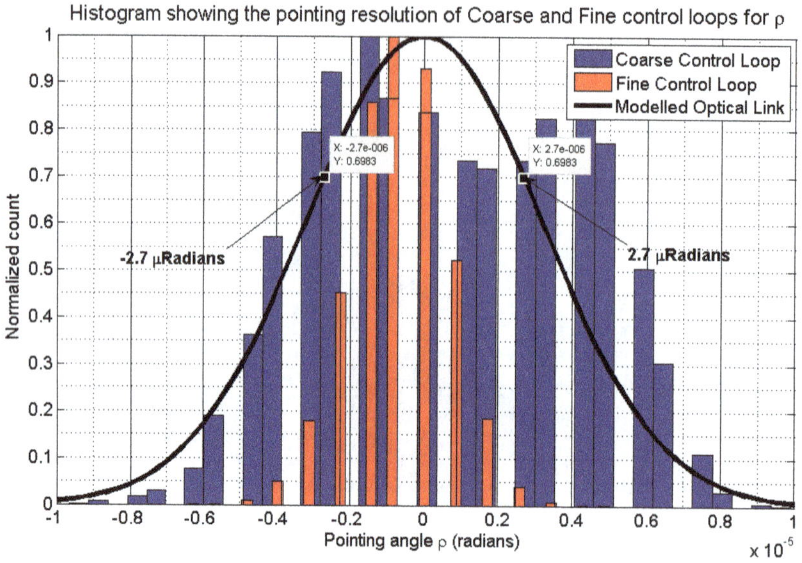

Figure 23. Coarse and fine pointing control loops plotted against the received optical power as a function of pointing angle error for ρ.

micro-meter, and sub micro-radian resolution in all six DOF over a larger displacement range than existing systems. This represents a significant step forward when compared with other mechanisms that utilise magnetic levitation, which, while achieving resolutions comparable to the PPM generally lack the displacement range.

The control strategy presented allows the combination of low cost lower resolution sensors for measuring the position within 6 DOF, with a higher cost greater resolution sensor to achieve an overall increase in accuracy and resolution of the system. This strategy can be utilised to reduce the cost of implementation while retaining the resolution requirements within a system.

The frictionless environment, and contact less nature of the PPM means that the positional accuracy will not degrade as a function of mechanical wear. This also means that no lubrication is required, which can be a significant issue in space craft mechanism. Although the lifetime and consistency of operation cannot be evaluated in the lifetime of this research, no degradation or inconsistency in functionality could be observed during the eight months of experimentation.

5. Acknowledgements

The authors would like to thank The Nuffield Foundation

for their support of this project through grant NAL/32791, and ESA for their contribution under a pump priming fund (FRSF09/16).

REFERENCES

[1] S. Arnon, S. Rotman and N. S. Kopeika, "Optimum Transmitter Optics Aperture for Satellite Optical Communication," *IEEE Transactions on Aerospace and Electronic Systems*, Vol. 34, No. 2, 1998, pp. 590-596. doi:10.1109/7.670339

[2] H. Wang and Q. Chu, "An Inline Coaxial Quasi-Elliptic Filter with Controllable Mixed Electric and Magnetic Coupling," *IEEE Transactions on Microwave Theory and Techniques*, Vol. 57, No. 3, 2009, pp. 667-673.

[3] G. Baister and P. Gatenby, "The SOUT Optical Intersatellite Communication Terminal," *IEEE Proceedings Optoelectronics*, Vol. 141, No. 6, 1994, pp. 345-355.

[4] G. Baister and C. Haupt, "The ISLFE Terminal Development Project Results from the Engineering Breadboard Phase", 20*th AIAA International Communication Satellite Systems Conference*, 12-15 May 2002.

[5] M. Toyoshima and K. Araki, "In-Orbit Measurements of Short Term Attitude and Vibrational Environment on the Engineering Test Satellite VI Using Laser Communication Equipment," *Optical Engineering*, Vol. 40, No. 5, 2001, pp. 827-832.

[6] D. Bamber and P. Palmer, "Attitude Determination through Image Registration Model and Test-Case for Novel Attitude System in Low Earth Orbit," *AIAA/AAS Astrodynamics Specialist Conference and Exhibit*, Keystone, 21-14 August 2006.

[7] S. Arnon and N. S. Kopeika, "Laser Satellite Communication Network-Vibration Effect and Possible Solutions," *Proceedings of the IEEE*, Vol. 85, No. 10, 1997, pp. 1646-1661. doi:10.1109/5.640772

[8] R. Seiler and C. Allegranza, "Mechanism Noise Signatures: Identification and Modelling," 13*th European Space Mechanisms and Tribology Symposium*, Vienna, 23-25 September 2009.

[9] S. Verma and W. J. Kim, "Six-Axis Nanopositioning Device with Precision Magnetic Levitation Technology," *IEEE/ASME Transactions on Mechatronics*, Vol. 9, No. 2, 2004, pp. 384-391.

[10] J. Seddon and A. Pechev, "3Dwheel: 3-Axis Low Noise, High-Bandwidth Attitude Actuation from a Single Momentum Wheel Using Magnetic Bearings," 23*rd Annual AIAA/USU Conference on Small Satellites*, Utah, 10-13 August 2009.

[11] Z. Ren and L. Stephens, "Laser Pointing and Tracking Using a Completely Electromagnetically Suspended Precision Actuator," *American Institute of Aeronautics and Astronautics, Journal of Guidance, Control and Dynamics*, Vol. 29, No. 5, 2006, pp. 1235-1238.

[12] Y. Chen and M. Wang, "Modeling and Controller Design of a Maglev Guiding System for Application in Precision Positioning," *IEEE Transactions on Industrial Electronics*, Vol. 50, No. 3, 2003, pp. 493-506.

[13] F. Ayoub, S. Leprince, R. Binet, K. Lewis and O. Aharonson, "Influence of Camera Distortions on Satellite Image Registration and Change Detection Applications", *Geoscience and Remote Sensing Symposium*, Vol. 2, No. 1, 2008, pp. II-1072-II-1075.

[14] R. McKay and M. Macdonald, "Non-Keplerian Orbits Using Low Thrust, High ISP Propulsion Systems," 60*th International Astronautical Congress*, Daejeon, 12-16 October 2009.

[15] T. Frame and A.Pechev, "Optical Communications Terminals with Precise Pointing," *Proceedings of the* 13*th European Space Mechanisms and Tribology Symposium*, Vienna, 23-25 September 2009.

Reducing Refractive Index Variations in Compression Molded Lenses by Annealing

Bo Tao[1], Lianguan Shen[1], Allen Yi[2], Mujun Li[1], Jian Zhou[1]
[1]Department of Precision Machinery and Precision Instrumentation, University of Science and Technology of China, Hefei, Anhui, China
[2]Department of Integrated Systems Engineering, the Ohio State University, Columbus, Ohio, USA
Email: taoboq@mail.ustc.edu.cn

ABSTRACT

Compression molding of glass optics is gradually becoming a viable fabrication technique for high precision optical lenses. However, refractive index variation was observed in compression molded glass lenses, which would contribute to image quality degradation. In this research, annealing experiments were applied to control the refractive index variation in molded glass lenses. The refractive index variations pre and post annealing experiment in molded lenses were measured by an experiment setup based on Mach-Zehnder interferometer. The experimental results showed that the refractive index variation can be controlled providing that a proper cooling process is applied during cooling.

Keywords: Refractive Index; Mach-Zehnder Interferometer; Optical; Annealing; Compression Molding

1. Introduction

Compression molding is a thermal forming process for precision glass optics [1, 2]. However, refractive index variation was induced in glass during cooling when the glass material went through its glass transition region [3-5]. On the other hand, the variation of refractive index in a molded glass lens will introduce distortion to the wave-front passing through the glass lens, which leads to image quality degradation. In order to ensure a proper optical performance of thermally formed glass lens, it is important to reduce the degree of refractive index variation in the molded glass lens.

Annealing has been studied for improving the quality of glass [6-9]. Through annealing, glass can achieve homogeneous refractive index. Annealing of the compression molding of glass lenses was investigated in this research. In order to identify the effects of annealing on refractive index variations, computer tomography method was employed to measure the refractive index distribution in the glass lenses. The measurement was conducted by using an optical setup based on Mach-Zehnder interferometer [4, 10]. By comparing experimental results of the glass lens pre and post annealing, the results showed that refractive index variation was reduced.

2. Design of Experiments

Glass lenses studied in this research were molded in a commercial glass molding machine (GMP-211V). Thermal forming was carried out at 684 °C. After forming, the temperature was maintained until stresses in glass caused by pressing were completely released. Two steps cooling process was followed: the glass lens was cooled to 520 ℃ at a rate of 0.8 ℃/s and then to 200 ℃ at a rate of 1.6 ℃/s. When the glass lens was cooled to 200 ℃, the lens was taken out of the molding machine and cooled naturally at room temperature. **Figure 1** is the illustration of the glass lens. BK7 was chosen as the glass material. The properties of BK7 are shown in **Table 1**.

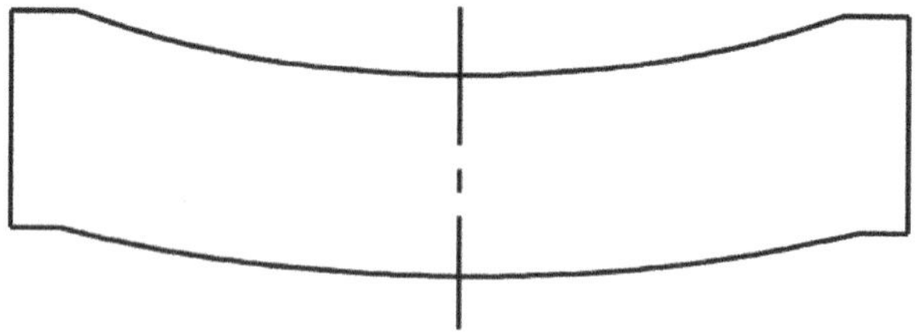

Figure 1. Illustration of the compression molded glass lens.

Table 1. Properties of BK7 glass [11].

Material Properties	BK7
Elastic modulus, E [MPa]	82,500
Poisson's ratio, ν	0.206
Transition temperature, T_g [℃]	557
Refractive index, n	1.5148

2.1. Refractive Index Measurement

Figure 2 shows a schematic of the experiment setup used to measure the refractive index in glass lenses. To avoid refraction, tested lens was placed in a box filled with refractive index matching liquid of BK7. He-Ne laser was used as the light source. In the experiment setup, beam splitter 1 divides the laser beam into two. One is used as reference beam, the other one goes through the lens under test. These two beams interfere with each other at beam splitter 2. Fringe pattern carried with refractive index distribution in the lens was captured by a CCD camera.

Figure 3 is the fringe pattern of the molded glass lens at one direction. The fringe pattern was analyzed by a 2D Fourier transform technique [12] and a least-square phase unwrapping method [13] for unwrapped phase. Because of the axisymmetric property of the glass lens, it is sufficient to reconstruct the refractive index distribution in the glass lens through only one fringe pattern. With the unwrapped phase, 3 dimensional (3D) refractive index distribution of the glass lens relative to the refractive index of matching liquid can be reconstructed using the filtered back-projection method [14].

$$n(x,y,z) = \frac{p(x,y,z)\lambda}{2\pi d} \quad (1)$$

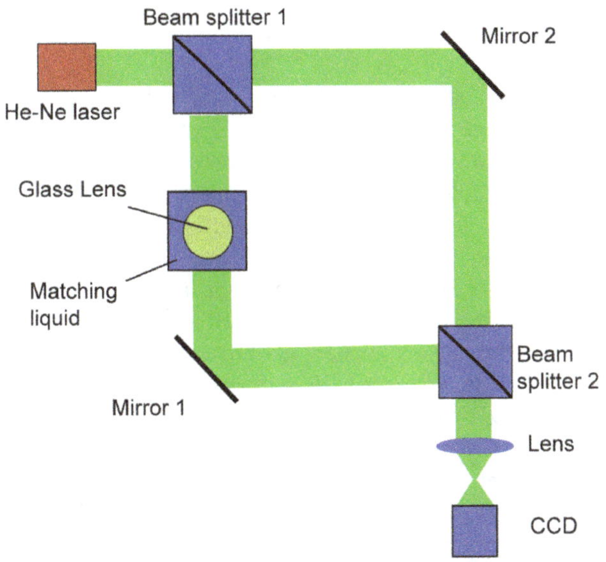

Figure 2. Schematic of the experiment setup for refractive index measurement.

Figure 3. Fringe pattern of the compression molded glass lens before annealing at one direction.

where, $p(x,y,z)$ is the reconstructed phase distribution at point (x,y,z), d is the pixel size in the test lens of the interferogram. The reconstructed refractive index in the middle section of the compression molded lens relative to the matching liquid is shown in **Figure 4**.

2.2. Annealing

Annealing experiments were conducted in a commercial furnace (Grieve, BF-12128-HT). **Figure 5** illustrates the time-temperature history of the annealing experiments.

At first, the molded glass lens was heated to 560 ℃, slightly higher than the glass's transition temperature, and soaked for 10 minutes. After soaking, the glass lens was cooled to 500 ℃ at a rate of 1 ℃/min. Then the furnace was turned off. The maximum cooling rate was about 1.29 ℃/min after turning off the furnace, because there was no force cooling system. Once the temperature decreased to 150 ℃, the glass lens was taken out of the furnace and cooled to room temperature naturally. In the experiments, the glass lens under thermal treatment was placed on a ceramic plate in the furnace with the concave side facing down.

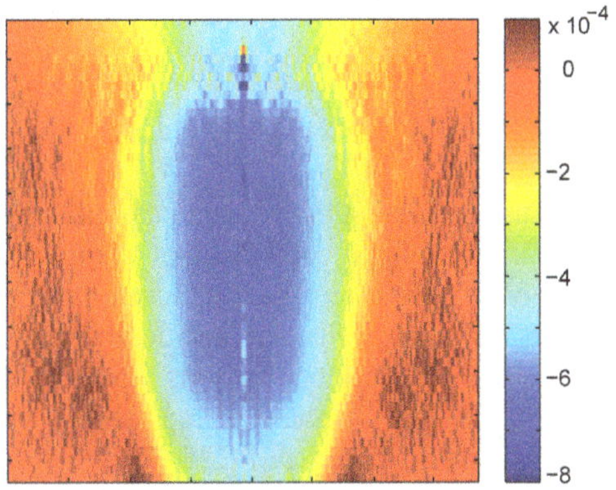

Figure 4. Refractive index distribution of middle section of the molded glass lens relative to the matching liquid.

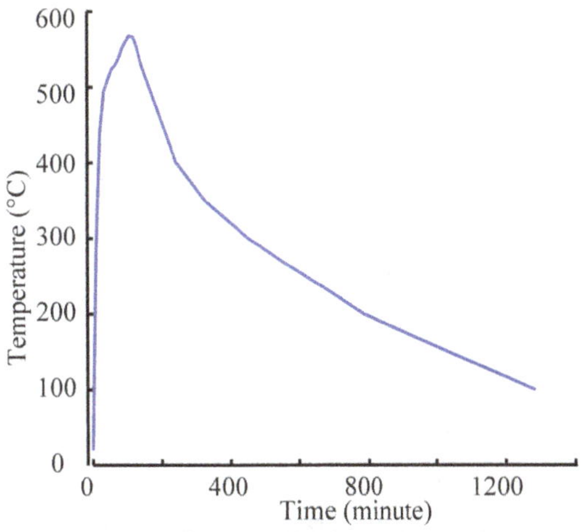

Figure 5. Time-temperature history of annealing.

3. Results and Discussion

Refractive index variations of the molded glass lens pre and post annealing were both measured. **Figure 6** shows the average refractive index variations in the middle section of the glass lens along radial direction pre and post annealing experiment. The maximum refractive index variation was reduced about 4 × 10^{-4} which was more than half of the maximum variation before annealing.

The relations between refractive index n and density ρ of the glass material can be described by Lorentz–Lorenz equation [15]:

$$\frac{n^2-1}{n^2+2}=\frac{4\pi}{3}\frac{N_A\rho\alpha}{M} \tag{2}$$

where N_A is the Avogadro number, α is the mean polarization and M is the molar weight. Differentiating Equation (2), the relations between refractive index change dn and density change $d\rho$ can be obtained by:

$$\frac{dn}{d\rho}=\frac{(n^2-1)(n^2+2)}{6n\rho} \tag{3}$$

Substituting volume for the density, refractive index change Δn can be calculated from the volume change ΔV and the original volume V_0:

$$\Delta n=\frac{(n^2-1)(n^2+2)}{6n}\left(\frac{-\Delta V}{V_0+\Delta V}\right) \tag{4}$$

After annealing, residual stresses inside the glass lens were released. Values of the coefficient of thermal expansion (CTE) of the glass lens were different at different cooling rates due to the behavior of structural relaxation when the glass went through its glass transition region. As such, the volume change was imported by both stresses relaxation and the changes of CTE [16]:

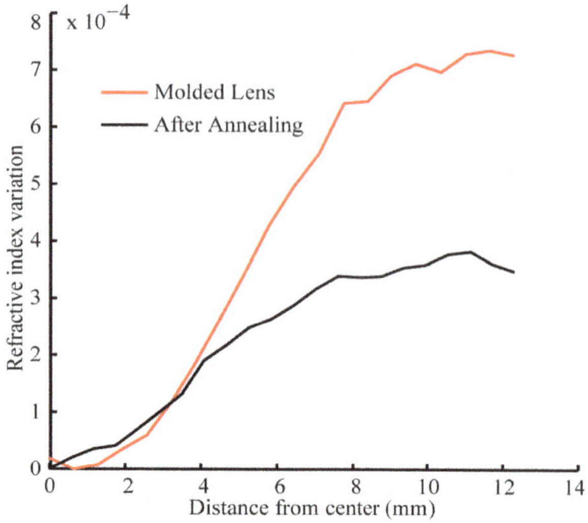

Figure 6. Comparison of the refractive index variations in a molded glass lens pre and post annealing.

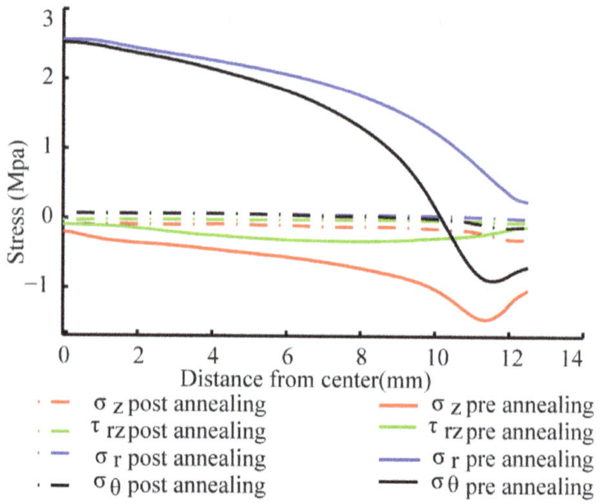

Figure 7. Residual stresses in the glass lens pre and post annealing in cylindrical coordinate.

$$\Delta V=V_o\left[3\beta_h-3\beta_c\right]+V_o\left[\frac{1-2v}{E}(\sigma_{11}+\sigma_{22}+\sigma_{33})\right] \tag{5}$$

where, $\beta_h=\int_{T_1}^{T_2}\alpha_h(T)dT$ and $\beta_c=\int_{T_2}^{T_1}\alpha_c(T)dT$.

α_h and α_c are CTE during heating and cooling, respectively. T_1 is room temperature and T_2 is soaking temperature. v is Poisson's ratio and E is elastic modulus. σ_{11}, σ_{22} and σ_{33} are the changes of normal stresses along the axes.

The residual stresses inside the glass lens can be measured by a circular polariscope based on the property of birefringence when the glass is stressed [17-19]. The maximum residual stress in the compression molded glass lens was about 3 Mpa in cylindrical coordinate system. **Figure 7** shows residual stresses in the middle section of the glass lens pre and post annealing. The residual stresses were significantly released after annealing. Substituting the maximum changes of residual stresses, E and v of the glass lens into Equation (5), the volume change induced by stress relaxation can be calculated, ΔV = 6 × 10^{-5} V_0. The maximum refractive index change induced by stresses relaxation was 4 × 10^{-5}.

Therefore the refractive index change was mainly caused by the changes of CTE. Basically, slower cooling rates yield lower volumes. In order to get lower refractive index variation, lower cooling rate should be applied.

4. Conclusions

Refractive index variations in compression molded glass lenses were investigated using an experiment setup. The refractive index variation was induced during the cooling due to structural relaxation of the glass material. Consequently the refractive index variation in the glass lens can induce wave-front distortion. To control the refractive index variation, annealing of the glass lenses were conducted and the results demonstrated that the refractive index variations were significantly reduced. This research demonstrated that CTE played a critical role in process optimization for precision compression molded glass optics.

5. Acknowledgements

The material is partially based on work supported by National Science Foundation under Grants No. CMMI 0547311. Any opinions, findings, and conclusions or recommendations expressed in this material are those of the authors and do not necessarily reflect the views of the National Science Foundation. The work is also supported by National Natural Science Foundation of China (No. 51075381). Bo Tao acknowledges the financial support from China Scholarship Council.

REFERENCES

[1] R. O. Maschmeyer, C. A. Andrysick, T. W. Geyer, H. E. Meissner, C. J. Parker and L. M. Sanford, "Precision Molded-Glass Optics," *Applied Optics*, Vol. 22, No. 16, 1983, pp. 2413-2415. doi:10.1364/AO.22.002413

[2] A. Y. Yi and A. Jain, "Compression Molding of Aspherical Glass Lenses – A Combined Experimental and Numerical Analysis," *Journal of the American Ceramic Society*, Vol. 88, No. 3, 2005, pp. 579-586. doi:10.1111/j.1551-2916.2005.00137.x

[3] U. Fotheringham, A. Baltes, P. Fischer, P. Höhn, R. Jedamzik, C. Schenk, C. Stolz and G. Westenberger, "Refractive Index Drop Observed After Precision Molding of Optical Elements: A Quantitative Understanding Based on the Tool– Narayanaswamy –Moynihan model," *Journal of the American Ceramic Society*, Vol. 91, No. 3, 2008, pp. 780-783.doi:10.1111/j.1551-2916.2007.02238.x

[4] W. Zhao, Y. Chen, L. G. Shen and A. Y. Yi, "Investigation of the Refractive Index Distribution in Precision Compression Glass Molding by Use of 3D Tomography," *Measurement Science and Technology*, Vol. 20, No. 5, 2009, p. 055109. doi:10.1088/0957-0233/20/5/055109

[5] L. J. Su, Y. Chen, A. Y. Yi, F. Klocke and G. Pongs, "Refractive Index Variation in Compression Molding of Precision Glass Optical Components," *Applied Optics*, Vol. 47 No. 10, 2008, pp. 1662-1667. doi:10.1364/AO.47.001662

[6] J. Wolfe and R. Chipman, "Reducing Symmetric Polarization Aberrations in A Lens by Annealing," *Optics Express*, Vol. 12, No. 15, 2004, pp. 3443-3451. doi:10.1364/OPEX.12.003443

[7] H. R. Lillie and H. N. Ritland, "Fine Annealing of Optical Glass," *Journal of the American Ceramic Society*, Vol. 37, No. 10, 1954, pp. 466-473. doi:10.1111/j.1151-2916.1954.tb13978.x

[8] H. E. Hagy, "Fine Annealing of Optical Glass for Low Residual Stress and Refractive Index Homogeneity," *Applied Optics*, Vol. 7, No. 5, 1968, pp. 833-835. doi:10.1364/AO.7.000833

[9] N. M. Brandt, "Annealing of 517.645 Borosilicate Optical Glass: I, Refractive Index," *Journal of the American Ceramic Society*, Vol. 34, No. 11, 1951, pp. 332-338. doi:10.1111/j.1151-2916.1951.tb13480.x

[10] W. Zhao, Y. Chen, L. G. Shen and A. Y. Yi, "Refractive index and Dispersion Variation in Precision Optical Glass Molding by Computed Tomography," *Applied Optics*, Vol. 48, No. 9, 2009, pp. 3588-3595. doi:10.1364/AO.48.003588

[11] Schott Glass Inc., Products and Applications, www. schott. de

[12] M. Takeda, H. Ina and S. Kobayashi, "Fourier- Transform Method of Fringe-Pattern Analysis for Computer-Based Topography and Interferometry," *Journal of the Optical Society of America*, Vol. 72, No. 1, 1982, pp. 156-160. doi:10.1364/JOSA.72.000156

[13] D. C. Ghiglia and L. A. Romero, "Robust Two-Dimensional Weighted and Unweighted Phase Unwrapping That Uses Fast Transforms and Iterative Methods," *Journal of the Optical Society of America*, Vol. 11, No. 1, 1994, pp. 107-117. doi:10.1364/JOSAA.11.000107

[14] G. T. Herman, "Fundamentals of Computerized Tomography: Image Reconstruction from Projections," *Springer*, London 2009. doi:10.1007/978-1-84628-723-7

[15] H. N. Ritland, "Relation between Refractive Index and Density of A Glass at Constant Temperature," *Journal of the American Ceramic Society*, Vol. 38, No. 2, 1955, pp. 86-88. doi:10.1111/j.1151-2916.1955.tb14581.x

[16] L. J. Su and A. Y. Yi, "Investigation of the Effect of Coefficient of Thermal Expansion on Prediction of Refractive Index of Thermally Formed Glass Lenses Using FEM Simulation," *Journal of Non-Crystalline Solids*, Vol. 357, No. 15, 2011, pp. 3006-3012. doi:10.1016/j.jnoncrysol.2011.04.005

[17] Anton, A. Errapart, H. Aben and L. Ainola, "A Discrete Algorithm of Integrated Photoelasticity for Axisymmetric Problems," *Experimental Mechanics*, Vol. 48, No. 5, 2008, pp. 613-620. doi:10.1007/s11340-008-9121-9

[18] A. Errapart, H. Aben, L. Ainola and J. Anton, "Photoelastic Tomography for the Measurement of Thermal and Residual Stresses in Glass," *9th International Congress on Thermal Stresses*, Budapest, Hungary, 2011.

[19] H. Aben and A. Errapart, "A Non-Linear Algorithm of Photoelastic Tomography for the Axisymmetric Problem," *Experimental Mechanics*, Vol. 47, No. 6, 2007, pp. 821-830. doi:10.1007/s11340-007-9057-5

In-Line Chromatic Dispersion Measurement for NRZ and RZ Signals Using a Novel RF Spectrum Phase Detection Technique

Guozhou Jiang, Ying Mei
College of Educational Information & Technology, Hubei Normal University, Huangshi, China
Email: jgz_hust@163.com

ABSTRACT

In this paper, a novel method of in-line chromatic dispersion (CD) measurement is proposed, we theoretically and experimentally demonstrated the CD measurement of 10-Gbit/s NRZ and RZ signals by a novel RF spectrum phase detection technique, this is performed using in-band tone monitoring RF, electrically down-converted to direct current (DC) or a low intermediate-frequency (IF) of less than 1MHz through electronic mixing with local oscillator (LO) of 2.4 GHz. The measurement provides a large CD measuring range with good accuracies (±2000 ± 35 ps/nm), and independent of the bit-rate and data format. In addition, the use of electronic mixer and low-speed detectors makes it cost effective for in-line CD measurement.

Keywords: Radio Frequency Photonics; Frequency Modulation; Fiber Measurements; Fiber Optics Links; Subsystems; Modulation; Ultrafast Processes in Fibers

1. Introduction

With the increased demand for large data capacity, higher data rates have become an essential requirement of next generation light-wave system. It has been recognized that at data rates higher than 10 Gb/s, it will be necessary to provide tunable dispersion compensators (TDC) in order to accommodate uncertainties and dynamic chromatic dispersion (CD) variations in link, various tunable dispersion compensators have been demonstrated [1]. For such devices to be working effectively, an appropriate residual CD measuring and feedback signal must be obtained from the data stream passing through the TDC. Various feedback signals are possible including, bit-error rate (BER) derived from forward error correction algorithms, eye monitoring e.g [2], and schemes that rely on monitoring the RF spectrum of the detected bit stream after a fast photo-detector e.g [3], but these feedback signals can't indicate the value of residual CD, and it is necessary to adjust TDC till obtain a perfect state of feedback signals, that need a lot of time for adjusting, if residual CD value changed, the process must be iterated.

So some researchers have focused on researching of residual CD monitoring, and there have been several approaches of CD monitoring recently demonstrated, including asynchronous sampling and histogram evaluation [4-8], electrical dispersion equalizer [9], self-phase modulation, four-wave mixing, and cross-phase modulation (XPM) in optical fibers [10-12], radio-frequency (RF) tone measurement [13,14]. However, some of these approaches tends to require either high-speed components (e.g., oscilloscope, detector, RF spectrum analyzer, or analog-to-digital converter), a tunable DLI to decode phase information into amplitudes, or high data input power.

2. Principle of CD Measurement

The model of this technique is depicted in **Figure 1**. This technique is based on electrical mixing with orthogonal I-Q procedure.

The incoming signal $E_{DSB}(t)$ is a modulated dual sideband (DSB) signal, two single sideband (SSB) tunable band-pass optical filter is used to distill upper and lower sideband of the incoming signal, that the upper sideband signal and lower sideband signal are respectively given by:

$$E_L(t) = \alpha\sqrt{I_0}\cos(\omega_0 t) + \beta\sqrt{I_0}\cos((\omega_0 - \omega_d)t + \varphi_L), \quad (1)$$

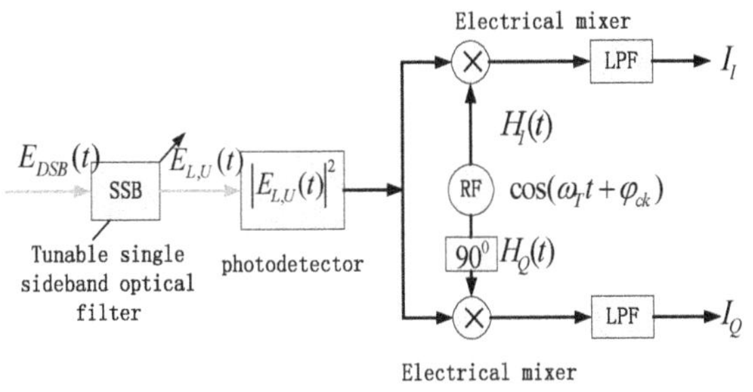

Figure 1. Principle of CD measurement based on RF spectrum phase detection using electrical mixing.

$$E_U(t)=\alpha\sqrt{I_0}\cos(\omega_0 t)+\beta\sqrt{I_0}\cos((\omega_0+\omega_d)t+\varphi_U). \quad (2)$$

After the photo-detector and the electrical mixing, the in-phase and in-quadrature received signals $I_{I,Q}$ for the LSB and USB are given by:

$$I_{IU}=|E_U(t)|^2\times H_I(t), \quad (3)$$

$$I_{QU}=|E_U(t)|^2\times H_Q(t), \quad (4)$$

$$I_{IL}=|E_L(t)|^2\times H_I(t), \quad (5)$$

$$I_{QL}=|E_L(t)|^2\times H_Q(t), \quad (6)$$

where $H_I(t)$ and $H_Q(t)$ are the electrical mixer transfer functions for the in-phase and in-quadrature signals $I_{I,Q}$, respectively given by:

$$H_I(t)=\cos(\omega_T t+\varphi_{ck}), \quad (7)$$

$$H_Q(t)=\cos\left(\omega_T t+\varphi_{ck}+\frac{\pi}{2}\right) \quad (8)$$

where ω_T is the RF modulation frequency of the LO signal and the φ_{ck} is its relative phase to the data clock.

When $\omega_T=\omega_d$, the detected signals I_{IU}, I_{IL}, I_{QL}, I_{QU} after low pass filter (LPF) are determined from Equations (9)-(12)

$$I_{IL}=\frac{\alpha\beta I_0}{2}\cos(\varphi_L+\varphi_{ck}), \quad (9)$$

$$I_{QL}=\frac{\alpha\beta I_0}{2}\cos\left(\varphi_L+\varphi_{ck}+\frac{\pi}{2}\right), \quad (10)$$

$$I_{IU}=\frac{\alpha\beta I_0}{2}\cos(\varphi_U+\varphi_{ck}), \quad (11)$$

$$I_{QU}=\frac{\alpha\beta I_0}{2}\cos\left(\varphi_U+\varphi_{ck}+\frac{\pi}{2}\right), \quad (12)$$

The relative phase of the carrier and each sideband for the LSB and USB (φ_L and φ_U) are given by:

$$\varphi_L=\mathrm{Arctg}\left(\frac{I_{QL}}{I_{IL}}\right)-\varphi_{ck}, \quad (13)$$

$$\varphi_U=\mathrm{Arctg}\left(\frac{I_{QU}}{I_{IU}}\right)-\varphi_{ck}. \quad (14)$$

It is noted that the phase difference $\Delta\varphi$ is independent of LO signal's phase φ_{ck}, and $\Delta\varphi$ is given by:

$$\Delta\varphi=\varphi_U-\varphi_L=\mathrm{Arctg}\left(\frac{I_{QU}}{I_{IU}}\right)-\mathrm{Arctg}\left(\frac{I_{QL}}{I_{IL}}\right). \quad (15)$$

The GVD is then given by:

$$GVD=\frac{2\pi c\Delta\varphi}{\lambda^2\omega_T^2}, \quad (16)$$

3. Experimental Implementation

Experimental demonstration was performed using the setup of **Figure 2**. The transmitter comprises a tunable laser (TL) operating at 1550 nm with 10MHz line width, the part of PRBS generating 10 Gbit/s pseudorandom bit sequence (PRBS), and pulse generator generats 66% NRZ and RZ pulse shape respectively, the optical carrier was modulated with a 10 GHz NRZ/RZ PRBS of length $2^{15}-1$ through Mach-Zenhder modulator (MZM). The fiber under test (FUT) comprises single mode fiber (G.652) and dispersion compensation fiber. An erbium-doped fiber amplifier (EDFA) was used to compensate the fiber loss. At the receiver, a tunable optical band-pass filter with 3dB bandwidth of 0.6 nm was used to eliminate the redundant amplified spontaneous emission (ASE) noise, the Mach-Zenhder interferometer (MZI) with FSR (Free Spectrum Range) of 20 GHz was used to distill upper and lower signal band, the optical signal of upper and lower band were respectively detected by analog detector with 3dB bandwidth of 3 GHz, the output RF electrical signal was split into I and Q channels for mixing with local oscillator (LO), and I channel LO signal has $\pi/2$ phase difference with Q channel LO signal. In this experiment we used ADL5382 to actualize mixing RF with LO signal, the LO frequency is 2.4 GHz. The analog-digital-converters (ADC) with 14

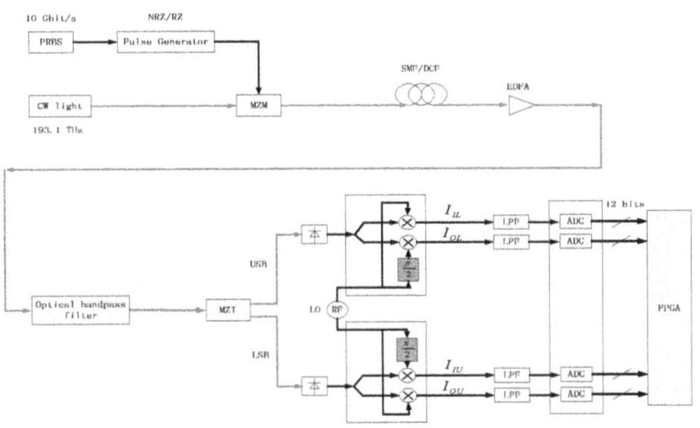

Figure 2. Experimental setup of CD measurement for NRZ and RZ systems.

bit-width and 20 MHz sample rate were used to sampling the output analog intermediate frequency (IF) signals and converting to digital signals, signals processing was performed in Field Programmable Gate Array (Xilinx: XC4VLX15).

4. Experimental Results

The positive chromatic dispersion was added to the signal using five spools of SMF of 20, 40, 60, 80, 100 km corresponding, respectively, to 335, 670, 1005, 1340 and 1675 ps/nm. The negative chromatic dispersion was added to the signal using four spools of DCF of 3, 6, 9, 12 km corresponding, respectively, to −420, −840, −1260, −1680 ps/nm. The OSNR was varied with a variable noise loading stage using an ASE source. The OSNR was maintained at a 20.5 dB level for all CD measurements. The CD measurement was tested without introducing any DGD. **Figure 3** shows the experimental results of the NRZ format signal. **Figure 4** shows the experimental results of the RZ format signal.

The precision of the dispersion measurement mainly depends on the OSNR. **Figures 5(a)** and **(b)** shows the detected phase corresponding to different OSNR under the CD of 0 ps/nm. In measurement, the result of NRZ link is more precise than RZ link under the same OSNR.

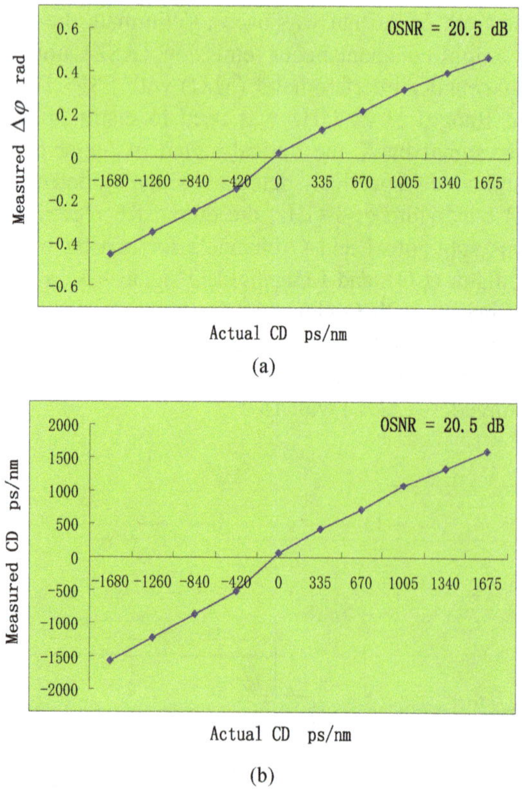

Figure 3. (a) Measured value of $\Delta\varphi$ corresponding to variable CD (LO = 2.4 GHz); (b) Measured CD corresponding to actual CD (LO = 2.4 GHz).

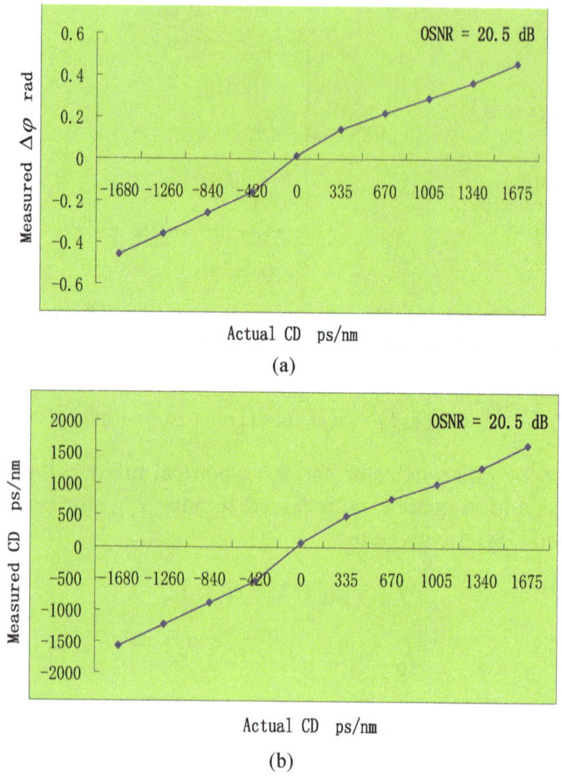

Figure 4. (a) Measured value of $\Delta\varphi$ corresponding to variable CD (LO = 2.4 GHz); (b) Measured CD corresponding to actual CD (LO = 2.4 GHz).

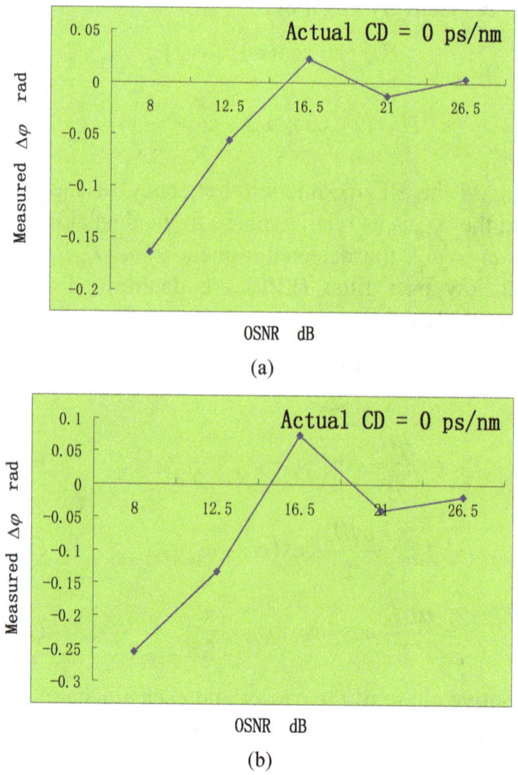

Figure 5. Detected phase corresponding to different OSNR for 10 G, (a) NRZ; (b) RZ.

5. Conclusion

In this experiment, CD measuring technique for independent of data rates and formats has been developed, the performance of the technique was experimentally and respectively assessed for 10 Gbit/s NRZ and RZ systems. The measurement range and error achieved with the proposed method, based on monitoring a 2.4 GHz in-band tone, was ±2000 ± 35 ps/nm for CD. In condition of 2.4 GHz LO, the maximum measurement range may up to ±9000 ps/nm, the LO is tunable that adjust it up for CD range. The technique was shown to operate in single wavelength system, it can also be used in WDM system by using a tunable optical filter (TOF) to achieve multi-wavelength channels CD measurement in serial mode, or by using array waveguide gates (AWG) to achieve all wavelength channels CD measurement in parallel mode. The minimum acquisition-time is found to be a trade-off between the required measurement accuracy and the monitoring speed. The 5 ms acquisition time, required to obtain the accuracy stated above, is suitable for measuring application in dynamic optical networks with reconfiguration times greater than that, such as ASON, OBS or OPS systems. Detuning of the MZI optical filter from the optimum position by up to 5 GHz does not have any effect on the phase measurement for the CD measuring. In addition, the use of electronic mixer for spectra down-conversion and the use of low-speed detectors make it potentially cost effective for multi-channel operation.

REFERENCES

[1] B. J. Eggleton, A. K. Ahuja, P. S. Westbrook, J. A. Rogers, P. Kuo, T. N. Nielsen and B. Mikkelsen, "Integrated Tunable Fiber Gratings for Dispersion Management in High-Bit Rate Systems," *Journal of Lightwave Technology*, Vol. 18, No. 10, 2000, pp. 1418-1432. doi:10.1109/50.887194

[2] N. Liu, W. D. Zhong, Y. J. Wen and Z. H. Li, "New Transmitter Configuration for Subcarrier Multiplexed DPSK Systems and Its Applications to Chromatic Dispersion Monitoring," *Optics Express*, Vol. 15, No. 3, 2007, pp. 839-844. doi:10.1364/OE.15.000839

[3] H. Ohta, S. Nogiwa, Y. Kawaguchi and Y. Endo, "Measurement of 200 Gbit/s Optical Eye Diagram by Optical Sampling with Gain-Switched Optical Pulse," *Electronics Letters*, Vol. 36, No. 8, 2000, pp. 737-739. doi:10.1049/el:20000538

[4] Z. Q. Pan, Q. Yu, Y. Xie, S. A. Havstad, A. E. Willner, D. S. Starodubov and J. Feinberg, "Chromatic Dispersion Monitoring and Automated Compensation for NRZ and RZ Data Using Clock Regeneration and Fading without Adding Signaling," *Optical Fiber Communication Conference and Exhibit*, Anaheim, 17-22 March 2001, Article ID: WH5-1-3.

[5] S. D. Dods, T. B. Anderson, K. Clarke, M. Bakaul and A. Kowalczyk, "Asynchronous Sampling for Optical Performance Monitoring," *Optical Society of America* (*OFC*). Anaheim, 25-29 March 2007, pp. OMM5-1-3.

[6] B. Kozicki, A. Maruta and K. I. Kitayama, "Transparent Performance Monitoring of RZ-DQPSK Systems Employing Delay-Tap Sampling," *Journal of Optical Networking*, Vol. 6, No. 11, 2007, pp. 1257-1269. doi:10.1364/JON.6.001257

[7] B. Kozicki, A. Maruta and K. I. Kitayama, "Experimental Demonstration of Optical Performance Monitoring for RZ-DPSK Signals Using Delay-Tap Sampling Method," *Optics Express*, Vol. 16, No. 6, 2008, pp. 3566-3576. doi:10.1364/OE.16.003566

[8] Z. H. Li and G. F. Li, "Chromatic Dispersion and Polarization-Mode Dispersion Monitoring for RZ-DPSK Signals Based on Asynchronous Amplitude-Histogram Evaluation," *Journal of Lightwave Technology*, Vol. 24, No. 7, 2006, pp. 2859-2866. doi:10.1109/JLT.2006.876089

[9] Z. H. Li, Z. Jian, L. H. Cheng, Y. F. Yang, C. Lu, A. P. T. Lau, C. Y. Yu, H. Y. Tam and P. K. A. Wai, "Signed Chromatic Dispersion Monitoring of 100 Gbit/s CS-RZ DQPSK Signal by Evaluating the Asymmetry Ratio of Delay Tap Sampling," *Optics Express*, Vol. 18, No. 3, 2010, pp. 3149-3157. doi:10.1364/OE.18.003149

[10] W. Chen, F. Buchali, X. W. Yi, W. Shieh, J. S. Evans and R. S. Tucker, "Chromatic Dispersion and PMD Mitigation at 10 Gb/s Using Viterbi Equalization for DPSK and DQPSK Modulation Formats," *Optics Express*, Vol. 15, No. 9, 2007, pp. 5271-5276. doi:10.1364/OE.15.005271

[11] P. S. Westbrook, B. J. Eggleton, G. Raybon, S. Hunsche, T. H. Her, "Measurement of Residual Chromatic Dispersion of a 40-Gb/s RZ Signal via Spectral Broadening," *IEEE Photonics Technology Letters*, Vol. 14, No. 3, 2002, pp. 346-348. doi:10.1109/68.986808

[12] S. P. Li and D. V. Kuksenkov, "A novel dispersion monitoring technique based on four-wave mixing in optical fiber," *IEEE Photonics Technology Letters*, Vol. 16, No. 3, 2004, pp. 942-944. doi:10.1109/LPT.2004.823751

[13] T. Luo, C. Y. Yu, Z. Q. Pan, Y. Wang, J. E. McGeehan, M. Adler and A. E.Willner, "All-Optical Chromatic Dispersion Monitoring of a 40-Gb/s RZ Signal by Measuring the XPM-Generated Optical Tone Power in a Highly Nonlinear Fiber," *IEEE Photonics Technology Letters*, Vol. 18, No. 2, 2006, pp. 430-432. doi:10.1109/LPT.2005.862359

[14] Y. K. Lizé, L. C. Christen, J. Y. Yang, P. Saghari, S. R. Nuccio, A. E. Willner and R. Kashyap, "Independent and Simultaneous Monitoring of Chromatic and Polarization-Mode Dispersion in OOK and DPSK Transmission," *IEEE Photonics Technology Letters*, Vol. 19, No. 1, 2007, pp. 3-5. doi:10.1109/LPT.2006.888039

22

Generation of Feedback-induced Chaos in a Semiconductor Ring Laser

Xin Zhang, Guohui Yuan, Zhuoran Wang
School of Optoelectronic Information, University of Electronic Science and Technology of China, Chengdu, Sichuan, China
Email: yuanguohui@uestc.edu.cn

ABSTRACT

A scheme for chaotic signal generation in a semiconductor ring laser (SRL) with optical feedback is presented. Part of the output is returned to the SRL, resulting in chaotic oscillation.

Keywords: (140.1540) Chaos; (140.5960) Semiconductor Lasers; (140.0140) Lasers and Laser Optics; (060.0060) Fiber Optics and Optical Communications

1. Introduction

It is previously demonstrated that semiconductor lasers are widely used in the chaotic Optical communications as chaotic carrier source [1-2]. As a special case of semiconductor laser, semiconductor ring lasers (SRLs) can also be utilized to chaotic communication systems.

In this paper, we demonstrate the chaotic signal generation in a SRL with an optical feedback. Simulated results indicate the existence of chaotic oscillation in the SRL with appropriate disturbance.

2. Feedback-induced Chaos Scheme

The feedback-induced scheme for the generation of chaos is based on a SRL with a feedback waveguide as shown in **Figure 1**. Part of the output of the SRL is injected back to its cavity after a certain time delay, which induces chaotic oscillation in the SRL with appropriate feedback parameters. Simultaneously, the lasing direction of the drive SRL is set to the clockwise as a result of mode competition.

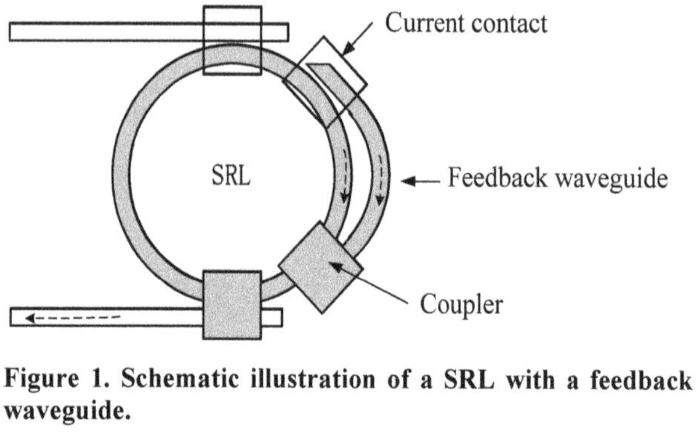

Figure 1. Schematic illustration of a SRL with a feedback waveguide.

The rate equations of the SRL with optical feedback are described here as [3]:

$$\frac{dE_1}{dt}=\frac{1}{2}\left(\Gamma v_g G(N-N_0)\left(1-\varepsilon_s E_1^2-\varepsilon_c E_2^2\right)-\frac{1}{\tau_P}\right)E_1 + \frac{K_f}{\tau_{in}}E_1(t-\tau)\cos\left(\omega_1\tau+\phi_1(t)-\phi_1(t-\tau)\right) \quad (1)$$

$$\frac{d\phi_1}{dt}=\frac{1}{2}\alpha\left(\Gamma v_g G(N-N_0)\left(1-\varepsilon_s E_1^2-\varepsilon_c E_2^2\right)-\frac{1}{\tau_P}\right) -(\omega_1-\varpi_{th})-\frac{K_f}{\tau_{in}}\frac{E_1(t-\tau)}{E_1}\sin\left(\omega_1\tau+\phi_1(t)-\phi_1(t-\tau)\right) \quad (2)$$

$$\frac{dE_2}{dt}=\frac{1}{2}\left(\Gamma v_g G(N-N_0)\left(1-\varepsilon_s E_2^2-\varepsilon_c E_1^2\right)-\frac{1}{\tau_P}\right)E_2 \quad (3)$$

$$\frac{d\phi_2}{dt}=\frac{1}{2}\alpha\left(\Gamma v_g G(N-N_0)\left(1-\varepsilon_s E_2^2-\varepsilon_c E_1^2\right)-\frac{1}{\tau_P}\right) -(\omega_2-\varpi_{th}) \quad (4)$$

$$\frac{dN}{dt}=\frac{\eta_i I}{eV}-\frac{N}{\tau_s}-v_g G(N-N_0) \times\left(\left(1-\varepsilon_s E_1^2-\varepsilon_c E_2^2\right)E_1^2+\left(1-\varepsilon_s E_2^2-\varepsilon_c E_1^2\right)E_2^2\right) \quad (5)$$

where E is the electric field amplitude, Φ is the phase, and N is the carrier density. The subscript *1* and *2* account for the clockwise and counter-clockwise directions of the SRL; τ is the delay time of the feedback light. I is the injection current of SRL; K_f is the feedback coefficient, the ratio of the feedback light to the light in SRL controlled by the bias current of the couplers. The detailed parameters of SRL are described in [3].

3. Simulation Results

The dynamics of the SRL with optical feedback depend

on the adjustable system parameters including the delay time of the feedback light τ, and the bias injection current I of the SRL, the feedback coefficient Kf. In this paper, τ is set to 179 fs, which means that the feedback waveguide is about 15μm longer than that of the corresponding part of the resonant cavity of the SRL. We focus on the effect of the feedback coefficient and the bias injection current of the SRL on the nonlinear system.

It is well known that any system containing at least one positive lyapunov exponent is defined to be chaotic and the larger the magnitude of the positive lyapunov exponent is, the more chaotic the system is. The map of largest lyapunov exponent of the system is presented in **Figure 2** as a function of the feedback coefficient and the bias injection current of the SRL, which is approximately computed based on the classic Wolf's algorithm [4]. It is clear that the system is chaotic for the most part of the region in **Figure 2**. **Figure 3** shows that the chaotic output from the SRL when the feedback coefficient 0.25 and the bias current of the SRL is 110 mA, where the largest lyapunov exponent is about 0.14. **Figure 3(a)** is the Random-like time series and **Figure 3(b)** is the impulse-like autocorrelation, which also indicates a chaotic system.

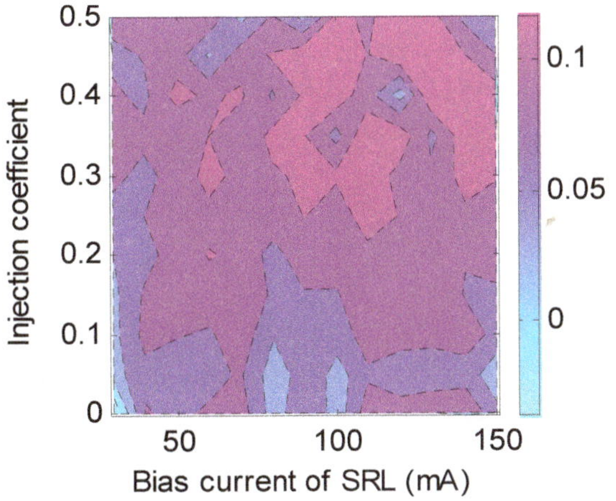

Figure 2. Largest lyapunov exponent map as a function of the feedback coefficient and the bias current of the SRL.

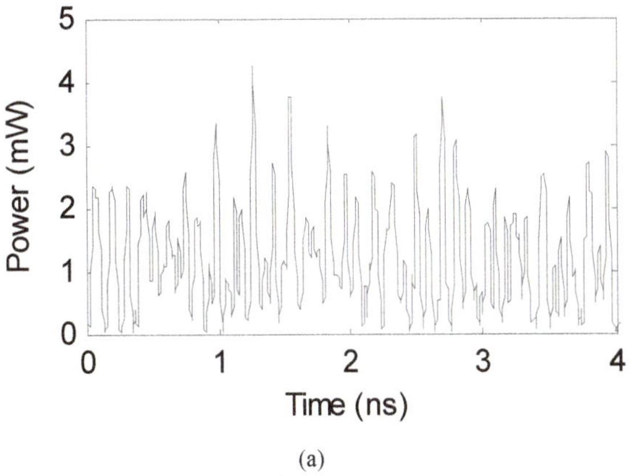

(a)

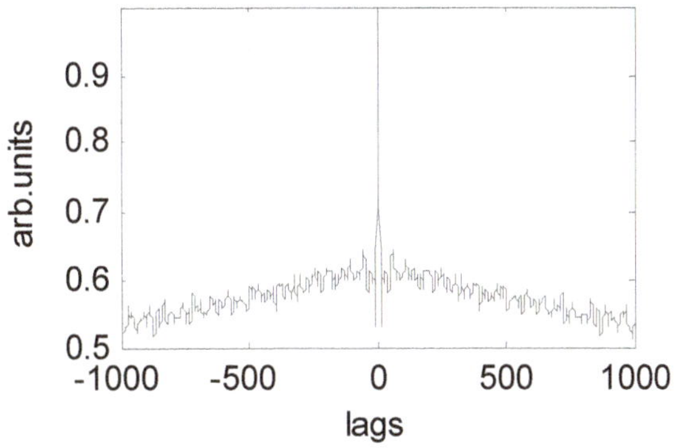

(b)

Figure 3. Time series (a) and autocorrelation (b) of SRL when the feedback coefficients is 0.25 and the bias current of the SRL is 110mA.

4. Conclusions

The generation of chaotic signal in a SRL with an optical feedback is proposed in this paper. The positive lyapunov exponent map, time series and autocorrelation of SRL indicate the occurring of chaotic oscillation in our nonlinear system with suitable system parameters, which paves the way for the utilization of SRLs in the chaotic Optical communication systems.

5. Acknowledgements

This work was sponsored in part by the National Natural Science Foundation of China under Grant 61107061, Grant 61107088, and Grant 61090393, Program for New Century Excellent Talents in University under Grant NCET-12-0092, Specialized Research Fund for the Doctoral Program of Higher Education (SRFDP) under Grant 20100185120016, Project of international sci-tech cooperation and exchange research of Sichuan Province under Grant 2012HH0001, the Scientific Research Foundation for the Returned Overseas Chinese Scholars of State Education Ministry 2012GJ002, the State Key Laboratory of Electronic Thin Films and Integrated Devices under Grant KFJJ201112, and State Key Laboratory on Integrated Optoelectronics under Grant 2011KFB008.

REFERENCES

[1] N. Jiang, W. Pan, L. S. Yan, B. Luo, S. Y. Xiang, L. Yang, D. Zheng and N. Q. Li, "Chaos Synchronization and Communication in Multiple Time-Delayed Coupling Semiconductor Lasers Driven by a Third Laser," *IEEE Journal of Selected Topics in Quantum Electronics*, Vol. 17, 2011, pp. 1220-1227.

[2] J. M. Liu, H. F. Chen and S. Tang, "Optical- Communication Systems Based on Chaos in Semiconductor Lasers," *IEEE Transactions on Circuits and Systems*, Vol. 48, 2001, pp. 1475-1483.

[3] G. Yuan and S. Yu, "Bistability and Switching Properties of Semiconductor Ring Lasers With External Optical In-jec- tion," *IEEE Journal of Quantum Electronics*, Vol. 44, 2008, pp. 41-48. doi:10.1109/JQE.2007.909523

[4] A. Wolf, J. B. Swift, H. L. Swinney and J. A. Vastano, "De-Termining Lyapunov Exponents from a Time Se-ries," *Physica D.,*Vol. 16D, 1985, pp. 285-317. doi:10.1016/0167-2789(85)90011-9

23

Preparation and Microstructural, Structural, Optical and Electro-Optical Properties of La Doped PMN-PT Transparent Ceramics

Fernando Andrés Londono Badillo, Jose Antonio Eiras, Flavio Paulo Milton, Ducinei Garcia
Ferroelectric Ceramics Group, Physics Department, Federal University of São Carlos, São Carlos, Brazil
Email: flondono@df.ufscar.br

ABSTRACT

Transparent relaxor ferroelectric ceramics of the system lanthanum modified lead magnesium niobate have been investigated for a variety of electro-optic properties that could make these materials alternatives to (Pb,La)(Zr,Ti)O_3. However, a study that relates the to properties in function stoichiometric formula, has not been analyzed heretofore. Therefore, in this work the effect of A-site substitution of La^{3+} in the characterization microstructural, structural, optical and electro-optical on $(1-x)\left[Pb_{(1-3/2y)}La_y\left(Mg_{1/3}Nb_{2/3}\right)O_3\right]-xPbTiO_3$ and $(1-z)\left[(1-x)Pb\left(Mg_{1/3}Nb_{2/3}\right)O_3+xPbTiO_3\right]+zLa_2O_3$ has been performed. It was observed that the properties according to the stoichiometric formula and the PT had a maximum whose behavior was related to the addition of lanthanum in each stoichiometries.

Keywords: Ferroelectric; Optics; Transmittance; Electro-Optical

1. Introduction

Lead magnesium niobato $Pb\left(Mg_{1/3}Nb_{2/3}\right)O_3$ (PMN), as one of the most widely investigated relaxor ferroelectric with a perovskites structure, was first synthesized in the late 1950s [1]. Initially, PMN was prepared by the conventional mixed oxide method, where pyrochlore phase was inevitably produced. In order to synthesize stoichiometric perovskite PMN ceramics, Swartz and Shrout [2] proposed a columbite precursor method, where the intermediate reaction of the formation of pyrochlore phase was bypassed, which results in the stabilization of perovskites structure as compared to the mixed oxide method. Recently, a novel methodology was devised to stabilize perovskites structure by adding stable normal $PbTiO_3$ (PT). The formation of the solid solution increases the tolerance factor and electronegativity difference, leading to the stabilization of the perovskites structure and the enhancement of dielectric property of the relaxor ferroelectric [3].

A study of the optical, electrical and electro-optical properties of PMN-PT ceramics in function of the stoichiometrie is of interest both for possible insight into the physical nature of relaxor ferroelectrics as well as for making practical extension of its several present applications to include usage in electro-optical devices. Present applications take advantages of PMN-PT singularly excellent dielectric, low thermal expansion, and high electrostrictive properties [4]. From the viewpoint of crystal chemistry, the substitution of Ti^{4+} ions for the complex $\left(Mg_{1/3}Nb_{2/3}\right)^{4+}$ ions on the B-site of the perovskite structure in the $(1-x)Pb\left(Mg_{1/3}Nb_{2/3}\right)O_3-xPbTiO_3$ (PMN-PT) system leads to the outstanding properties of the PMN-PT ceramics that exhibit excellent electrical and electro-optical performance, which make them promising applications in multilayer capacitors, piezoelectric transducers and actuators and optical devices [5,6]. However, perovskite structure are extremely difficult to fabricate reproducibly without the appearance of stable pyrochlore phase, which always exist in the PMN-PT ceramics and significantly deteriorates such properties as the dielectric property [7].

The optical transmittance as function of wavelength was studied in $Pb\left(Mg_{1/3}Nb_{2/3}\right)_{0.62}Ti_{0.38}O_3$ single crystal by Wan *et al.* It was found that the crystal is transparent in the visible region and rolls off in near 450 nm. Using the Senarmont compensador method, the effective electro-optic coefficient r_c = 42.8 pm/V was also obtained [8]. Some effect and applications, such as photorefractive, Second-Harmonic Generation (SHG), electro-optic and elasto-optical devices, are based on the large optical property coefficient of the materials [9,10]. The knowledge of factors that can modify these properties is desirable to find new application of the relaxor and electro-optical materials. In this work, ceramics in the PMN-PT system were prepared by doping with lanthanum. La addition to PMN-PT has been shown to promote densification and through hot uniaxial pressing, optically transparent materials have been achieved, allowing the determination of various optic and electro optic properties [11,12].

It is the purpose of this article, to report the microstructural, structural, electrical, optical and electro-optical properties according to the stoichiometric formula of lanthanum doped PLMN-XPT and PMN-XPT:La with $0.11 \leq x \leq 0.15$.

2. Experimental Procedure

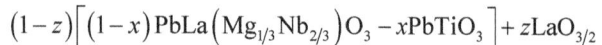

The powder was synthesized by the columbite or two-stage calcining method [2]. The batch formulae were $(1-x)\left[Pb_{(1-y)}La_y\left(Mg_{1/3}Nb_{2/3}\right)O_3\right]-xPbTiO_3$ PLMN-XPT, and $(1-z)\left[(1-x)PbLa\left(Mg_{1/3}Nb_{2/3}\right)O_3-xPbTiO_3\right]+zLaO_{3/2}$ PMN-XPT:La, with $y = 0.01$. $z = 0.01$, $X = 100\ x$ and $0.11 \leq x \leq 0.15$. The starting materials were lanthanum oxide, La_2O_3 (Aldrich, >99% purity), niobium oxide, Nb_2O_5 (Alfa Aesar 99.9% purity), magnesium carbonate hydroxide pentahydrate, $(MgCO_3)4\cdot Mg(OH)_2\cdot 5H_2O$ (Aldrich 99% purity), lead oxide, PbO (MGK 99%) and titanium oxide, TiO_2 (Alfa Aesar, 99.8% purity) powders. The $(MgCO_3)4\cdot Mg(OH)_2\cdot 5H_2O$ was carried up to 1100°C, for 4 h, to drive off CO_2 and H_2O and obtain the correct amount of MgO for a stoichiometric reaction with Nb_2O_5. In the first stage, MgO and Nb_2O_5 powders were ball-milled and prereacted at 1100°C for 4 h, in air, to form the columbite phase ($MgNb_2O_6$). In the second stage, the synthesized $MgNb_2O_6$ (MN) was ball-milled in isopropanol, for 24 h, with appropriate amounts of PbO, TiO_2 and La_2O_3 and heated at 900°C, for 4 h, in oxygen atmosphere, at a controlled pressure of 200 kPa. The calcined powders were pressed into pellets 10 mm in diameter and 10 mm in thickness with the addition of an appropriate amount of polyvinyl alcohol (PVA) binder. The pellets were sintered in hot uniaxial pressing in O_2 atmosphere followed for 4 h at 1220°C, and 6 MPa.

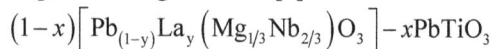

The phases and the structural parameters were determined by X-ray diffraction (XRD) using a Rigaku diffractometer, with CuKα radiation. The lattice parameters were calculated by least squares from the positions of the diffraction peaks. Apparent densities were determined by the Archimedes method. The microstructural features of the samples were investigated by scanning electron microscopy (SEM), using a Jeol JSM 5800 LV. The grain sizes were calculated from the SEM images of the polished thermally etched surfaces by the linear-intercept method. For dielectric measurement, gold electrodes were deposited on both faces of the disk samples (5 mm diameter and ~1.0 mm thick). The relative permittivity, ε was measured by impedance analysis (HP4194A). The transmittance was measured in a spectrophotometer (Micronal-B582), at wavelengths ranging from 200 to 1000 nm for optically polished samples, 600 μm thick. The method used in the electro-optical characterization, determination of the Pockels and Kerr coefficients, is based in the Senarmont configuration illustrated in the **Figure 1** [13]. The birefringence medium (sample) is located between two linear polarizer's, P_1 and P_2, where the P_2 polarizer is known as analyzer, whit polarizer axis forming angles of ±45° in relation to principal axis of the sample, this definite by the electric-field direction applied. Between the sample and the analyzer is located a $\lambda/4$ plate used to compensation the effects caused by the natural birefringence of the medium. Then the analyzer is positioned in the angle β that optimizes the relation between the birefringence and the applied electric-field. In the Equation (1) is presented the relation between birefringence and electric-field for medium with linear and quadratic electro-optic response.

$$\Delta n(E) = -1/2n^3\left(rE + RE\right) \quad (1)$$

where Δn is the birefringence induced for the electric-field E, n is the natural birefringence of the medium, r and R are the Pockels and Kerr electro-optic coefficients, respectively.

3. Results and Discussion

The relative density of PLMN-XPT and PMN-XPT:La at different PT concentration are greater than 96% of the theoretical density as can be seen in **Figure 2**, which is especially suitable for electronic industry application [14]. The relative density of the two system exhibits the tendency to decrease when PT concentration is increased. This can be attributed the loss of PbO (control atmosphere of PbO was not used in this work) [15]. As well can be seen the relative densities of PLMN-XPT ceramics is greater for all PT concentrations when compared with PMN-XPT:La. This possibly associated with the less occupation of La^{3+} in Pb^{2+} sites, implying in the appearance of vacancies in the A and B sites in different quantities [16] (La^{3+} atoms are lighter than Pb^{2+} atoms, and PMN-XPT:La have major number of La atoms).

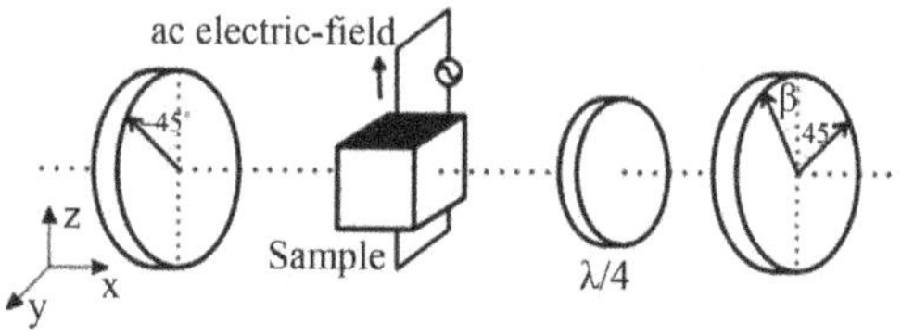

Figure 1. Illustration of the experimental Senarmont method for characterization of the transversal electro-optical effect.

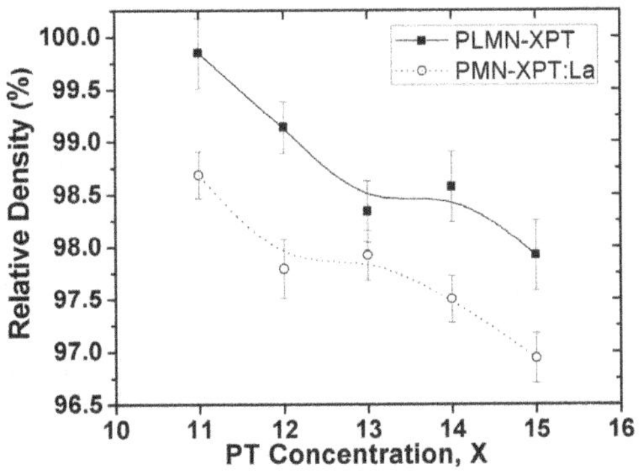

Figure 2. Variation of relative density as a function of PT concentration of the PLMN-XPT and PMN-XPT:La ceramics with $0.11 \leq x \leq 0.15$.

The variation of density is considered as relating to mass loss induced by evaporation of PbO or to variation of temperature of densification relating to the increase of PT, this changes the grain size, porosity could appear which decreases the relative density of ceramics. However the slight decrease of relative density of PLMN-XPT and PMN-XPT:La ceramics can be attributed to the variation of microstructure and to evaporation of lead at elevated sintering temperatures.

The high densification of PLMN-XPT and PMN-XPT:La ceramics is further confirmed by SEM observation, which is shown in **Figure 3**. All ceramics exhibited similar microstructure as can be seen in the micrograph of polished and thermally attacked surface of PLMN-13PT (shown here).

The variation of grain size with the increase of PT concentration is shown in **Figure 4**. An increase in PT concentration resulted in a decrease in grain size and a corresponding slight decrease in density [17]. The grain size in PMN-XPT:La is lesser for all concentration when compared to PLMN-XPT because the major quantities of La^{3+} resulted in reduction of grain size as observed by Kim *et al.* in La^{3+} doped PMN-PT ceramics [18].

The XRD patterns of sintered ceramics with different compositions are presented in **Figure 5**, where a complete perovskite structure (JCPDS 391488) is formed and the pyrochlore or other second phases were not detected.

The permittivity (real, ε' and imaginary, ε'') was measured at various temperatures in the frequency range 1 kHz - 1MHz. In the **Figure 6** can be seen a typical examples of measurements realized in this work. All the samples exhibit typical relaxor behavior [19,20], with the magnitude of ε' decreasing with increasing frequency, and the maximum permittivity real, ε'_{max} shifting to higher temperatures, as shown in the PLMN-12PT ceramics (**Figure 6**). The data of all ceramic analyzed in this work can be seen in the **Table 1**.

Variations in the shape of ε' in function of T curves can be also directly related to the diffuseness coefficient,

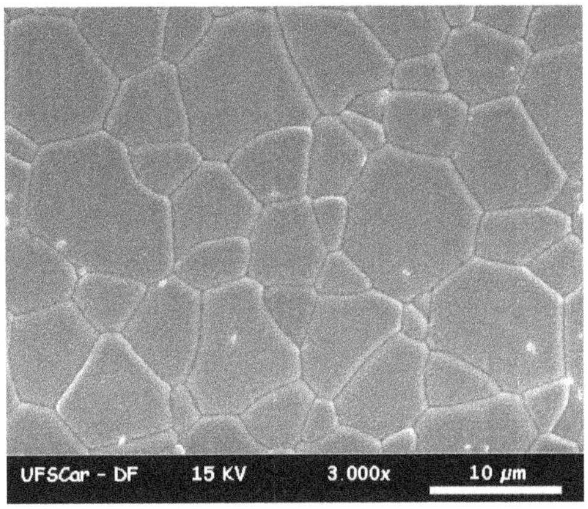

Figure 3. SEM image of surfaces polished and thermally attacked of PLMN-13PT ceramic.

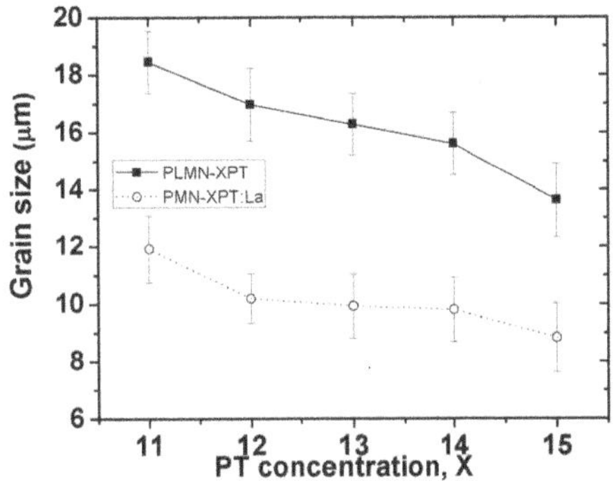

Figure 4. Grain size in function of PT concentration of the PLMN-XPT and PMN-XPT:La ceramics with $0.11 \leq x \leq 0.15$.

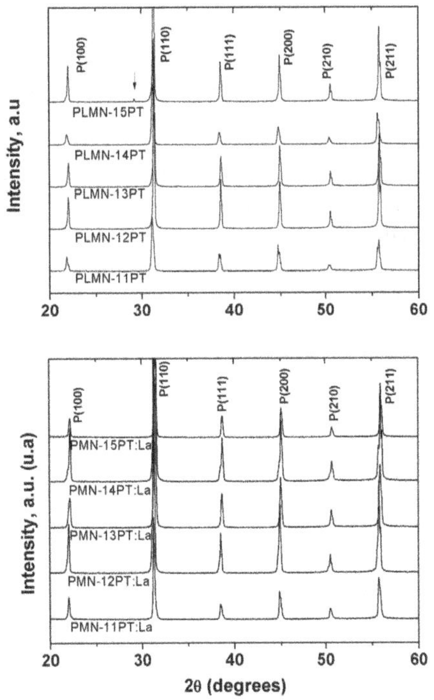

Figure 5. XRD patterns of PLMN-XPT and PMN-XPT:La ceramics with $0.11 \leq x \leq 0.15$.

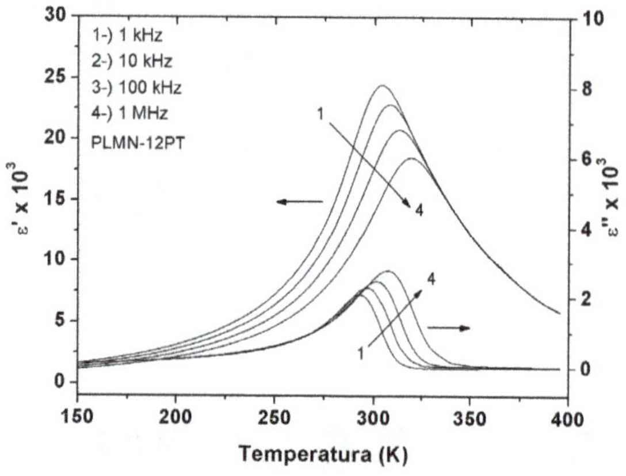

Figure 6. Permittivity real, ε' and permittivity imaginary, ε'' at various frequencies in function of temperature for PLMN-XPT and PMN-XPT:La ceramics with $0.11 \le x \le 0.15$.

Table 1. Dielectric properties of PLMN-XPT and PMN-XPT:La with $0.11 \le x \le 0.15$ at 1 kHz.

Ceramic	ε'_{max}	T_{max} (K)	Δ
PLMN-11PT	27,000	293	1.66
PLMN-12PT	26,800	300	1.65
PLMN-13PT	25,200	305	1.63
PLMN-14PT	29,000	308	1.53
PLMN-15PT	28,000	317	1.55
PMN-11PT:La	25,000	297	1.62
PMN-12PT:La	24,700	299	1.62
PMN-13PT:La	25,100	308	1.62
PMN-14PT:La	23,000	318	1.60
PMN-15PT:La	30,000	319	1.52

δ in **Table 1**, calculated from the Santos Eiras equation [21], which for relaxors reflect the level of disorder in term of the diffusivity of the ε'_{max}. As can be deduced from **Table 1**, the value of ε'_{max} for PMN-XPT:La^{3+} are smaller that PLMN-XPT, inferring that lanthanum additions result in the reduction in ε'_{max}, while increasing the degree of ordering, as observed for Kim *et al.* [18].

The transmission spectrum as a function of wavelength (200 nm - 1100 nm) of PLMN-XPT and PMN-XPT:La ceramics with $0.11 \le x \le 0.15$ and 630 μm thickness is presented in **Figure 7**. For both of these stoichiometries formula the percentage of transmitted light begins to rise abruptly at just below 380 nm and then increases only gradually with wavelength about 500 nm. This gradual increase in transmittance continues into the near IR at least through 1100 nm without any noticeable absorption band being observed. This is similar to what was observed for most crystals with oxygen-octahedral perovskites structure [22-24]. From the transmission characterization, we can see that the optical absorption is very small at higher wavelength range. In general, optical transmission relates to reflection loss and scattering loss. Using the refractive indices measured by Mchenry *et al.* [12], e by the Fresnel expression, the reflection loss of the light at two surfaces was calculated about 20%. Beyond of reflection loss is observed loss by scattering which can be attributed the dispersion by ferroelectrics domain (spontaneous birefringence), increases in the porosity (by loss of PbO) and precipitation of spurious phases. In general the addition of PT increase the transmittance both the PLMN-XPT and PMN-XPT:La, nevertheless exist a concentration optimal of PT in 14%. As previously mentioned the increase of content of PT promotes the densification of the samples, favoring the transmittance, though the sample with PMN-XPT:La present less transmission in comparison with the PLMN-XPT in function of wavelength. This effect is attributed the replacement of lanthanum ions in the sites of the lead, that increased the number of vacancies reducing the density and consequently affecting the transmittance.

In the **Table 2** can be seen the values for the electro-optic Kerr coefficients as a function of PT concentration and stoichiometric formula, which were determined under frequency of 200 Hz, at room temperature. The values of electro-optical coefficients here obtained are similar the reported in literature for La doped PMN-PT ce-

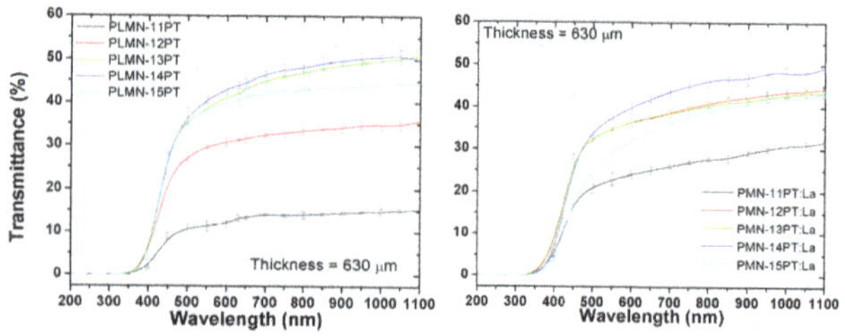

Figure 7. Percent transmission as a function of wavelength for optically polished PLMN-XPT and PMN-XPT:La with $0.11 \le x \le 0.15$ and 630 μm thickness.

Table 2. Electro-optical coefficient, *R* of PLMN-XPT and PMN-XPT:La with $0.11 \le x \le 0.15$ calculated under frequency of 200 Hz.

Ceramic	R ($\times 10^{-16}$ m^2/V^2)
PLMN-11PT	8.0
PLMN-12PT	14.1
PLMN-13PT	23.7
PLMN-14PT	10.1
PLMN-15PT	10.3
PMN-11PT:La	7.1
PMN-12PT:La	10.3
PMN-13PT:La	13.0
PMN-14PT:La	9.2
PMN-15PT:La	6.1

ramics [12,25] and greater to commercial PLZT ceramics [26]. The Kerr coefficients of the PLMN-XPT ceramics are higher when compared with PMN-XPT:La ceramics because of their major transmittance, since that the phase induced by electro-optical effect is directly proportional intensity of light transmission [27].

4. Conclusion

The ceramics obtained in this study showed high dense and homogeneous microstructure, independent of the stoichiometric which were obtained. The La addition to the PLMN-XPT and PMN-XPT:La relaxor ferroelectric system was found to promote densification and inhibit grain growth, however the PMN-XPT:La ceramic to present smaller density in relation to other composition. The transmittance characterization revealed a transmission of about 50% from 0.45 μm to 1.1 μm, presenting a lower transmission in the samples of PMN-XPT:La. The quadratic electro-optic coefficients, calculated from measurements of birefringence in this work to PMN-XPT and PLMN-XPT ceramics are up to two times higher than the maximum reported for PLZT. Thus confirming the potential use of these system as substitutes for PLZT in electro-optical applications.

5. Acknowledgements

To CAPES, FAPESP and CNPq for the financial support. To Dr. Y. P. Mascarenhas (São Carlos Physics Institute at the University São Paulo), for the use of the XRD laboratory facilities. And to Mr. Francisco José Picon for the technical support.

REFERENCES

[1] G. A. Smolenskii and A. I. Agranovskaya, "Dielectric Polarization and Losses of Some Complex Compounds," *Soviet Physics—Technical Physics*, Vol. 3, No. 7, 1958, pp. 1380-1382.

[2] S. L. Swartz and T. R. Shrout, "Fabrication of Perovskite Lead Magnesium Niobate," *Materials Research Bulletin*, Vol. 17, No. 10, 1982, pp. 1245-1250. doi:10.1016/0025-5408(82)90159-3

[3] T. R. Shrout and A. Halliyal, "Preparation of Lead-Based Ferroelectric Relaxors for Capacitors," *The Bulletin of the American Ceramic Society*, Vol. 66, No. 4, 1987, pp. 704-711.

[4] S. J. Jang, K. Uchino, S. Nomura and L. E. Cross, "Electrostrictive Behavior of Lead Magnesium Niobate Based Ceramic Dielectrics," *Ferroelectrics*, Vol. 27, No. 1, 1980, pp. 31-34. doi:10.1080/00150198008226059

[5] Y. H. Chen, K. Uchino, M. Shen and D. Viehland, "Substituent Effects on the Mechanical Quality Factor of $Pb(Mg_{1/3}Nb_{2/3})O_3$-$PbTiO_3$ and $Pb(Sc_{1/2}Nb_{1/2})O_3$-$PbTiO_3$ Ceramics," *Journal of Applied Physics*, Vol. 90, No. 3, 2001, pp. 1455-1458. doi:10.1063/1.1379248

[6] L. B. Kong, J. Ma, W. Zhu and O. K. Tan, "Rapid Formation of Lead Magnesium Niobate-Based Ferroelectric Ceramics via a High-Energy Ball Milling Process," *Materials Research Bulletin*, Vol. 37, No. 3, 2002, pp. 459-465. doi:10.1016/S0025-5408(01)00823-6

[7] E. M. Jayasingh, K. Prabhakaran, R. Sooraj, C. Durgaprasad and S. C. Sharma, "Synthesis of Pyrochlore Free PMN-PT Powder by Partial Oxalate Process Route," *Ceramics International*, Vol. 35, No. 2, 2009, pp. 591-596. doi:10.1016/j.ceramint.2008.01.022

[8] X. M. Wan, H. Q. Xu, T. H. He, D. Lin and H. S. Luo, "Optical Properties of Tetragonal $Pb(Mg_{1/3}Nb_{2/3})_{0.62}$-$TiO_{0.38}$ Single Crystal," *Journal of Applied Physics*, Vol. 93, No. 8, 2003, pp. 4766-4768. doi:10.1063/1.1561991

[9] A. S. Bhalla, R. Guo, A. S. Cross, G. Burns, F. H. Dacol and R. R. Neurgaonkar, "Glassy Polarization in the Ferroelectric Tungsten Bronze $(Ba,Sr)Nb_2O_6$," *Journal of Applied Physics*, Vol. 71, No. 11, 1992, pp. 5591-5595. doi:10.1063/1.350537

[10] A. S. Bhalla, R. Guo, A. S. Cross, G. Burns, F. H. Dacol and R. R. Neurgaonkar, "Measurements of Strain and the Optical Indices in the Ferroelectric $Ba_{0.4}Sr_{0.6}Nbz$: Polarization Effects," *Physical Review B*, Vol. 36, No. 4, 1987, pp. 2030-2035. doi:10.1103/PhysRevB.36.2030

[11] F. A. Londono, J. A. Eiras and D. Garcia, "New Transparent Ferroelectric Ceramics with High Electro-Optical Coefficients: PLMN-PT," *Cerâmica*, Vol. 57, No. 344, 2011, pp. 404-408.

[12] D. A. Mchenry, J. Giniewicz, S. J. Jang, A. Bhalla and T. R. Shrout, "Optical Properties of Hot Pressed Relaxor Ferroelectrics," *Ferroelectrics*, Vol. 93, No. 1, 1989, pp. 351-359. doi:10.1080/00150198908017367

[13] R. Go, "Ferroelectric Properties of Lead Barium Niobate Compositions near the Mosphotropic Phase Boundary," Ph.D. Theses, Pennsylvania State University, Pennsylvania, 1990.

[14] C. Ding, *et al.*, "Phase Structure and Electrical Properties of $0.8Pb(Mg_{1/3}Nb_{2/3})O_3$ Relaxor Ferroelectric Ceramics Prepared by the Reaction-Sintering Method," *Physica Status Solidi (A)*, Vol. 207, No. 4, 2010, pp. 979-985. doi:10.1002/pssa.200925377

[15] G. S. Snow, "Fabrication of Transparent Electrooptic PLZT Ceramics by Atmosphere Sintering," *Journal of the American Ceramic Society*, Vol. 56, No. 2, 1973, pp. 91-96. doi:10.1111/j.1151-2916.1973.tb12365.x

[16] S. M. Gupta and D. Viehland, "Role of Charge Compensation Mechanism in La-Modified $Pb(Mg_{1/3}Nb_{2/3})O_3$-$PbTiO_3$ Ceramics: Enhanced Ordering and Pyrochlore Formation," *Journal of Applied Physics*, Vol. 80, No. 10, 1996, pp. 5875-5883. doi:10.1063/1.363581

[17] N. Kim, D. A. McHenry, S. J. Jang and T. R. Shrout, "Fabrication of Optically Transparent Lanthanum Modified $Pb(Mg_{1/3}Nb_{2/3})O_3$ Using Hot Isostatic Pressing," *Journal of the American Ceramic Society*, Vol. 73, No. 4, 1990, pp. 923-928. doi:10.1111/j.1151-2916.1990.tb05137.x

[18] N. Kim, W. Huebner, S. J. Jang and T. R. Shrout, "Dielectric and Piezoelectric Properties of Lanthanum-Modified Lead Magnesium Niobium-Lead Titanate Ceram-

ics," *Ferroelectrics*, Vol. 93, No. 1, 1989, pp. 341-349. doi:10.1080/00150198908017366

[19] L. E. Cross, S. J. Jang, R. E. Newnham, S. Nomura and K. Uchino, "Large Electrostrictive Effects in Relaxor Ferroelectrics," *Ferroelectrics*, Vol. 23, No. 1, 1980, pp. 187-191. doi:10.1080/00150198008018801

[20] K. Uchino, S. Nomura, L. E. Cross, S. J. Jang and R. E. Newnham, "Electrostrictive Effect in Lead Magnesium Niobate Single Crystals," *Journal of Applied Physics*, Vol. 51, No. 2, 1979, pp. 1142-1145. doi:10.1063/1.327724

[21] I. A. Santos and J. A. Eiras, "Phenomenological Description of the Diffuse Phase Transitiion Inferroelectrics," *Journal of Physics: Condensed Matter*, Vol. 13, No. 50, 2001, pp. 11733-11740. doi:10.1088/0953-8984/13/50/333

[22] X. M. Wan, H. S. Luo and X. Y. Zhao, "Refractive Indices and Linear Electro-Optic Properties of $(1-x)Pb(Mg_{1/3}Nb_{2/3})O_3-xPbTiO_3$ Single Crystals," *Applied Physics Letters*, Vol. 85, No. 22, 2004, pp. 5233-5235. doi:10.1063/1.1829393

[23] C. J. He, L. H. Luo and X. Y. Zhao, "Optical Properties of Tetragonal Ferroelectric Single Crystal Lead Magnesium Niobate Lead Titanate," *Journal of Applied Physics*, Vol. 100, No. 1, 2006, Article ID: 013112. doi:10.1063/1.2217487

[24] F. A. Londono, J. Eiras and D. Garcia, "Optical and Electro-Optical Properties of $(Pb,La)TiO_3$ Transparent Ceramics," *Optical Materials*, Vol. 34, No. 8, 2012, pp. 1310-1313. doi:10.1016/j.optmat.2012.02.020

[25] K. Uchino, "Ferroelectric Devices," Marcel Dekker, New York, 2000.

[26] K. Uchino, "Electro-Optic Ceramics and Their Display Applications," *Ceramics International*, Vol. 21, No. 5, 1995, pp. 309-315. doi:10.1016/0272-8842(95)96202-Z

[27] A. Yariv and P. Yeh, "Optical Waves in Crystals: Propagation and Control of Laser Radiation," John Wiley & Sons, New York, 1984.

24

Three-Photon Absorption in ZnO Film Using Ultra Short Pulse Laser

Raied K. Jamal, Mohammed T. Hussein, Abdulla M. Suhail
Department of Physics, College of Science, University of Baghdad, Baghdad, Iraq
Email: raiedkamel@yahoo.com, Abdulla_shl@yahoo.com, mohammedtaki97@yahoo.com, hani_saka@yahoo.com

ABSTRACT

The three-photon absorption (3PA) in nanostructure wide-band gap ZnO semiconductor material is observed under high intensity femtosecond Titanium-Sapphire laser of 800 nm wavelength excitation. The ZnO film was prepared by chemical spray pyrolysis technique with substrate temperature of 400˚C. The optical properties concerning the absorption, transmission, reflection, Raman and the photoluminescence spectra are studied for the prepared film. The structure of the ZnO film was tested with the X-ray diffraction and it was found to be a polycrystalline with recognized peaks oriented in (002), and (102). The measured of three photon absorption coefficient was found to be about 0.0166 $cm^3/Gwatt^2$, which is about five times higher than the bulk value. The fully computerized z-scan system was used to measure the nonlinear coefficients from the Gaussian fit of the transmitted laser incident.

Keywords: Multiphoton Processes; ZnO Nanocrystalline; Nonlinear Optics

1. Introduction

The ZnO nanostructures have many applications in gas sensors, UV detectors and solar cells [1-3]. This material has some additional advantages compared to other large band-gap semiconductors; for example, its large exciton binding energy (about 60 meV) which is three times the binding energies of ZnSe and GaN [4]. This allows a stable exciton distribution and achieves efficient excitonic emission at room temperature. The optical and electrical properties of ZnO nanostructures are studied at different preparation techniques and at different substrate materials [5,6]. The pumping of ZnO crystalline nanofilm with near-infrared femtosecond radiation pulses, enhances the nonlinear interaction between the ultrahigh intensity applied optical field and the ZnO nanostructures [7,8]. The nonlinear interaction leads to the simultaneous absorption of two or more photons of subband gap energy. The absorption is through a virtual-states that assists the inter band transitions. This transition produces electron-hole pairs in the excited states and, subsequently, the band-edge emission via their radiative recombination [9]. The two photon absorption (2PA) in semiconductor nanocrystals (NCs) has been widely investigated [8,10], while the research effort on their three-photon absorption (3PA) is limited [11]. The three photon absorption was observed in ZnO and ZnS crystals when pump with lasers of ultra excitation irradiance which was more than 40 GW/cm^2 [12-15]. Three-photon absorption in nanostructure wide-Band gap semiconductor ZnO using femtosecond laser for thin film was studied by [16].

In this work the three photon absorption in ZnO nanostructure illuminated by intensity 110 GW/cm^2 Ti-Sapphire laser is observed. The work concentrates on the effect of the nanostructure on the nonlinear parameters through the studying of the 3PA coefficient and the laser threshold pumping power. Simple mathematical relations are developed to describe the dependence of the fluorescent emission on the pumping laser intensity. The three photon absorption coefficient was calculated from the experimental measurements.

2. Experimental Work

The ZnO film was prepared by chemical spray pyrolysis technique. The film was deposited on quartz substrates heated to 400˚C. The spray solution is prepared by mixing Zinc Chloride ($ZnCl_2$) at 0.1 M with distillated water. The above mixture solution was placed in the flask of the atomizer and spread by controlled nitrogen gas flow on the heated substrates. The chemical spray pyrolysis experimental setup is similar to the standard unit fully described by [17]. The spraying time was controlled by adjustable a solenoid valve. The heated substrate was left for 12 sec after each spraying run to give time for the deposited ZnO layer to be dry. In order to get film of proper thickness many layers deposited of ZnO are required. The optimum experimental conditions for obtaining homogeneous ZnO film at 400°C are determined by the spraying time, the drying time and the flashing gas pressure. The schematic representation of the spray system is given in **Figure 1**. The possible reaction of the spray chemical on the heated substrate is yielding for the following reaction:

$$ZnCl_2 + H_2O \xrightarrow[400^\circ C]{Heat} ZnO\ (film/substrate) + 2HCl \quad (1)$$

During the chemical reaction, gas and water vapor is obtained from this reaction due to the high temperature of the substrate. At end of reaction a white precipitates remain from the reaction as a nanofilm of ZnO as shown in **Figure 2**. The topography study of the prepared nanofilm surface was carried on using Scanning Electron Microscopy (SEM) type ULTRA 55 with different magnification; as shown in **Figure 3**. The X-ray diffraction (XRD) pattern of ZnO film was recorded by XDR 2000. The X-ray diffractometer use copper tube radiation line of wavelength 1.54Å in 2θ range from 20° to 60°. The scanning rate was 1 deg/min.

The UV-VIS-NIR absorption, transmission and reflection spectra of the sample were recorded by Hitachi U-4100 spectrometer covering the spectral range 200 - 1100 nm. The photoluminescence spectrum (PL) was studied using SL1174 spectrophotometer in the range 300 - 900 nm.

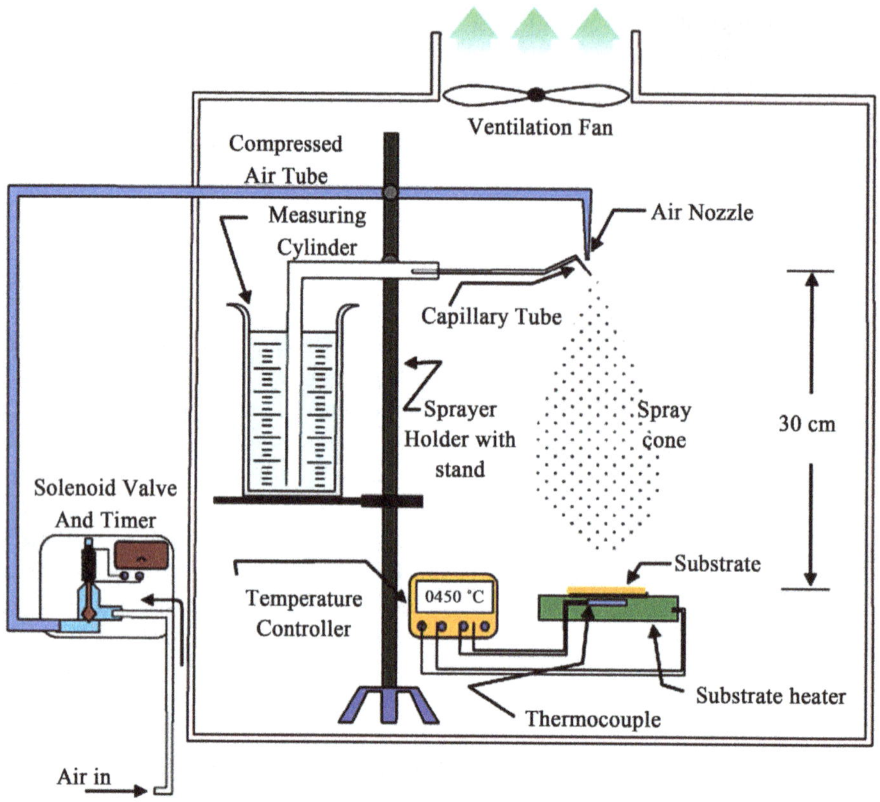

Figure 1. Schematic representation of the spray system.

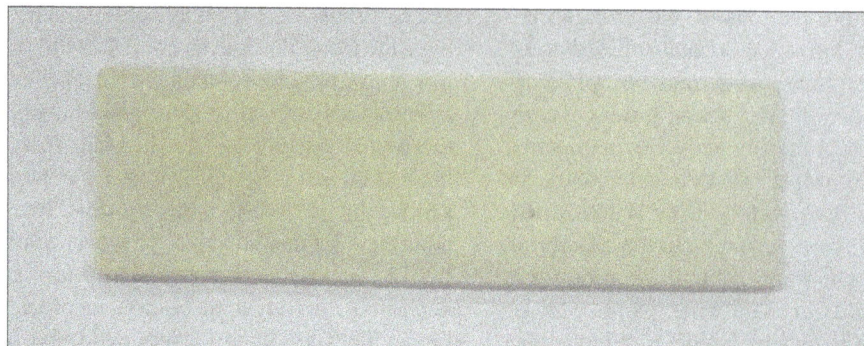

Figure 2. A photo of spray pyrolyzed ZnO nanofilm on glass samples.

Figure 3. (a)-(e) represent SEM images of ZnO nanofilm at different magnification power and (f) for a sample thickness.

The nonlinear absorption study at the near resonant regime was carried out using single beam femtosecond *z*-scan technique. The *z*-scan setting is illustrated by schematic diagram shown in **Figure 4**. A femtosecond laser of pulse duration 60 fs and of average power 0.165 W was used as a laser source. The pulse duration was measured by autocorrelation system and the energy was measured by pyroelectric energy prob model type (PDA36A), covering the rang 350 - 1100 nm from THORLABS. The beam profile was adjusted by spatial filter leading to spatial intensity profile near Gaussian with beam quality of $M^2 \approx 1.79$. The laser beam was focused by a lens of 15 cm focal length to produce a waist of 22.5 μm. The sample was translated along the beam axis (*z*-axis) through the Rayleigh distance 2000 μm. The distance from lens to aperture was 53 cm and beam diameter at detector was 16.4 mm.

3. Results and Discussion

The topography study of the prepared film shows the formation of the ZnO nanostructure and the film thickness was around 2 μm. as reveled by Scanning Electron Microscopy (SEM) type ULTRA 55 with different magnification; as shown in **Figure 3**. These figures showed nanocrystals of size ~(100 - 200 nm). The sample was scanned in all zones before the picture was taken. The micrographs reveled that the particles were hexagonal in shape. This indicated that ZnO nanocrystal were grown with c-axis orientation and almost perpendicular to the substrate surface as observed in the scanning electron microscope (SEM) image.

The XRD pattern of ZnO film prepared with 2 μm thickness is illustrated in **Figure 3**. The spectrum indicates that the ZnO film is a polycrystalline structure. The observed values of the XRD peaks are compared with American Society for testing and Material (ASTM) data for hexagonal zinc oxide. It can be noticed from the XRD pattern the strongest peak observed at Bragg's angle $2\theta = 34.38°$ can be attributed to the (002) plan of the hexagonal ZnO and the interplanar distance can be determined by Bragg's law, $d_{002} = 0.26$ nm. The (102) peak is also observed at $2\theta = 47.44°$ but this peak is much lower intensity than the (002) peak. The figure shows broad peaks which give evidence of the nanostructure formation. Using the width of (002) peak which appears at angle 34.38° on 2θ scale in Scherrer's formula [18]:

$$d = 0.94\lambda / \beta \cos\theta \tag{2}$$

where d is the average crystalline grain size, λ is the wavelength, β represents the full width at half maximum (FWHM) in radian and θ is the Bragg diffraction angle in degree. The size of the formed nanoparticles was found to be about 46 nm, this value is close from the value that measured by SEM. Using the width of (102) peak which appears at angle 47.44° on 2θ scale the size of the formed nanoparticles was found to be about 26 nm. The interplaner distance can be determined by Bragg's law $d_{102} = 0.19$ nm.

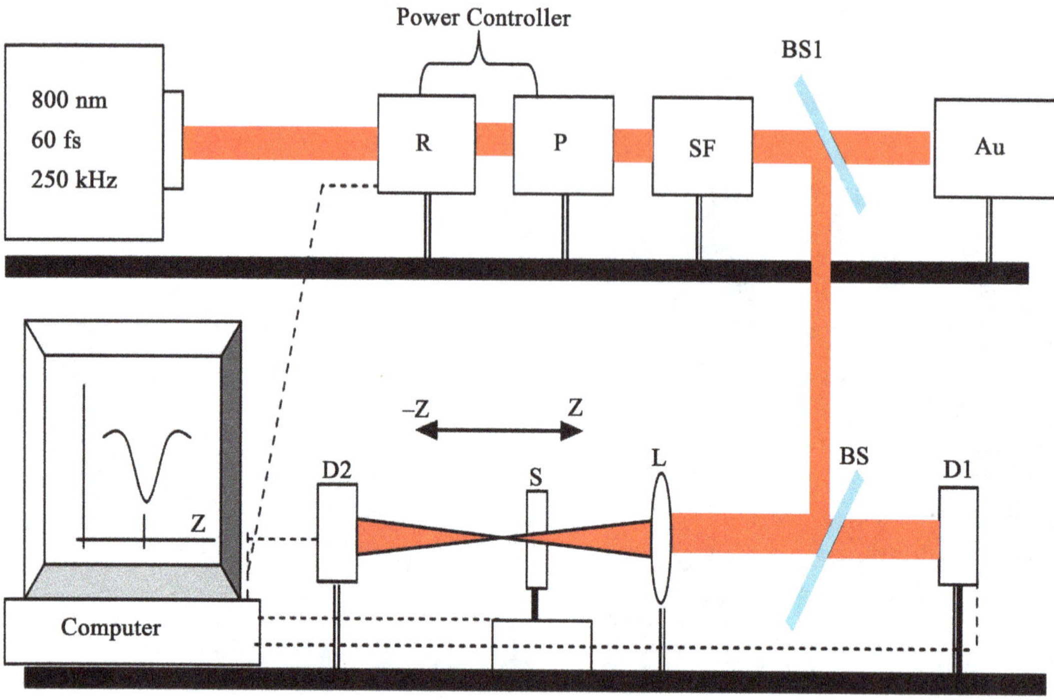

Figure 4. Schematic of the z-scan setup recording the nonlinear absorption, R-rotator, P-polarizer, SF-spatial filter, BS1, BS2-beam splitter, Au-autocorrelation, D1,D2-detectors, L-lens, S-sample.

The optical properties of ZnO film which were prepared on quartz substrate have been studied in this work. The absorption spectrum of the ZnO film is shown in **Figure 5**. This spectrum shows low absorption in the visible and infrared regions; however, the absorption in the ultraviolet region is high. This result were in agreement with the absorption spectrum which obtained by other workers. The energy band gap of ZnO film was estimated using Tauc equation which can be written as:

$$(\alpha hv) = A(hv - Eg)^n \qquad (3)$$

where A is a constant, α absorption coefficient, hv the photon energy (Eg) the band gap, $n = 1/2$ for the direct transitions. Referring to the data extracted from the absorption spectrum in **Figure 6**, the absorption coefficient (α) was calculated as a function of wavelength. Assuming allowed transition; the dependence of $(\alpha hv)^2$ on (hv) is plotted as in **Figure 7**. The extrapolation of the linear part of the plot $(\alpha hv)^2 = 0$, gives rise on estimation of the energy gap value of the prepared ZnO film. The value of the energy gap was found to be about 3.3 eV. This value was in a good agreement with values mentioned by other works.

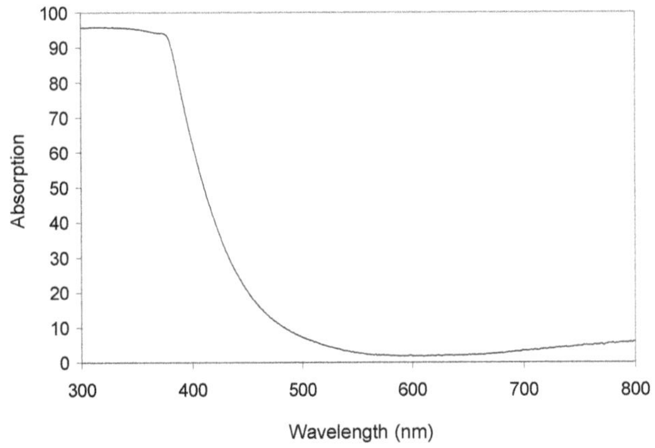

Figure 5. UV-VIS absorption spectra of ZnO nanofilm.

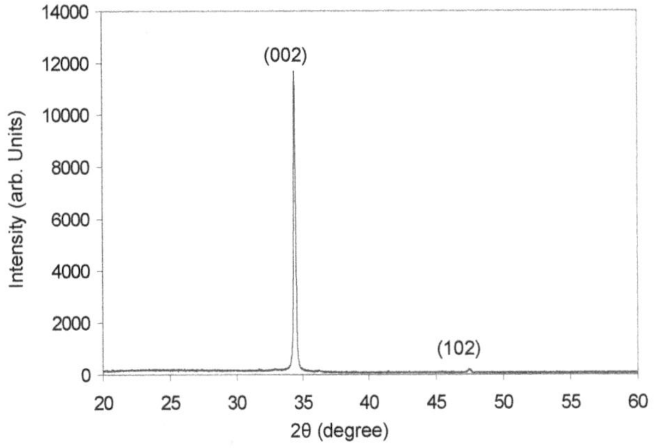

Figure 6. XRD pattern of ZnO nanofilm deposited on a glass substrate at 400˚C.

The optical transmittance spectrum of the ZnO film is shown in **Figure 8**. It can be noticed from this figure that the transmittance is high in the visible and infrared regions and low transmission in ultraviolet region.

The optical reflection spectrum of the ZnO film is shown in **Figure 9**. It can be noticed from this figure that the maximum reflection at 470 nm and minimum value was 390 nm. The optical fluorescence spectrum of the

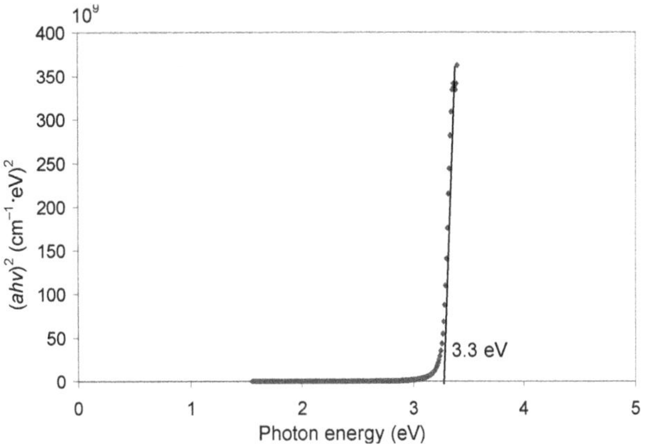

Figure 7. Plot of $(ahv)^2$ versus photon energy hv for ZnO nanonanofilm.

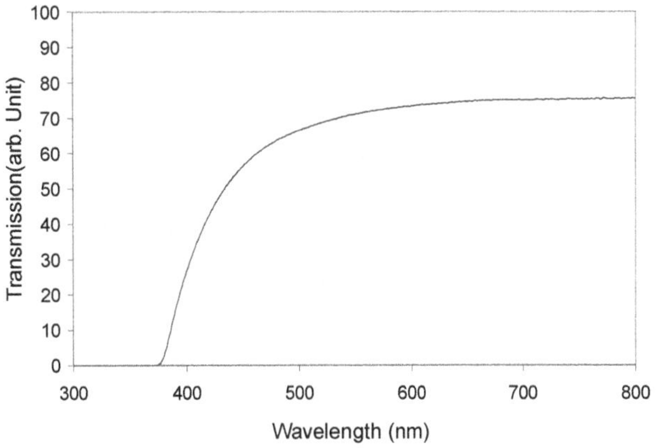

Figure 8. Optical transmission spectra of ZnO nanofilm.

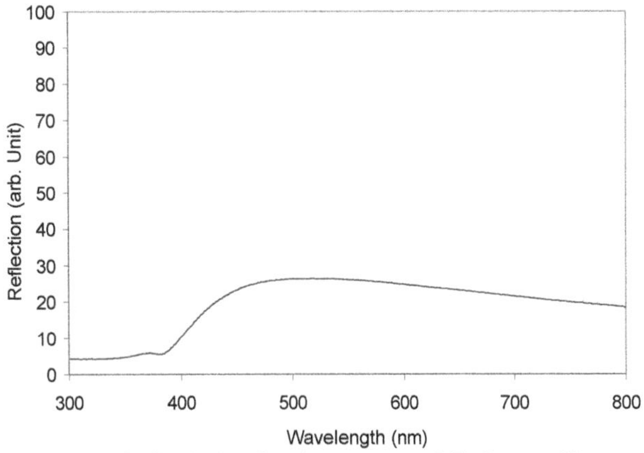

Figure 9. Optical reflection spectra of ZnO nanofilm.

ZnO film illuminated by 320 nm UV line is shown in **Figure 10**. The spectrum displays two major luminance peaks at 380 nm and around 556 nm. The first peak is due to the energy gap transmission which corresponds to 3.28 eV. The second peak is due to the excitonic emission; and it is in a good agreement with the results measured by many other authors [19,20]. The broad green emission peak that is dominated at 556 nm (≈2.23 eV) is a good evidence for the exciton formation in ZnO with a binding energy of 60 meV. The high binding energy enables the finding of the exciton at room temperature. The intensity at the 556 nm peak is higher than that found around 380 nm peak. This is because the band-to-band transition was quenched by the defect states. The same behavior was observed by [21].

The Raman active zone-center optical phonons predicated by the group theory are $A_1 + 2E_2 + E_1$. The phonons of A_1 and E_1 symmetry are polar phonons and, hence, exhibit different frequencies for the transverse- optical (TO) and longitudinal optical (LO) phonons. In wurtzite ZnO crystals, The non-polar phonon modes with symmetry E_2 have two frequencies, E_2(high) is associated with oxygen atoms and E_2(low) is associated with Zn sublattice. The described phonon modes have been reported in the Raman scattering spectra of bulk ZnO [22,23]. In **Figure 11**, we present a typical non-resonant Raman scattering spectrum from ZnO nanocrystal obtain under the 633 nm non-resonant excitation on of He-Ne laser. For comparison, the phonons peaks observed in bulk ZnO crystal are summarized in **Table 1**.

In our experiment, Raman spectra were recorded at room temperature under the 633 nm excitation and laser power 75 mW, the E_2(low) peak is red shifted by ~3 cm^{-1} from its bulk value, E_2(high) peak is red shift by ~2 cm^{-1}, but the peak E_1(LO) at 591 cm^{-1} corresponding to E_1(LO). The A_1(TO), E_1(TO), A_1(LO) peaks being absent in this sample.

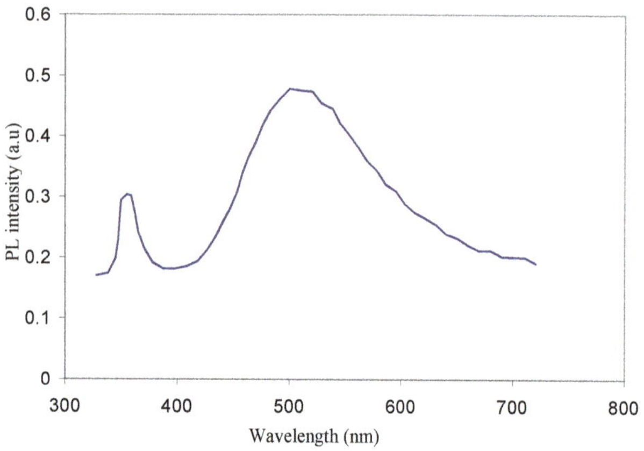

Figure 10. Photoluminescence emission spectrum of ZnO nanofilm.

There are three possible mechanisms for the phonon peak shift in Raman spectra of nanostructures. The first one is spatial confinement within the quantum dot (nanocrystal) boundaries. The second one is related to the phonon localization by defect. Nanocrystal or quantum dots, produced by chemical methods or by the molecular-beam epitaxy, normally have more defect than corresponding bulk crystal. In order elucidate the possible of the peak shifts we carried out simple calculations in the framework of the H. Richer [24] and I. H. Chambell [25] phenomenological models.

The z-scan transition curve at 110 GW/cm^2 excitation intensity is recorded for the nanoparticles ZnO film and it shown in **Figure 12**, Where the average power was 0.165 W, Repetition rate (RR) was 250 kHz, Pulse energy E_p was 6.6×10^{-7} J, Pulse duration (FWHM) was 60 fs, Peak power (PP) was 10.34×10^6 W.

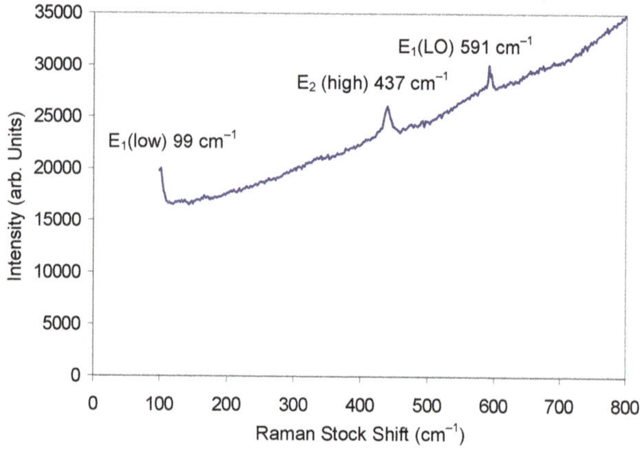

Figure 11. Non-resonant Raman spectra from ZnO nanocrystal under 633 nm excitation and 75 mW for He-Ne laser.

Table 1. Raman active phonon mode frequencies (in cm^{-1}) for bulk ZnO.

E_2(low)	A_1(TO)	E_1(TO)	E_2(high)	A_1(LO)	E_1(LO)
102	379	410	439	574	591

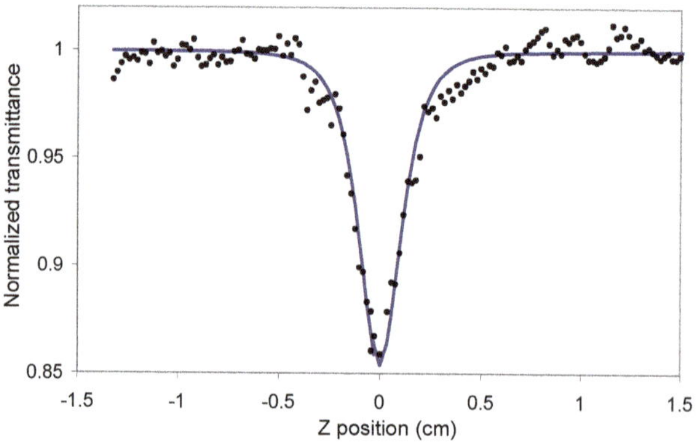

Figure 12. OA Z-scan measured at irradiance 110 GW/cm^2, wavelength 800 nm, a pulse duration 60 fs and repetition rate of 250 kHz. The solid line is the fitting curve employing the z-scan theory, described in the text, on 3PA.

The rate equations model describing the transitions between the valance band and the conduction band in our semiconductor sample illuminated by Titanium-Sapphire femtosecond laser of 800 nm wavelength can be written as:

$$\frac{dN_h}{dt} = -\frac{\alpha I^3}{3\hbar\omega} + \frac{N_e}{\tau_i} \tag{4}$$

$$\frac{dN_e}{dt} = \frac{\alpha I^3}{3\hbar\omega} - \frac{N_e}{\tau_i} \tag{5}$$

where N_h and N_e are the population in the valance band and the conduction band respectively. α is the three photon absorption coefficient, $\hbar\omega$ is the energy of the IR pump Titanium-Sapphire laser photon, I is the laser intensity at the sample position and τ_i is the life time of the excited level in the conduction band. The Gaussian laser beam intensity is given by:

$$I = I_o\left[\frac{\omega_o^2}{\omega^2(z)}\right]\exp\left[-\frac{2r^2}{\omega^2}\right]\exp\left[-\frac{t^2}{\tau_p^2}\right] \tag{6}$$

where I_o is the laser intensity at the waist of the beam and τ_p is the laser pulse duration. Since the measuring intensity was carried out at the focal point, in the Z-coordinate, thus the intensity in Equation (4) can be reduced to $I = I_o$. In order to find the relation between the power of the out put emitted fluorescent pulse and the intensity of the Titanium-Sapphire laser pump, Equation (5) can be arranged as:

$$\frac{dN_e}{dt} + \frac{N_e}{\tau_i} = \frac{\alpha I_o^3}{3\hbar\omega} \tag{7}$$

The solution of Equation (7) can be written as:

$$N_e = \frac{\alpha\tau_i I_o^3}{3\hbar\omega}\left[1 - e^{-\frac{\tau_i}{\tau_p}}\right] \tag{8}$$

Since the pulse duration of the pump laser (τ_p) is in femtosecond and the life time of the excited state (τ_i) in picosecond [26], the Equation (8) can be simplified as:

$$N_e = \frac{\alpha\tau_p}{3\hbar\omega} I_o^3 \tag{9}$$

Multiply both sides of the above equation by the volume (V) of the illuminated part of the ZnO sample and arranged the terms, Equation (9) can be written as:

$$\frac{3\hbar\omega \cdot N_e \cdot V}{\tau_p} = P_f = \alpha \cdot V \cdot I_o^3 \tag{10}$$

where P_f is the power of the emitted fluorescent from the ZnO sample pumped by the Titanium-Sapphire laser. Introducing the quantum yield which include the probability of the exciton self-trapping and the efficiency of the detection unit, Equation (10) can be written as:

$$P_f = \eta\alpha V I_o^3 \tag{11}$$

In our experimental setting described before, the focal spot are about 1590 μm^2 and the ZnO film has 2 μm thickness, thus the illuminated volume is about 3180 μm^3 and η was estimated to by 10^{-5} in our experimental set up. The value of (a) is found from the measurements of P_f at different values of I_o, and by applying Equation (11), it was found to by about of 0.0142 cm^3/GWatt^2. The result which is not far away from the calculate result from Gaussian fit technique as follow:

The normalized energy transmittance for 3PA of the open aperture z-scan is given by R. L. Sutherland [27]:

$$T(z) = \frac{1}{\pi^{1/2} p_o} \times \int_{-\infty}^{\infty} \ln\left\{\left[1 + p_o^2\exp\left(-2x^2\right)\right]^{1/2} + p_o\exp(-x^2)\right\}dx \tag{12}$$

where $P_o = \left(2\gamma I_o L'_{eff}\right)^{1/2}$, $I_o = I_{oo}/\left(1 + z^2/z_o^2\right)$ is the excitation intensity at the position z, $z_o = \pi\omega_o^2/\lambda$ where z_o is the Rayleigh range, ω_o is the minimum beam waist at focal point ($z = 0$), λ is the laser free-space wavelength, $L'_{eff} = \left[1 - \exp\left(-2a_o L\right)\right]/2a_o$ is the effective sample length for 3PA processes; L is the sample length and a_o is the linear absorption coefficient. The open aperture z-scan graph is always normalized to linear transmittance *i.e.*, transmittance at large values of $|z|$. If $P_o < 1$ Equation (12) can be expanded in a Taylor series as:

$$T = \sum_{m=1}^{\infty}(-1)^{m-1}\frac{p_o^{2m-2}}{(2m-1)!(2m-1)^{1/2}} \tag{13}$$

Furthermore, if the higher order terms are ignored, the transmission as a function of the incident intensity is given by R. L. Sutherland [27]:

$$T = 1 - \frac{\gamma I_o^2 L'_{eff}}{3^{3/2}} \tag{14}$$

The 3PA coefficient can be extracted from the best fit for Equation (14). The sold curve in **Figure 12** is the best fit for Equation (14). The Equation (14) shows clearly that the depth of the absorption dip is linearly proportional to the 3PA coefficient γ, but the shape of the trace is primarily determined by the Rayleigh range of the focused Gaussian beam. The fitted value of γ is on the order of 0.0166 cm^3/Gwatt^2. This value is five times of magnitudes higher than the value observed with bulk ZnO sample. The natural logarithm of the $(1 - T)$ values are plotted as a function of natural logarithm of the incident intensity I_o in **Figure 13**. The curve can be reasonably fitted with a straight line with slope of 1.99, this indicate that the 3PA was occur in ZnO pump by 800 nm laser source of 60 fs pulse duration.

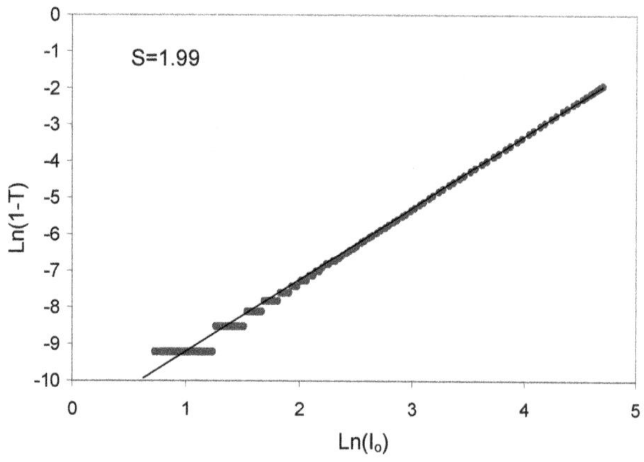

Figure 13. Plot of Ln(1 − *T*) versus Ln(I_o) at 800 nm wavelength, the solid line is the example of the linear at 800 nm with slope $s = 1.99$.

4. Conclusion

The three photon absorption has been observed in ZnO nanocrystalline prepared by chemical method upon illuminating it by femtosecond Titanium-Sapphire laser. The fully computerized *z*-scan system was used to measure the nonlinear absorption coefficient of the prepared samples. The value of the measured nonlinear coefficient was found to be five times higher than the bulk value. The enhancement of the nonlinear coefficient was attributed to the formation of the nanocrystallites of ZnO and to the existence of the exciton in the prepared film.

5. Acknowledgements

This work has been carried out in the physics Department, School of Engineering and Applied Sciences, Harvard University. The authors would like to thanks Mazur Research Group in Harvard University for their help through this work. Thank also to Christopher C. Evans, Jonthan D. B. Bradley, and Eric Mazur for their interest, guide and useful discussion. We thanks also the Ministry of Higher Education in the republic of Iraq for support this work.

REFERENCES

[1] Y. W. Zhu, *et al.*, "Multiwalled Carbon Nanotubes Beaded with ZnO nanoparticles for Ultrafast Nonlinear Optical Switching," *Advanced Materials*, Vol. 18, No. 5, 2006, pp. 587-592. doi:10.1002/adma.200501918

[2] K. Ramanathan, J. Keane and R. Noufi, "Properties of High-Efficiency CIGS Thin-Film Solar Cells," *NREL/CP*, Vol. 520, 2005, Article ID: 37404.

[3] C. L. Rhodes, S. Lappi, D. Fischer, S. Sambasivan, J. Genzer and S. Franzen, "Characterization of Monolayer Formation on Aluminum-Doped ZincOxide Thin Films," *American Chemical Society*, Vol. 24, No. 2, 2008, pp. 433-440.

[4] H. Li, *et al.*, "Microstructural Study of MBE-Grown ZnO Film on GaN/Sapphire (0001) Substrate," *Central European Journal of Physics*, Vol. 6, No. 3, 2008, pp. 638-642. doi:10.2478/s11534-008-0032-2

[5] T. Sato, *et al.*, "Production of Transition Metal-Doped ZnO Nanoparticles by Using RF Plasma Field," *Journal of Crystal Growth*, Vol. 275, No. 1-2, 2005, pp. 983-987. doi:10.1016/j.jcrysgro.2004.11.152

[6] S. Singh and M. S. R. Rao, "Structure and Physical Properties of Undoped ZnO and Vanadium Doped ZnO Films Deposited by Pulsed Laser Deposition," *Journal Nanoscience and Nanotechnology*, Vol. 8, No. 5, 2007, pp. 2575-2577.

[7] Ü. Özgür, *et al.*, "A Comprehensive Review of ZnO Materials and Devices," *Journal of Applied Physics*, Vol. 98, No. 4, 2005, Article ID: 041301. doi:10.1063/1.1992666

[8] J.-H. Lin, *et al.*, "Two-Photon Resonance Assisted Huge Nonlinear Refraction and Absorption in ZnO Thin Films Institute of Electro-Optical Engineering," *Journal of Applied Physics*, Vol. 97, No. 3, 2005, Article ID: 033526. doi:10.1063/1.1848192

[9] S. J. Bentley, *et al.*, "Three-Photon Absorption for Nanosecond Excitation in Cadmium Selenide Quantum Dots," *Optical Engineering*, Vol. 46, No. 12, 2007, Article ID: 128003. doi:10.1117/1.2823156

[10] E. W. Van Stryiand, M. Sheik-Bahae, A. A. Said and D. J. Hagan, "Characterization of Nonlinear Optical Absorption and Refraction," *Progress in Crystal Growth and Characterization of Materials*, Vol. 27, No. 3-4, 1993, pp. 279-311. doi:10.1016/0960-8974(93)90026-Z

[11] J. He, W. Ji and J. Mi, "Three-Photon Absorption in Water-Soluble ZnS Nanocrystal," *Applied Physics Letters*, Vol. 88, No. 18, 2006, Article ID: 181114. doi:10.1063/1.2198823

[12] B. Gu, *et al.*, "Three-Photon Absorption Saturation in ZnO and ZnS Crystals," *Journal of Applied Physics*, Vol. 103, No. 7, 2008, Article ID: 073105. doi:10.1063/1.2903576

[13] S. Pearl, *et al.*, "Three Photon Absorption in Silicon for 2300 - 3300 nm," *Applied Physics Letters*, Vol. 93, No. 13, 2008, Article ID: 131102. doi:10.1063/1.2991446

[14] A. Penzkofer and W. Falkenstein, "Three Photon Absorption and Subsequent Excited-State Absorption in CdS," *Optics Communications*, Vol. 16, 1976, pp. 247-250.

[15] M. G. Vivas, T. Shih, T. Voss, E. Mazur and C. R. Mendonca, "Nonlinear Spectra of ZnO: Reverse Saturable, Two- and Three-Photon Absorption," *Optics Express*, Vol. 18, No. 9, 2010, pp. 9628-9633. doi:10.1364/OE.18.009628

[16] A M. Suhail, H. J. Kbashi and R. K. Jamal, "Three-Photon Absorption in Nanostructure Wide-Band Gap Semiconductor ZnO Using Femtosecond Laser," *Modern Applied Science*, Vol. 5, No. 6, 2011, pp. 199-210. doi:10.5539/mas.v5n6p199

[17] M. R. Islam and J. Podde, "Optical Properties of ZnO Nanofiber Thin Film Grown by Spray Pyrolysis of Zinc Acetate Precursor," *Crystal Research and Technology*, Vol. 44, No. 3, 2009, pp. 286-292.

doi:10.1002/crat.200800326

[18] A. L. Patterson, "The Scherrer Formula for X-Ray Particle Size Determination," *Physical Reviews*, Vol. 56, No. 10, 1939, pp. 978-982. doi:10.1103/PhysRev.56.978

[19] K. Yim and C. Lee, "Optical Properties of Al-Doped ZnO Thin Films Deposited by Two Different Sputtering Methods," *Crystal Research and Technology*, Vol. 41, No. 12, 2006, pp. 1198-1202. doi:10.1002/crat.200610749

[20] C. X. Xu, *et al.*, "Growth and Spectral Analysis of ZnO Nanotubes," *Journal of Applied Physics*, Vol. 103, No. 9, 2006, Article ID: 094303. doi:10.1063/1.2908189

[21] D. D. Wang, J. H. Yang, L. L. Yang, Y. J. Zhang, J. H. Lang and M. Gao, "Morphology and Photoluminescence Properties of ZnO Nanostructures Fabricated with Different Given Time of Ar," *Crystal Research and Technology*, Vol. 43, No. 10, 2008, pp. 1041-1045. doi:10.1002/crat.200800109

[22] K. Alim, V. A. Fonoberov, M. Shamsa and A. A. Balandin, "Micro-Raman Investigation of Optical Phonons in ZnO Quantum Dots," *Journal of Applied Physics*, Vol. 97, No. 12, 2005, Article ID: 124313. doi:10.1063/1.1944222

[23] N. Ashkenov, B. N. Mbenkum, C. Bundesmann, V. Riede, M. Lorenz, D. Spenmann and E. M. Kaidashev, "Infrared Dielectric Functions and Phonon Modes of High-Quality ZnO Film," *Journal of Applied Physics*, Vol. 93, No. 1, 2003, pp. 126-133. doi:10.1063/1.1526935

[24] H. Richer, Z. P. Wany, "The One Phone Raman Spectrum in Microcrystalline Silicon," *Solid State Communications*, Vol. 39, No. 5, 1981, pp. 625-629. doi:10.1016/0038-1098(81)90337-9

[25] I. H. Chambell and P. M. Fanchet, "Three Effects of Microcrystal Size and Shape on the One Phonon Raman Spectra of Crystalline Semiconductors," *Solid State Communications*, Vol. 58, 1986, p. 739.

[26] J. He, *et al.*, "Three-Photon Absorption in ZnO and ZnS Crystals," *Optics Express*, Vol. 13, 2005, pp. 9235-9247.

[27] R. L. Sutherland, D. G. McLean and S. Kirkpatrick, "Handbook of Nonlinear Optics," 2nd Edition, Reserved and Expanded, New York, 2003.

Multi-TBaud Optical Coding Based on Superluminal Space-to-Time Mapping in Long Period Gratings

Reza Ashrafi, Ming Li, José Azaña
Institut National de la Recherche Scientifique–Énergie, Matériaux et Télécommunications, Montréal, Québec, Canada
Email: ashrafi@emt.inrs.ca

ABSTRACT

A novel time-domain ultra-fast pulse shaping approach for multi-TBaud serial optical communication signal (e.g. QPSK and 16-QAM) generation based on the first-order Born approximation in feasible all-fiber long-period gratings is proposed and numerically demonstrated.

Keywords: Pulse Shaping; Fiber Optics Components; All-optical Devices; Ultra-fast Processing

1. Introduction

Fiber and integrated-waveguide grating structures have been widely investigated for ultrafast optical pulse shaping and processing applications, including generation and detection of high-speed complex data streams in telecommunication systems [1,2]. The advantages of these solutions are associated with their intrinsic compact, low-loss all-fiber/waveguide implementations, e.g. in contrast to widely used programmable linear waveshapers based on bulk-optics configurations (involving diffraction gratings and spatial modulators) [3]. In particular, there has been an important body of work on the use of short-period (Bragg) fiber/waveguide gratings (BGs) for ultra-fast optical coding, namely generation of customized temporal optical data streams under different amplitude and/or phase coding schemes [1,2]. These solutions are particularly interesting for applications requiring the generation of time-limited data streams (composed of a few consecutive symbols), such as for optical code-division multiple access (OCDMA) and optical label-switching communications [1,2]. Long-period fiber gratings (LPGs) have recently attracted a great deal of interest for linear optical pulse shaping and processing applications [4]. However, to date, there are very few published works on their potential for general optical coding operations; some interesting LPG designs have been recently reported [5] but they are limited to the synthesis of temporally symmetric, binary intensity-only (on-off-keying, OOK) optical codes.

As a general rule, in optical grating-based linear filters, to achieve a faster temporal signal, a smaller spatial feature is required in the coupling-coefficient (grating apodization) profile. Previous studies in counter-directional coupling structures [6,7], e.g. fiber/waveguide BGs, have revealed that under the first-order Born approximation (*i.e.* weak-coupling conditions), the output time-domain optical field complex envelope variation follows the spatial variation of the complex coupling coefficient. This phenomenon, referred to as space-to-time mapping, provides a very straightforward mechanism to synthesize optical waveforms (*e.g.* coded communication data streams) with prescribed complex temporal shapes. However, in BGs, the ratio (v) between the resolution of the mentioned variations in space (Δz) and time (Δt) is necessarily lower than the propagation speed of light in vacuum (c) [8], *i.e.* $v = \Delta z/\Delta t < c$, (see the case of BG in **Figure 2** and the given numerical example in **Table 1**). Considering a typical achievable sub-mm resolution for fiber grating apodization profiles, fiber BG pulse shapers/ coders are thus limited to resolutions of at least several picoseconds [1,2,7].

This work focuses on the use of the first-order Born approximation in co-directional coupling filters, particularly LPGs. As illustrated in **Figure 1(a)**, similarly to the case of BGs [6,7], under weak-coupling conditions, the grating complex (amplitude and phase) apodization profile can be directly mapped into the LPG filter's temporal impulse response [8,9]. In contrast to the BG case, the space-to-time mapping speed ($v = \Delta z/\Delta t$) in LPG filters can be made much higher than the propagation speed of light in vacuum. As illustrated in **Figure 2**, this superluminal space-to-time mapping speed in LPGs enables the synthesis of waveforms with temporal features several orders of magnitude faster than those achievable by BGs

*This research was supported in part by the Natural Sciences and Engineering Research Council of Canada (NSERC), and le Fonds Québécois de la Recherche sur la Nature et les Technologies (FQRNT).

(assuming the same spatial resolution in the grating apodization profile). In this work, we numerically demonstrate the straightforward use of this phenomenon for ultrafast optical coding applications, particularly for generation of customized serial optical communication streams under any desired complex coding format (*e.g.* QPSK and 16-QAM modulation formats in the examples reported here), well in the TBaud range (femtosecond resolutions) using readily feasible LPG designs, *e.g.* with grating apodization resolutions above the millimeter range.

2. Theory of Superluminal Space-to-Time Mapping in LPGs

Our theoretical derivations on the superluminal space-to-time mapping phenomenon in LPGs rely on the standard coupled-mode equations for the case of co-directional coupling. The mathematical details of these derivations will be reported elsewhere [9].

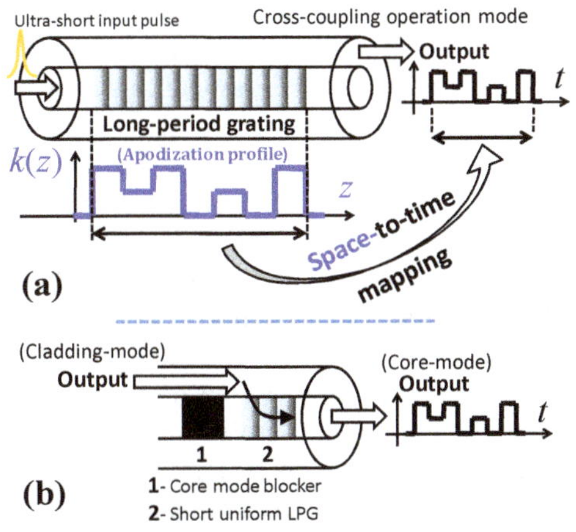

Figure 1. (a) Schematic of the proposed ultra-fast pulse shaping/coding approach based on superluminal space-to-time mapping in LPGs; (b) Illustration of a previously demonstrated fiber-optic approach [4] to transfer the cross-coupling signal from the fiber cladding-mode into the fiber core-mode by concatenating (1) a core-mode blocker and (2) a short, strong uniform LPG.

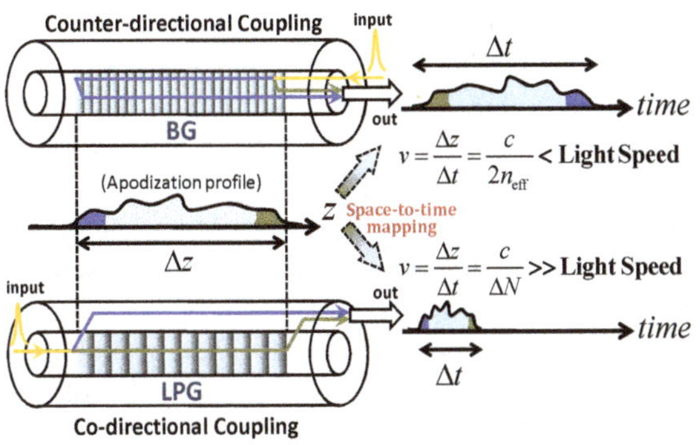

Figure 2. Comparison of the two pulse shaping approaches based on space-to-time mapping in fiber BGs and LPGs.

The LPG coupling coefficient (apodization) profile, i.e. $k(z)$ in **Figure 1**, is a complex function defined as $k(z) = |k(z)|\exp[j\varphi(z)]$. The magnitude $|k(z)|$ depends on the amplitude of the refractive index modulation along the LPG length, as illustrated in **Figure 3**. The grating discrete phase-shifts and grating period changes along the LPG length are accounted for in the phase term of the coupling coefficient, *i.e.* $\varphi(z)$. Some single phase-shifted gratings are also illustrated in **Figure 3** aimed to induce the corresponding discrete jumps in the phase of the coupling coefficient profile, *i.e.* $\varphi(z)$. Our theoretical studies [8,9] have shown that under weak-coupling strength conditions (*i.e.* strictly, cross-coupling power spectral response peak < 10%), the complex envelope of the temporal impulse response (let us call it $h(t)$) of the cross-coupling transfer function, *i.e.* core-to-cladding transfer function in fiber LPGs, is approximately proportional to the variation of the complex coupling coefficient $k(z)$, as a function of the grating length z after a suitable space-to-time scaling [8,9]. In particular, the space-to-time mapping speed (v), is obtained as $v = c/\Delta N$, where $\Delta N = (n_{\mathrm{eff1}} - n_{\mathrm{eff2}})$, and n_{eff1} and n_{eff2} are the effective refractive indices of the two coupled-modes around the wavelength of interest. Mathematically,

$$h(t) \propto \left\{ |k(z)| e^{j\varphi(z)} \right\}_{z=t\cdot c/\Delta N} \quad (1)$$

Clearly, ΔN can be designed to be much smaller than 1, and consequently the resulting speed (v) can be made significantly higher than the speed of light in vacuum. This superluminal space-to-time mapping speed is also significantly higher than the corresponding (subluminal) speed in the case of BG devices, *i.e.* $v = c/(2n_{\mathrm{eff}})$, where n_{eff} is the average effective refractive index of the propagating mode in the grating, see the comparison in **Figure 2**. This is the key to design optical pulse shapers (*e.g.* coders) based on LPGs with impulse responses having several orders of magnitude faster temporal features than their counter-directional filter counterparts (BGs).

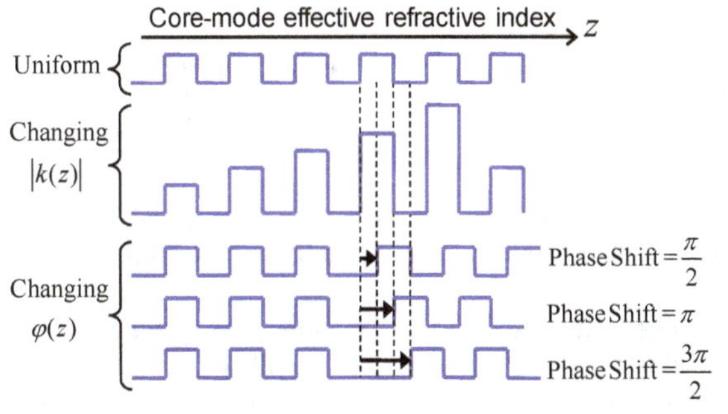

Figure 3. Illustration of variations on the amplitude and phase of the coupling coefficient profile, *i.e.* $|k(z)|$ and $\varphi(z)$ respectively, along the LPG length. For the phase change examples, some single phase-shifted gratings to generate the corresponding discrete jumps in $\varphi(z)$ are illustrated.

Notice that the LPG's cross-coupling operation mode can be practically implemented based on either integrated- waveguide technology (by simply inducing the coupling between two physically separated waveguides [10]) or a fiber-optic approach [4]. **Figure 1(b)** shows a schematic of a previously demonstrated all-fiber approach for implementation of the cross-coupling operation mode in LPGs [4], *i.e.* to ensure that both the input and output signals are carried by the fiber core mode. A core-mode blocker and a short broadband uniform LPG can be used for undistorted transference of the desired output signal from the cladding mode into the core mode. designations.

3. Numerical Comparison between BG-Based and LPG-Based Pulse Coders

Let us assume a fiber BG working in reflection and a fiber LPG working in the cross-coupling operation mode, both made in standard single-mode fiber (Corning SMF28), see **Figure 4**. The grating period for the LPG is assumed to be Λ = 430 μm, which corresponds to coupling of the fundamental core mode into the LP06 cladding mode at a central wavelength of 1550 nm. The BG has a period of 528 nm, corresponding to a Bragg wavelength of 1550nm. The average effective refractive index of the propagating mode in the BG is n_{eff} = 1.4684 and for the LPG: n_{eff1} = 1.4684 and n_{eff2} = 1.4648 [11-13]. **Table 1** shows the estimated space-to-time mapping speeds for these two examples. Let us further assume that the two considered BG and LPG devices have the same length of 10cm and they are both identically spatially-apodized for a target optical OOK bit stream pattern generation, as shown in **Figure 4**.

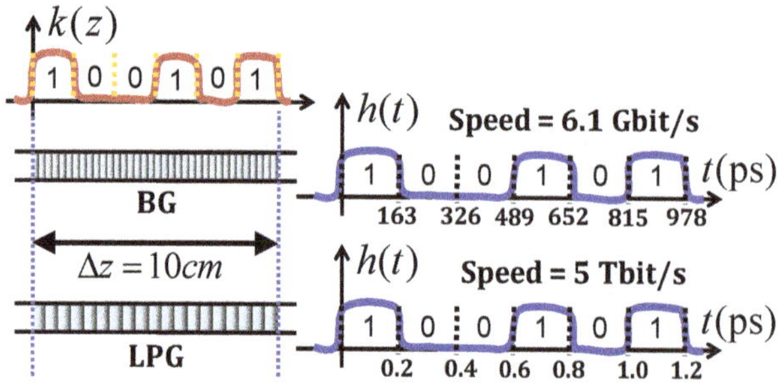

Figure 4. Comparison of the two OOK pulse-coding approaches based on space-to-time mapping in BGs and LPGs.

Table 1. The estimated space-to-time mapping speed for the considered BG and LPG made in SMF28 fiber.

	Space-to-time mapping speed
BG	$V = c/(2n_{eff}) = 1.022 \times 10^8$ (m/s)
LPG	$V = c/(n_{eff1} - n_{eff2}) = 833.3 \times 10^8$ (m/s)

In both cases, the amount of peak coupling coefficient is assumed to be low enough to satisfy weak-coupling conditions. Based on the space-to-time mapping theory, by launching an ultra-short optical pulse into the considered optical filters, the target bit stream patterns (*i.e.* $h(t)$ in **Figure 4**) will be generated at the filters' output port. As expected from the different space-to-time mapping speeds, the bit rate of the generated bit stream pattern by the LPG device should be nearly 1,000 faster than that generated by the BG filter.

4. Numerical Simulations

Using coupled-mode theory combined with a transfer-matrix method [13], we have numerically simulated two different LPG designs for generation of two 8-symbol optical QPSK and 16-QAM signals, each with a speed of 4TBaud (4TBaud), from an input ultra-short optical Gaussian pulse with a (full width at 10% of the peak amplitude) duration of 100 fs. **Figure 5** shows the results of these numerical simulations. The LPG design parameters are those defined above and the input optical pulse is assumed to be centered at the LPG resonance wavelength of 1550 nm. In the numerical simulations, the following wavelength dependence has been assumed for the effective refractive indices of the two interacting (coupled) modes [12]: $n_{eff1}(\lambda) = 1.4884 - 0.031547\lambda + 0.012023\lambda^2$ for the core-mode and $n_{eff2}(\lambda) = 1.4806 - 0.025396\lambda + 0.009802\lambda^2$ for the LP06 cladding-mode, where $1.2 < \lambda < 1.7$ is the wavelength variable in μm.

Figsures 5(a) and **(b)** show the designed amplitude and phase grating-apodization profiles for the target QPSK and QAM coding operations, respectively. The grating designs are relatively straightforward and simple, just being spatial-domain mapped versions of the respective targeted complex time-domain optical data streams. In particular, **Figrues 5(g)** and **(h)** show the amplitude and phase profiles of the time-domain waveforms at the outputs of the simulated LPG designs, demonstrating accurate generation of the targeted 4TBaud data streams, as per the coding formats defined in **Figures 5(c)** and **(d)**, respectively, in excellent agreement with the inscribed grating-apodization profiles.

Notice that considering the superluminal space-to-time mapping scaling value in the designed LPG (~833.3 × 10^8 m/s), each symbol time period of 250 fs corresponds to a fairly large spatial period of ~2.07 cm. As anticipated, time resolutions in the femtosecond regime (e.g. for the inter-symbol amplitude transitions and discrete phase jumps) can be achieved based on readily feasible millimeter grating spatial resolutions. The spectral responses of the two designed LPG filters are shown in **Figures 5(e)** and **(f)**, respectively. It is worth noting the intrinsic complexity of these responses (also for the phase, not shown here), which would make it very challenging for implementation using a frequency-based optical filter design approach, *e.g.* such as using conventional programmable linear wave-shapers.

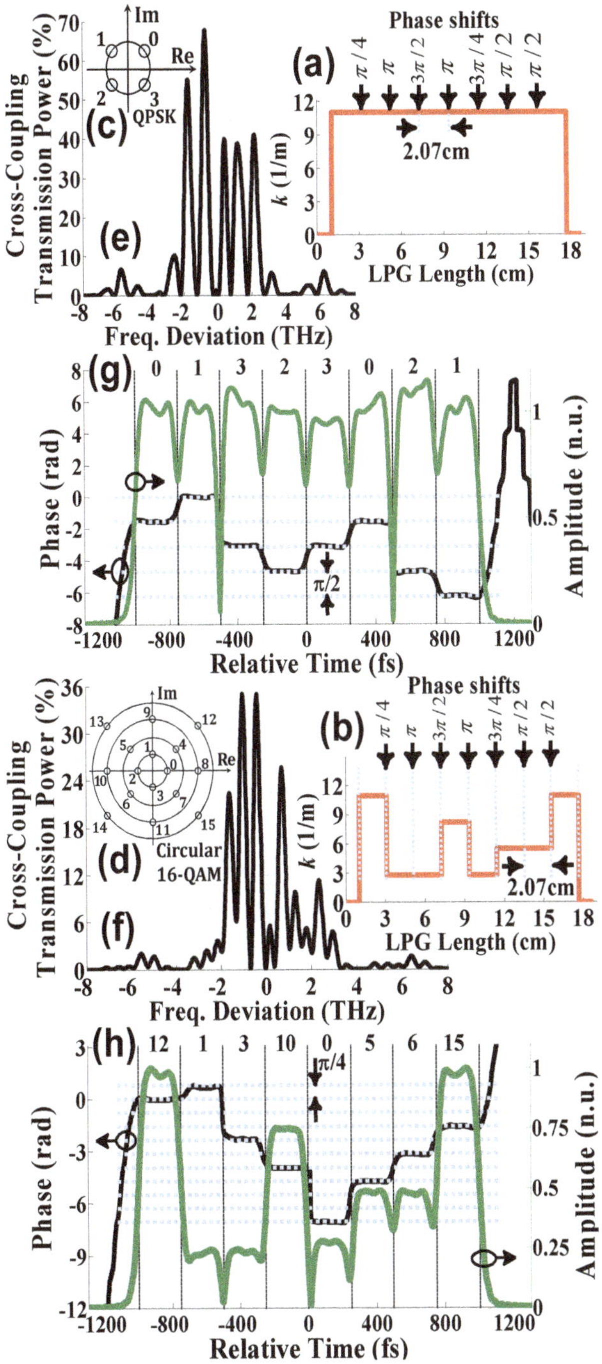

Figure 5. Simulation results of the two designed LPGs (a,b) to generate 8-symbol optical QPSK (c) and 16-QAM (d) data stream patterns, *i.e.* "0"1"3"2"3"0"2"1" and "12"1"3" 10"0"5"6"15" respectively, with a speed of 4TBaud from an input (full width at 10% of the peak amplitude) 100fs optical Gaussian pulse. *(e,f)* The corresponding spectral power responses of the designed LPGs. *(g,h)* The corresponding output temporal amplitude and phase responses.

5. Conclusions

We have proposed and numerically demonstrated a novel time-domain pulse shaping approach for synthesizing THz-bandwidth linear optical filters with arbitrary ultrafast temporal impulse responses based on the first-order Born approximation in LPGs. The proposed technique is particularly useful for generation of multi-TBaud serial optical communication data streams under complex (PSK, QAM *etc.*) coding formats using readily feasible and simple LPG designs, *e.g.* with spatial resolutions above the millimeter range. The corresponding matched-filtering devices for efficient decoding and detection of the generated data streams could be also designed and implemented using this same LPG approach.

REFERENCES

[1] P. C. Teh, M. Ibsen, J. H. Lee, P. Petropoulos and D. J. Richardson, "Demonstration of A Four-Channel WDM/OCDMA System Using 255-Chip 320-Gchip/s Quarternary Phase Coding Gratings," *IEEE Photonics Technology Letters*, Vol. 14, No. 2, 2002, pp. 227-229. doi:10.1109/68.980530

[2] L. M. Rivas, M. J. Strain, D. Duchesne, A. Carballar, M. Sorel, R. Morandotti and J. Azaña, "Picosecond Linear Optical Pulse Shapers Based on Integrated Waveguide Bragg Gratings," *Optics Letters*, Vol. 33, No. 21, 2008, pp. 2425-2427. doi:10.1364/OL.33.002425

[3] A. M. Weiner and A. M. Kan'an, "Femtosecond Pulse Shaping for Synthesis, Processing, and Time-to-Space Conversion of Ultrafast Optical Waveforms," *IEEE Journal of Selected Topics in Quantum Electronics*, Vol. 4, No. 2, 1998, pp. 317-331.

[4] R. Slavík, M. Kulishov, Y. Park and J. Azaña, "Long-Period-Fiber-Grating-Based Filter Configuration Enabling Arbitrary Linear Filtering Characteristics," *Optics Letters*, Vol. 34, No. 7, 2009, pp. 1045-1047. doi:10.1364/OL.34.001045

[5] S. J. Kim, T. J. Eom, B. H. Lee and C. S. Park, "Optical Temporal Encoding/Decoding of Short Pulses Using Cascaded Long-Period Fiber Gratings," *Optics Express*, Vol. 11, No. 23, 2003, pp. 3034-3040. doi:10.1364/OE.11.003034

[6] H. Kogelnik, "Filter response of nonuniform almost-periodic structures," *Bell System Technical Journal*, Vol. 55, No. 1, 1976, pp. 109-126.

[7] J. Azaña and L. R. Chen, "Synthesis of Temporal Optical Waveforms by Fiber Bragg Gratings: A New Approach Based on Space-to-Frequency-to-Time Mapping," *Journal of the Optical Society of America B*, Vol. 19, No. 11, 2002, pp. 2758-2769.doi:10.1364/JOSAB.19.002758

[8] R. Ashrafi, M. Li and J. Azaña, "Femtosecond Optical Waveform Generation Based on Space-to-Time Mapping in Long Period Gratings," *IEEE Photonics Conference*, 2012, pp. 104-105.

[9] R. Ashrafi, M. Li, S. LaRochelle and J. Azaña, "Superluminal Space-to-Time Mapping in Grating-Assisted

Co-Directional Couplers," *Optics Letters*, Vol. 21, No. 5, 2013, pp. 6249-6256. doi:10.1364/OE.21.006249

[10] J. Jiang, C. L. Callender, J. P. Noad and J. Ding, "Hybrid Silica/Polymer Long Period Gratings for Wavelength Filtering and Power Distribution," *Applied Optics*, Vol. 48, No. 26, 2009, pp. 4866-4873. doi:10.1364/AO.48.004866

[11] R. Kritzinger, D. Schmieder and A. Booysen, "Azimuthally Symmetric Long-Period Fibre Grating Fabrication with A TEM_{01}-Mode CO_2 Laser," *Measurement Science and Technology*, Vol. 20, No. 3, 2009, p. 034004. doi:10.1088/0957-0233/20/3/034004

[12] M. Smietana, W. J. Bock., P. Mikulic and J. Chen, "Increasing Sensitivity of Arc-induced Long-Period Gratings—Pushing the Fabrication Technique Toward Its Limits," *Measurememt Science and Technology*, Vol. 22, No.1,p.015201,2011.doi:10.1088/0957-0233/22/1/015201

[13] T. Erdogan, "Fiber Grating Spectra," *Journal of Lightwave Technology*, Vol. 15, No. 8, 1997, pp. 1277-1294. doi:10.1109/50.618322

26

Guiding of Waves between Absorbing Walls

Dmitrii Kouznetsov, Makoto Morinaga
Institute for Laser Science, University of Electro-Communications, Tokyo, Japan
Email: {dima, morinaga}@ils.uec.ac.jp

ABSTRACT

Guiding of waves between parallel absorbing walls is considered. The principal mode is constructed; its absorption is estimated. The agreement with previous results about reflection of waves from absorbing walls is discussed. Roughly, the effective absorption of the principal mode is proportional the minus third power of the distance between walls, minus 1.5d power of the wavenumber and minus 0.5d power of the local absorption of the wave in the wall. This estimate is suggested as hint for the design of the atomic waveguides, and also as tool for optimization of attenuation of the amplified spontaneous emission (and suppression of parasitic oscillations) in high power lasers.

Keywords: Zeno Effect; Atom Optics; Waveguides; Suppression of Amplified Spontaneous Emission

1. Introduction

The consideration of reflection of waves from absorbing walls had been stimulated by the experiments with ridged mirrors [1] and their interpretation in terms of the Zeno effect [2,3]. The Zeno approximation [2] showed good agreement with experiments in wide range of parameters [3,4]; it describes reflection of waves of any origin. In particular, it applies to the atomic waves and to the optical waves.

In addition to the atom optics (discussed in [2]), the reflection and guiding of waves may affect the suppression of the Amplified Spontaneous Emission (ASE) that is considered as serious problem [5-9]. At the scaling-up the power, the efficient suppression of ASE becomes more important, and the unwanted guiding of ASE by the absorber (which is supposed to suppress it) might take place. The estimates of the conditions of such a guiding is necessary tool for the design of powerful devices.

In this article, the wave function of a particle (atom, photon) between two absorbing walls is constructed. The effective absorption of such a mode is estimated and compared to the previous results. The results are expected to have applications in both the atom optics (wanted guiding of neutral particles by the absorbing walls) and the design of powerful lasers (estimates of conditions of the unwanted guiding of the ASE by the walls that are supposer to absorb the ASE, and optimization of suppression of the ASE).

This article is organized in the following way:

In Section 2, the phenomenological absorbing Schrödenger equation is suggested. The case with absorbing walls corresponds to the pure anti Hermitian potential. The frequency of decay is denotes with γ. The special system of units is used in such a way that $\hbar = 1$ and the energy of the particle in vacuum is assumed to be square of its wavenumber.

In Section 3, the special case of uniform absorption is considered; this gives relations between parameters of the wave function and the physical quantities that can be determined experimentally. Such physical quantities are energy ω of the particle and its absorpfion s in the material of the wall; this quantities are assumed to be independent parameters.

In Section 4, the case of channeling is considered; parameters of the principal mode (transversal wavenumber p, decay q and the damping α) are defined and expressed through the holomorphic function $\operatorname{cosc}(z) = \cos(z)/z$, its inverse function $\operatorname{acosc} = \operatorname{cosc}^{-1}$ and $\operatorname{acosq}(z) = \operatorname{acosc}(z\exp(\mathrm{i}\pi/4))$; properties of these functions are discussed and the efficient C++ implementations are indicated.

In Section 5, the example of the principal mode is considered; the real and imaginary parts of the wave functions are built through the trigonometric function of complex argument and complex exponential for the damping $\alpha = 1/4$. The amplitude and phase of the principal mode are shown for $\alpha = 1/4$ and for $\alpha = 1/50$.

In Section 6, the asymptotic behavior of parameters at small damping is considered; this is realistic case of good guiding conditions.

In Section 7, the estimates for the effective absorption of the principal mode are compared with the previous results; the agreement is interpreted as confirmation of the validity of the analysis.

In Section 8, the special case is suggested for physically realistic parameters of the experimental conditions of realization of the guiding of waves between absorbing walls.

In Section 9 (Conclusion), the asymptotic estimate of the effective absorption A of the guided mode through the real and imaginary parts c, s of the wavenumber in the wall is suggested. This main result is expected to be confirmed (or rejected) by the physical experiments with photons, atoms or any other wave of any origin.

2. Schrödinger Wave and the Absorption

In this case the dimension less Schrödinger equation is considered in the paraxial approximation.

For simplicity, in this section the special system of units is used such that $\hbar = 1$ and mass of the atom is half. Then, the Schrödinger equation for the wave function $\Psi = \Psi(x, z, t)$ can be written as follows

$$i\partial\Psi/\partial t = -\partial^2\Psi/\partial x^2 - \partial^2\Psi/\partial z^2 + U(x)\Psi \qquad (1)$$

where the potential $U = U(x)$ depends only on the transversal coordinate x (and does not depend on time z in the direction of propagation).

In the simplest approximation, the entangling with numerous degrees of freedom of scattered (or relaxed) atom can be taken into account with non Hermitian potential. Such an approximation is considered, in particular, in the interpretation of the quantum reflection in therms of the Zeno effect [2]. For the wave guide, the absorption correspond to complex values of $U(x)$ at $x > X$, where X is half-width of the channel between the mirrors.

Consider the quasi-monochromatic solution, assuming the exponential dependence on t; let

$$\Psi(x, z, t) = \exp(-\mathrm{i}\omega t)\psi(x, z) \qquad (2)$$

Then, instead of (1) we get the stationary Schrödinger equation

$$\omega\psi = -\partial^2\psi/\partial x^2 - \partial^2\psi/\partial z^2 + U(x)\psi \qquad (3)$$

below, the two cases of the solution are considered: for

$$U(x) = -\mathrm{i}\gamma = \text{const} \qquad (4)$$

and for

$$U(x) = -\mathrm{i}\gamma\,\text{UnitStep}(|x| - d) \qquad (5)$$

where UnitStep is the conventional unit step function implemented in various programming languages including Mathematica, and $d \in \mathbb{R}$ is constant, that has sense of half-width of the channel that confines the particle.

3. Uniform Absorption

In order to understand the physical sense to the constant γ, consider the case of the uniform absorption of the plane wave at U by (4); let ψ does not depend on the first argument, id est, $\psi(x, z) = f(z)$. Then, for $f = f(x)$ we get the equation

$$f'' + (\omega + \mathrm{i}\gamma)f = 0 \qquad (6)$$

At positive values of γ, the decaying in the direction z solution has form

$$f = \exp(\mathrm{i}kz) \qquad (7)$$

where $k \in \mathbb{C} : \{\Re(k) > 0, \Im(k) > 0\}$ is solution of equation

$$k^2 = \omega + \mathrm{i}\gamma \qquad (8)$$

Assuming $\omega > 0$, $\gamma > 0$, wavenumber k can be expressed as follows:

$$k = \sqrt{\omega + \mathrm{i}\gamma} \qquad (9)$$

For the compactness of notations, let $k = c + \mathrm{i}s$, where c and s are real parameters. Then

$$c^2 - s^2 = \omega \qquad (10)$$

$$2cs = \gamma \qquad (11)$$

Alternatively, we may consider the absorption of wave in the medium, id est, $\Im(k)$ as initial parameter. Making estimates for the ridged atomic mirror with distance T between ridges, the absorption by intensity can be approximated as $1/T$, and the absorption by amplitude is of order of $1/(2T)$

$$s = \Im(k) = 1/(2T) \qquad (12)$$

Real part of k is determined by the mass and the energy of the particle we intent to reflect or to guide, or just $2\pi n/\lambda$, where λ is vacuum wavelength of the waves (perhaps, ASE) that could be guided, and n is the refraction index of the medium.

In order to avoid additional guiding by the step of the refraction index, the real part of the refraction index of the wall is supposed to be matched to that of the central region; but some reflection still may happen due to the imaginary part. Then, the real and imaginary parts of the swiare o wavenumber k can be expressed as follows:

$$\omega = \Re(k^2) = \Re(k)^2 - \Im(k)^2 = c^2 - s^2 \qquad (13)$$

$$\gamma = \Im(k^2) = 2\Re(k)\Im(k)) = 2sc \qquad (14)$$

Parameter ω has sense of the energy of the particle, and also sense of the square of wavenumber, while γ has sense of the decay rate at the time scale, if a wave would be uniform in the space. For the estimates, the term with $\Im(k)$ in (13) can be neglected; however it is kept in the deduction that may be applied not only to atoms, but also to other kinds of waves (optical, acoustical, waves on the surface of a liquid, etc.) For optics (both atom optics and conventional optics), the typical case is of low absorption; waves rather propagate than absorb, to,

$$\frac{s}{c}=\left|\frac{\Im(k)}{\Re(k)}\right|\ll 1 \quad (15)$$

In this case, for the estimate of the primary parameters, we may use the approximation

$$\omega \approx \Re(k)^2 \approx c^2 \quad (16)$$

and treat $c=\Re(k)$ and $s=\Im(k)$ as initial parameters of the model describing the absorption of the wave inside the walls.

On the other hand, one may consider as "given" the energy ω of the particle and the absorption $\Im(k)$. Then,

$$c=\Re(k)=\sqrt{\Im(k)^2+\omega} \quad (17)$$

$$\gamma=2\Im(k)\sqrt{\Im(k)^2+\omega} \quad (18)$$

In the following consideration, parameters $\omega\in\mathbb{R}$ and $\gamma\in\mathbb{R}$ and $k\in\mathbb{C}$ are supposed to be known. These parameters determine behavior of the wave inside the absorbing wall.

For the atom optics, id est, for the atom wave, the absorption $s=\Im(k)$ should be positive. For the optical wave, in principle, the absorption may be negative (gain medium), but the amplified spontaneous emission (unavoidable in the gain medium) limits the application of the formalism to very short distance of propagation. For this reason, the channeling in the pumped region may have more applications. In such a way, this section gives the sense to the parameters ω and γ that appear in the Equations (3) and (5), that describe the channeling of a particle by the potential U by (5). This channeling is considered in the next section. Then, the effective propagation constant $\kappa\in\mathbb{C}$ is estimated in terms of k and half-width d of the channel.

4. Channeling

For the case of potential U by (5), search the solution ψ in the following form:

$$\psi(x,z)=f(x)\,\mathrm{e}^{\mathrm{i}\kappa z} \quad (19)$$

where $\kappa\in\mathbb{C}$ is constant.

For the experimental realization, $B=\Re(\kappa)$ is expected to be of order of $c=\Re(k)$, determined in the previous section. As for the imaginary part, $A=\Im(\kappa)$ is expected to be small compared to $s=\Im(k)$, and $B=\Im(\kappa)$ is expected to decrease as $s=\Im(k)$ increases, allowing the interpretation in terms of the Zeno effect [2]: the stronger is the absorption in the region of the "observation" (id est, $|x|>d$), the better is the channeling.

Substitution of (19) to (3) gives the equation for $f=f(x)$ in the following form:

$$f''+\left(\omega-\kappa^2+\mathrm{i}\gamma\,\mathrm{UnitStep}(|x|-d)\right)f=0 \quad (20)$$

Search the solution of (20) as the combination of the cosinusoidal and the exponential, let

$$f(x)=\begin{cases}\cos(px), & |x|\le d\\ r\exp\left(-q(|x|-d)\right), & |x|\ge d\end{cases} \quad (21)$$

where $r\in\mathbb{C}$, $\kappa\in\mathbb{C}$, $p\in\mathbb{C}$, $q\in\mathbb{C}$ are constant parameters. From the physical reasons (almost free propagation inside the channel), parameter κ is expected to be of order of wavenumber k from the previous section.

The substitution of (21) into (20) determines that

$$\kappa^2+p^2=\omega=k^2 \quad (22)$$

$$\kappa^2-q^2=\omega+\mathrm{i}\gamma \quad (23)$$

From the physical reasons, it is expected that $|q|$ is small compared to $k=\sqrt{\omega}$ and, therefore, $|q|\ll|\kappa|$.

The continuity of $f(x)$ at $x=\pm d$ and the continuity of f' give the relations

$$\cos(pd)=r \quad (24)$$

$$-p\sin(pd)=-qr \quad (25)$$

The four Equations (22), (23), (24), (25) allows to express new parameters k,p,q,r in terms of the already defined parameters ω and γ.

Subtraction of (23) from (22) gives

$$p^2+q^2=-\mathrm{i}\gamma \quad (26)$$

Dividing of (25) by (24) gives

$$p^2\tan(pd)^2=q^2 \quad (27)$$

The combination of (26) and (27) gives

$$\left(1+\tan(pd)^2\right)p^2=-\mathrm{i}\gamma \quad (28)$$

Using the relation $1+\tan(\alpha)^2=1/\cos(\alpha)^2$, Equation (28) can be written as

$$\frac{(pd)^2}{\cos(pd)^2}=-\mathrm{i}\gamma d^2 \quad (29)$$

then

$$\frac{\cos(pd)^2}{(pd)^2}=\frac{\mathrm{i}}{\gamma d^2} \quad (30)$$

This equation can be written as follows:

$$\operatorname{cosc}(pd)^2 = \frac{\mathrm{i}}{\gamma d^2} \tag{31}$$

where for all complex $z \neq 0$,

$$\operatorname{cosc}(z) = \frac{\cos(z)}{z} \tag{32}$$

Complex map of function cosc is shown in the left hand side of figure 0; properties of this function and its inverse function $\operatorname{acosc} = \operatorname{cosc}^{-1}$ are described in TORI [10]. The name of function cosc is chosen in analogy with well established name of function sinc [11], defined with

$$\operatorname{sinc}(z) = \frac{\sin(z)}{z} \tag{33}$$

As usually, the name of the inverse function is created adding prefix "a" or "arc".

Equation for one of solutions with $\Re(\operatorname{acosc}(pd)) > 0$ can be written as follows:

$$\operatorname{cosc}(pd) = \sqrt{\frac{\mathrm{i}}{\gamma d^2}} \tag{34}$$

Equation (34) can be "inverted", giving

$$pd = \operatorname{acosc}\left(\sqrt{\frac{\mathrm{i}}{\gamma d^2}}\right) = \operatorname{acosc}\left(\frac{\exp(\mathrm{i}\pi/4)}{d\gamma^{1/2}}\right) = \operatorname{acosc}(\exp(\mathrm{i}\pi/4)\alpha) = \operatorname{acosq}(\alpha) \tag{35}$$

where $\operatorname{acosc} = \operatorname{cosc}^{-1}$ is inverse function of cosc. Damping $\alpha \in \mathbb{R}$ is dimensionless parameter determining the efficiency of channeling,

$$\text{Damping} = \alpha = \frac{1}{\gamma^{1/2} d} = \frac{1}{\left(2\Re(k)\Im(k)\right)^{1/2} d} = \frac{1}{2d\sqrt{sc}} \tag{36}$$

Properties of function acosc are described at TORI; the efficient implementation in C++ is suggested [10]. The complex map of acosc is show in the right hand side of **Figure 1** with levels of constant real part and levels of constant imaginary part. Behavior of real and imaginary parts of function acosq of real argument is shown in fthe left hand side of **Figure 2**.

The decay rate of mode in the region with absorption is determined with parameter q. Once p is determined, from Equation (27),

$$qd = pd\tan(pd) = \operatorname{acosc}\left(\mathrm{e}^{\mathrm{i}\pi/4}\alpha\right)\tan\left(\operatorname{acosc}\left(\mathrm{e}^{\mathrm{i}\pi/4}\alpha\right)\right) = \operatorname{acosq}(\alpha)\tan(\operatorname{acosq}(\alpha)) = \operatorname{acosqq}(\alpha) \tag{37}$$

Function acosqq is shown in the right hand part of **Figure 2**.

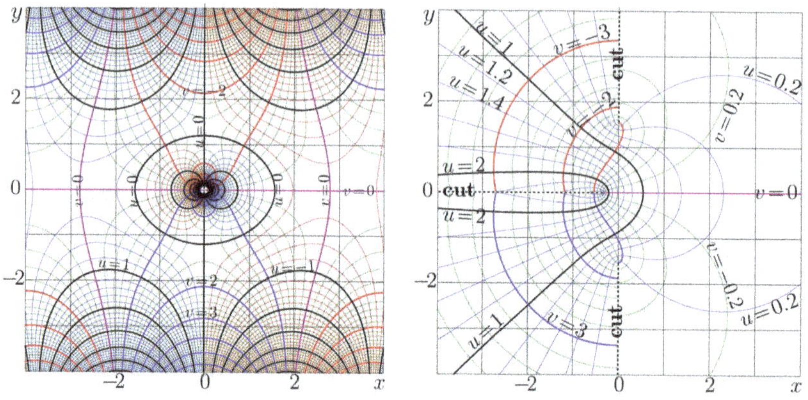

Figure 1. Complex map of function $f = \operatorname{cosc}(x + iy)$ by (32), left, and that of $f = \operatorname{acosc}(x + iy)$ by [10], right, in the x, y plane. Levels $u = \Re(f) = \text{const}$ and levels $v = \Im(f) = \text{const}$ are shown.

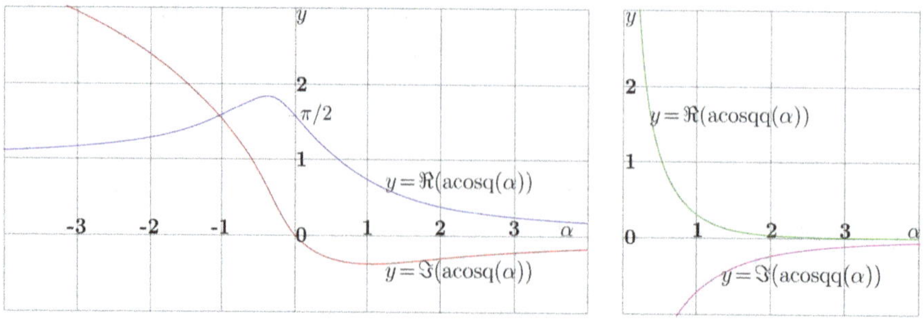

Figure 2. $y = \Re(\operatorname{acosq}(\alpha))$ and $y = \Im(\operatorname{acosq}(\alpha))$ by (35), left; $y = \Re(\operatorname{acosqq}(\alpha))$ and $y = \Im(\operatorname{acosqq}(\alpha))$ by (35), right.

5. Assembling of Mode, Example

The damping α by (36) determines the properties of the mode with given distance $2d$ between walls and given real and imaginary parts of the wavenumber k that determines the propagation of wave in the material of the wall. With tools defined above, the principal mode of wave guided between absorbing walls is expressed with Equation (21). Parameters p, q and r are defined with Equations (35), (37), (24).

As an example of the assembling of such a mode, the case $\alpha = 1/4$ is presented in **Figure 3**; the real and imaginary parts of the components of function f are plotted versus dimensionless product xd. In this case,

$$pd \approx 1.30652013112871 - 0.20108562381528\mathrm{i} \quad (38)$$

$$qd \approx 2.63604614403057 - 2.93518290714528\mathrm{i} \quad (39)$$

$$r \approx 0.26650956316919 + 0.19541505906974\mathrm{i} \quad (40)$$

The amplitude and phase of the mode f for $\alpha = 1/4$ and $\alpha = 1/50$ are shown in **Figure 4**. for $\alpha = 1/50$, the parameters are

$$pd \approx 1.54859380700524 - 0.02159861746934\mathrm{i} \quad (41)$$

$$qd \approx 35.33791574606151 - 35.37182445894684\mathrm{i} \quad (42)$$

$$r \approx 0.02220587422227 + 0.02159497306720\mathrm{i} \quad (43)$$

With functions ArcCosq and ArcCosqq implemented in TORI through [10], one can easy assemble the principal mode for other values of the damping parameter α with minimal modification of the codes supplied there.

6. Application to Atomic Waves and the Asymptotic

The effective absorption of a guided mode is one of most important parameters of any waveguide. This section consider the case of low damping and, correspondently, strong channeling.

According to (19) the effective absorption is determined by parameter κ,

$$\text{Effective Absorption} = v = A = \Im(\kappa) \quad (44)$$

From Equation (22),

$$\kappa = \sqrt{\omega - p^2} \quad (45)$$

The real and imaginary parts of κ determine the effective wavenumber and absorption of the guided mode "exactly" in the mathematical sense. As for the physical applications, the case of strong guiding (and low absorption) is of interest. This case corresponds to the small values of the damping parameter $\alpha \ll 1$, and the asymptotic behavior of the absorption of the guided mode is considered in this section.

For the strong guiding, the propagation constant κ can be expanded as follows:

$$\kappa = \sqrt{\omega p^2} \approx \sqrt{\omega}\left(1 - \frac{p^2}{2\omega}\right) \approx \sqrt{\omega} - \frac{p^2}{2\sqrt{\omega}} \quad (46)$$

then, the absorption A of the mode can be expressed as follows:

$$A = \Im(\kappa) \approx -\Re(p)\Im(p)/\sqrt{\omega} \quad (47)$$

In order to provide the flux of probability from the center of mode to the absorbing walls, the imaginary part of the transversal wavenumber should be negative, $\Im(p) < 0$. The expansion of funciton acocq at zero gives:

$$\operatorname{acosq}(z) = \frac{\pi}{2} - \frac{\pi}{2}\mathrm{e}^{\mathrm{i}\pi/4} z + O(z^2) \quad (48)$$

This gives the approximation for the transversal wavenumber p in the following form:

$$pd \approx \frac{\pi}{2} - \frac{\pi}{2}\mathrm{e}^{\mathrm{i}\pi/4}\alpha \quad (49)$$

and the estimate for the effective absorption

$$A \approx \frac{\pi}{2d} \times \frac{\pi}{d\sqrt{2}}\alpha \times \frac{1}{\sqrt{\omega}} \quad (50)$$

Then, $\alpha = \dfrac{1}{d\sqrt{\gamma}}$ by (36) and $\gamma \approx 2s\sqrt{\omega} \approx 2sc$ should be used, giving $\alpha \approx \dfrac{1}{d\sqrt{2sc}}$. Then, the effective absorption of the mode

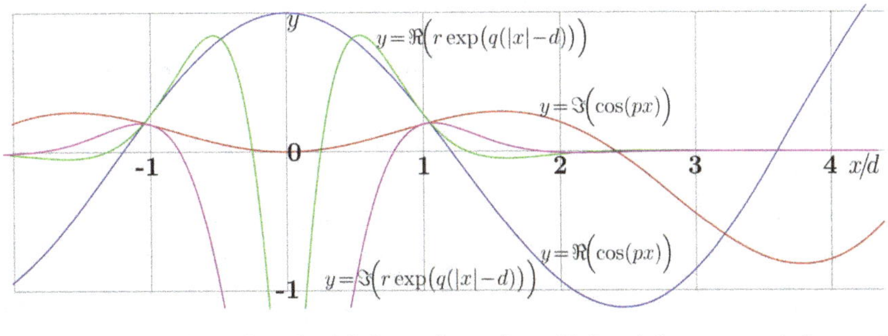

Figure 3. Combination of mode (21) from the cosinusoidal and the exponents for $\alpha = 1/4$.

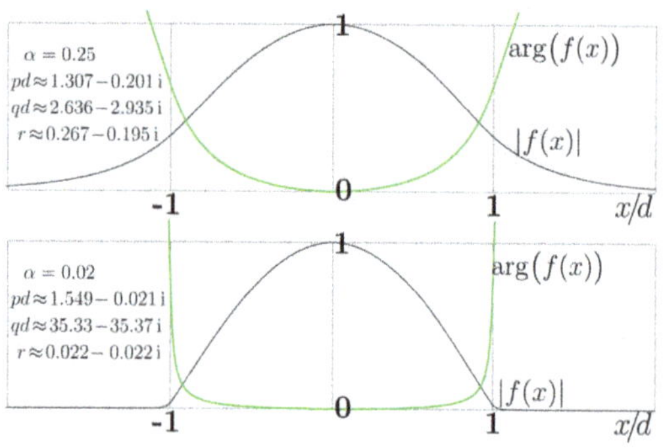

Figure 4. Amplitude and phase of the mode *f* by (21) for α = 1/4 and for α = 1/50.

$$A \approx \frac{\pi^2}{\sqrt{8}\,d^2} \times \frac{1}{d\sqrt{2sc}} \times \frac{1}{c} = \frac{\pi^2}{4\,d^3 c^{3/2} s^{1/2}} \tag{51}$$

where c has sense of wavenumber, and s is the absorption in the wall.

In the similar way, the highest modes can be constructed. For the *m*th transversal mode, the transversal wavenumber scales proportionally to n; and the absorption of mode scales proportionally to m^2.

7. Comparison to Previous Results

The absorption of mode can be interpreted also in terms of the multiple reflection of guided wave from the walls. The coefficient of reflection r_{zeno} is estimated in the description of the ridged mirrors in terms of the Zeno effect [2],

$$r = r_{\text{zeno}} = \frac{\sqrt{\sqrt{1/\chi^4+1}+1} - \sqrt{2}}{\sqrt{\sqrt{1/\chi^4+1}+1} + \sqrt{2}} \approx \exp\left(-\sqrt{8}\chi\right) \tag{52}$$

where

$$\chi = \sqrt{KL}\theta \tag{53}$$

K is wavenumber and can be replaced to c; while L is distance between idealized absorbers that can be approximated as $1/(2s)$, and θ is the grazing angle. (Notation p of [2] is not used here, to keep letter p denoting the transversal wavenumber; so, in (52) and (53), notation χ is used instead.)

For the good channeling conditions, the effective absorption can be approximated with

$$a = \frac{1 - r_{\text{zeno}}}{2d/\theta} \tag{54}$$

where d is half-width of the channel and $\theta = p/K$ is ratio of the transversal wavenumber

$$p = \frac{\pi}{2d} \tag{55}$$

to the wavenumber K. At the reflection of wave from a ridged mirror, $\theta = p/K$ plays role of the grazing angle.

Substitution of (52) and (55) into (54) gives the following expression for the absorption

$$a \approx \sqrt{8KL}\,\frac{\theta^2}{2d} \tag{56}$$

The grazing angle can be approximated with

$$\theta \approx \frac{\pi}{2dK} \tag{57}$$

giving the estimate for the efficient absorption

$$a \approx \sqrt{8KL}\,\frac{\pi^2}{8d^3K^2} \approx \sqrt{\frac{L}{2K^3}}\,\frac{\pi^2}{2d^3} \tag{58}$$

For the comparison to the previous result, L should be replaced to $1/(2s)$ and K should be replaced to c, giving the absorption by probability

$$a \approx \frac{\pi^2}{4\,d^3 c^{3/2} s^{1/2}} \tag{59}$$

This expression should be compared to (51).

In the first approximation, the consideration of the multiple reflection from absorbing walls and the consideration of mode guided between the absorbing walls give the same prediction about effective absorption of this mode. The consideration of the multiple reflection from absorbing walls and the consideration of mode guided between the absorbing walls give the same prediction about effective absorption of this mode.

8. Numerical Example

Consider the application of the estimate (59) for the guiding of the realistic laser beam. Assume, the absorption in the walls

$$s = \frac{0.5}{35\text{ cm}} \approx 0.143\text{ cm}^{-1} \tag{60}$$

Following the ideology of the Zeno interpretation of absorbing walls [2], such an absorption may be approximated with series of slits separated by distance 35 mm.

Let the wavenumber is

$$c = \frac{2\pi}{1.064\,\mu} \approx 5.9\,\mu^{-1} \tag{61}$$

Let the halfwidth of the channel

$$d = 500\,\mu \tag{62}$$

This gives the estimate for the absorption of guided modes,

$$a \approx 0.73\,m^2 \cdot \text{m}^{-1} \tag{63}$$

For the principal mode ($m = 1$), after to propagate distance $Z = 35\text{ cm}$, the attenuation factor is of order of

$$\mathrm{AF} = \exp(-aZ) \approx \exp(-0.73 \times 0.35) \approx 78\% \quad (64)$$

that means, that the most of the initial power of the guided mode is still delivered. As for the second mode, its attenuation

$$\mathrm{AF} = \exp(-4aZ) \approx \exp(-1.02) \approx 36\% \quad (65)$$

that means significant dicrimination of the second mode.

Using the approximation of the set of absorbers as a continuous medium [2], the example above may correspond to transfer of near infra-red light through the set of 10 slits separated with distance 35 mm. Roughly, the amplitude of field after the set of slits can be approximates with the cosinusoidal profile. However, the presence of the highest modes, as well as the diffraction of the tails of the mode on the edges should make the similarity qualitative. Similar result one may expect to observe at propagation of light through the set of pinholes of radius d. The similarity with the idealized cosinusoidal or Besseliean mode should improve at the increase of number of slits or pinholes; a hundred of silts or pinholes may be sufficient to get the quantitative agreement with the idealized cosinusoidal or Besselian profile.

The accurate consideration of the discrete character of the absorbing walls, as well as construction of the mode for the case with circular symmetry may be continuation of this work. For the paraxial case, the estimates are universal and are not sensitive to the origin of waves. In particular, the results are expected to apply to the electromagnetic waves as well as to the cold atoms, exhibiting the wave properties.

9. Conclusions

Guiding of wave of any origin between absorbing walls is considered. Wavenumber c and the amplitude absorption s of wave in the wall, and the half-width d of the channel are considered as given parameters.

The dimensionless damping parameter $\alpha = 1/\left(2d\sqrt{sc}\right)$ by (36) is suggested to characterize the scale of the effect.

The first (principal) mode (21) with lowest absorption is explicitly constructed. The transversal wavenumber p of the mode is expressed through the function acosc of complex argument; properties of this function are described and the numerical implementation is supplied [10]. The propagation constant $\kappa = B + \mathrm{i}A$ is expressed with Equation (45); the asymptotic estimate (51) of the absorption A of the mode is suggested. The estimate agrees with that on the base of the Zeno reflection of the waves from the absorbing medium reported earlier [2].

The estimates above are important in the design of the suppression of the amplified spontaneous emission (ASE) in the high power lasers. The guiding of modes by the absorption walls happens whenever the engineers want this effect or not. Similar estimate is valid for the highest modes. For the mth transversal mode, the asymptotic estimate is suggested for the absorption

$$A = \frac{\pi^2 m^2}{4\, d^3 c^{3/2} s^{1/2}} \quad (66)$$

through the half-width d of the channel, wavenumber c and absorption s of wave in the walls.

Similar estimate (with slightly higher absorption) correspond to the case with circular symmetry, that can be treated in the similar way; the mode is expressed with the Bessel function, parameter d plays role of the radius of the channel. At small value of damping α by (36), the transversal wavenumber p is almost real.

The result should be useful in both, wanted guiding of cold neutral particles by their detection (absorption) and the efficient suppression of the unwanted guiding of waves, for example, ASE in powerful optical amplifiers, and optimization of the ASE absorbers.

10. Acknowledgements

Authors are grateful to Dr. Hilmar Oberst and Prof. Fujio Shimizu and Prof. Kazuko Shimizu for the collaboration.

REFERENCES

[1] F. Shimizu and J. Fujita, "Giant Quantum Reflection of Neon Atoms from a Ridged Silicon Surface," *Journal of the Physical Society of Japan*, Vol. 71, No. 1, 2002, pp. 5-8. doi:10.1143/JPSJ.71.5

[2] D. Kouznetsov and H. Obrest, "Reflection of Waves from a Ridged Surface and the Zeno Effect," *Optical Review*, Vol. 12, No. 5, 2005, pp. 363-366. doi:10.1007/s10043-005-0363-9

[3] D. Kouznetsov and H. Oberst, "Scattering of Waves at Ridged Mirrors," *Physical Review A*, Vol. 72, No. 1, 2005, p. 013617. doi:10.1103/PhysRevA.72.013617

[4] D. Kouznetsov, H. Oberst, K. Shimizu, A. Neumann, Y. Kuznetsova, J.-F. Bisson, K. Ueda and S. R. J. Brueck, "Ridged Atomic Mirrors and Atomic Nanoscope," *Journal of Physics B*, Vol. 39, No. 7, 2006, pp. 1605-1616. doi:10.1088/0953-4075/39/7/005

[5] J. Itatani, J. Faure, M. Nantel, G. Mourou and S. Watanabe, "Suppression of the Amplified Spontaneous Emission in Chirped-Pulse-Amplification Lasers by Clean High-Energy Seed-Pulse Injection," *Optics Communications*, Vol. 148, No. 1-3, 1998, pp. 70-74.

[6] H. Yagi, J. F. Bisson, K. Ueda and T. Yanagitani, "$Y_3Al_5O_{12}$ Ceramic Absorbers for the Suppression of Parasitic Oscillation in High-Power Nd:YAG Lasers," *Journal of Luminescence*, Vol. 121, No. 1, 2006, pp. 88-94. http://www.sciencedirect.com/science/article/pii/S0022231305002504

[7] K. Ertel, C. Hooker, S. J. Hawkes, B. T. Parry and J. L.

Collier, "ASE Suppression in a High Energy Titanium Sapphire Amplifier," *Optics Express*, Vol. 16, No. 11, 2008, pp. 8039-8049. doi:10.1364/OE.16.008039

[8] L. A. Hackel, T. F. Soules, S. N. Fochs, M. D. Rotter and S. A. Letts, "A Method for Suppressing ASE and Parasitic Oscillations in a High Average Power Solid-State Laser," Patent US7463660, 2008. http://www.google.com/patents/US20050254536

[9] A. K. Sridharan, S. Saraf, S. Sinha and R. L. Byer, "Zig-zag Slabs for Solid-State Laser Amplifiers: Batch Fabrication and Parasitic Oscillation Suppression," *Applied Optics*, Vol. 45, No. 14, 2006, pp. 3340-3351. doi:10.1364/AO.45.003340

[10] http://tori.ils.uec.ac.jp/TORI/index.php/ArcCosc

[11] http://mathworld.wolfram.com/SincFunction.html

27

Dynamics of Entanglement in the Cavity with Nonlinear Medium

Shaojiang Du, Hairan Feng
Physics and Information Engineering Department, Jining University, Qufu, China
Email: dsjsd@126.com, hairanfeng@mail.sdu.edu.cn

ABSTRACT

We investigate the evolution of initial entangled two-atom in cavity with single-mode light field. Using the method of negative eigenvalue of partially transposed matrix we analysis the evolution of the entanglement of the two-atom in a field of number state and Kerr-media environment and find that entanglement sudden death phenomenon occurs in the number of particles field. When the atoms interact with the Kerr medium, we obtain that the phenomenon of sudden death can be eliminated in the particle-number field, and the entanglement of two-atom oscillates around a high-value.

Keywords: Quantum Optics; Kerr Medium; Negativity; Entanglement Sudden Death

1. Introduction

Quantum information science, which mainly includes quantum computer and quantum communication, has increasingly evolved as a new object. In the past few years, quantum information has made a surprise progress both in theoretical and experimental fields. It has created many miracles, such as absolute secure quantum key, quantum dense coding, quantum teleportation, and so on. Recently the theory of quantum information has been widely used to every branch of physics, which has acelerated the development of information science and technology. Jaynes-Cummings mode (J-C mode) and Tavis-Cummings model (T-C mode) could in deal the quantum optics and quantum information problems effectively. In special condition, they have analytic solutions, so they have theoretical and real meanings, and have been afforded more and more attention. Now the study of quantum entanglement becomes more important, such as the quantum teleportation due to quantum entanglement, which acts as quantum channel to join different space sites [1-5]. Yu and Eberly [6] found Entanglement Sudden Death, it made widely concern about the theory [7,8] and experiment [9-11]. In this letter we analysis the two-atom entanglement characteristics in Numericall state field and Kerr-media environment.

[1]Foundation of Natural Science Foundation of China (11147019) and Jining University (Grant No. 2010QNKJ02).

2. Model and Method

2.1. Theoretical Model

One of the initial entangled two-atom interact with a cavity in the single-mode light field. We can use the J-C mode to describe the action of atom and light field. The Hamiltonian of this system can be described as

$$H=\frac{1}{2}\sum_{i=A,B}\omega_i\sigma_i^z+\omega a^+a+g\left(\sigma_A^+a+\sigma_A^-a^+\right) \quad (1)$$

Here ω_1,ω_2 is essential transition frequency of atom and ω is the frequency of light field; g is coupling constant of atom and light field. $\sigma^z,\sigma^+,\sigma^-$ is spin operator of atom that $\sigma_A^z=|e_A\rangle\langle e_A|-|g_A\rangle\langle g_A|$, $\sigma_B^z=|e_B\rangle\langle e_B|-|g_B\rangle\langle g_B|$, $\sigma_A^+=|e_A\rangle\langle g_A|$ and $\sigma_A^-=|g_A\rangle\langle e_A|$, which $|e\rangle,|g\rangle$ is excited state and base state of atom.

In this letter we discussed the entanglement state of two-atom. The initial state can be written as

$$\Psi(0)=\left(\cos\theta|e_A\rangle|g_B\rangle+\sin\theta|g_A\rangle|e_B\rangle\right)\otimes|n\rangle \quad (2)$$

At any time in the interaction picture, the state of system is depicted as

$$\psi(t)=C_1(t)|e_A,e_B,n-1\rangle+C_2(t)|e_A,g_B,n\rangle+C_3(t)|g_A,e_B,n\rangle+C_4(t)|g_A,g_B,n+1\rangle$$

In the interaction picture, the Schrödinger equation of

system is

$$i\frac{\partial|\psi(t)\rangle}{\partial t}=H_I|\psi(t)\rangle \quad (3)$$

Using the initial condition we can get

$$\begin{aligned}C_1(t)&=-\mathrm{i}\sin\theta\sin\sqrt{n}gt\\C_2(t)&=\cos\theta\cos\sqrt{n+1}gt\\C_3(t)&=\sin\theta\cos\sqrt{n}gt\\C_4(t)&=-\mathrm{i}\cos\theta\sin\sqrt{n+1}gt\end{aligned} \quad (4)$$

2.2. Calculation

We analysis the two-atom entanglements characteristics in a vacuum field by using the partially transposed matrix of negative Eigen values (negativity) method [12]. It is defined as

$$N=-2\sum_i\lambda_i^- \quad (5)$$

Which λ_i^- is negative Eigen values of the partially transposed matrix ρ^{T}, and $N\in[0,1]$. By this calculation method we get the result as follows.

$$N=\sqrt{|C_1|^4+|C_4|^4+4|C_2|^2|C_3|^2-2|C_1|^2|C_4|^2}-|C_1|^2-|C_4|^2 \quad (6)$$

The evolution of entanglement degree of the system is described as **Figure 1**.

2.3. The Effect of Kerr Medium

When the cavity with single-mode numerical light field fills with Kerr medium, the Hamiltonian of the system can be written as

$$\begin{aligned}H&=\frac{1}{2}\left(\sum_{i=1}^{2}\omega_i\sigma_i^z\right)+\omega a^+a\\&+g\left(\sigma_A^+a+\sigma_A^-a^+\right)+\chi a^+aa^+a\end{aligned} \quad (5)$$

which χ describes the strength of interaction between Kerr medium and light field.

As the initial state is

$$\Psi(0)=\left(\cos\theta|e_A\rangle\langle g_B|+\sin\theta|g_A\rangle\langle e_B|\right)\oplus|n\rangle \quad (6)$$

At any time in the interaction picture, the state of system is depicted as

$$\begin{aligned}\psi(t)&=C_{k1}(t)|e_A,e_B,n-1\rangle+C_{k2}(t)|e_A,g_B,n\rangle\\&+C_{k3}(t)|g_A,e_B,n\rangle+C_{k4}(t)|g_A,g_B,n+1\rangle\end{aligned} \quad (7)$$

2.4. Calculation and Discussion

The two-atom entanglement characteristics in a vacuum field, using the partially transposed matrix of negative eigenvalues (negativity) method [12], it is defined as

$$N=-2\sum_i\lambda_i^- \quad (8)$$

The evolution of entanglement degree is described as **Figures 1** and **2**.

According to the picture we can see that the entanglement degree of the two-atom declines seriously in the number of particles field; sometimes sudden death of entanglement occurs (as it is shown in **Figure 1**). The regions of entanglement sudden death increase by the increasing particle number. The reason is that the action between atom and light field destroys the entanglement of the atoms. For the further study, we find that Kerr medium can erase the phenomenon of sudden death entanglement in numerical light field when two entangled atoms enter-act in the environment of Kerr medium and single-mode numerical light field (as it is shown in **Fig-**

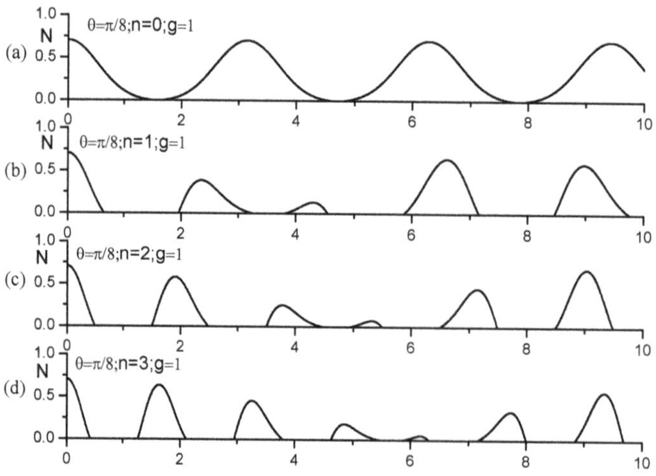

Figure 1. Entanglement time evolution property in number field.

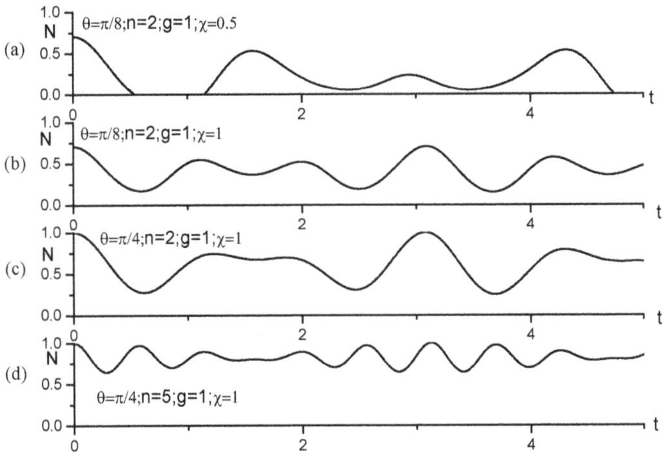

Figure 2. The influence of Kerr medium in the entanglement evolution.

ure 2). Further more, it can make the entangled atoms keep high entanglement degree, and can make the evolution of them near the maximal entanglement regions.

3. Conclusion

From the above discussion, we find that the entanglement degree decreases seriously when the entangled atoms are in the number of particles field. Sometimes entanglement sudden death phenomenon occurs. Especially we obtain that the phenomenon of entanglement sudden death can be eliminated in the particle-number field, and the entanglement of two-atom oscillates around a high-value state under a certain condition when the light field is filled with the Kerr medium. It means that Kerr medium could change the regions of disentanglement and could improve entanglement degree.

4. Acknowledgements

Thanks for the discussion with Doctor Zhang Yingjie and associate professor Han feng. Otherwise this work was supported by the National Natural Science Foundation of China (10774088, 10947006) of Doctor Yunjie Xia and Zhongxiao Man.

REFERENCES

[1] T. Pellizzari, S. Cardiner, J. Cirac, *et al.*, *Physical Review Letters*, Vol. 75, 1995, pp. 3788-3791. doi:10.1103/PhysRevLett.75.3788

[2] F. Giraldi and P. Girgolini, *Physical Review A*, Vol. 64, 2001, Article ID: 032310. doi:10.1103/PhysRevA.64.032310

[3] S. B. Zheng and G. C. Guo, *Physical Review Letters*, Vol. 85, 2000, pp. 2392-2395. doi:10.1103/PhysRevLett.85.2392

[4] A. Rauschenbeutel, *et al.*, *Physical Review Letters*, Vol. 83, 1999, pp. 5166-5169. doi:10.1103/PhysRevLett.83.5166

[5] D. Bouwmeester, J. W. Pan, M. Daniell, *et al.*, *Physical Review Letters*, Vol. 82, 1999, pp. 1345-1349. doi:10.1103/PhysRevLett.82.1345

[6] G. Vidal and R. F. Werner, *Physical Review A*, Vol. 65, 2002, Article ID: 032314. doi:10.1103/PhysRevA.65.032314

[7] Z. X. Man, Y. J. Xia and N. B. An, *Journal of Physics B*, Vol. 41, 2008, Article ID: 085503. doi:10.1088/0953-4075/41/8/085503

[8] T. Yu and J. H. Eberly, *Physical Review Letters*, Vol. 93, 2006, Article ID: 140403. doi:10.1103/PhysRevLett.97.140403

[9] M. P. Almeida, F. de Melo, M. Hor-Meyll, A. Salles, S. P. Wallborn, P. H. Souto Ribeiro and L. Davidovich, *Science*, Vol. 316, 2007, pp. 579-582. doi:10.1126/science.1139892

[10] J. S. Xu, C. F. Li, M. Gong, X. B. Zou, C. H. Shi, G. Chen and G. C. Guo, *Physical Review Letters*, Vol. 104, 2010, Article ID: 100502. doi:10.1103/PhysRevLett.104.100502

[11] J. Laurat, K. S. Choi, H. Deng, C. W. Chou and H. J. Kimble, *Physical Review Letters*, Vol. 99, 2007, Article ID: 180504. doi:10.1103/PhysRevLett.99.180504

[12] G. Vidal and R. F.Werner, *Physical Review A*, Vol. 65, 2002, Article ID: 032314. doi:10.1103/PhysRevA.65.032314

[13] D. F. Walls and G. J. Milbum, "Quantum Optics," Springer-Verlag, New York, 1995, pp. 15-18

[14] G. J. Milburn and S. L. Braunstein, *Physical Review A*, Vol. 60, 1999, pp. 937-955. doi:10.1103/PhysRevA.60.937

[15] S. Parker, S. Bose and M. B. Plenio, *Physical Review A*, Vol. 61, 2000, Article ID: 32305.

[16] L.-M. Duan, G. Gieske, J. I. Cirac, *et al.*, *Physical Review Letters*, Vol. 84, 2000, pp. 4002-4005. doi:10.1103/PhysRevLett.84.4002

[17] L.-M. Duan, G. Gieske, J. I. Cirac, *et al.*, *Physical Review A*, Vol. 62, 2000, Article ID: 032304.

[18] Y.-J. Xia and G.-C. Guo, *Chinese Physics Letters*, Vol. 21, 2004, pp. 1877-1880.

[19] B. C. Sanders, S. M. Barnett and P. L. Knight, *Optics Communications*, Vol. 58, 1986, pp. 290-294. doi:10.1016/0030-4018(86)90453-0

[20] H. Y. Fan and H. R. Zaidi, *Optics Communications*, Vol. 68, 1988, pp. 143-149. doi:10.1016/0030-4018(88)90140-X

[21] F. G. Deng, *et al.*, *Chinese Physics Letters*, Vol. 19, 2002, pp. 893-898. doi:10.1088/0256-307X/19/2/327

[22] L. M. Duan, *et al.*, *Physical Review Letters*, Vol. 84, 2000, pp. 2722-2726. doi:10.1103/PhysRevLett.84.2722

Modeling of Complex Solitary Waveforms for Micro-Width Doped ZnO Waveguides

Rosmin Elsa Mohan[1], M. Sivakumar[2], K. S. Sreelatha[3]
[1]Amrita Vishwa Vidyapeetham, Kollam, India
[2]Amrita Vishwa Vidyapeetham, Coimbatore, India
[3]Govt.Polytechnic College, Kottayam, India
Email: rosminelsa@am.amrita.edu, r.m.sivakumar@gmail.com, kssreelatha@yahoo.com

ABSTRACT

The potential applications of metallic oxides as supporters of nonlinear phenomena are not novel. ZnO shows high nonlinearity in the range 600 - 1200 nm of the input wavelength [1]. ZnO thus make way to become efficient photoluminescent devices. In this paper, the above mentioned property of ZnO is harnessed as the primary material for the fabrication of waveguides. Invoking nonlinear phenomena can support intense nonlinear pulses which can be a boost to the field of communication. The modeling characteristics of undoped and doped ZnO also confirm the propagation of a solitary pulse [1]. An attempt to generalize the optical pattern of the doped case with varying waveguide widths is carried out in the current investigation. The variations below 6 um are seen to exhibit complex waveforms which resemble a continuum pulse. The input peak wavelength is kept constant at 600 nm for the modeling.

Keywords: Solitons; Nonlinear Optics; Doped ZnO Waveguides; Continuum

1. Introduction

ZnO has recently attracted wide interest for its unique properties and versatile applications in the fields of piezoelectric devices, light sensors, spintronics [2-10] and acoustic wave devices [11,12]. Nanostructured electrodes of ZnO have also been used as solar cells [13] with its physical and chemical properties that can be varied by adequate doping by cationic or anionic substitution. Doping with B or Mn decreased the resistivity [14,15] or introduced ferromagnetism [16] respectively. The optoelectronic properties are generally affected by impurities and defects. Impurity incorporation thus plays a dominant role in the possible applications of ZnO in the field of optoelectronics.

The effect of Ag as a Group 1 element is a candidate acceptor for ZnO. Ag doping could greatly increase the catalytic doping and photo activity in semiconductors [17,18]. The silver atoms may be incorporated into the lattice sites of ZnO only as the substitution of the Zn atom sites [19]. Thus the doping with Ag requires systematic investigation. The silver ions have novel applications of shifting the emission spectrum of doped ZnO beyond the UV-blue region making it a promising candidate for communication via the propagation of solitary pulses.

2. Theory

The Guiding Phenomena

For any waveguide a refractive index larger than the surroundings is needed. Planar waveguides allow confinement in one direction though diffraction may occur along the plane of the film. Fiber and channel waveguides allow cross-sectional dimensions with the size of confinement to be the order of the wavelength [20]. As a result higher intensities for a given input power can be supplied as the effective beam area is minimized in a waveguide. In effect, the guided wave field is maximum in the region of high index and decays with distance into the media of lower index. Nonlinearity can therefore be either in the core or the surrounding media. However dominance of the nonlinear phenomena with optimum efficiency is mostly seen in the core region.

The propagation constants with their corresponding eigen modes depend on the dimensions of the high-index (core) region, the geometry of the waveguide structure, and the refractive index of the wave guiding media. The modes, TE and TM with E or H in the plane of the surfaces, need to be orthogonal to one another and should occur in two unique polarizations. The two orthogonal modes dominate though the fields contain contribution from all three polarizations. Any degeneracy in the corresponding modes can cause birefringence which can be termed polarization preserving in that specification to support two orthogonally linearly polarized eigen modes [21].

Wave guiding can also introduce reduction in the spatial degrees of freedom which can in turn limit the propagation wave vectors to two dimensions in planar waveguides or to one dimension in fiber and channel guides. This greatly benefits nonlinear interaction associated with intensity dependant refractive index. Wave vector interactions that result can be accomplished by adjusting the lengths at which these wave vector interactions occur.

It is possible to excite the waveguide modes from the sides or from the ends of the waveguide. The angle of incidence is so chosen that only one mode is excited at a time for plane wave incidence. Optically aligning the waveguide to the incident beam allows almost all the guided-wave power to be launched in to lowest-order mode with appropriate polarization.

Many $\chi^{(2)}$ phenomena have been demonstrated in planar waveguides such as second-harmonic generation, difference frequency generation, optical parametric amplification and optical parametric oscillation. SHG has been widely studied of these, though restricted mainly to the field of integrated optics. As for the simplest case of SHG, a single fundamental guided wave is excited at $z = 0$ propagating to $z = L$ where it leaves the waveguide along with the second harmonic generated between 0 and L. The second harmonic power, $P(2\omega, z)$ is given in terms of the fundamental input power [21] $P(\omega, 0)$ by

$$P(2\omega, L) = (k_0 L)^2 \frac{d_{eff}^2}{n_{eff}^3} \frac{\sin^2\phi}{\phi^2} |K^2| P^2(\omega, L)$$

Here $P(2\omega, L)$ is the second harmonic power in terms of the fundamental input power $P(\omega, L)$ across the waveguide length L.

The power scaling of the Second Harmonic generation is enables characterization and generalization of the waveguides. Here the waveguide figure of merit is given by $\frac{d_{eff}^2}{n_{eff}^3}$ where n_{eff} is the effective index and d_{eff} the effective waveguide thickness.

Φ is the phase vector, for the simplest SHG condition in terms of the guided wave vectors, $\Phi = 0$, $\Phi = 0.5(\beta_2 - 2\beta_1)$ where 1 and 2 refer to the fundamental and harmonic respectively.

The overlap integral [21], a concept unique to waveguides is given by

$$K = \int_{-\infty}^{\infty} dx \int_{-\infty}^{\infty} dy \frac{d_{ijk}}{d_{eff}} e_i(2\omega) e_j(\omega) e_k(\omega) \cdot f_i(x, y) f_j(x, y) f_k(x, y)$$

where the terms govern the *product* of the field distributions across the waveguide; the latter if negative reduces the value of K in effect.

The overlap integral is usually small for the field distribution modes. It seems that the existence of modes with different values of the effective index facilitate phase matching. In this case, the overlap integral is extremely reduced or even zero.

In planar waveguides, the angles at which the mode intersections associated with phase matching occurs must be very small so as to satisfy the minimum thickness for the film as required for phase matching. This condition equally lets the birefringence and material dispersion to be quite small [20,21].

3. Doped ZnO on Silica Substrates: Modeling in the Dispersive Regime

The propagation characteristics of the guided wave are obtained provided the guided-wave field satisfies the proper boundary conditions at the interface of two different media (*i.e.*, tangential electric and magnetic-field vectors must be continuous across the boundary) and necessary radiation conditions.Along a straight line path, every component of the electromagnetic wave that propagates may be represented as $f(u, v)e^{-i\beta z}e^{i\omega t}$ [10]' where z is chosen as the propagation direction and u, v are orthogonal coordinates in a transverse plane. β is the propagation constant and ω is the frequency of the wave. However, the fundamental property of a planar waveguide is the relation between the number and nature of the waveguide modes propagated and its refractive index [22].

For nonlinear waveguides, integral representations for the longitudinal electric and magnetic fields satisfy the appropriate wave equations and all the necessary boundary conditions. By approximate expansions of these fields and employing the analytic continuation technique [23], the relevant integral equations may be reduced to linear algebraic equations which may be solved to obtain the propagation constants.

The field mode distributions of doped ZnO in the 800 - 1200 nm of the input wavelength have shown increased nonlinear effects. Experimental analysis of Ag doped ZnO has revealed interesting changes in physical and chemical properties at the nano scale such as crystallinity, optical transmittance, absorption and refraction patterns etc. [24, 25]. The doped waveguides can be used for making inexpensive optical devices. The field modes for a doped ZnO waveguide structure, $n_2 = 2.099$, show dispersive behavior in accordance with linear losses and two photon absorption around 1000 nm. However, we have considered variations with waveguide width of 0.6 micrometer and less for an input wavelength of within 600 nm. This was so chosen so as to minimize the dispersive effects and oscillatory behavior of solitary pulses beyond these dimensions [26].

4. Solitary Pulses in Doped ZnO: The Route to Supercontinuum

The interplay of dispersion and nonlinear self-action in wave dynamics has been a major area of interest across many branches of physics since the Fermi-Pasta-Ulam work. Localized nonlinear waves have been often referred to as solitary waves, however today the term' soliton' has been extended over the nonintegrable cases as well. Optical solitons in fibers have been researched much over the years as potential information carriers (Mollenauer and Gordon 2006; Agrawal in 2007). Octave wide spectral broadening was later observed in the beginning of the 21st century which was extensively studied and came to be known as Supercontinuum.

The first experiments of Supercontinuum inadvertently marked the presence of solitons in the process. A fiber with high nonlinearity and the GVD point close to the pump wavelength has large potential for harnessing. Fibers with silica cores (~1 - 5 µm) of a few microns in diameter have been studied extensively with a variety of sources. Investigations with Femtosecond pulses with wavelength around 800 nm (Ranka *et al.*, 2000) and with nanosecond microchirp lasers close to 1 µm (Stone and Knight 2008) gave much promising results. The dispersion profile in the latter exhibited intense supercontinua extending towards the shorter "bluer" wavelengths (Harbold *et al.*, 2002; Efimov *et al.*, 2004).

The modeling of ZnO and the doped structure (Ag-ZnO) confirm the possible passage of a solitary pulse [1]. The solitons show a self-consistency, characteristic of its self-guided nature, to be a mode of the linear waveguide it induces [1,2]. The length of the passage may be extended using a doped form of the initial waveguide. The field mode distributions are found to vary with the input power and the wavelength. Solitons if incorporated in waveguides can revolutionize optical communication systems with their ability to carry information over long distances without a change in shape.

In the current investigation, an index contrast of glass (1.456) to that of air in addition to the increased nonlinearity of the material of the waveguides provides high modal confinement and significant contribution to dispersion. The width of the waveguide structures are varied within and below the 6 µm scale when a continuum pulse is seen to propagate (**Figures 1(a)-(c)**) with variation in the refractive index governing the waveguide width. The width variations show a continuum spectra for the silver doped ZnO structure which confirms the possibility of a supercontinuum in these structures. The propagating pulses retain a continuous solitonic path thus enabling

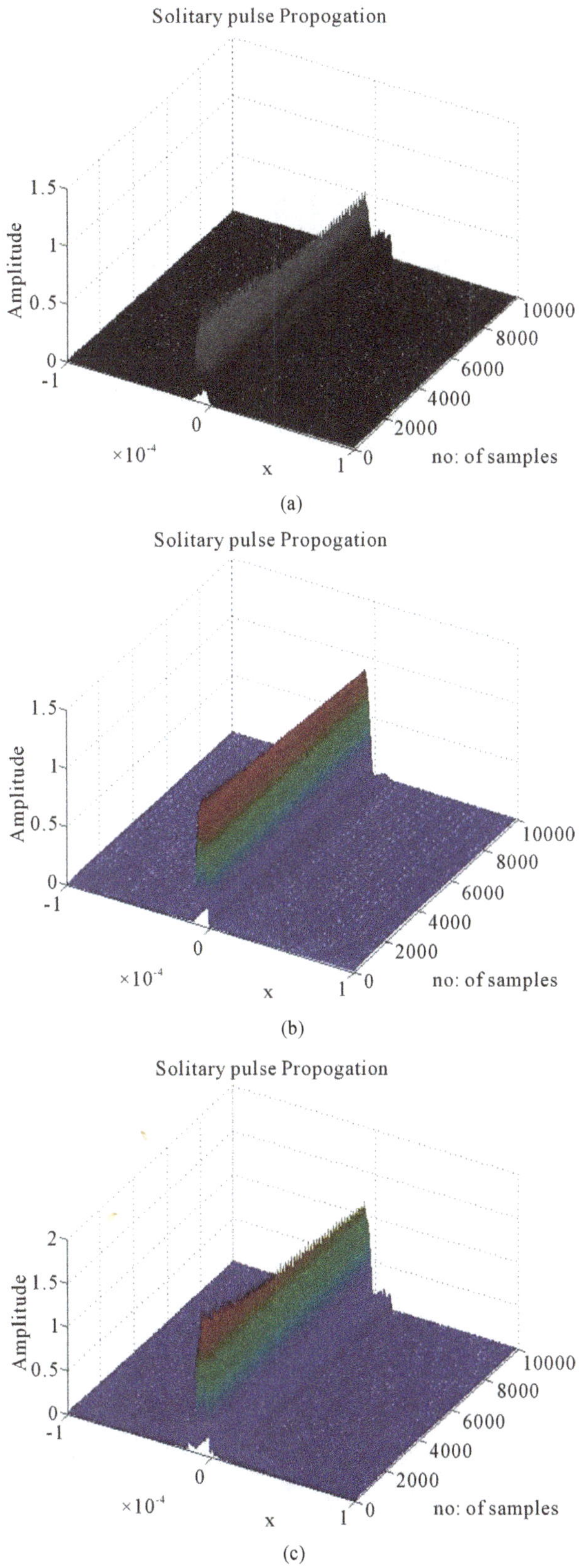

Figure 1. (a) Solitary propagation for *w* = 1 um at 600 nm peak input wavelength; (b) for *w* = 3 um at 600 nm peak input wavelength; (c) for *w* = 5 um at 600 nm input wavelength.

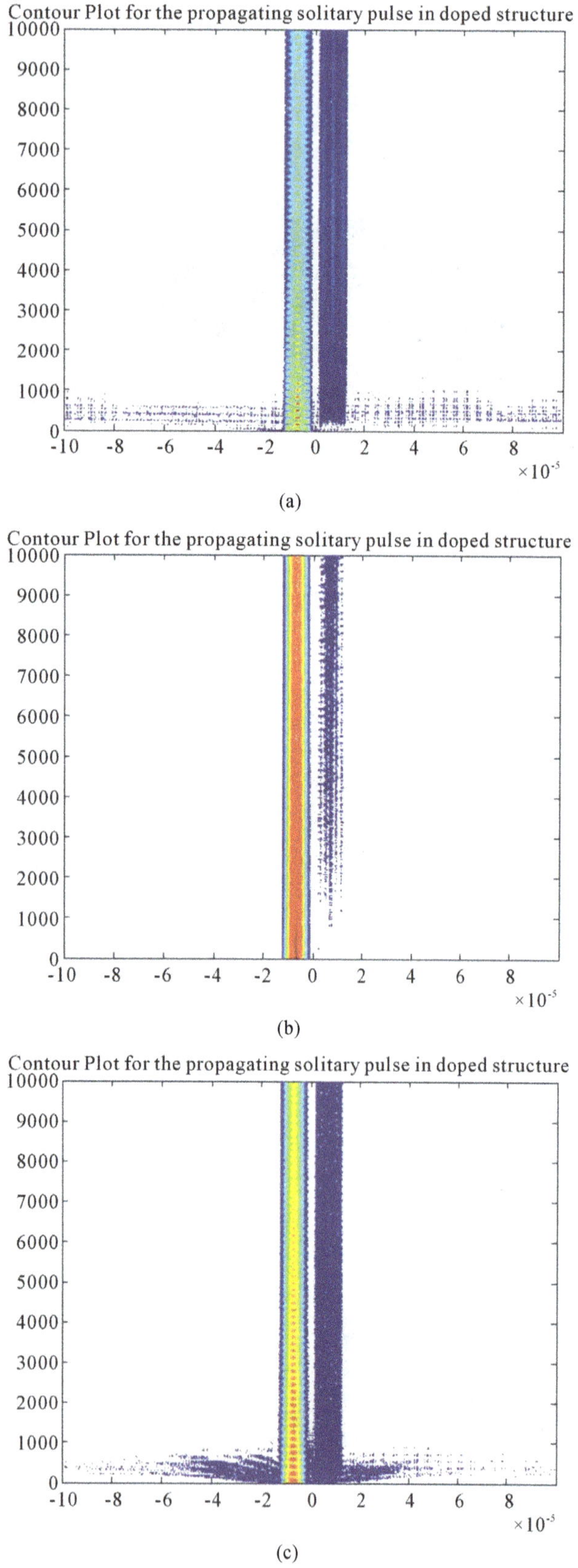

Figure 2. Contour plots for the propagating continua pulses at (a) *w* = 1 um (b) 3 um (c) 5 um for the input wavelength of 600 nm.

the basic need of low losses in communication. The refractive index of the doped structure (n_2 = 2.099) is particularly important with relevance to the fact that it prevents the dispersive spreading of the radiation waves within the short wavelength range of the continuum. The change in refractive index is seen to exert an inertial force which further ensures a dispersionless propagation of the radiation (**Figures 2(a)-(c)**). Sub-wavelength diameter nanowires have proved to offer effective nonlinearities and interesting dispersion profiles [27]. Such structures enabled ultra-efficient octave spanning for nanosecond and femto second input pulses [28].

In the **Figures 1** and **2** it can seen that an initial pulse, which has an input wavelength of 600 nm, takes form as a solitary pulse when it propagates through the waveguide length. The soliton stability is improved as the width of wave guide is increased which can be directly seen from the figures. The amplitude of the pulse can be optimized using the refractive index variations.

5. Conclusion

Earlier studies carried out in this regard had confirmed solitons for an input wavelength between 800 nm and 1000 nm [1]. In the present study we have varied the waveguide width within a micro range thereby varying the nature of the continuum passing through the waveguide. The doped ZnO waveguide widths below 6 um were considered in the present investigation. The onset of a solitary pulse which was confirmed to be stable without dispersion around 600 nm was exploited. The variation in the solitary propagation resembles closely to continuum propagation.

6. Acknowledgement

The author would like to thank University Grants Commission for JRF under the MANF scheme.

REFERENCES

[1] R. E. Mohan, K. S. Sreelatha, M. Sivakumar and A. Krishnasree, "Modeling of Doped ZnO Waveguides for Nonlinear Applications," *Advanced Materials Research*, Vol. 403-408, 2011, pp. 3753-3757. 10.4028/www.scientific.net/AMR.403-408.3753

[2] K. Nomura, H. Ohta, K. Ueda, T. Kamiya, M. Hirano and H. Hosono, "Thin-Film Transistor Fabricated in Single-Crystalline Transparent Oxide Semiconductor," *Science*, Vol. 300, No. 5623, 2003, pp. 1269-1272. doi:10.1126/science.1083212

[3] T. Nakada, Y. Hirabayashi, T. Tokado, D. Ohmori and T. Mise, "Novel Device Structure for Cu(In,Ga)Se_2 Thin Film Solar Cells Using Transparent Conducting Oxide Back and Front Contacts," *Solar Energy*, Vol. 77, No. 6, 2004, pp. 739-747.

[4] S. Y. Lee, E. S. Shim, H. S. Kang, S. S. Pang, *et al.*,

"Fabrication of ZnO Thin Film Diode Using Laser Annealing," *Thin Solid Films*, Vol. 437, No. 1, 2005, pp. 31-34.

[5] R. Könenkamp, R. C. Word and C. Schlegel, "Vertical Nanowire Light-Emitting Diode," *Applied Physics Letters*, Vol. 85, No. 24, 2004, pp. 6004-6006. doi:10.1063/1.1836873

[6] S. Trolier-McKinstry and P. Muralt, "Thin Film Piezoelectrics for MEMS," *Journal of Electroceramics*, Vol. 12, No. 1-2, 2004, pp. 7-17. doi:10.1023/B:JECR.0000033998.72845.51

[7] Z. L. Wang, X. Y. Kong, Y. Ding, P. Gao, W. L. Hughes, R. Yang and Y. Zhang, "Semiconducting and Piezoelectric Oxide Nanostructures Induced by Polar Surfaces," *Advanced Functional Materials*, Vol. 14, No. 10, 2004, pp. 943-956. doi:10.1002/adfm.200400180

[8] M. S. Wagh, L. A. Patil, T. Seth and D. P. Amalnerkar, "Surface Cupricated SnO_2-ZnO Thick Films as a H_2S Gas Sensor," *Materials Chemistry and Physics*, Vol. 84, No. 2-3, 2004, pp. 228-233. doi:10.1016/S0254-0584(03)00232-3

[9] Y. Ushio, M. Miyayama and H. Yanagida, "Effects of Interface States on Gas-Sensing Properties of a CuO/ZnO Thin-Film Heterojunction," *Sensors and Actuators B: Chemical*, Vol. 17, No. 3, 1994, pp. 221-226.

[10] H. Harima, "Raman Studies on Spintronics Materials Based on Wide Bandgap Semiconductors," *Journal of Physics: Condensed Matter*, Vol. 16, No. 48, 2004, pp. S5653-S5660. doi:10.1088/0953-8984/16/48/023

[11] S. J. Pearton, W. H. Heo, M. Ivill, D. P. Norton and T. Steiner, "Dilute Magnetic Semiconducting Oxides," *Semiconductor Science and Technology*, Vol. 19, No. 10, 2004, pp. R59-R74. doi:10.1088/0268-1242/19/10/R01

[12] Ü. Özgür, Y. I. Alivov, C. Liu, A. Teke, M. A. Reshchikov, S. Doğan, V. Avrutin, S.-J. Cho and H. Morkoc, "A Comprehensive Review of ZnO Materials and Devices," *Journal of Applied Physics*, Vol. 98, No. 4, 2005, Article ID: 041301.

[13] O. Lupan, S. Shishiyanu, L. Chow and T. Shishiyanu, "Nanostructured Zinc Oxide as Sensors by Successive Ionic Layer Adsorption and Reaction Method and Rapid Photothermal Processing," *Thin Solid Films*, Vol. 516, No. 10, 2008, pp. 3338-3345. doi:10.1016/j.tsf.2007.10.104

[14] A. A. Ibrahim and A. Ashour, "ZnO/Si Solar Cell Fabricated by Spray Pyrolysis Technique," *Journal of Materials Science: Materials in Electronics*, Vol. 17, No. 10, 2006, pp. 835-839.

[15] W. W. Wenas and S. Riyadi, "Carrier Transport in High-Efficiency ZnO/SiO_2/Si Solar Cells," *Solar Energy Materials and Solar Cells*, Vol. 90, No. 18-19, 2006, pp. 3261-3267.

[16] B. J. Lokhande, P. S. Patil and M. D. Uplane, "Studies on Structural, Optical and Electrical Properties of Boron Doped Zinc Oxide Films Prepared by Spray Pyrolysis Technique," *Physica B: Condensed Matter*, Vol. 302-303, 2001, pp. 59-63.

[17] P. Sharma, A. Gupta, K. V. Rao, F. J. Owens, R. Sharma, R. Ahuja, J. M. O. Guillen, B. Johansson, G. A. Gehring, "Ferromagnetism above Room Temperature in Bulk and Transparent Thin Films of Mn-Doped ZnO," *Nature Materials*, Vol. 2, No. 10, 2003, pp. 673-677. doi:10.1038/nmat984

[18] C. A. Gouvêa, F. Wypych, S. G. Moraes, N. Durán and P. Peralta-Zamora, "Semiconductor-Assisted Photodegradation of Lignin, Dye, and Kraft Effluent by Ag-Doped ZnO," *Chemosphere*, Vol. 40, No. 4, 2000, pp. 427-432.

[19] R. H. Wang, J. H. Z. Xin, Y. Yang, H. F. Liu, L. M. Xu and J. H. Hu, "The Characteristics and Photocatalytic Activities of Silver Doped ZnO Nanocrystallites," *Applied Surface Science*, Vol. 227, No. 1-4, 2004, pp. 312-317. doi:10.1016/j.apsusc.2003.12.012

[20] A. N. Gruzintsev, V. T. Volkov and E. E. Yakimov, *Semi Conductors*, Vol. 37, 2003, pp. 275-279.

[21] G. I. Stegeman and R. H. Stolen, "Waveguides and Fibers for Nonlinear Optics," *Journal of the Optical Society of America B*, Vol. 6, No. 4, 1989, pp. 652-662.

[22] S. C. Rashleigh and R. H. Stolen, "Preservation of Polarization in Single-Mode Fibers," *Laser Focus*, Vol. 19, No. 5, 1983, pp. 155-161.

[23] L. H. Yin, Q. Lin and G. P. Agrawal, "Dispersion Tailoring and Soliton Propagation in Silicon Waveguides," *Optics Letters*, Vol. 31, No. 9, 2006, pp. 1295-1297.

[24] M. Foster and A. Gaeta, "Ultra-Low Threshold Supercontinuum Generation in Subwavelength Waveguides," *Optics Express*, Vol. 12, No. 14, 2004, pp. 3137-3143. doi:10.1364/OPEX.12.003137

[25] R. Gangwar, S. P. Singh and N. Singh, "Soliton Based Optical Communication," *PIER*, Vol. 74, No. 3, 2007, pp. 157-166. doi:10.2528/PIER07050401

[26] G. P. Agrawal, "Nonlinear Fiber Optics," 4th Edition, Academic Press, Waltham, 2001

[27] A. M. Zheltikov, "The Physical Limit for the Waveguide Enhancement of Nonlinear-Optical Processes," *Optics and Spectroscopy*, Vol. 95, No. 3, 2003, pp. 410-415. doi:10.1134/1.1613005

[28] A. Suryanto "Numerical Study of Spatial Soliton Propagation in a Triangular Waveguide with Nonlocal Nonlinearity," *Proceedings of the Third International Conference on Mathematics and Natural Sciences* (*ICMNS* 2010), Bandung, 23-25 November 2010.

29

The Oscillating Universe Theory (To the Unified Field Theory)

Evgeny V. Chensky
Modern Science Institute, SAIBR, Moscow, Russia
Email: echensky@yandex.ru

ABSTRACT

This paper represents model of oscillating universe theory. We try to realize model of both electromagnetic waves and spectrum of elementary particles from the unified point of view. Consideration of problems of the gravitational optics and dark matter is developing from the solid crystal model for the vacuum. The vacuum is represented as a three-dimensional crystal lattice matter with a very small lattice period, much less than 10^{-26} cm. The oscillators are located at the nodes of an infinite lattice. It is shown that an infinite set of equations to describe the coupled oscillations of moving oscillators converges to a system of twelve equations. We have obtained the combined equations for a multicomponent order parameter in the form of the electric and magnetic vacuum polarization, which defines the spectrum and symmetry of normal oscillations in the form of elementary particles. Two order parameters—a polar vector and an axial vector—had to be introduced as electrical and magnetic polarization, correspondingly, in order to describe dynamic properties of vacuum. Vacuum susceptibility has been determined to be equal to the fine structure constant α. Unified interaction constant g for all particles equal to the double charge of Dirac monopole has been found ($g = e/\alpha$, where e—electron charge). The fundamental vacuum constants are: g, α, parameters of length ξ_e, ξ_n and parameters of time τ_e, τ_n for electron and nucleon oscillations, correspondingly. Energy of elementary particles has been expressed in terms of the fundamental vacuum parameters, light velocity being equal to $c = \xi_e/\tau_e = \xi_n/\tau_n$. The term mass of particle has been shown to have no independent meaning. Particle energy does have physical sense as wave packet energy related to vacuum excitation. Exact equation for particle movement in the gravitational field has been derived, the equation being applied to any relatively compact object: planet, satellite, electron, proton, photon and neutrino. The situation has been examined according to the cosmological principle when galaxies are distributed around an infinite space. In this case the recession of galaxies is impossible, so the red shift of far galaxies' radiation has to be interpreted as the blue time shift of atomic spectra; it follows that zero-energy, and consequently electron mass are being increased at the time. Since physical vacuum has existed eternally, vacuum parameters can be either constant, or oscillating with time. It is the time oscillation of the parameters that leads to the growth of electron mass within the last 15 billion years and that is displayed in the red shift; the proton mass being decreased that is displayed in planet radiation.

Keywords: Gravitational Optics; Crystal Model of Vacuum; Electromagnetism and Particles Physics; Universe Evolution Modeling; Unified Field Theory

1. Introduction

During recent years the science about cosmology has been in rather difficult situation. On one hand, observations of star dynamics in galaxies and of galaxies in clasters show substantial deviation of rotation velocities from Kepler's law; this proves the existence of additional matter (dark matter) which participates in gravitational interaction [1-3]. On the other hand, more careful examinations of the red shift in the nearer space at the distances of 10^5 - 10^7 light years as well as observation of supernova outburst [4,5] show that velocity of the Universe expansion increases with time, and this in turn requires introduction of additional dark energy with anti-gravitational properties. Thus, a contradiction arises. Practically, in one and the same point it is necessary to introduce both dark matter creating additional gravitational field and dark energy having anti-gravitation. Since there is no doubt about the facts above, their interpretation must be revised.

At the present time there are two mutually exclusive points of view. First, despite very distinctive spatial non-homogeneity of matter, observations show that at the distances of about 10^9 light years (cell of homogeneity) matter is distributed in the space quite homogeneously. Besides, the cosmological principle suggests that these homogeneous cells should cover the entire infinite space. Second, the red shift discovered by Hubble, which he interpreted as Doppler's principle related to the galaxies expansion, made Friedman's model of expanding Universe quite necessary. From Hubble's empirical law that determines dependence of velocity of galaxies on the distance $v = Hr$, we can suppose existence of a singularity at a certain time. Since velocity of expansion of the galaxies cannot exceed the light velocity c, it follows from the relation $c = HR = HcT_0$, that there is quite a definite size of the Universe growing with time $R = cT_0$, here $-T_0$ is the singularity offset counted from the present moment; Hubble's constant equal to $H = 1/T_0$ decreases with time; however, observations show, that the value H, on the contrary, increases with time.

If we interpret the existence of a singularity as a Big Bang, we have to bear in mind that the explosion is a phase transition from a metastable state into another more stable state accompanied with release of energy. Before the phase transition, this energy is homogeneously distributed around the space. They sometimes say: explosion power is equivalent to e.g. one kilogram of trotyl; it is obvious that two kilograms of trotyl give off right twice as much energy as one kilogram does. Besides, the phase transition does not begin with the singularity but with the nucleation of a new phase whose size exceeds the critical radius. In this case energy is released in accordance with broadening the new phase at the expense of the phase edge motion. Since the average energy density of the entire matter in vacuum is approximately 0.008 erg/m^3, this very energy should be released at the phase transition of each cubic meter of vacuum. It is difficult to imagine, however, that electrons and protons could be created out of this homogeneously distributed in space energy, and, besides, in exactly equal quantities. An explosion of a hydrogen bomb in vacuum can serve as a model of a hot Universe. The hydrogen bomb is a local object in a metastable state. There is a mixture of light and heavy nuclei under the temperature of several million degrees at the moment of detonation. According to D'Alambert equation, the electromagnetic pulse and the neutrino pulse will start to disperse with the light velocity. Following electromagnetic pulse relativistic electrons will fly and then light, and heavy nuclei. In a second, the electromagnetic pulse will reach the Moon area and nothing will stay at the point of explosion. Thus, the examined case is also far from the Friedman's model of expanding Universe.

In order to somehow reconcile the model of the infinite matter distribution in space with that of the expanding Universe, Milne offered the following reasoning [6]. If we mentally specify a sphere of a definite size in a matter homogeneously distributed around an infinite space, then external layers of the sphere due to their spherical symmetry have no influence on the sphere dynamics. Therefore, we can ignore the external layers and consider the Universe as a sphere of a definite size that precisely coincides with the Friedman's model. However, this statement is a mistake. The thing is that with matter being homogeneously distributed about the entire infinite space, the gravitational potential follows the condition of the translational invariance: $U(\boldsymbol{r}) = \text{const}$. We may consider this constant to be equal to zero, therefore, a gravitational potential only arises at deviation of a matter distribution from an average value. For that reason the equation for the potential can be written as follows:

$$\Delta U(\boldsymbol{r}) = 4\pi\gamma\left(\rho(\boldsymbol{r}) - \rho_0\right). \qquad (1)$$

Here ρ_0 is an average density of matter. From Equation (1) we can see that it is not necessary to search for dark energy as the density is both the gravitating and the anti-gravitating matter in the form of $\rho(\boldsymbol{r})$ and ρ_0.

On the other hand, if we mentally specify a sphere of radius R with the density of matter ρ_0 and ignore the external matter, we come to another equation for the potential:

$$\Delta U(\boldsymbol{r}) = 4\pi\gamma\begin{cases}\rho_0, & r < R;\\ 0, & r > R.\end{cases}$$

This equation has the following solutions:

$$U(r) = \begin{cases}-2\pi\gamma\rho_0 R^2\left(1 - \dfrac{r^2}{3R^2}\right), & r < R;\\ -\dfrac{4\pi\gamma\rho_0 R^3}{3r}, & r > R.\end{cases} \qquad (2)$$

Similar expressions can be used for determining gravitational potentials of planets, stars and galaxies in a form of the sum of the potentials of stars with their specific location. However, for the scales comparable with the size of homogeneity cell and bigger, we come to an obviously non-physical result: the potential in any arbitrary point depends on the radius of a sphere which we mentally specify out of the entire infinite space. Thus, any result depending on the mentally specified radius of the sphere, including the radius of the visible part of the Universe, is physically incorrect.

For instance, we can determine the circular orbital velocity v_1 for the Universe of radius R on the sphere surface from the equality of centripetal and centrifugal forces:

$$R\dot{\varphi}^2 = \frac{\partial U(r)}{\partial r}\Big|_{r=R} = \frac{4\pi\gamma\rho_0 R}{3},$$

$$v_1 = R\dot{\varphi} = R\sqrt{\frac{4\pi\gamma\rho_0}{3}}$$

If v_1 is equal to the light velocity c, we obtain the following expression for the critical matter density in the Universe:

$$\rho_c = \frac{3c^2}{4\pi\gamma R^2} = \frac{3H^2}{4\pi\gamma}.$$

This corresponds to the condition $R = r_g$ when r_g is a gravitational radius. Therefore, with the definite choice for R we may come to the conclusion that the Universe is a black hole, while, as it follows from the cosmological principle at the scales comparable with the radius of the visible part of the Universe, the gravitational potential has no specific features and its average value is zero.

The same situation takes place when we consider the influence of a pressure on the dynamics of the expanding Universe. For instance, if we take a big vessel with a gas, mentally specify a sphere of radius R in it, and ignore the gas surrounding the sphere, we can state that the gas will broaden and get cool at the expense of the internal pressure. This may remind the model of the expanding Universe. Remember, however, that the specified sphere is surrounded with the same gas at the same pressure; that is why there will be neither broadening nor cooling. Thus, for the infinite Universe both an average gravitational potential and an average pressure are constant; besides, since the expanding dynamics is influenced by the equal to zero gradients of these variables, there cannot be neither expanding nor compression. An infinite system can only stratify according to the energy density and we really observe this stratification on giant scales from the value less than 10^{-9} erg/cm^3 for an inter-galaxy space to the value over 10^{39} erg/cm^3 for nuclear energy.

Nevertheless, within the frames of the cosmological principle there is a problem, the so called photometric paradox. The thing is that at present time when stars and galaxies radiate light in the entire infinite space, we can introduce an average luminosity L of a unit volume, provided that the densities of a luminous flux intensity at the distance r from a single volume is equal to $j = L/4\pi r^2$. The integral over the sphere of radius R gives the total flux intensity equal to $J = RL$; it follows that with R approaching infinity the flux intensity must approach infinity as well. Practically, however, we see rather a low sky luminosity. This is the photometric paradox.

In fact, by calculating the intensity, we must take into consideration the retardation effects. The flux that comes to a certain point $(r = 0)$ at a certain time $(t = 0)$ radiates at different moments depending on the distance:

$$J = \int j\left(t = -\frac{r}{c}\right)\mathrm{d}\boldsymbol{r} = c\int_{-\infty}^{0} L(t)\,\mathrm{d}t \tag{3}$$

Expression (3) shows that the flux coming from the deep Universe will be finite if $L(t)$ at longer t decreases faster than $1/t$.

Besides, we can divide the entire flux observed at any point of the infinite space into two parts: the flux J_{vis} of a visible part of space $R = cT_0$, $T_0 \sim 15 \times 10^9$ years and the relict flux J_{rel} radiating from the spots with $r > R$:

$$J = J_{vis} + J_{rel} = \sum_{n=1}^{N} J_n + \int_{-\infty}^{-T_0} L(t')\,\mathrm{d}t'$$

Here summing was carried out over a countable number of galaxies in the visible part of the Universe. Thus, from the expressions given above it follows that the Universe must be non-stationary, not due to an expansion of galaxies', but at the expense of a variation of physical vacuum parameters. Since the relict radiation corresponds to the temperature 3K, the Universe had such a temperature long ago. The one but not the only feature of a non-stationarity is the red shift of atomic spectra that we can interpret as the blue temporal shift of both characteristic Bohr energy and all atomic energy levels correspondingly, at the expense of variation of physical vacuum parameters. Observations show that the characteristic Bohr frequency depends on time and increases with time. By introducing a frequency of an arbitrary atomic level, we obtain the following expression for the Hubble's constant:

$$\frac{\mathrm{d}}{\mathrm{d}t}\left(\frac{\omega(t)}{\omega(t=0)}\right) = H(t) \tag{4}$$

Both $\omega(t)$ and $H(t)$ are monotonously increasing functions. The latest observations of the flashes of far supernova [4,5] show temporal growth of H. It is senseless to explain this situation using space-time properties.

Speaking about space-time properties is quite the same as judging about wine quality by the curvature of a bottle surface. Dilettantes are often attracted by the appearance of the vessel, while connoisseurs pay attention to its contents, conservation conditions, and temporal changes. We should regard space like a vessel with the only feature: its volume is infinite [7]. Its internal properties are to be discussed.

2. Hidden Parameters of Vacuum

We should proceed from the experimental fact that the energy and the pulse of any elementary particle are:

$$\varepsilon_k = \hbar\omega_k; \quad \boldsymbol{p} = \hbar\boldsymbol{k} \tag{5}$$

Here ω_k—frequency for electron, proton, photon and

neutrino, correspondingly, we expressed as follows:

$$\omega_{ek}=\sqrt{\omega_{e0}^2+c^2k^2};\quad \omega_{pk}=\sqrt{\omega_{p0}^2+c^2k^2};\quad \omega_{\gamma,\nu;k}=ck \quad (6)$$

The unified formula for the energy of any elementary particle points to the existence of the universal interaction for fields related to each particle. Besides, the two oscillation branches with the energy gap observed in the excitation spectrum prove an existence of a certain set of discrete oscillators whose interaction causes normal oscillations with frequencies ω_k. In fact, we can represent vacuum as a crystal object of a cubic or hexagonal symmetry with a very small lattice period, much less than 10^{-26} cm. We can estimate the upper limit of a lattice period by the maximum particle energy in cosmic rays equal to 10^{21} eV that corresponds to the wave vector of 10^{26} cm^{-1}. The vacuum ground state is the equilibrium position of all oscillators; these are the points of equilibrium forming crystal lattice related to the absolute coordinate system. Under the deviation of an oscillator from the equilibrium position, a dipole moment arises. For the scales exceeding the lattice period we can introduce a macroscopic order parameter as an electric polarization of vacuum:

$$\boldsymbol{P}(\boldsymbol{r})=\frac{4\pi}{\delta V}\sum_i \boldsymbol{d}_i.$$

Suppose, there are two branches of normal oscillations of field P that we can call electron and nucleon modes. The Hamiltonian for electron and nucleon modes written in the unified form, is:

$$H=\frac{1}{8\pi}\int\left(\tau_{e,n}^2\dot{\boldsymbol{P}}^2+\xi_{e,n}^2(\nabla\boldsymbol{P})^2+\sigma\boldsymbol{P}^2\right)\mathrm{d}\boldsymbol{r} \quad (7)$$

For electronic and nucleonic parts, we introduced the parameters of time τ_e, τ_n and length ξ_e, ξ_n that characterize the kinetic and gradient energy of the fields. Besides, we introduced a dimensionless parameter of an elastic coefficient σ corresponding to the reciprocal susceptibility common to both modes. These are the latent parameters of vacuum and the available experimental data are sufficient to determine them.

By using the minimal action principle for the Lagrange function equal to the difference of the kinetic energy—the first member of expression (7), and the potential energy—the second and the third members of (7), we obtain the equation of motion for six independent normal oscillations $P_{ex}, P_{ey}, P_{ez}, P_{nx}, P_{ny}, P_{nz}$:

$$\left(\sigma+\xi_{e,n}^2\left(\frac{\partial^2}{\partial x^2}+\frac{\partial^2}{\partial y^2}+\frac{\partial^2}{\partial z^2}\right)-\tau_{e,n}^2\frac{\partial^2}{\partial t^2}\right)P_{e,n,x,y,z}=0 \quad (8)$$

By setting up the following solutions:

$$P_{x,y,z}=a_{x,y,z}\exp\left(i(\boldsymbol{kr}+\omega t)\right), \quad (9)$$

we obtain the spectrum for normal oscillations:

$$\omega_{e,n;k}=\pm\sqrt{\omega_{e,n;0}^2+c^2k^2};$$
$$\omega_{e,0}=\frac{\sqrt{\sigma}}{\tau_e};\quad \omega_{n,0}=\frac{\sqrt{\sigma}}{\tau_n};\quad c=\frac{\xi_e}{\tau_e}=\frac{\xi_n}{\tau_n}. \quad (10)$$

Therefore, we can represent physical vacuum as some coherent state with the natural frequency standards in the form of homogeneous polarization oscillations about an absolute coordinate system

$$P_{e,n;x,y,z}(\boldsymbol{r},t)=a_{e,n;x,y,z}\exp\left(i\omega_{e,n;0}t\right)$$

with the absolute time, homogeneous around the entire space $t(\boldsymbol{r})\equiv t_{abs}$.

The situation, however, becomes more complicated, since the electrical vacuum polarization generates the following electric charge:

$$div\boldsymbol{P}=-4\pi\rho_e. \quad (11)$$

Here ρ_e is the electric charge density, while the polarization is determined by both electron and nucleon modes $\boldsymbol{P}=\boldsymbol{P}_e+\boldsymbol{P}_n$. This results in an additional long-range Coulomb interaction between the normal oscillations $P_{ex}, P_{ey}, P_{ez}, P_{nx}, P_{ny}, P_{nz}$

$$U=\frac{1}{2}\iint\frac{\rho_e(\boldsymbol{r})\rho_e(\boldsymbol{r}')}{|\boldsymbol{r}-\boldsymbol{r}'|}\mathrm{d}\boldsymbol{r}\mathrm{d}\boldsymbol{r}' \quad (12)$$

For simplicity, we consider normal oscillations inside the electronic modes. We dimensionlize coordinates and time. We express new variables like this: $t\Rightarrow t/\tau_e$; $r\Rightarrow r/\xi_e$; velocity being in terms of $c=\xi_e/\tau_e$. It makes sense to specify a dimensional value for the electric polarization in the terms of the electron charge:

$$\boldsymbol{P}\Rightarrow\frac{e}{\xi_e^2}\boldsymbol{P},$$

after that the electron field action reduces to form (see (13) below):

Here i and j run over x, y, z and we carried out summation over repeated indices. By varying action S over the values $P_i, \dot{P}_i, \partial P_i/\partial x_j$, we come to the following system

$$S=\frac{e^2}{8\pi c}\int\mathrm{d}t\int\mathrm{d}\boldsymbol{r}\left\{\dot{P}_i\dot{P}_i-\frac{\partial P_i}{\partial x_j}\frac{\partial P_i}{\partial x_j}-\sigma P_iP_i-\frac{1}{4\pi}\int\frac{\dfrac{\partial P_i(\boldsymbol{r},t)}{\partial x_i}\dfrac{\partial P_j(\boldsymbol{r}',t)}{\partial x_j'}}{|\boldsymbol{r}-\boldsymbol{r}'|}\mathrm{d}\boldsymbol{r}'\right\} \quad (13)$$

of the integral-differential equations:

$$\left(\frac{\partial^2}{\partial t^2}-\Delta+\sigma\right)P_i(\boldsymbol{r},t)+\frac{1}{4\pi}\int\frac{(x_i-x_i')}{|\boldsymbol{r}-\boldsymbol{r}'|^3}\frac{\partial P_j(\boldsymbol{r}')}{\partial x_j'}\mathrm{d}\boldsymbol{r}'=0 \quad (14)$$

Consider the solutions in the form of plane waves:

$$P_i(\boldsymbol{r},t)=P_i\exp\left(i(\boldsymbol{kr}+\omega t)\right) \quad (15)$$

For plane waves, Equation (14) reduce to the form:

$$\left(\sigma+k^2-\omega^2\right)P_i+\frac{k_ik_j}{k^2}P_j=0 \quad (16)$$

By making the determinant of the Equation (16) equal to zero, we obtain the oscillation spectrum (see (17) below):

Equation (17) transforms to

$$\left(\sigma+k^2-\omega^2\right)^2\cdot\left(\sigma+1+k^2-\omega^2\right)=0 \quad (18)$$

Thus, from (18) we obtain the normal spectrum of the oscillations; from Equation (16), we obtain the form of the oscillations:

$$\begin{aligned}&\omega_k=\pm\sqrt{\sigma+1+k^2};\ k_yP_x=k_xP_y,\\&k_zP_x=k_xP_z,\ k_zP_y=k_yP_z;\\&\omega_k=\pm\sqrt{\sigma+k^2};\\&k_xP_x+k_yP_y+k_zP_z=0.\end{aligned} \quad (19)$$

Expression (19) allow the definition of the general properties of the normal oscillations for vacuum fields linked by the long-range Coulomb interaction. The Laplacian operator in Equations (14) requires that the polarization components be eigenfunctions of this operator:

$$\Delta P_i=\beta^2P_i \quad (20)$$

The result of the Coulomb interaction is that the oscillations of the polarization are divided into two classes: longitudinal $\boldsymbol{P}_1$ with $rot\boldsymbol{P}_1=0$ and lateral $\boldsymbol{P}_2$ with $div\boldsymbol{P}_2=0$, according to the Helmholtz theorem $\boldsymbol{P}_1=\nabla\Phi$; $\boldsymbol{P}_2=rot\boldsymbol{A}$, here Φ and $\boldsymbol{A}$ are scalar and vector potentials. Longitudinal oscillations provide a depolarizing electric field $\boldsymbol{E}=-\boldsymbol{P}_1$, which meets the following condition:

$$div\boldsymbol{E}=4\pi\rho_e.$$

For lateral (transverse!) oscillations, the depolarizing field equals to zero. As a result, the frequencies for longitudinal and lateral oscillations are different.

The problem, however, is that for linear homogeneous differential equations we may take into consideration both eigenfunctions and eigenvalues, while the amplitude of the eigenfunctions remains arbitrary. Suppose, an eigenfunction specifies the configuration of the excitation; though the excitation energy and pulse are the integrals of motion, and yet they can have arbitrary meanings. Nevertheless, in practice we can see that energy of any excitation has quite a definite meaning both for light quantum and for any elementary particle. Therefore, within the framework of homogeneous equations it is impossible to realize the origin and the physical meaning of the Planck constant.

For linear systems, the amplitude of oscillations turns out to be quite definite under the external force; then we can express the solution by means of the Green function, which meets the homogeneous equation and has quite definite amplitude. Non-homogeneous equations are necessary for the following reasons. We know from the theory of many-body systems that, if a system consists of discrete particles, the correlation effects substantially decrease the ground state, and local states such as polarons can occur. Therefore, we pass to consideration of the ground state taking into account correlation effects.

From an endless number of particles forming a crystalline vacuum state we examine one particle as a point unit source $Q_e=\delta(\boldsymbol{r})$, which generates longitudinal electric field defined by equations:

$$div\boldsymbol{E}_0=4\pi\delta(\boldsymbol{r});\ rot\boldsymbol{E}_0=0;\ \boldsymbol{E}_0=\frac{\boldsymbol{r}}{r^3}.$$

Thereafter, we can write the interaction energy of the point source with vacuum fields as follows:

$$U_0=-\frac{g}{4\pi}\boldsymbol{E}_0\boldsymbol{P}. \quad (21)$$

Here g is the constant of interaction between the point unit field and vacuum fields; it is convenient to express this constant in a normalized form: $g=g_1e$. By varying the Lagrange function over $\boldsymbol{P}$, we obtain a non-homo-

$$\begin{vmatrix}\left(\sigma+k^2+\frac{k_x^2}{k^2}-\omega^2\right) & \frac{k_xk_y}{k^2} & \frac{k_xk_z}{k^2}\\ \frac{k_xk_y}{k^2} & \left(\sigma+k^2+\frac{k_y^2}{k^2}-\omega^2\right) & \frac{k_yk_z}{k^2}\\ \frac{k_xk_z}{k^2} & \frac{k_yk_z}{k^2} & \left(\sigma+k^2+\frac{k_z^2}{k^2}-\omega^2\right)\end{vmatrix}=0 \quad (17)$$

geneous equation for polarization:

$$\left(\sigma - \Delta + \frac{\partial^2}{\partial t^2}\right) \boldsymbol{P} = g_1 \boldsymbol{E}_0(\boldsymbol{r}, t); \quad (22)$$

Divergence of the left and the right parts of the Equation (22) results in the expression for the induced charge density, related to the electrical polarization for the case when a source is moving with velocity $\boldsymbol{v}$:

$$\left(\sigma - \Delta + \frac{\partial^2}{\partial t^2}\right) \rho_e(\boldsymbol{r}, t) = -g_1 \delta(\boldsymbol{r} - \boldsymbol{v}t). \quad (23)$$

It is obvious from (23) that the induced charge density is the Green function for a point source that fulfills the homogeneous equation over the entire space except one point; but due to this point, the function acquires quite definite values over the entire space.

At first, we consider a particular solution of Equation (23). Fourier-transformation over coordinates results in the Fourier-harmonics for the induced charge density in vacuum:

$$\rho_{ek} = -\frac{g_1}{\sigma + k^2 - (\boldsymbol{k}\boldsymbol{v})^2} \quad (24)$$

Here the corresponding coordinate dependence of the induced charge density for the case, when velocity lies in z-axis, is:

$$\rho(\boldsymbol{r}, t) = -\frac{g_1}{(2\pi)^3} \iiint \frac{\exp\left(i\left(k_x x + k_y y + k_z (z - vt)\right)\right)}{\sigma + k_x^2 + k_y^2 + k_z^2 (1 - v^2)} dk_x dk_y dk_z$$

Here, it is convenient to proceed to the new integration variables

$$k_x' = k_x;\; k_y' = k_y;\; k_z' = k_z\sqrt{1 - v^2},$$

in addition, to a new coordinate system:

$$x' = x;\; y' = y;\; z' = \frac{z - vt}{\sqrt{1 - v^2}}.$$

After that, the induced charge density expressed in dimensional units transforms to the equation:

$$\rho(\boldsymbol{r}, t) = \frac{1}{\sqrt{1 - v^2}} \rho'(r'),\; \rho'(r') = -\frac{g}{4\pi \xi_e^2} \frac{1}{r'} \exp\left(-\frac{r'\sqrt{\sigma}}{\xi_e}\right), \quad (25)$$

From (25) we can see that the characteristic dimension of the polarization charge is a definite value equal to the correlation radius or the Compton length of electron: $r_{c,e} = \xi_e / \sqrt{\sigma} = c/\omega_{e0}$. The polarization charge moving relative to the absolute coordinate system, in accordance with the Lorentz transformation, is deformed in such a way that its dimensions decrease along the direction of motion

$$r_{ex} = r_{ey} = r_{c,e};\; r_{ez} = r_{c,e}\sqrt{1 - v^2}.$$

Total polarization charge as an integral over the entire space is proportional to the constant of interaction g and the vacuum susceptibility σ^{-1}:

$$q = \int \rho(\boldsymbol{r}, t) \mathrm{d}\boldsymbol{r} = -\frac{g}{\sigma} \quad (26)$$

The total polarization charge does not depend on the particle velocity that we can interpret as the law of conservation of charge.

We can find a scalar potential for the motionless source from the expression:

$$\Phi(r) = \sum_k \frac{4\pi}{k^2} \rho_k \exp(i\boldsymbol{k}\boldsymbol{r}) = 4\pi g \sum_k \frac{\exp(i\boldsymbol{k}\boldsymbol{r})}{k^2(\sigma + k^2)} = \frac{g}{\sigma} \frac{1}{r}\left(1 - \exp\left(-\frac{r}{r_e}\right)\right) \quad (27)$$

The polarization for the electron is similar to that of the proton within an accuracy of a charge sign:

$$\boldsymbol{P}_{e,n}(\boldsymbol{r}) = \frac{g}{\sigma} \frac{\boldsymbol{r}}{r^3}\left[1 - \left(1 + \frac{r}{r_{c;e,n}}\right) \exp\left(-\frac{r}{r_{c;e,n}}\right)\right]. \quad (28)$$

They only differ in the characteristic wavelength $r_{c,e}$ and $r_{c,n}$. The main feature of the solution for the polarization (28) is an absence of divergence at a point $r = 0$ that leads to the finite value of the particle energy.

Therefore, we can see that the vacuum polarization results in decrease of the source energy by U_0, both electronic and nucleonic modes having the same form:

$$U_{0e,n} = -g\Phi_{e,n}(r = 0) = -\frac{g^2}{\sigma r_{c;e,n}} \quad (29)$$

Non-homogeneous Equation (23) defines two parameters: the polarization charge q and the radius of a charge localization $r_{c;e,n}$.

In order to determine vacuum parameters, we require that the polarization charge, both for proton and electron, be equal to the electron charge, whereas the particle energy must be equal to the ionization energy of a source out of a potential energy well, which the source creates for itself:

$$q = \frac{g}{\sigma} = e; \qquad \hbar\omega_{0,e} = \frac{g^2}{\sigma r_{c,e}};\; \hbar\omega_{0,n} = \frac{g^2}{\sigma r_{c,n}}. \quad (30)$$

By adding the definition of the fine structure constant $\alpha = e^2/\hbar c$ to the latter equations, we obtain the equality $g_1 = \sigma = \alpha^{-1} \approx 137$. It follows that the vacuum polarizability σ^{-1} equals to the fine structure constant α, whereas the constant of the interaction of a point source with vacuum fields equals to the Dirac monopole charge $g = e/\alpha$ [8].

We can define the correlation radius and the fundamental frequency for electronic and nucleonic normal mode as follows:

$$r_{c,e} = \xi_e \sqrt{\alpha}, \quad \omega_{e,0} = \frac{1}{\tau_e \sqrt{\alpha}};$$

$$r_{c,n} = \xi_n \sqrt{\alpha}, \quad \omega_{n,0} = \frac{1}{\tau_n \sqrt{\alpha}}.$$

As a result, the rest energy in form (30) reduces to a quite transparent form:

$$\varepsilon_{e,0} = \hbar\omega_{e0} = \frac{eg}{r_{c,e}}; \ \varepsilon_{n,0} = \hbar\omega_{n0} = \frac{eg}{r_{c,n}} \tag{31}$$

It follows that the point source with the interaction constant g polarizes vacuum and induces charges with dimension $r_{c,e}$ for electronic and $r_{c,n}$ nucleonic modes. The electric field energy for both proton and electron is equal to $e^2/r_{c,e}$ and to $e^2/r_{c,n}$, correspondingly, and it turns out to be 137 times less than the energy related to the electrical polarization. We should notice that the solution for the polarization (28) is formed by three modes of normal vacuum oscillations P_x, P_y, P_z, each creating a charge equal to $e/3$:

$$\frac{1}{4\pi}\int \frac{\partial P_x}{\partial x} d\boldsymbol{r} = \frac{1}{4\pi}\int \frac{\partial P_y}{\partial y} d\boldsymbol{r} = \frac{1}{4\pi}\int \frac{\partial P_z}{\partial z} d\boldsymbol{r} = \frac{e}{3}.$$

Now we consider the structure of the fields in the excited state. The excited state corresponds to the generation of a source at a certain time. Suppose, a point source is generated at time $t = 0$ under the initial conditions for polarization:

$$\boldsymbol{P}(\boldsymbol{r},t)\big|_{t=0} = 0; \quad \frac{\partial \boldsymbol{P}(\boldsymbol{r},t)}{\partial t}\Big|_{t=0} = 0.$$

In this case, the general solution of Equation (22) consists of a particular solution (28) and two fundamental solutions of the homogeneous wave equation:

$$\boldsymbol{P}(\boldsymbol{r},t) = \boldsymbol{P}^{(0)}(\boldsymbol{r}) + \boldsymbol{P}^{(+)}(\boldsymbol{r})\exp(i\omega_0 t) + \boldsymbol{P}^{(-)}(\boldsymbol{r})\exp(-i\omega_0 t) \tag{32}$$

By taking into account initial conditions, we can reduce the solution (32), both for electron and nucleon, to the form:

$$\boldsymbol{P}_{e,n}(\boldsymbol{r},t) = 2e\frac{\boldsymbol{r}}{r^3}\left[1 - \left(1 + \frac{r}{r_{c;e,n}}\right)\exp\left(-\frac{r}{r_{c;e,n}}\right)\right]\sin^2\left(\omega_{e,n;0}t\right) \tag{33}$$

The characteristic feature of the solution above is that the electrical polarization for both electron and proton, covers the entire infinite space and oscillates synchronously with the frequency $\omega_{e,n;0}$.

The solution (33), however, contains a substantial disadvantage: such wave packet cannot move in space, it is a typical standing wave. Impossibility of motion is caused by the fact that the phase velocity of different harmonics $v_f = \omega_k/k$ changes from infinity to the light velocity c, whereas the group velocity $v_g = \partial\omega_k/\partial k$ changes from zero to c.

In a general case, the solution for the polarization for a wave packet moving with velocity $\boldsymbol{v}$ should have a soliton form:

$$\boldsymbol{P}(\boldsymbol{r},t) = \boldsymbol{P}(\boldsymbol{r} - \boldsymbol{v}t) f(t) \tag{34}$$

A similar property is natural for the solution of a one-dimensional D'Alambert equation that fulfills the condition of deviation from a state of equilibrium for a flexible infinite string $u(x,t) = u(x \pm ct)$. A possibility of motion without changing the form is directly connected to a linear excitation spectrum in k-space $\omega_k = ck$. For two- and three-dimensional cases, the solution of the D'Alambert equation substantially differs from the one-dimensional one. An excitation generated in some point starts spreading (propagating) at velocity c in the form of concentrated circles for two-dimensional case, and in the form of concentrated spheres for the three-dimensional case. The propagation of radio waves strictly follows the three-dimensional D'Alambert equation, which proceeds from the Maxwell equations. Radio waves, however, are a multiquantum process. Nevertheless, a single quantum, while having wave properties, yet behaves like a particle. The thing is that a light quantum radiated by an excited atom at a distant star can cover million years without spreading dispersion. After colliding with a similar atom on the Earth, the light quantum transfers into a similar state of excitation. Therefore, there must be a solution of a soliton type for a light quantum in the form (34), which gives the origin of ray optics.

Analysis shows that it is impossible to obtain such a spectrum in a three-dimensional isotropic space for one order parameter. Following strictly the terminology, we should consider electromagnetic oscillations as coupled oscillations of a two-component order parameter in the form of an electric and magnetic polarization of vacuum.

Suppose a magnetic polarization with the same Hamiltonian, as that for the electric polarization (7) is possible to appear in vacuum:

$$H = \frac{1}{8\pi}\int\left(\tau_{e,n}^2 \dot{\boldsymbol{M}}^2 + \xi_{e,n}^2 (\nabla \boldsymbol{M})^2 + \sigma \boldsymbol{M}^2\right) d\boldsymbol{r} \tag{35}$$

We define a magnetic order parameter, as well as an electric polarization, through the sum of the elementary magnetic moments:

$$M = \frac{4\pi}{\delta V}\sum \mu_i.$$

Practice shows that electric and magnetic dipole moments create, correspondingly, electric and magnetic fields, similar in configuration:

$$E(r) = \frac{3(r-r')(d(r-r'))}{|r-r'|^5} - \frac{d}{|r-r'|^3};$$
$$H(r) = \frac{3(r-r')(\mu(r-r'))}{|r-r'|^5} - \frac{\mu}{|r-r'|^3}. \quad (36)$$

It follows that for a similar distribution of the electric and magnetic polarization, electric and magnetic fields will be similar as well. We can reduce expressions (36) to the form:

$$E(r) = -\nabla\Phi_d(r);\ \Phi_d = d\nabla_{r'}\Phi_0(r-r');$$
$$H(r) = -\nabla\Phi_\mu(r);\ \Phi_\mu = \mu\nabla_{r'}\Phi_0(r-r'), \quad (37)$$

Here Φ_0 is the potential Coulomb function for a unit source $\Phi_0 = 1/|r-r'|$. Under an arbitrary distribution of the electric and magnetic polarization, scalar potentials (37) acquire the form:

$$\Phi_P = \frac{1}{4\pi}\int P(r')\nabla_{r'}\Phi_0(r-r')dr'$$
$$= -\frac{1}{4\pi}\int \frac{divP(r')}{|r-r'|}dr' = \int \frac{\rho_e(r')}{|r-r'|}dr';$$
$$\Phi_M = \frac{1}{4\pi}\int M(r')\nabla_{r'}\Phi_0(r-r')dr'$$
$$= -\frac{1}{4\pi}\int \frac{divM(r')}{|r-r'|}dr' = \int \frac{\rho_\mu(r')}{|r-r'|}dr', \quad (38)$$

It follows that the sources of the electric field are the electric charges defined by the relation $divP = -4\pi\rho_e$, whereas the sources of the magnetic field are the magnetic charges defined by the relation $divM = -4\pi\rho_\mu$. As a result, the electric and magnetic fields meet the conditions:

$$divE = 4\pi\rho_e;\ divH = 4\pi\rho_\mu.$$

Energy of the electric and magnetic fields turns up to be 137 times less than that of the electric and magnetic polarization, correspondingly. The configurations of the electric and magnetic fields are similar under the similar distribution of the electric and magnetic polarization. For example, if we create a homogeneous electric polarization P in a full-sphere, then it causes generation of the depolarizing electric field inside the sphere $E = -P/3$; therefore, the depolarization coefficient for a sphere is equal to $1/3$. The situation is the same with a spherical magnet: $H = -M/3$. Generation of the magnetic field also leads to the long-range Coulomb interaction between the normal oscillations: $M_{ex}, M_{ey}, M_{ez}, M_{nx}, M_{ny}, M_{nz}$.

We can define the electric current with the expression:

$$4\pi j_e = \dot{P} + rotM.$$

The continuity equation follows from here:

$$divj_e + \frac{\partial\rho_e}{\partial t} = 0.$$

Now we show in what way the interaction between the electric and magnetic polarization provides the solution of the soliton type. We add the interaction energy of currents to the Hamiltonians (7) and (35) in the form:

$$U = \frac{\xi_e\tau_e}{4\pi}\int(\dot{P}\cdot rotM - \dot{M}\cdot rotP)dr \quad (39)$$

From there we obtain the combined equations for a plane polarized electromagnetic wave:

$$\left(\sigma - \Delta + \frac{\partial^2}{\partial t^2}\right)P_x(r,t) - 2\frac{\partial}{\partial t}\frac{\partial}{\partial z}M_y(r,t) = 0;$$
$$-2\frac{\partial}{\partial t}\frac{\partial}{\partial z}P_x(r,t) + \left(\sigma - \Delta + \frac{\partial^2}{\partial t^2}\right)M_y(r,t) = 0. \quad (40)$$

By setting up the solutions in the form:

$$P_x(r,t) = P_x\exp(i(kr+\omega t));$$
$$M_y(r,t) = M_y\exp(i(kr+\omega t)), \quad (41)$$

we can obtain the system of equations:

$$(\sigma + k^2 - \omega^2)P_x - 2\omega k_z M_y = 0;$$
$$-2\omega k_z P_x + (\sigma + k^2 - \omega^2)M_y = 0. \quad (42)$$

The compatibility condition for the Equations (42) leads to the equation:

$$\begin{vmatrix} \sigma + k^2 - \omega^2 & -2\omega k_z \\ -2\omega k_z & \sigma + k^2 - \omega^2 \end{vmatrix} = 0,$$

This gives the spectrum of normal oscillations:

$$\omega = \pm k_z \pm \sqrt{\sigma + k_x^2 + k_y^2 + 2k_z^2},$$

After that, the solutions for the electric and magnetic polarization transmitting with the light velocity reduce to the soliton form:

$$P_x(r,t) = a\cdot\exp(i(k_x x + k_y y + k_z(z \pm t)))\exp\left(i\left(\pm\sqrt{\sigma + k_x^2 + k_y^2 + 2k_z^2}\cdot t\right)\right);$$
$$M_y(r,t) = a\cdot\exp(i(k_x x + k_y y + k_z(z \pm t)))\exp\left(i\left(\pm\sqrt{\sigma + k_x^2 + k_y^2 + 2k_z^2}\cdot t\right)\right).$$

We proceed from the supposition that the electron radiates a light quantum; then from a wide range of possible solutions we should choose a solution compatible with the own field of the electron. Since the light quantum propagating along z-axis has a wave vector k_z, we can specify a Fourier-harmonic k_z from the scalar potential (27) which defines the field of the electron. In the cylindrical coordinate system, the Fourier-harmonic for the scalar potential becomes:

$$\Phi_{k_z} = \int_{-\infty}^{\infty} \Phi \exp(-ik_z z)\,dz = q\left(K_0(rk_z) - K_0\left(r\sqrt{2\sigma + k_z^2}\right)\right).$$

Here K_0 is the Macdonald function. We express the electric polarization along the x-axis as $P_x = \partial\Phi/\partial x$. Thus, for a plane-polarized wave compatible with the field of the electron and fulfilling the system of Equations (40), we obtain the solution for the electric polarization:

$$P_x = q\left(k_z K_1(rk_z) - \sqrt{2\sigma + k_z^2}\,K_1\left(r\sqrt{2\sigma + k_z^2}\right)\right) \times \cos\varphi \exp\left(i\left(k_z(z-t) \pm \sqrt{2\left(\sigma + k_z^2\right)}\cdot t\right)\right) \quad (43)$$

This solution is a quasi-one-dimensional infinite monochromatic wave propagating at the light velocity along the z-axis and interacting with the similar magnetic polarization M_y. In the transversal direction, the monochromatic wave (43) is localized with the dimension equal to the wavelength, since the Macdonald function at big values of argument approximately equals:

$$K_1(x) \approx \sqrt{\frac{\pi}{2x}}\exp(-x).$$

This precisely corresponds to the experiment, as it is impossible to localize a light ray more than the light wavelength.

Therefore, from the values, which we consider as fundamental $e, \hbar, c, m_e, m_p$, we go over to the set of values, which characterize properties of physical vacuum $\alpha, g, \xi_e, \xi_p, \tau_e, \tau_p$ under the additional condition: $\xi_e/\tau_e = \xi_p/\tau_p = c$. In connection with this, we must change the concepts of mass and matter.

Wave equations can only be applied to the material medium having definite dynamic properties, so the idea of physical vacuum means that the entire infinite space is filled with a definite matter. The particles that we observe—electrons, protons, photons—these are excitations of vacuum in the form of wave packets, which are eigenfunctions of the united system of twelve equations. From the point of view of wave mechanics, we can characterize a wave packet with energy, momentum, angular momentum and oscillation amplitude; specifically for the electric polarization, we define the amplitude by the electric charge. For a multi-component order parameter, the form or symmetry of oscillations is important. In this connection, the concept of a particle mass does not have independent meaning. Researchers introduced the values of mass and charge, as well as Planck constant for particles, in different periods of time and so far, they have considered these values as independent ones. As we showed above, charge quantization and existence of Planck constant are the consequences of correlation effects related to discreteness of physical vacuum. Now it makes sense to study the concept of mass for a wave packet.

From practice we know that, if we describe the particle oscillation spectrum with the expression: $\omega_k = \sqrt{\omega_0^2 + c^2k^2}$ then the particle velocity is equal to the group velocity:

$$\boldsymbol{v} = \boldsymbol{v}_g = \frac{\partial\omega_k}{\partial\boldsymbol{k}} = \frac{c^2\boldsymbol{k}}{\sqrt{\omega_0^2 + c^2\boldsymbol{k}^2}} \quad (44)$$

If we express the wave vector $\boldsymbol{k}$ from (44) through the group velocity, we obtain the value of frequency in the form:

$$\omega_k = \sqrt{\omega_0^2 + c^2k^2} = \frac{\omega_0}{\sqrt{1 - \frac{v^2}{c^2}}} \quad (45)$$

Multiplying terms of (45) by $\hbar$, we come to the relativistic expression for the particle energy:

$$E(v) = \frac{\hbar\omega_0}{\sqrt{1 - \frac{v^2}{c^2}}} = \frac{E_0}{\sqrt{1 - \frac{v^2}{c^2}}} \approx E_0 + \frac{E_0}{2}\frac{v^2}{c^2} = E_0 + \frac{mv^2}{2} \quad (46)$$

The expression for the particle mass $m = E_0/c^2$ follows from the latter Equation (46), the concept of mass being not necessary if we specify velocity in terms of light velocity.

The examples given below illustrate how to express some known values in terms of vacuum parameters:

De Broglie wavelength:

$$\lambda_{\partial,e} = \frac{\hbar}{m_e v} = \frac{\hbar}{p} = \frac{\hbar}{\hbar k} = \frac{1}{k};\quad \lambda_{\partial,n} = \frac{\hbar}{m_n v} = \frac{1}{k}.$$

Here we have to consider k, the particle wave vector, as a quantum number independent of vacuum parameters.

Compton wavelength:

$$\lambda_{K,e} = \frac{\hbar}{m_{e,0}c} = \frac{c}{\omega_{0,e}} = r_{c,e} = \xi_e\sqrt{\alpha};$$

$$\lambda_{K,n} = \frac{\hbar}{m_{n,0}c} = \frac{c}{\omega_{0,n}} = r_{c,n} = \xi_n\sqrt{\alpha}$$

Classical radius of electron:

$$r_{0,e} = \frac{e^2}{m_{0,e}c^2} = r_{c,e}\alpha = \xi_e \alpha^{3/2}.$$

Bohr radius:

$$a_B = \frac{\hbar^2}{m_{0,e}e^2} = \frac{r_{c,e}}{\alpha} = \frac{\xi_e}{\sqrt{\alpha}} \quad (47)$$

Bohr energy:

$$E_B = \frac{m_{0,e}e^4}{\hbar^2} = m_{0,e}c^2\alpha^2 = \frac{eg}{r_{c,e}}\alpha^2 \quad (48)$$

By making Bohr energy equal to photon energy,

$$E_B = \hbar\omega_{\gamma,B} = \hbar c k_{\gamma,B}$$

we obtain γ-quantum wavelength, which corresponds to Bohr energy

$$\lambdabar_{\gamma,B} = \frac{1}{k_{\gamma,B}} = \frac{\hbar^3 c}{m_{0,e}e^4} = \frac{r_{c,e}}{\alpha^2} = \frac{\xi_e}{\alpha^{3/2}}.$$

We can express Rydberg constant through vacuum parameters:

$$R = \frac{m_{0,e}e^4}{2\hbar^3} = \frac{1}{2}\omega_{0,e}\alpha^2 = \frac{\alpha^{3/2}}{2\tau_e}.$$

It follows from the above expressions that fine structure constant characterizes not only the fine structure of the hydrogen atomic spectrum but the entire lengths hierarchy of the quantum mechanics as well. It is easy to see that characteristic lengths form a geometrical progression:

$$r_{0,e} = \alpha r_{c,e} = \alpha^2 a_B = \alpha^3 \lambdabar_{\gamma,B}.$$

All the lengths contain neither Planck constant, nor mass, nor charge of electron. In this connection, it makes sense to express the Schrödinger equation through the natural parameters of physical vacuum.

The Hamiltonian for the Schrödinger equation for a hydrogen atom looks like this:

$$H = -\frac{\hbar^2}{2m_{e,0}}\nabla^2 - \frac{e^2}{r}$$

In this expression, we take the fundamental constants $\hbar, m_{e,0}, e$, which specify the characteristic parameters of a hydrogen atom (47, 48), as independent; however, as we demonstrated above, none of these constants ought to be taken as a fundamental one.

We can write the Hamiltonian of the electron in the nuclear field of a hydrogen atom in a different form:

$$H = \hbar\sqrt{\omega_{0,e}^2 + c^2k^2} - \frac{e^2}{r} \quad (49)$$

The Planck constant expressed through the electron charge reduces (49) to the form

$$H = \frac{e^2\omega_{0,e}}{\alpha c}\sqrt{1 + \frac{c^2k^2}{\omega_{0,e}^2}} - \frac{e^2}{r} = \frac{eg}{r_{c,e}}\left(\sqrt{1 + r_{c,e}^2k^2} - \frac{\alpha r_{c,e}}{r}\right) \quad (50)$$

Here, it is convenient to use the dimensionless length $r \Rightarrow r/r_{c,e}$, the dimensionless wave vector $\boldsymbol{k} \Rightarrow \boldsymbol{k}\cdot r_{c,e}$ and the dimensionless time $t \Rightarrow \omega_{0,e}t$. We express the energy in terms of the electron rest energy $eg/r_{c,e}$:

$$H = \sqrt{1+k^2} - \frac{\alpha}{r} \approx 1 + \frac{1}{2}k^2 - \frac{\alpha}{r} \quad (51)$$

The particle velocity is equal to the group velocity of the wave packet $\boldsymbol{v} = \partial\omega_k/\partial\boldsymbol{k} = \boldsymbol{k}/\sqrt{1+k^2} \approx \boldsymbol{k}$ and we express it in terms of light velocity. Approximate expressions correspond to the case of a low velocity $k \approx v \ll 1$. We can regard the value k^2 in the approximate expression (51) as the eigenvalue of the Laplacian operator; then we may reduce (51) to the equation for the eigenfunction and the eigenvalue:

$$H\psi = \varepsilon\psi; \quad H = -\frac{1}{2}\nabla^2 - \frac{\alpha}{r} \quad (52)$$

From (52) it follows that the Schrödinger equation only contains one dimensionless small parameter α of a physical vacuum susceptibility. The fundamental function ψ of a free electron in Cartesian coordinates is equal to $\exp(i\boldsymbol{kr})$; we express the eigenvalue by the equality: $\varepsilon = k^2/2$.

Now we find out the Bohr quantization conditions for a hydrogen atom. The circular motion of electron around an atomic nucleus is defined by the equality of centrifugal and centripetal forces:

$$r\dot{\varphi}^2 = \frac{\alpha}{r^2}. \quad (53)$$

Bohr assumed a quantization of adiabatic invariants:

$$\oint p_i dq_i = nh$$

For the circular motion, the latter relation reduces to the form:

$$pr = n\hbar$$

Externally, it looks as if a quantum of action existed, that provides quantization of a pulse moment. However, by taking into consideration the pulse $p = \hbar k$, we come to the cyclic boundary conditions for a wave vector:

$$kr = n.$$

It follows that Planck constant has nothing to do with forming the wave function. Since $k = v = r\dot{\varphi}$, we can add to Equation (53):

$$r^2\dot{\varphi} = n$$

From where we can obtain the energy, radius and velocity at the stationary Bohr orbits:

$$\varepsilon_n = -\frac{\alpha^2}{2n^2};\ r_n = \frac{n^2}{\alpha};\ v_n = \frac{\alpha}{n}.$$

That accurately corresponds to the relations (47, 48).

Compton scattering, which we regard as one of the evidences proving existence of quantum of action, proceeds from the laws of conservation of energy and momentum for electron and γ- quantum:

$$m_{e,0}c^2 + \hbar\omega_\gamma = \frac{m_{e,0}c^2}{\sqrt{1-\frac{v^2}{c^2}}} + \hbar\omega'_\gamma;$$
$$\hbar\boldsymbol{k}_\gamma = \frac{m_{0,e}\boldsymbol{v}}{\sqrt{1-\frac{v^2}{c^2}}} + \hbar\boldsymbol{k}'_\gamma. \quad (54)$$

It follows from (54), that we can specify the wave vector of a scattered light by the relation:

$$k'_\gamma = \frac{k_\gamma}{1+\frac{\hbar k_\gamma}{m_{e,0}c}(1-\cos\theta)}, \quad (55)$$

Here θ is the angle between vectors $\boldsymbol{v}$ and $\boldsymbol{k}'$; besides, there is a length parameter $\lambda\!\!\!^{-}_{K,e} = \hbar/m_{e,0}c$ where we take the values $\hbar, m_{e,0}, c$ as the fundamental ones. However, by taking into consideration the fact, that relations (5) and (6) define the spectrum of a particle, we can reduce combined Equations (54) to the form:

$$\omega_{e,0} + ck_\gamma = \sqrt{\omega_{e,0}^2 + c^2k^2} + ck'_\gamma;$$
$$\boldsymbol{k}_\gamma = \boldsymbol{k} + \boldsymbol{k}'_\gamma,$$

It follows that the scattering characteristic is defined neither by the Planck constant nor by the electron mass, but by the space and frequency resonance for the wave packets; scattering being submitted to the same formula (55) with the Compton length $\lambda\!\!\!^{-}_{K,e} = c/\omega_{e,0} = r_{c,e} = \xi_e\sqrt{\alpha}$ equal to the correlation radius.

Once in his days Planck supposed that radiation and absorption of light should proceed by quanta. Later this brilliant supposition was confirmed. After that, scientists had only to examine the properties of electron responsible for light radiation and absorption in a quantum way. Albert Einstein, however, considered something different. Since we can observe light quanta, then light is quantized due to existence of quantum of action; the question "Why?" being quite inappropriate here since physical mechanism for quantization of action just does not exist. We can only say that these are the properties of spacetime. We just substitute one senseless statement by another one. Nevertheless, proceeded from the fact that electron radiates and absorbs light per quanta, a planetary model of electron is suggested by itself. The electron rest energy equals to: $\varepsilon_{0,e} = \hbar\omega_{0,e} = eg/r_{c,e}$. We can write γ- quantum energy in a similar way: $\varepsilon_\gamma = \hbar\omega_\gamma = egk_\gamma$. Since the photon spin equals to $\hbar$, then, by representing it in the form of the orbital moment $s = pr = \hbar k_\gamma r_\gamma = \hbar$, we come to quite transparent cyclic conditions for the radius of photon orbit $k_\gamma r_\gamma = 1$. After that, the photon energy reduces to the form: $\varepsilon_\gamma = eg/r_\gamma$. We can obtain such an energy as follows: use the solution for the electron polarization in the form (28), set it up into Hamiltonian (7) and integrate over space from infinity to the radius r_γ. Therefore, the nature of fields for photon and electron is the same. By radiating photon, an electron takes off some part of its polarization coat, the intrinsic energy of the electron being reduced.

3. Gravitational Optics

In the previous part we showed that all particles can be considered as excitations of physical vacuum; they are the solutions of the unified system of equations for coupled oscillations of the multicomponent order parameter $(\boldsymbol{P},\boldsymbol{M})$. That is why we can be sure to a certain degree that all particles similarly contribute to the gravitational interaction, particle energy being the interaction parameter. Now we write down the standardized form of the Hamiltonian for a particle in the gravitational field caused by a massive body of mass m_1

$$H = \varepsilon(\boldsymbol{p}) - \frac{\gamma m m_1}{r} \quad (56)$$

We express the particle mass through energy $m = \varepsilon(\boldsymbol{p})/c^2$; after that the Hamiltonian (56) reduces to the form:

$$H = \varepsilon(\boldsymbol{p})\left(1-\frac{r_g}{r}\right);\ r_g = \frac{\gamma m_1}{c^2}. \quad (57)$$

Here r_g is the gravitational radius which scales gravitational potential of a massive body. For an arbitrary potential, the Hamiltonian has the form:

$$H = \varepsilon(p)(1-\Phi(\boldsymbol{r})).$$

In general case, particle energy is defined by the following expression:

$$\varepsilon(\boldsymbol{p}) = \sqrt{\varepsilon_0^2 + c^2\boldsymbol{p}^2} \quad (58)$$

In addition, particle velocity equals to

$$\boldsymbol{v} = \frac{\partial\varepsilon(\boldsymbol{p})}{\partial\boldsymbol{p}} \quad (59)$$

From the coordinate system (x,y,z,t) we proceed to a new time $t \Rightarrow ct$ and, in Equation (58)—to a new momentum $c\boldsymbol{p} \Rightarrow \boldsymbol{p}$; then the particle velocity does not depend on the chosen scales of length and time, but becomes a dimensionless value expressed in terms of light velocity:

$$\boldsymbol{v}=\frac{\partial\varepsilon(\boldsymbol{p})}{\partial\boldsymbol{p}}=\frac{\boldsymbol{p}}{\sqrt{\varepsilon_0^2+\boldsymbol{p}^2}} \tag{60}$$

The second equation that defines the particle motion in the gravitational field looks like this:

$$\dot{\boldsymbol{p}}=-\nabla H=\varepsilon(\boldsymbol{p})\nabla\Phi \tag{61}$$

By taking into account equation (60), we can rewrite (61) as follows:

$$\frac{\mathrm{d}}{\mathrm{d}t}(\boldsymbol{v}\varepsilon(p))=\varepsilon(p)\nabla\Phi(\boldsymbol{r}) \tag{62}$$

For low velocities we can substitute value $\varepsilon(p)$ by an approximate expression ε_0; after that Equation (62) reduces to that of Newton's mechanics:

$$\ddot{\boldsymbol{r}}=\nabla\Phi(\boldsymbol{r}).$$

Based upon this equation, Albert Einstein affirmed that the inertial mass and the gravitating mass are equivalent. This statement, however, is incorrect. An accurate equation of motion (62) is transformed to:

$$\ddot{\boldsymbol{r}}=\nabla\Phi(\boldsymbol{r})-\boldsymbol{v}(\boldsymbol{v}\nabla\Phi(\boldsymbol{r})) \tag{63}$$

It follows, that particle inertia depends on the direction of motion. It is interesting to note that the intrinsic (internal) energetic properties of a particle are lost in the equation of motion (63). This means that we can apply the obtained equation to any relatively compact object. It can be a planet, a satellite, an electron, a proton, a photon, a neutrino—all the same.

Bearing in mind (60), we reduce (61) as follows:

$$\frac{\boldsymbol{p}\dot{\boldsymbol{p}}}{\varepsilon_0^2+p^2}=\boldsymbol{v}\cdot\nabla\Phi(\boldsymbol{r}),$$

From the latter equation we obtain the integral of motion in two different forms:

$$\begin{aligned}&\varepsilon(p)\exp(-\Phi)=\text{const};\\&(1-v^2)\exp(2\Phi)=\text{const}.\end{aligned} \tag{64}$$

Now we examine the motion in the Coulomb potential with the Hamiltonian (57). In a centrally symmetrical field motion develops in a plane crossing the centre of a massive body; therefore, we can re-write Equation (63) for the plane (x,y):

$$\begin{aligned}\ddot{x}&=-\frac{r_g}{r^3}(x-\dot{x}(x\dot{x}+y\dot{y}));\\ \ddot{y}&=-\frac{r_g}{r^3}(y-\dot{y}(x\dot{x}+y\dot{y})).\end{aligned} \tag{65}$$

In the polar coordinate system Equations (65) becomes

$$\begin{aligned}\ddot{r}-r\dot{\varphi}^2&=-\frac{r_g}{r^2}(1-\dot{r}^2);\\ r\ddot{\varphi}+2\dot{r}\dot{\varphi}&=\frac{r_g}{r}\dot{r}\dot{\varphi}.\end{aligned} \tag{66}$$

The second equation in (66) can be integrated easily; after that we obtain the integral of motion corresponding to the momentum conservation law:

$$r^2\dot{\varphi}\exp\left(\frac{r_g}{r}\right)=\text{const}. \tag{67}$$

At the beginning, we consider a circular motion: $\ddot{r}=\dot{r}=0$. Then, the first equation of (66) leads to

$$r\dot{\varphi}^2=\frac{r_g}{r^2};\quad \dot{\varphi}=\frac{r_g^{1/2}}{r^{3/2}}, \tag{68}$$

This exactly coincides with the results of Kepler's problem, the first space velocity on the orbit of radius r being equal to

$$v_1=r\dot{\varphi}=\sqrt{\frac{r_g}{r}} \tag{69}$$

Consequently, the first space velocity attains to the light velocity at $r=r_g$.

Further, we consider an arbitrary motion relative to a heavy centre. Let us assume that at time t = 0, a particle has coordinates $(r=r_0,\varphi=0)$, complete velocity v_0 and azimuth velocity $v_{\varphi0}=r_0\dot{\varphi}_0$. From the integrals of motion (64, 67) it follows:

$$\begin{aligned}&1-v^2=1-\dot{r}^2-r^2\dot{\varphi}^2=(1-v_0^2)\exp\left(2\left(\frac{r_g}{r_0}-\frac{r_g}{r}\right)\right);\\&r^2\dot{\varphi}=r_0^2\dot{\varphi}_0\exp\left(\frac{r_g}{r_0}-\frac{r_g}{r}\right).\end{aligned} \tag{70}$$

From the system of equations (70) we obtain the equation that combines φ and r:

$$\mathrm{d}\varphi=\frac{v_{\varphi0}\mathrm{d}r}{r\sqrt{\left(\frac{r}{r_0}\right)^2\left(\exp\left(2\left(\frac{r_g}{r}-\frac{r_g}{r_0}\right)\right)-1+v_0^2\right)-v_{\varphi0}^2}} \tag{71}$$

Now proceed to a new variable $\rho=r/r_0$ and new parameters of the problem:

$$\beta=r_g/r_0;\ v_{\varphi0}=\sqrt{\beta(1+\delta)}. \tag{72}$$

We examine the situation when the radial velocity at the starting point is zero. It follows that $v_0=v_{\varphi0}$; after this, Equation (71) acquires the form

$$\mathrm{d}\varphi=\frac{\sqrt{\beta(1+\delta)}\mathrm{d}\rho}{\rho\sqrt{\rho^2\left[\exp\left(2\beta\left(\frac{1}{\rho}-1\right)\right)-1+\beta(1+\delta)\right]-\beta(1+\delta)}} \tag{73}$$

Here, the parameter δ defines a deviation from the circular motion in an orbit. We apply the Equation (73) to the Solar system. The gravitational radius of the Sun equals to 1.5 km. The radius of the terrestrial orbit is 1.5 × 10^8 km, the radius of Mercury orbit is 0.5 × 10^8 km, the radius of the solar sphere is 6. 96 × 10^5 km. The parameter β in (73) is equal to $\beta_E = 10^{-8}$ for the Earth planet; to $\beta_M = 3\times10^{-8}$ for the Mercury; and to $\beta_S = 2.1\times10^{-6}$ for the Sun surface. It follows that the circular orbital velocity of the Earth is $v_{\varphi E} = \sqrt{\beta_E} = 10^{-4}$. In dimensional terms the circular velocity of the Earth equals to $10^{-4}c = 30\ \text{km/sec}$ The Mercury moves in an elliptic orbit according to (73), where the value $\beta \ll 1$. Second order expansion in series of the exponent (73) leads to the equation

$$\varphi = \int_1^\rho \frac{d\rho\sqrt{1+\delta}}{\rho\sqrt{-(1-\delta-2\beta)\rho^2 + 2(1-2\beta)\rho - (1+\delta-2\beta)}}, \tag{74}$$

It enables to obtain the orbit path

$$\rho = \frac{1+\delta-2\beta}{1-2\beta+\delta\cos\left(\varphi\sqrt{\frac{1+\delta-2\beta}{1+\delta}}\right)} \tag{75}$$

We can find the complete revolution of the path from the condition:

$$\varphi\sqrt{\frac{1+\delta-2\beta}{1+\delta}} = 2\pi$$

Consequently, the angle gain over one revolution of the path is

$$\varphi \approx 2\pi\left(1+\frac{\beta}{1+\delta}\right).$$

The century displacement of the Mercury perigee means that while the Earth makes 100 revolutions around the Sun, the Mercury makes the number of revolutions equal to $100(\beta_M/\beta_E)^{3/2}$. From here we obtain

$$\Delta\varphi = 100\frac{2\pi\beta_M}{1+\delta_M}\left(\frac{\beta_M}{\beta_E}\right)^{3/2} = 21'' \tag{76}$$

The value $\delta_M = 0.2$ is the eccentricity of the Mercury elliptic orbit.

From (75) we can obtain the condition when an elliptic orbit transforms into a parabolic path:

$$1-2\beta-\delta = 0.$$

It follows that the second space velocity is a little less than that of Kepler's problem and is equal to

$$v_2 = v_{\varphi 0} = \sqrt{\beta(1+\delta)} = \sqrt{2\beta(1-\beta)} = v_1\sqrt{2(1-v_1^2)} \tag{77}$$

Further, we consider the motion of a photon or a neutrino in a gravitational field. In this case, for the equation of motion (71), it is necessary to assume $v_0 = v_{\varphi 0} = 1$. Then, Equation (71) leads to

$$\varphi = \int_1^\rho \frac{d\rho}{\rho\sqrt{\rho^2\exp\left(2\beta\left(\frac{1}{\rho}-1\right)\right)-1}} \tag{78}$$

In order to calculate the complete angle of displacement φ_β for a light beam passing a gravitating mass, we move to a new variable $\xi = 1/\rho$ and, as a result, we obtain:

$$\varphi_\beta = \int_0^1 \frac{2d\xi}{\sqrt{\exp(2\beta(\xi-1))-\xi^2}} \tag{79}$$

Integral (79) is divergent at $\beta \to 1$. It proceeds from the fact that at the gravitational radius a photon has a stationary orbit. For $\beta \ll 1$.

$$\varphi_\beta \approx \pi + 2\arcsin\beta, \tag{79a}$$

it follows that the deviation of a light beam moving, for example, along the Sun surface is $0.86''$. The only stationary orbit for a photon is $r = r_g$ that corresponds to the parameter $\beta = 1$. The slightest deviation from unit makes a photon either leave for infinity, or fall down to the centre. **Figure 1** illustrates a photon getting off a stationary orbit.

Now we study a radial motion which we can determine from the integrals of motion (68). Under the given input conditions of the coordinate and velocity directed along the radius, and by using the integrals of motion (68), we obtain the energy of a photon moving away from the

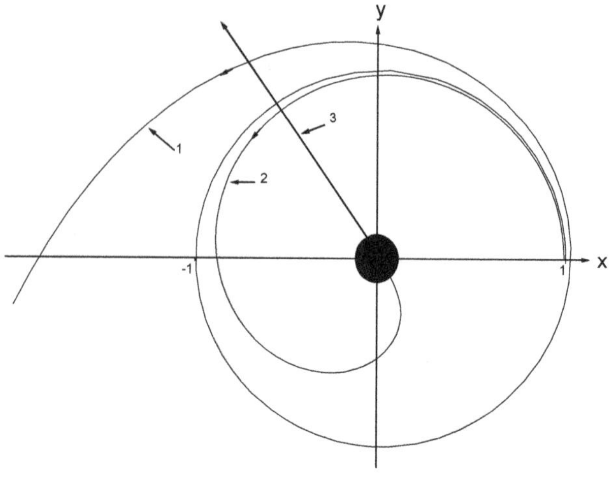

Figure 1. The paths of a photon at different initial conditions. Curve 1 exhibits the photon leaving for infinity at the input condition $\beta = 0.9999$. Curve 2 shows the photon falling down to the centre at $\beta = 1.01$ Arrow 3 displays the photon radially leaving for infinity from under the gravitational radius.

centre:

$$\varepsilon(p) = p = p_0 \exp\left(-\beta\left(1-\frac{1}{\rho}\right)\right) \tag{80}$$

It follows that the photon crosses freely the gravitational radius and at the infinity the photon energy equals to:

$$p = p_0 \exp(-\beta) = p_0 \exp\left(-\frac{r_g}{r_0}\right) \tag{81}$$

The radial velocity of the photon remains constant:

$$v_x = \frac{p_x}{p};\quad v_y = \frac{p_y}{p};\quad v^2 = v_x^2 + v_y^2 = \frac{p_x^2 + p_y^2}{p^2} = 1.$$

The velocity of particles with non-zero mass is defined by the equation:

$$\left(1-v^2\right)\exp\left(2\frac{r_g}{r}\right) = \left(1-v_0^2\right)\exp\left(2\frac{r_g}{r_0}\right) \tag{82}$$

From (82) we can define the second space velocity:

$$v_0^2 = v_2^2 = 1-\exp(-2\beta), \tag{83}$$

it follows from (83) that at the initial velocity $v_0 > v_2$, any particle crosses freely the gravitational radius and leaves for infinity. Note that the first space velocity v_1 equals to $\sqrt{\beta}$. A circular orbit is steady under the condition that $v_2 > v_1$. From the equation

$$\beta = 1-\exp(-2\beta)$$

we define the boundary of stability for circular orbits $\beta = \beta_c \approx 0.796812$. Circular orbits are only stable to small disturbances under the condition $\beta < \beta_c$. This situation is described by the equation of motion (73) where we can consider the value of δ as a disturbance of a circular orbit. It follows from (73) that for $\beta > \beta_c$ any small value $\delta > 0$ makes a particle leave for infinity along the path similar to that shown on **Figure 1** (curve 1). Under the disturbance $\delta < 0$, a particle falls down to the centre and as well leaves for infinity along the curve similar to 2, 3 on **Figure 1**.

Since all bodies in the Solar system obey the same equation of motion (66), we can measure time in terms of any periodical process that occurs in the Solar system; for example, in terms of revolution of the Earth around the Sun. Further, since we can calculate the periods of revolution for any bodies beforehand, time in the entire Solar system runs similarly. Moreover, we extend the time over the entire visible part of the Universe; and we are quite right when we measure time in billions of years, whereas we measure distance in billions of light years.

Therefore, following Newton, we can repeat that a particle moves uniformly and straight until no force is applied. Following Galilee, we can say that under the same initial conditions in the gravitational field all particles move along the same paths. For example, under the same initial conditions an ultra relativistic proton moves in the same path as a photon does. However, Einstein's statement that time runs differently in each lift does not have any physical meaning, since every electron covers the entire infinite space (33) and simultaneously interacts with all particles in the Universe.

4. Problem of the Dark Matter

In the previous section we introduced the concept that it is the total particle energy $\varepsilon(\boldsymbol{p})$ which plays the key part in the gravitational interaction, but not the rest mass, as it is usually considered. This fact substantially changes the estimations of the matter quantity participating in the gravitational interaction. For example, the protons whose energy achieves 10^{21} eV in cosmic rays create a gravitational potential 10^{12} times higher than that for protons on the Earth whose energy is 10^9 eV. The situation is similar for neutrino. The mean energy of neutrino emitted by neutron beta decay is about 10^6 eV; whereas zero energy of neutrino, which we usually take into account for gravitational interaction, is estimated by value of 10 eV. Consequently, neutrino contributes into the gravitational interaction 10^5 times more. Photons having zero mass are not considered as carriers of the gravitational interaction at all. Deviation from the straight motion for a photon is caused by the Einstein deflection effect. This point of view contradicts elementary physics. The thing is that, if two bodies exist at positions $\boldsymbol{r}_1$ and $\boldsymbol{r}_2$, and interact according to the law $U(\boldsymbol{r}_1 - \boldsymbol{r}_2)$, then their momenta $\boldsymbol{p}_1$ and $\boldsymbol{p}_2$ follow the equations

$$\dot{\boldsymbol{p}}_1 = -\nabla_{r_1} U(\boldsymbol{r}_1 - \boldsymbol{r}_2);\quad \dot{\boldsymbol{p}}_2 = -\nabla_{r_2} U(\boldsymbol{r}_1 - \boldsymbol{r}_2). \tag{84}$$

Since

$$\nabla_{r_1} U(\boldsymbol{r}_1 - \boldsymbol{r}_2) = -\nabla_{r_2} U(\boldsymbol{r}_1 - \boldsymbol{r}_2), \tag{85}$$

then, as a consequence of (84 – 85), follows the law of total momentum conservation:

$$\frac{\mathrm{d}}{\mathrm{d}t}(\boldsymbol{p}_1 + \boldsymbol{p}_2) = 0;\quad \boldsymbol{p}_1 + \boldsymbol{p}_2 = \text{const}. \tag{86}$$

Thus, the distortion of the trajectory for a photon passing e.g. the Sun shows (demonstrates) the variation of its momentum; it follows from the law of the total momentum conservation that the momentum of the Sun changes by the same amount. We can make an obvious conclusion: if a photon is attracted to a massive body, then the massive body is attracted to the photon to the same extent. Therefore, photons, like any other particles, participate in the gravitational interaction, interaction intensity being proportional to the proper intrinsic energy of the particle: $\varepsilon = \hbar\omega = p$.

The azimuth velocity of stars in galaxies is about 100 -

200 km/sec. That is why, the dark matter elements belonging to a certain galaxy at first sight may seem to have the same velocities. Hence, all relativistic particles, such as photons, neutrino, and cosmic rays, are beyond our consideration; as a result, practically none of the observed particles can create an additional gravitational field. In this connection an idea arises that there are heavy cold particles contributing only to the gravitational interaction; they are called dark matter.

However, a possible alternative point of view exists. First, we examine a simple example. A charged ion of a hydrogen atom creates a Coulomb potential where localized states for an electron are formed. Filling up one of the localized states makes the hydrogen atom electrically neutral, as the nuclear field is completely screened by an electron. On the other hand, if we insert a proton into a metal where there is a sea of free electrons, the localized state does not occur, but this time the nuclear field is screened by free electrons. The trajectory of each electron is distorted near the nucleus so much, that, as a result, electron density increases exactly to the same extent and it screens the nuclear field completely. A positive charge interacts with all free electrons of metal in a Coulomb way and attracts them.

Any heavy body attracts all free particles of a cosmic space by the gravitational interaction in a Coulomb way as well. Nevertheless, there is a significant difference between these two processes. Free electrons of metal are attracted to a positive charge, begin repulsive from each other, as a consequence, the electrical field of the positive charge is screened by electrons. The situation is quite opposite with the gravitational interaction. A massive body attracts particles from the surrounding space. Due to this attraction the total gravitational potential increases, thereby increasing the particle attraction even more. A positive feedback or antiscreening arises that can lead to the system instability. As an illustration, we examine the both situations: screening of an electrical field by free electrons in metal and antiscreening of a gravitational field by free particles (any) in cosmic space.

An external charge with harmonics $\rho_{ext}(k)$ placed into a metal creates a real charge $\rho_i(k)$ defined as a sum of external and induced charges:

$$\rho_i(k) = \rho_{ext}(k) + \rho_{res}(k) \tag{87}$$

We can express the induced charge through the polarizability of electrons in metal $\rho_{res}(k) = -\chi(k)\rho_i(k)$.

Here $\chi(k) = k_{TF}^2/k^2$; k_{TF} is a characteristic wave vector calculated using a Thomas-Fermi approximation [9]. As a result, we obtain

$$\rho_i(k) = \frac{\rho_{ext}(k)}{1+\chi(k)} = \frac{k^2}{k_{TF}^2 + k^2}\rho_{ext}(k); \quad k_{TF}^2 = \frac{4k_F}{\pi a_B}, \tag{88}$$

Here k_F is a Fermi momentum in metal. It follows from (88) that a Coulomb potential of a point charge q, for example, is transformed into a screened potential:

$$\Phi_{ext} = \frac{q}{r}; \quad \Phi_i = \frac{q}{r}\exp(-k_{TF}r).$$

Now we consider a situation rather close to the gravitational interaction. Suppose, the entire space is filled up with neutral particles that have some homogeneous density ρ_0 and interact according to the law of gravitation. If any density fluctuation $\rho_{ext}(k)$ occurs in the space, then, owing to the gravitational interaction, all other particles begin to adjust to this density; there-after we can re-write the real density in the form:

$$\rho_i(k) = \rho_{ext}(k) + \rho_{res}(k) = \rho_{ext}(k) + \chi_\gamma(k)\rho_i(k) \tag{89}$$

On the analogy of a free electrons susceptibility, we imagine a gravitational susceptibility $\chi_\gamma(k)$ like this: $\chi_\gamma = k_0^2/k^2$. Here k_0 depends on the value of ρ_0 and on the distribution function of the particle velocity. Afterwards, the real density acquires the form:

$$\rho_i(k) = \frac{k^2}{k^2 - k_0^2}\rho_{ext}(k)$$

This causes gravitational instability of the system relative to the long-wave density fluctuations. As fluctuations develop, slow particles, which compose a small part of an average density, are pulled out of the surrounding space and transformed into clusters of matter in form of stars and galaxies. Fast relativistic particles remain free and continue to participate in creating an additional gravitational field. We denote clusters of a cool matter in form of stars and galaxies having finite motion as $\rho_{cold}(k)$. The remainder relativistic particles in form of cosmic rays, photons and neutrino create additional non-homogeneous matter density due to the trajectory distortion $\rho_{rel}(k) = \chi_{rel}(k)\rho_i(k)$. Here $\chi_{rel}(k)$ is the gravitational polarizability of the relativistic particles. Thus, the total density is equal to

$$\rho_i(k) = \frac{\rho_{cold}(k)}{1-\chi_{rel}(k)}.$$

Being on Earth, we have no possibility to scan the distribution of a total energy over the entire space. However, judging from the fact that the azimuth velocity of stars moving away from the centre of galaxy remains nearly constant, the total gravitational potential must have the form:

$$\Phi_i(r) = \eta \ln\left(\frac{r}{R_c}\right),$$

Here η is a dimensionless parameter, R_c —a gravitational size of a space belonging to a certain galaxy.

Provided that the centrifugal and centripetal forces are equal

$$r\dot{\varphi}^2=\frac{\partial \Phi_i(r)}{\partial r}=\frac{\eta}{r}$$

we come to the expression for the circular velocity:

$$v=r\dot{\varphi}=\sqrt{\eta}.$$

At the star velocity being approximately equal to 200 km/sec we obtain the value $\eta=8\times10^{-7}$. From the expression for the potential and with the aid of the Poisson equation we obtain the space distribution density of matter:

$$\Delta\Phi_i=\frac{\eta}{r^2}=\frac{4\pi\gamma}{c^2}\rho_i(r);$$

$$\rho_i(r)=\frac{c^2}{4\pi\gamma}\frac{\eta}{r^2}.$$

The space integral of density provides a value of mass inside a sphere of radius R:

$$M_i=\frac{\eta c^2 R}{\gamma}=\frac{v^2 R}{\gamma}.$$

For our Galaxy having the size of about $R=5\times10^4$ light years and velocity of $v=200$ km/sec we obtain $M_i=10^{45}$ *gr*. Mass of cool matter is estimated by value $M_{cold}=4\times10^{44}$ *gr*., therefore, mass of a relativistic matter is comparable with that of the cool one $M_{rel}\approx M_{cold}$. Thus, as a result of the trajectory distortion for relativistic particles, an additional nonhomogeneous distribution of relativistic matter occurs and, consequently, an additional gravitational potential as well. That is why, there is no need to search for a mystical dark matter; relativistic energy is quite sufficient to create an additional gravitational field. Moreover, emission of radiation by stars and galaxies as well as supernova outburst lead to the constant growth of relativistic energy in space. So, observations of the azimuth stellar motion both in galaxies and galaxies in clusters point to the existence of an additional gravitational field. Since azimuth and radial motion follows from the general equation of motion, for example in form (63), the radial motion is submitted to the same additional gravitational attraction; for this reason, there is no dark energy to create antigravitation [10,11]. Thus, if red shift is related to recession of galaxies, then a contradiction arises, because galaxies have to scatter with acceleration but, judging from the azimuth motion, this is impossible.

It is more natural to consider atomic spectra of far stellar radiation to be time dependent as a consequence of time dependence of physical vacuum parameters. Since atomic levels are proportional to Bohr energy, and Bohr energy, in turn, is proportional to the rest energy of electron $(\varepsilon_B=\alpha^2\varepsilon_{0e})$ we can affirm that the electron mass increases with time; this means that vacuum parameters for the electron oscillation branch $\tau_e(t)$ and $\xi_e(t)$ decrease with time. The Hubble constant can be defined from the following expression:

$$H(t)=\frac{\mathrm{d}}{\mathrm{d}t}\left(\frac{m_{e0}(t)}{m_{e0}(t=0)}\right)=\frac{\mathrm{d}}{\mathrm{d}t}\left(\frac{\tau_e(t=0)}{\tau_e(t)}\right)$$
$$=\frac{\mathrm{d}}{\mathrm{d}t}\left(\frac{\omega_{0e}(t)}{\omega_{0e}(t=0)}\right). \qquad (90)$$

The red shift indicates that the Hubble constant is a monotonically growing time function and at the present moment it equals to $2.5\cdot10^{-18}$ sec^{-1}.

Nowadays the Universe is in a metastable state, energy emission transitions occurring in two opposite directions. On one hand, nucleosynthesis of light nuclei—takes place, which is the source of stellar energy. On the other hand, nuclear disintegration of heavy nuclei (natural radioactivity)—occurs, as well with energy emission. From today's point of view, nuclear fusion looks quite natural as there is a binding energy between nuclei; moreover, the binding energy on one nucleus increases with the growth of atomic number up to iron. Creation of heavier elements turns out to be less gainful; in this connection it is a surprise that heavy elements, up to uranium, exist on Earth. Nuclei of uranium are in metastable state. If we launched a piece of uranium towards the Sun, the uranium nuclei, under neutron bombardment, would decompose into lighter fragments. This means that uranium cannot occur on Sun. Deposits of uranium on Earth, however, prove that the Earth is an earlier formation than the Sun. Chemical composition of the Earth principally differs from that of the Sun. Sun consists of 75% hydrogen, 24% helium and a negligibly small amount of heavier elements, whereas Earth consists of 32% iron, 30% oxygen and a noticeable amount of heavier elements up to uranium. Heavy elements existing on Earth, as well as the red shift, point out to nonstationarity of physical vacuum parameters. Heavy elements could only occur on Earth when they were energetically gainful; variations of physical vacuum parameters led to the transition of heavy nuclei into a metastable state. A further evidence of nonstationarity of physical vacuum parameters is that not only stars, but also planets emit energy; moreover, volcanic activity, similar to that on the Earth, is still being observed on Jupiter satellites. It is known that the Jupiter emits twice as much energy as it receives from the Sun. We can express the Jupiter energy emission via the Hubble constant. From the law of conservation of energy it follows:

$$\frac{\mathrm{d}}{\mathrm{d}t}\left(M_J c^2\right)+L_J=0. \qquad (91)$$

Here $M_J c^2$—is Jupiter energy of 1.8×10^{51} erg, and

L_J is the integral emission flux of 6.5×10^{25} erg/sec. By dividing both parts of (91) into Jupiter energy, and considering that the Jupiter only consists of hydrogen, we can reduce (91) to form

$$\frac{d\varepsilon_{0e}/dt + d\varepsilon_{0n}/dt}{\varepsilon_{0e} + \varepsilon_{0n}} + l_J = 0. \quad (92)$$

Here l_J is specific luminosity of the Jupiter which is equal to 3.6×10^{-26} sec^{-1}. Taking into consideration the definition of the Hubble constant (90), we can re-write Equation (92) as follows:

$$\frac{d\varepsilon_{0n}/dt}{\varepsilon_{0n}} = -\frac{\varepsilon_{0e}}{\varepsilon_{0n}} H(t) - l_J, \quad (93)$$

All values at the right part of (93) are known, therefore,

$$\frac{d\varepsilon_{0n}}{dt} \approx -\frac{d\varepsilon_{0e}}{dt}.$$

It can be seen from this expression, that the eigen frequencies of electron and proton vary practically with the same rate, with their values approaching each other. Let us present, for the illustration, a table of a specific luminosity of a number of objects (**Table 1**).

Table 1 shows that the specific luminosity of the Sun is maximum, which is natural due to thermonuclear fusion. The specific luminosity of Jupiter and Saturn are the same, due to their similar composition of light atoms of hydrogen and helium. The specific luminosity of the Earth is much less than that of Jupiter. This can be attributed to the fact that the Earth is composed of heavier elements, and its nuclear energy is determined not only by the energy of the protons, but the binding energy between protons and neutrons.

So, the red shift shows that the rest energy of the electron is growing with time, whereas emission of radiation by planets indicates that the rest energy of the proton is decreasing. The total change of the energy for the electron and proton is so great, that it leads to planet heating and emission of radiation.

Since physical vacuum has existed eternally, the values, which characterize the vacuum, can only be of two types: either time independent constants, or oscillating functions. The fundamental values are general for both electron and nuclear modes $g, \alpha, c = \xi_e/\tau_e = \xi_n/\tau_n$ seem to be thought as constant values; however, we have to consider as time dependent the values, which are characteristic either for an electron mode only by ξ_e, τ_e, or for a nuclear one—by ξ_n, τ_n. At the present moment electron and nuclear frequencies are moving towards each other.

Table 1. A specific luminosity of a number of objects.

	$E = Mc^2$ Erg	L Erg/sec	$l = L/E$ 1/sec
Sun	1.8×10^{54}	3.86×10^{33}	2.14×10^{-21}
Earth	5.4×10^{48}	3.58×10^{20}	6.6×10^{-29}
Jupiter	1.8×10^{51}	6.48×10^{25}	3.6×10^{-26}
Saturn	5.13×10^{50}	1.85×10^{25}	3.6×10^{-26}

Finally, we pay attention to one more mechanism of a gravitational instability. Not coincidentally, there has been some cause for concern so far, that microscopic black holes are possible to occur under the experimental research with Large Hadron Collider in CERN. The thing is that the gravitational attraction between particles grows with increasing particle energy, whereas, the electrical repulsion remains constant due to the law of conservation of charge. In this connection we examine two protons which are speeded up to the certain energy in an accelerator. We write down the Hamiltonian for two protons, taking into account an electrical and gravitational interaction:

$$H = \varepsilon_1 + \varepsilon_2 + \frac{e^2}{|\mathbf{r}_1 - \mathbf{r}_2|} - \frac{\gamma m_1 m_2}{|\mathbf{r}_1 - \mathbf{r}_2|}.$$

Since masses of the particles are proportional to their energy $m_1 = \varepsilon_1/c^2$, $m_2 = \varepsilon_2/c^2$,, then, under the following condition

$$\frac{\gamma \varepsilon_1 \varepsilon_2}{c^4} > e^2$$

the gravitational attraction turns out to exceed the electrical repulsion. Consequently, the gravitational collapse may occur when the particle energy amounts to

$$\varepsilon > \varepsilon_c = \frac{ec^2}{\sqrt{\gamma}} = 10^{27} \text{ eV}.$$

Maximum particle energy in cosmic rays reaches 10^{21} eV. The value of energy expected at the accelerator in CERN is 7×10^{15} eV that is eleven orders less than the critical value. That is why the microscopic black holes are impossible to appear in the accelerator. From the expression for the critical energy, we can define the specific wave vector and the corresponding de Broglie wave length:

$$\varepsilon_c = c\hbar k_c = \frac{c\hbar}{\lambda_c} = \frac{ec^2}{\gamma^{1/2}}$$

It follows:

$$\lambda_c = \frac{e\sqrt{\gamma}}{\alpha c^2} = \sqrt{\frac{\gamma\hbar}{\alpha c^3}} = 1.88\times10^{-32} \text{ cm}$$

Planck introduced the specific length by reason of dimension:

$$r_{Pl} = \sqrt{\frac{\gamma\hbar}{c^3}}.$$

It is easy to see that the specific length λ_c can be expressed via Planck length as follows:

$$\lambda_c = \frac{r_{Pl}}{\sqrt{\alpha}}$$

Thus, the considerations above allow attaching a physical sense to the Planck length, which defines the most probable value for the lattice constant of physical vacuum; here k_c specifies the edge of the Brillouin zone in k—space, and ε_c—the width of the allowed energy region.

5. Conclusion

In presented paper we try to consider problems of the gravitational optics and dark matter developing from the crystal model for the vacuum. The vacuum is represented as a three-dimensional crystal lattice matter with a very small lattice period, much less than 10^{-26} cm. The oscillators are located at the nodes of an infinite lattice. It is shown that an infinite set of equations to describe the coupled oscillations of moving oscillators converges to a system of twelve equations. We have obtained the combined equations for a multicomponent order parameter in the form of the electric and magnetic vacuum polarization, which defines the spectrum and symmetry of normal oscillations in the form of elementary particles. Thus, our model for vacuum is represented as a material medium in which dynamical properties of the crystal specify the spectrum of elementary particles. As we can see from the consideration presented above, the new theory allows describing with a single point of view both the electromagnetic waves and the spectrum of elementary particles as well. Namely we come to a unified field theory. We have obtained the combined equations for a multicomponent order parameter in the form of the electric and magnetic vacuum polarization, which defines the spectrum and symmetry of normal oscillations in the form of elementary particles. We have restored the fundamental parameters of physical vacuum, such as: susceptibility for the electric and magnetic polarization (equal to the constant of fine structure), parameters of length and time for the electron and nuclear branches of the oscillations, correspondingly. We have shown that the charge quantization is directly connected to discreteness of vacuum consisting of particles with the interaction constant equal to the double charge of a Dirac monopole. Elementary particles are excitations of vacuum in a form of wave packets of a soliton type. We have obtained an exact equation of motion for a particle in a gravitational field. Energy defines both gravitational interaction and particle inertia, inertia being of an anisotropic value; that is why the statement, that the inertial and gravitational masses are equivalent, is not correct. We have examined the situation when galaxies are distributed over the entire infinite space according to the cosmological principle. In this case recession of galaxies is impossible; therefore, the red shift of radiation emitted by far galaxies must be interpreted as the blue time shift of atomic spectra. As a consequence, it follows that both rest energy and mass of electron are increasing now. Since physical vacuum exists eternally, vacuum parameters can be either constant or oscillating with time. These are time oscillations of $\xi_e(t), \xi_n(t)$ and $\tau_e(t), \tau_n(t)$ which have caused electron mass growth within recent 15 billion years, inducing red shift; on the contrary, pro- ton mass decreases, responsible for emission of radiation by planets.

REFERENCES

[1] F. Zwicky, "Die Rotverschiebung von Extragalaktischen Nebeln," *Helvetica Physica Acta*, Vol. 6, 1933, pp. 110-127.

[2] K. C. Freeman, "On the Disk of Spiral and Galaxies," *Astrophysical Journal*, Vol. 160, 1970, p. 811. http://dx.doi.org/10.1086/150474

[3] J. A. Tyson, F. Valdes, J. F. Jarvis and A. P. Mills Jr., "Galaxy Mass Distribution from Gravitational Light Deflection," *Astrophysical Journal*, Vol. 281, 1984, pp. L59-L62.

[4] A. G. Riess, A. V. Filippenko, P. Challis, A. Clocchiatti, A. Diercks, P. M. Garnavich, R. L. Gilliland, C. J. Hogan, J. Saurabh, R. P. Kirshner, B. Leibundgut, M. M. Phillips, D. Reiss, B. P. Schmidt, R. A. Schommer, R. C. Smith, J. Spyromilio, C. Stubbs, N. B. Suntzeff and J. Tonry, "Observational Evidence from Supernovae for an Accelerating Universe and a Cosmological Constant," *Astronomical Journal*, Vol. 116, No. 3, 1998, p. 1009. http://dx.doi.org/10.1086/300499

[5] S. Perlmutter, G. Aldering, G. Goldhaber, R. A. Knop, P. Nugent, P. G. Castro, S. Deustua, S. Fabro, A. Goobar, I. M. Hook, M. Y. Kim, J. C. Lee, N. J. Nunes, R. Pain, C. R. Pennypacker and R. Quimby, "Measurements of {OMEGA} and {LAMBDA} from 42 High-Redshift Supernovae," *Astrophysical Journal*, Vol. 517, No. 2, 1999, p. 565. http://dx.doi.org/10.1086/307221

[6] E. Q. Milne, "A Newtonian Expending Universe," *Quarterly Journal of Mathematics*, Vol. 5, No. 1, 1934, pp. 64-72.

[7] E. V. Chensky, "New Consideration of Problems of Gravitational Optics and Dark Matter Based on Crystal Model of Vacuum," *Journal of Electromagnetic Analysis & Applications*, Vol. 2, No. 8, 2010, pp. 495-512. http://dx.doi.org/10.4236/jemaa.2010.28066

[8] P. A. M. Dirac, "Quantised Singularities in the Electromagnetic Field," *Proceedings of the Royal Society A*, Vol. 133, No. 821, 1931, pp. 60-72. http://dx.doi.org/10.1098/rspa.1931.0130

[9] D. Pines, "Elementary Excitations in Solid," W. A. Bejamin, New York, Amsterdam, 1963.

[10] A. D. Chernin, "Dark Energy and Universal Antigravitaition," *Physics-Uspekhi*, Vol. 51, No. 3, 2008, pp. 253-282. http://dx.doi.org/10.1070/PU2008v051n03ABEH006320

[11] V. N. Lukash and V. A. Rubakov, "Dark Energy: Myths and Reality," *Physics-Uspekhi*, Vol. 51, No. 3, 2008, pp. 283-290. http://dx.doi.org/10.1070/PU2008v051n03ABEH006567

Appendix

A1. Determination of the Order Parameters and its Conjugate Forces

In the study presented in the paper above and in [7], an attempt is made to create a non-local field theory to describe elementary particles, which should be taken as the elementary excitations of a particular environment, which is the physical vacuum. Since the discovery of the possibility of creation of particles and anti-particles in the vacuum, it became apparent that the vacuum is not empty space, but the environment with specific dynamic properties. It follows from this that there must be an absolute coordinate system tied to the medium (which is the model of vacuum as the three-dimensional rigid crystal lattice structure) in which the photons as excitations of the medium, propagate with the speed of light. So Einstein's postulate that there is no absolute coordinate system is flawed. Therefore, the second postulate of Einstein's that the "speed of light is independent of the source," is wrong. This situation occurs precisely when the signal is the excitation of the medium, the dynamic properties of which determine the speed. For instance, the speed of sound radiated by the flying aircraft, is not dependent on the aircraft speed v. Notice that the relative velocity of the wave front emitted forward, is equal $c-v$, and for the wave radiated back, the relative velocity is $c+v$. Mathematical equations have the property that if two physical processes are subject to the same equation, the solutions of this equation after renormalizetion of parameters apply to both the first and to the second physical processes. The main feature of any oscillatory process is the existence of equilibrium points, deviation from which induces an oscillatory process. Actually, these points of balance represent the absolute coordinate system in which the excitations, such as photons, travel at the speed of light. Therefore, the following postulate of Einstein "the speed of light in any inertial frame is the same", contradicts the previous postulate. It should be noted that the Lorentz transformations have appeared before the theory of relativity, and they have brought out to explain the negative result of the Michelson experiments for the coordinate system associated with the Earth, which moves relative to the absolute coordinate system tied to the physical vacuum. In the present work it is shown that the electron cloud of the charge moving relative to the absolute coordinate system is indeed deformed in strict accordance with the Lorentz transformations.

The necessity to introduce the well-defined order parameters for the vacuum environment follows from the terminology that we use. For example, in an electrically insulating medium, which is the physical vacuum, the electric charge can only be the result of vacuum polarization:

$$div\boldsymbol{P} = -4\pi\rho. \quad (1)$$

Here, $\boldsymbol{P}(\boldsymbol{r},t)$ the electric polarization of the vacuum, and the quantity $\rho(\boldsymbol{r},t)$ determines the density of the electric charge and is a quite standard definition. Expression (1) is not an equation, but the condition for the onset of the charge density. In the expression (1) a vector $\boldsymbol{P}(\boldsymbol{r},t)$ is an order parameter, which consists of three arbitrary functions:

$$\boldsymbol{P}(\boldsymbol{r},t) = \boldsymbol{e}_x P_x(\boldsymbol{r},t) + \boldsymbol{e}_y P_y(\boldsymbol{r},t) + \boldsymbol{e}_z P_z(\boldsymbol{r},t). \quad (2)$$

The density of the electric charge, $\rho(\boldsymbol{r},t)$ in turn, generates a force, conjugate to the order parameter $\boldsymbol{P}(\boldsymbol{r},t)$, which is called the electric field $\boldsymbol{E}$. It is possible to recover the electric field distribution for an arbitrary distribution of the electric charge density. From the Coulomb interaction between point charges:

$$U = \frac{e_1 e_2}{r} \quad (3)$$

we get:

$$\Phi(\boldsymbol{r}) = \int \frac{\rho(\boldsymbol{r}')\mathrm{d}\boldsymbol{r}'}{[\boldsymbol{r}-\boldsymbol{r}']}; \qquad \boldsymbol{E} = -\nabla\Phi, \quad (4)$$

from which follows the interaction energy:

$$U = \frac{1}{2}\int \rho(\boldsymbol{r})\Phi(\boldsymbol{r})\mathrm{d}\boldsymbol{r} = \frac{1}{8\pi}\int \boldsymbol{E}^2\mathrm{d}\boldsymbol{r}, \quad (5)$$

In addition, from the relationships (4) we have an equality:

$$div\boldsymbol{E} = 4\pi\rho, \quad (6)$$

which is included in the Maxwell equations. However, this is nothing more than a definition for the average density of a set of point charges. From (6) it does not follow neither the character of the Coulomb interaction, nor the interaction energy. Actually we use expressions (3) - (5), from which, after summing over all the charges we obtain both the electric field configuration and the interaction energy.

In the study of any dynamical system, is necessary first of all to determine the order parameter and its conjugate force. For example, in the theory of elasticity the strain tensor is the order parameter and its conjugate force is called the stress tensor. In equilibrium, the strain tensor is related to the stress tensor, as a result of the total energy can be expressed either through the strain tensor, or through the stress tensor with help of the coefficient of elasticity. In dynamic processes, the equation of motion can be written only for the strain tensor (the order parameter), and the stress tensor is decomposed into its own force, returning the system to the equilibrium, and the external force, being determined by the external conditions. The equation of motion for the stress tensor cannot be written.

In electrodynamics, the electric and magnetic fields determine the force field, so this field Maxwell called the stress tensor. By introducing the force field, the force itself cannot exist on its own, because the force has to be applied to something, as the action equals reaction. Therefore, it is first necessary to determine the order parameters associated with these forces. There is not arbitrariness here. If the electric field is generated by the electric charge, then the electric charge in an electrically neutral environment can only be generated by the electric polarization. In ("The Classical Theory of Fields" by L. D. Landau and E. M. Lifshitz, A Course of Theoretical Physics, Pergamon Press, Vol. 2, 1971) it is stated that that classical electrodynamics is internally inconsistent. For example, the theory of relativity requires that the charge of the electron was a point-like, but the point charge has an infinite energy. Obviously, it is necessary to reject the requirements of the theory of relativity, since the charge cannot be point-like. The equation of motion for the force field cannot written, as if we introduce an additional force, then effects of self-interaction would lead to the internal contradictions. Thus, the equations of motion for the electromagnetic wave can be expressed only by the order parameters in the form of the electric and magnetic polarization, but not by the both force fields, electric and magnetic fields.

On the other hand, quantum mechanics implies that the free electron—is the plane wave, which is also false, since the electron charge—is a localized formation. Therefore, from the viewpoint of wave mechanics elementary particles have to be considered as stable wave packets with a well-defined spatial distribution.

The first attempt to associate the field character to the electron energy is related with the introduction of the classical radius of the electron charge in the form:

$$\varepsilon_e = mc^2 = \frac{e^2}{r_0} \tag{7}$$

Further studies, however, have shown that the different length is associated with electron, which Compton obtained from the scattering of photons by the free electrons:

$$r_{ce} = \frac{\hbar}{m_{0e}c} \tag{8}$$

Here $\hbar, m_{0e}, c$ are considered as fundamental and independent variables. However, this is not the case. Neither of these units is neither basic nor independent. The fact that from quantum mechanical expression for the electron energy

$$\varepsilon_{0e} = m_{0e}c^2 = \hbar\omega_{0e} \tag{9}$$

is seen that the Compton length can be expressed in terms of the characteristics of the vacuum environment

$$r_{ce} = \frac{\hbar}{m_{0e}c} = \frac{c}{\omega_{0e}}, \tag{10}$$

The parameters of the vacuum c and ω_{0e}, in turn, are expressed in terms of the fundamental parameters of vacuum:

$$c = \frac{\xi_e}{\tau_e};\ \omega_{0e} = \frac{\sqrt{\alpha}}{\tau_e};\ r_{ce} = \xi_e\sqrt{\alpha}. \tag{11}$$

The physical meaning of the Planck constant remained unclear despite the fact that this term Planck introduced more than a hundred years ago. However, from the point of view of the existence of the vacuum medium (which is based on the rigid three-dimensional crystal model), the physical meaning is easily recovered, if in the expression (9) for the electron energy the Planck constant is expressed in terms of the electron charge e and the fine structure constant α

$$\varepsilon_{0e} = m_{0e}c^2 = \hbar\omega_{0e} = \frac{e^2\omega_{0e}}{\alpha c} = \frac{e^2}{\alpha r_{ce}}, \tag{12}$$

in addition, exactly the same relationship holds for the proton:

$$\varepsilon_{0p} = m_{0p}c^2 = \hbar\omega_{0p} = \frac{e^2\omega_{0p}}{\alpha c} = \frac{e^2}{\alpha r_{cp}}. \tag{13}$$

These last expressions for the energy of the electron and proton acquire a quite transparent physical meaning. The fact that the electric charge is related not only the energy of the electric field, but also the energy of the electric polarization, which induces the charge. As a result, we get that energy consists of two parts:

$$U = \frac{1}{8\pi}\int \boldsymbol{E}^2 \mathrm{d}\boldsymbol{r} + \frac{1}{8\pi\chi}\int \boldsymbol{P}^2 \mathrm{d}\boldsymbol{r}. \tag{14}$$

The first term determines the energy of the electric field, and the second term—the energy of the vacuum polarization, and the quantity χ characterizes the dielectric susceptibility of the vacuum.

For certain classes of solutions, Equation (14) can be written as:

$$U = \frac{e^2}{r} + \frac{e^2}{\chi r}, \tag{15}$$

where r characterizes a localization of the charge e. The last expression shows that under the condition $\chi < 1$ the energy associated with the polarization becomes greater than the energy of the electric field. Comparing the second term in (15) with (12) and (13) for the energy of the electron and the proton, respectively, we see that these expressions are consistent, if we assume that the polarizability of the vacuum is the fine structure constant: $\chi = \alpha \approx 1/137$.

Thus, the energy of the electric field is 137 times less than the energy associated with the polarization of the vacuum.

Dirac has raised the question of the need to understand the physical meaning of the fine structure constant. In this paper we show that the fine structure constant is a fundamental characteristic and determines the susceptibility of the vacuum.

A2. About the Inertia and Wave Properties of Electrons

It looks like there is a myth about the equivalence of inertial and gravitational masses. Since misunderstandings with the problem of equivalence of inertial and gravitational mass, it makes sense to consider the inertia and the wave properties of electrons on the example of the hydrogen atom. Consider the Hamiltonian of a particle in an arbitrary potential as

$$H = \varepsilon(p) + U(\boldsymbol{r}) \tag{16}$$

from the conservation energy law we have:

$$\frac{\mathrm{d}H}{\mathrm{d}t} = \frac{\partial \varepsilon(p)}{\partial \boldsymbol{p}} \dot{\boldsymbol{p}} + \frac{\partial U}{\partial \boldsymbol{r}} \dot{\boldsymbol{r}} = 0 . \tag{17}$$

Since

$$\frac{\partial \varepsilon(p)}{\partial \boldsymbol{p}} = \boldsymbol{v} = \dot{\boldsymbol{r}}, \tag{18}$$

then Equation (17) can be rewritten as follows:

$$\boldsymbol{v}(\dot{\boldsymbol{p}} + \nabla U) = 0, \tag{19}$$

which yields the Hamilton motion equation:

$$\boldsymbol{v} = \frac{\partial \varepsilon(p)}{\partial \boldsymbol{p}}; \quad \dot{\boldsymbol{p}} = -\nabla U \quad . \tag{20}$$

The (20) is valid for any particle and any externa l potential. In the particular case of an electron in t he nuclear field, the Hamiltonian has the form:

$$H = \sqrt{m_{0,e}^2 c^4 + c^2 p^2} - \frac{e^2}{r}. \tag{21}$$

Here one can get rid of the notion of the mass of the electron and go to the expression, reflecting the wave properties of the electron

$$H = \hbar\sqrt{\omega_{0,e}^2 + c^2 k^2} - \frac{e^2}{r}. \tag{22}$$

Since we are interested in the frequency spectrum of the hydrogen atom, then, in accordance with the theory of self-similarity, we must select the characteristic frequency of the problem, and then all the other components can be represented in a dimensionless form. Let us divide both sides of (22) by $\hbar$, then select the frequency $\omega_{0,e}$, and then the Hamiltonian takes the form:

$$\frac{H}{\hbar} = \omega_{0,e}\left\{\sqrt{1 + \frac{c^2}{\omega_{0,e}^2} k^2} - \frac{\alpha}{r}\frac{c}{\omega_{0,e}}\right\}. \tag{23}$$

Here we took into account that $\hbar = e^2/\alpha c$. From (23) it is clear that it is convenient to proceed to dimensionless coordinate and time:

$$\boldsymbol{r}' = \boldsymbol{r}/r_{c,e}; \quad t' = \omega_{0,e} t; \quad \boldsymbol{k}' = \boldsymbol{k} r_{c,e},$$

after which the Hamiltonian takes the form, which is displayed in the peer-reviewed paper:

$$\frac{H}{\hbar} = \omega_{0,e}\left\{\sqrt{1 + k'^2} - \frac{\alpha}{r'}\right\}, \tag{24}$$

which implies that the entire frequency spectrum of the hydrogen atom is determined solely by its principal electronic frequency $\omega_{0,e}$ and the fine structure constant α, and α—is the main fundamental quantity that determines the susceptibility of vacuum. There is no mass of the electron here, nor the Planck's constant. The velocity becomes also dimensionless and is expressed in terms of the speed of light. This is because the Planck constant and the mass of the electron and the proton are determined by the electron charge e, which actually determines the amplitude fluctuations of the order parameter in the form of electric polarization of the vacuum, and for a system of harmonic oscillators frequency oscillations does not depend on the amplitude of oscillations, which is reflected in the expression (24). The velocity becomes also dimensionless and is expressed in terms of the speed of light and can be expressed in terms of the group velocity of the wave packet:

$$\boldsymbol{v}' = \frac{\boldsymbol{v}}{c} = \frac{\mathrm{d}\boldsymbol{r}'}{\mathrm{d}t'} = \frac{\partial \omega_{k'}}{\partial \boldsymbol{k}'} = \frac{\boldsymbol{k}'}{\sqrt{1 + k'^2}}. \tag{25}$$

The equations of motion (20) for the electron in the nuclear field with the Hamiltonian (24) are as follows (the primes in (24, 25) can now be omitted):

$$\boldsymbol{v} = \frac{\boldsymbol{k}}{\sqrt{1 + k^2}}; \quad \dot{\boldsymbol{k}} = \nabla\left(\frac{\alpha}{r}\right). \tag{26}$$

From the first equation, the wave vector $\boldsymbol{k}$ can be expressed in terms of the group velocity of the wave packet—$\boldsymbol{v}$:

$$\boldsymbol{k} = \frac{\boldsymbol{v}}{\sqrt{1 - v^2}}, \tag{27}$$

after which the second Equation (26) transforms into a form which can be expressed in two different ways

$$\frac{\mathrm{d}}{\mathrm{d}t}\left(\frac{\boldsymbol{v}}{\sqrt{1 - v^2}}\right) = \nabla\left(\frac{\alpha}{r}\right), \tag{28}$$

and

$$\frac{\mathrm{d}}{\mathrm{d}t}\left(\boldsymbol{v}\sqrt{1+k^2}\right)=\nabla\left(\frac{\alpha}{r}\right), \quad (29)$$

which are reflecting the inertia and the wave properties of the electron.

From Equation (28) it follows

$$\frac{\dot{\boldsymbol{v}}}{\sqrt{1-v^2}}+\frac{\boldsymbol{v}(\boldsymbol{v}\cdot\dot{\boldsymbol{v}})}{\left(1-v^2\right)^{3/2}}=\nabla\left(\frac{\alpha}{r}\right), \quad (30)$$

so it can be seen that the acceleration in the direction perpendicular to the velocity $(\dot{\boldsymbol{v}}\boldsymbol{v}=0)$ is proportional to the square of $\sqrt{1-v^2}$. In this case, the particle inertia is proportional to the energy.

However, in the direction along velocity $(\dot{\boldsymbol{v}}\parallel\boldsymbol{v})$, the acceleration is proportional to the factor $\left(1-v^2\right)^{3/2}$, which implies that the inertia is growing faster in the direction of motion than in the perpendicular direction.

The inertia of the particles is the easiest checked on the accelerator. In the ring accelerator, the centripetal force required to keep the particles on a specific trajectory, increases proportional to the factor $1/\sqrt{1-v^2}$, therefore, in this case inertia is proportional to the energy. In a linear accelerator, the equation of motion for a particle is as follows:

$$\ddot{x}=-\left(1-v^2\right)^{3/2}\frac{\partial U}{\partial x}. \quad (31)$$

Thus, the inertia of the particles (inertial mass) is the magnitude of the anisotropic, while the gravitational mass (energy divided by c^2) is a scalar quantity.

The wave properties of electrons can be easier analyzed using Equation (29), which after differentiation in time takes the form:

$$\dot{\boldsymbol{v}}\sqrt{1+k^2}+\frac{\boldsymbol{v}\left(\boldsymbol{k}\dot{\boldsymbol{k}}\right)}{\sqrt{1+k^2}}=\nabla\left(\frac{\alpha}{r}\right). \quad (32)$$

After taking into account the relations

$$\sqrt{1+k^2}=\frac{1}{\sqrt{1-v^2}};\ \frac{\boldsymbol{k}}{\sqrt{1+k^2}}=\boldsymbol{v};\qquad \dot{\boldsymbol{k}}=\nabla\left(\frac{\alpha}{r}\right) \quad (33)$$

we arrive at the equation:

$$\dot{\boldsymbol{v}}=\sqrt{1-v^2}\left[\nabla\left(\frac{\alpha}{r}\right)-\boldsymbol{v}\left(\boldsymbol{v}\nabla\left(\frac{\alpha}{r}\right)\right)\right]. \quad (34)$$

In polar coordinates (r,φ), the Equation (34) reduces to:

$$\ddot{r}-r\dot{\varphi}^2=-\frac{\alpha}{r^2}\sqrt{1-\dot{r}^2-r^2\dot{\varphi}^2}\left(1-\dot{r}^2\right); \quad (35)$$

$$2\dot{r}\dot{\varphi}+r\ddot{\varphi}=\frac{\alpha}{r}\dot{r}\dot{\varphi}\sqrt{1-\dot{r}^2-r^2\dot{\varphi}^2}. \quad (36)$$

Equation (35) is also displays anisotropy of the inertial properties of the electron. When considering a motion along the radius $(\dot{\varphi}=0)$, Equation (35) reduces to the form:

$$\frac{\ddot{r}}{\left(1-\dot{r}^2\right)^{3/2}}=-\frac{\alpha}{r^2}, \quad (37)$$

while for the circular motion $(\ddot{r}=\dot{r}=0)$ the Equation (35) looks as follows:

$$\frac{r\dot{\varphi}^2}{\sqrt{1-r^2\dot{\varphi}^2}}=\frac{\alpha}{r^2}. \quad (38)$$

Multiplying both sides of (38) by r and taking into account that $r\dot{\varphi}=v_\varphi$, we obtain the following equation:

$$\frac{v_\varphi^2}{\sqrt{1-v_\varphi^2}}=\frac{\alpha}{r}. \quad (39)$$

Expressing v_φ through k_φ with help of (33), we obtain:

$$\frac{k_\varphi^2}{\sqrt{1+k_\varphi^2}}=\frac{\alpha}{r}. \quad (40)$$

The Bohr quantization condition should be added to Equation (40), which he wrote down in the form of quantization of adiabatic invariants:

$$\int p_i dq_i=nh. \quad (41)$$

For the circular motion, the relation (40) is transformed to the form:

$$2\pi r p_\varphi=nh;\ \Rightarrow\ rp_\varphi=n\hbar. \quad (42)$$

The relations (42) can be interpreted as a quantization of the orbital angular momentum $\boldsymbol{s}=[\boldsymbol{rp}]$:

$$s_z=rp_\varphi=n\hbar. \quad (43)$$

At first glance it is a real feeling that there is a quantum of action $\hbar$, which leads to the quantization of the orbital angular momentum. However, if we recall that the momentum itself is in turn equal to $\hbar k_\varphi$, then we come to the standard cyclic condition that holds for any wave process:

$$rk_\varphi=n, \quad (44)$$

which means that in a stationary orbit one must fit an integer number of wavelengths $2\pi r/\lambda=n$. Therefore Planck's constant has nothing to do with formation of stationary orbits in the hydrogen atom. Solving Equations (32) and (36) we obtain:

$$k_{\varphi,n}=\frac{\alpha}{n\sqrt{1-\left(\frac{\alpha}{n}\right)^2}};\ r_n=\frac{n^2}{\alpha}\sqrt{1-\left(\frac{\alpha}{n}\right)^2} \quad (45)$$

As a result, the frequency spectrum of the hydrogen atom that follows from the Hamiltonian (16) takes the form:

$$\omega_n = \omega_{0,e}\sqrt{1-\left(\frac{\alpha}{n}\right)^2}. \tag{46}$$

Wherein the spectrum depends only on the intrinsic electron frequency and the fine structure constant—susceptibility of vacuum, α. The term "the fine structure constant" is not appropriate, since both intrinsic electron frequency, and the entire spectrum of the hydrogen atom depend on the quantity α. It is important that the frequencies do not depend on the Planck constant, nor on $e, m_{0,e}, c$. The first question about the influence of relativistic effects on the spectrum of the hydrogen atom is considered by Sommerfeld (A. Sommerfeld, Atomic spectra and spectral lines, 1931). He obtained a more general formula, considering elliptic trajectories, but there he used five (!) different quantities $\hbar, m_{0,e}, e, c, \alpha$. As a result, for frequency spectrum of hydrogen he received:

$$\omega_n = \frac{2R}{\alpha^2}\sqrt{1-\left(\frac{\alpha}{n}\right)^2}; \quad R = \frac{m_{0,e}e^4}{2\hbar^3} \tag{47}$$

Here R is the Rydberg constant, the expression of which had received by Bohr. However, if we recall the relation

$$m_{0,e}c^2 = \hbar\omega_{0,e}; \quad \alpha = \frac{e^2}{\hbar c}, \tag{48}$$

then from (47) we arrive at the expression:

$$\frac{2R}{\alpha^2} = \frac{m_{0,e}e^4}{\hbar^3\alpha^2} = \omega_{0,e}. \tag{49}$$

Such a monstrous conglomeration of symbols in the expression (49) shows an absolute lack of understanding of the nature of the wave processes that occur in the hydrogen atom, and more over the wave processes within the electron itself.

Now, coming back to the energy $E_n = \hbar\omega_n$ from tor the discrete frequencies (46), we then arrive at the expression:

$$E_n = \hbar\omega_n = \frac{eg}{r_{c,e}}\sqrt{1-\left(\frac{\alpha}{n}\right)^2}. \tag{50}$$

Here e reflects the amplitude of the oscillations of the electron fields, and $r_{c,e}$—is the spatial configuration of the electron charge, a $g = e/\alpha \approx 137e$—the constant interaction between point particles of the crystal (physical) vacuum.

Heisenberg's uncertainty principle relationships are considered as the fundamental basis of quantum mechanics:

$$\Delta\varepsilon \cdot \Delta t \approx h, \quad \Delta p \cdot \Delta x \approx h. \tag{51}$$

Here, however, the Planck constant is quite superfluous quantity. If we recall that $\varepsilon = \hbar\omega$, and $\boldsymbol{p} = \hbar\boldsymbol{k}$ then we go back to the usual relationships of wave mechanics:

$$\Delta\omega \cdot \Delta t \approx 2\pi, \quad \Delta k_x \cdot \Delta x \approx 2\pi. \tag{52}$$

And these relations for photon and electron have completely different physical meanings.

Consider a simple example. If we connect the antenna to the generator of electrical oscillations with frequency ω for a time interval T, then this will result in the radiation of a wave packet of length $L = cT$. Now, in order to study the frequency spectrum of the wave packet, by using a narrowband receiver, we see that the frequencies will be distributed in a certain range $\omega \pm \Delta\omega$, with the value $\Delta\omega$ determined by the relationship: $\Delta\omega \cdot T \approx 2\pi$. Accordingly, since $\omega = ck$, the uncertainty of the wave vector is determined by the relationship $\Delta k \cdot L \approx 2\pi$, which implies that the creation of a strictly monochromatic wave requires infinite time. Of course, the total energy and total momentum of the wave packet, as integrals of motion are well defined, and cannot have any relation to the uncertainty relation (43). The same applies to photons. During atomic transition from the excited state to the ground, to create a photon as a wave packet with a length L it takes some time. Because the radiation time is limited by the lifetime of the excited state, then at the passage of photons through a diffraction grating, we see broadening of the lines. All of the diffraction pattern is determined by three parameters: the period of the grating a, wavelength of the quantum λ, and the length of the wave packet of photon $L = cT$. In this case, the Planck constant has nothing to do to the diffraction pattern. Moreover, the energy is a constant of motion and it cannot be uncertain, in principle. Considering the γ—ray quantum structure as a thin filament of length L, it becomes clear the quantization condition for the angular momentum $\hbar\boldsymbol{k}$. At γ—ray quantum capturing by an electron—the orbit length should be equal to the wavelength so that each subsequent wave cycle would be in phase with the previous one—$2\pi r_\gamma = \lambda_\gamma = 2\pi/k_\gamma$, which implies the condition: $r_\gamma k_\gamma = 1$, and thus the value of the orbital angular momentum $s = r_\gamma p_\gamma = \hbar r_\gamma k_\gamma = \hbar$. Thus it becomes clear condition of quantization of the angular momentum.

Let us now consider how the uncertainty relations are manifest for the electron. If an electron is placed in a one-dimensional potential well with the size L, then the solution for the electron wave function is as follows:

$$\psi_n = A_n \sin\left(\frac{\pi n}{L}x\right)e^{i\omega_n t}. \tag{54}$$

Here, the wave vector k_n and frequency of ω_n acquire the well-defined values:

$$k_n = \frac{\pi n}{L}; \quad \omega_n = \pm\sqrt{\omega_{0,e}^2 + c^2 k_n^2}. \quad (55)$$

The solution (54) is realized for any wave process. For example, if a guitar string is fixed at points at a distance L, then it results in a wave vector quantization according to (55), with frequencies $\omega_n = \pm c_1 k_n$, where c_1 – the signal propagation velocity along the string, depending on the tension of the string. Any player knows that the frequency of the string vibrations depends on the tension of a string and distance between the fixing points of the string. The same situation arises in the piezoelectric resonator having a thickness L, its vibrations may be given in the form (54) with the frequencies $\omega_n = \pm c_s k_n$, where c_s—the sound velocity in the piezoelectric. Thus, the restriction of the signal in time leads to the frequency uncertainties and restriction of vibrations in space leads to the quantization of the wave vector and the vibrations with the well-defined frequencies. Therefore piezoelectric resonators are used, for example, to stabilize the operation of an electronic clock.

A characteristic feature of linear vibration systems is that the frequency does not depend on the amplitude, therefore the secular equation determining the spectrum of normal oscillations, does not contain the amplitude at all.

If the Laplace operator is present in the equation of motion (or a system of equations), then the solution can be conveniently represented in terms of the eigenfunctions of the Laplace operator

$$\Delta\psi = \beta^2\psi$$

with the eigenvalues β^2. We obtain a well-defined frequency to each eigenfunction, as a result, we have the spectrum of normal vibrations $\omega(\beta)$. In the Cartesian coordinate system, the eigenfunctions can be plane waves $\exp(i\boldsymbol{kr})$ with eigenvalues $\beta^2 = -(k_x^2 + k_y^2 + k_z^2)$. In the cylindrical coordinate system, eigenfunctions can be Bessel functions, in the spherical coordinate system can be employed spherical cylinder functions. The amplitude of each normal mode is arbitrary and depends on external conditions. For example, the guitarist finger deflects a string to some distance from the equilibrium position. The initial deflection can be expanded in the eigenfunctions (54):

$$\psi(x, t=0) = \sum_n A_n \sin\left(\frac{\pi n}{L}x\right). \quad (56)$$

Once the string was released, every normal mode begins its own motion

$$\psi(x, t) = \sum_n A_n \sin\left(\frac{\pi n}{L}x\right)\cos\left(\frac{\pi n}{L}c_1 t\right). \quad (57)$$

And the energy of each normal vibration is proportional to A_n^2, where A_n depends only on the initial conditions.

If the electron is considered as an elementary excitation of the vacuum, the dynamic properties of the elementary particles must be determined directly by the dynamic properties of vacuum. Let us make rough estimates. Let the vacuum has a dielectric susceptibility equal α, then the energy associated with the creation of the electric polarization is:

$$\varepsilon = \frac{1}{8\pi\alpha}\int P^2(\boldsymbol{r})\,d\boldsymbol{r}. \quad (58)$$

On the example of a one-dimensional string, we can see that the arbitrary force excites the entire spectrum of normal vibrations in the form (57). However, really only three normal modes in the form of spherical functions are associated with the electron:

$$P_x = q\frac{\sin\theta\cos\varphi}{r^2}; \quad P_y = q\frac{\sin\theta\sin\varphi}{r^2}; \quad P_z = q\frac{\cos\theta}{r^2}, \quad (59)$$

and are eigenfunctions of the Laplace operator

$$\Delta P_i = 0 \quad (60)$$

with the eigenvalues equal to zero.

Such fields can be excited only in the event that the external force itself satisfies the Laplace equation. In fact, we need to find out who plays the role of "guitarist" that plays on the vacuum fields. At this point we are led to the introduction of the crystal structure of the vacuum with particles having an intrinsic field and the corresponding interaction energy:

$$\boldsymbol{E}_0(\boldsymbol{r}) = g\frac{\boldsymbol{r}}{r^3}; \quad U = -\frac{1}{4\pi}\int \boldsymbol{P}(\boldsymbol{r})\boldsymbol{E}_0(\boldsymbol{r})\,d\boldsymbol{r}. \quad (61)$$

However, the self-energy of the field for the solutions in the form (59) is infinite, and therein lies the problem of divergence for point charges. The introduction of the gradient energy with the parameter length ξ_e and the kinetic energy of the field with the parameter of time τ_e leads to a blurring of the polarization charge to the size equal to the correlation radius $r_{ce} = \xi_e\sqrt{\alpha} = c/\omega_{0e}$. As a result, the polarization energy becomes finite and equal to

$$\varepsilon = \frac{q^2}{\alpha r_{ce}}, \quad (62)$$

the polarization charge is then determined by the relation $q = \alpha g$. Assuming $q = e$, we get $g = e/\alpha$, with the energy of the electron

$$\varepsilon = \hbar\omega_{0e} = \frac{e^2}{\alpha c}\omega_{0e} = \frac{e^2}{\alpha r_{ce}} = \frac{g^2\sqrt{\alpha}}{\xi_e}, \tag{63}$$

which implies that the Planck constant is directly related to the charge of the electron and reflects the polarization and dynamic properties of the vacuum fields, however, is by no means a universal fundamental constant. Vacuum fundamental constants are $\alpha, g, \xi_{e,n}, \tau_{e,n}$. Charge, mass, energy and momentum of electron and proton, and the Planck constant are indeed expressed in terms of the fundamental parameters of vacuum, but it does not follow from this that if we take an arbitrary scalar field (for example, photon field) in the form $\exp(i\boldsymbol{kr})$, then the momentum can be quantized $\boldsymbol{p} = \hbar\boldsymbol{k}$ and then proceed to the quantum field theory. There is no physical meaning to this.

In a vacuum, as a linear system, the electromagnetic oscillations can occur with any amplitude, and the quantum of light can have arbitrary momentum and the corresponding energy $\boldsymbol{p} = A\boldsymbol{k}_\gamma$; $\varepsilon = A\omega_\gamma$, but in the emission and absorption of light by an electron the laws of conservation of energy and momentum must be fulfilled:

$$\begin{aligned} &\hbar\sqrt{\omega_{0,e}^2 + c^2k^2} + Ack_\gamma = \hbar\sqrt{\omega_{0,e}^2 + c^2k'^2} + Ack'_\gamma, \\ &\hbar\boldsymbol{k} + A\boldsymbol{k}_\gamma = \hbar\boldsymbol{k}' + A\boldsymbol{k}'_\gamma. \end{aligned} \tag{64}$$

In addition, the spatial and frequency resonance conditions must be fulfilled:

$$\begin{aligned} &\sqrt{\omega_{0,e}^2 + c^2k^2} + ck_\gamma = \sqrt{\omega_{0,e}^2 + c^2k'^2} + ck'_\gamma, \\ &\boldsymbol{k} + \boldsymbol{k}_\gamma = \boldsymbol{k}' + \boldsymbol{k}'_\gamma, \end{aligned} \tag{65}$$

it follows then that $A = \hbar$.

Thus, as conjectured by Planck, electron emits and absorbs light by quanta, but it does not follow from this that light is quantized, moreover, that any field is quantized, as stated by Einstein. But it is connected *only* specific field of the electron energy distribution in space.

The introduction of vacuum parameters allows us to understand the physical mechanism of the appearance of the spin and magnetic moment of the electron. The fact is that the magnetic field configuration of the electron coincides with the magnetic field of the ring current. The magnetic moment associated with the charge, rotating on a circular orbit of radius r is equal to

$$\mu = \frac{evr}{2c}. \tag{66}$$

Comparing an expression (66) with the magnetic moment of the electron, that is equal to the Bohr magneton,

$$\mu_e = \frac{e\hbar}{2m_{0e}c} = \frac{er_{ce}}{2}. \tag{67}$$

We see that the magnetic moment of the electron charge is related to the rotation of the charge in a circular orbit with the speed of light with a size equal to the correlation radius r_{ce}. Consequently, the emergence of the electron spin associated with the movement of fields in a circular orbit.

Closer examination shows that the charge rotates in the orbit with a radius $r_{ce}/2$ which leads to the spin of $\hbar/2$. The magnetic moment in this case is two times greater than that given by the formula (66). This means that the magnetic moment of the electron consists of two components: a magnetic moment associated with the motion of the charge and the magnetic moment associated with the magnetization of vacuum.

A3. The Formation of the Energy Spectrum of the Massive Body

Consider the process of formation of the energy spectrum of massive bodies, consisting of a certain set of quantum particles. Ignoring the interaction between particles the energy of a massive body can be represented as the sum of the energies of each of the particles:

$$\varepsilon = \sum_{i=1}^{N}\varepsilon_i = \sum_{i=1}^{N}\hbar\omega_i(\boldsymbol{k}_i) = \sum_{i=1}^{N}\hbar\sqrt{\omega_{0i}^2 + c^2\boldsymbol{k}_i^2}. \tag{68}$$

Convenient to go to the four-dimensional coordinate system of the same dimension x, y, z, t, where $t = ct'$. In this case, the velocity of the particle $\boldsymbol{v} = \mathrm{d}\boldsymbol{r}/\mathrm{d}t$ is a dimensionless quantity and is expressed in units of the speed of light, then the energy of elementary particles can be written as:

$$\varepsilon_i = \sqrt{\varepsilon_{0i}^2 + \boldsymbol{p}_i^2}, \tag{69}$$

where $\varepsilon_{0i} = \hbar\omega_{0i}$ —zero-energy particles, $\boldsymbol{p}_i = \hbar c\boldsymbol{k}_i$ — momentum of a particle, also has dimension of energy. The speed of the particle is determined by the expression:

$$\boldsymbol{v}_i = \frac{\mathrm{d}\varepsilon_i(\boldsymbol{p}_i)}{\mathrm{d}\boldsymbol{p}_i} = \frac{\boldsymbol{p}_i}{\sqrt{\varepsilon_{0i}^2 + \boldsymbol{p}_i^2}}. \tag{70}$$

Of the last expression can be expressed momentum of a particle and energy as follows:

$$\begin{aligned} \boldsymbol{p}_i &= \frac{\varepsilon_{0i}\boldsymbol{v}_i}{\sqrt{1 - v_i^2}}; \\ \varepsilon_i &= \frac{\varepsilon_{0i}}{\sqrt{1 - v_i^2}}. \end{aligned} \tag{71}$$

For massive body can be considered that the velocities of all particles are the same and then $v_i = v$.

The total energy and total impulse can be written as:

$$\varepsilon = \sum_{i=1}^{N}\varepsilon_i = \frac{\sum_{i=1}^{N}\varepsilon_{0i}}{\sqrt{1 - v^2}} = \frac{\varepsilon_0}{\sqrt{1 - v^2}}; \tag{72}$$

$$p = \sum_{i=1}^{N} p_i = \frac{v}{\sqrt{1-v^2}} \sum_{i=1}^{N} \varepsilon_{0i} = \frac{\varepsilon_0 v}{\sqrt{1-v^2}}; \qquad (73)$$

Using the definitions of zero energy and total momentum of a massive body (72) and (73), we can write the expression:

$$\varepsilon_0^2 + \boldsymbol{p}^2 = \frac{\varepsilon_0^2}{1-v^2} = \varepsilon^2 \qquad (74)$$

Whence it follows that the energy spectrum of a massive body has the form:

$$\varepsilon = \sqrt{\varepsilon_0^2 + \boldsymbol{p}^2}\,, \qquad (75)$$

Despite the fact that (72), (73), (75) have a relativistic view, speed in them has a different physical meaning and is defined as the velocity relative to the absolute coordinate system tied to the medium—physical vacuum which presented as the model of the three-dimensional rigid crystal lattice structure.

Optical Design of OCT with Gapped Magnetic Ring*

Xiaoyi Su, Qifeng Xu
College of Electrical Engineering and Automation, Fuzhou University, Fuzhou, China
Email: xiaos_320@163.com

ABSTRACT

The mathematics and physics model of OCT (Optical Current Transformer) with a gapped magnetic ring is briefly discussed in the paper. And some proposals of how to select the magnetic materials and crystals and reduce the stress birefringence of the crystal are also put forward in the paper. Based on the above, an OCT with 1000 A rated current is designed by using the ANSOFT Maxwell tools.

Keywords: Ferromagnetic Collector Type; Optics Current Transformer; Magnetic Properties; Linear Birefringence

1. Introduction

Faraday magneto-optical effect is applied in optical current transformer (OCT). Based on its sensing methodology it can be divided into four types: fiber one, bulk glass one, solenoid one, as well as gapped magnetic ring one [1]. The one with a gapped magnetic ring utilizes a small piece of magneto-optic crystal and other optical elements to be placed in the gap of the magnetic ring wounding around a current carrying conductor to form an optical circuit. The gap magnetically induced by the current flowing in the conductor and the crystal are the "heart" of an optical current transformer head. Taken together, the magnetic properties and the optical properties should be taken into consideration in the design. This paper discusses the characteristics of the OCT with gapped magnetic ring. Additionally, reasons and procedures to select magnetic materials and crystals and mitigate stress birefringence are also represented specifically in the paper. Depending on the analysis above and the ANSOFT Maxwell tools, the optimized structure of an OCT with 1000A has been achieved.

2. The Characteristics of OCT with Gapped Magnetic Ring

When a linearly polarized light passes through a magnetic field paralleling to its propagation direction, a optical phase shift of the light called Faraday rotation angle will created due to the Faraday effect

$$\theta = V\int Hdl \tag{1}$$

where V is the Verdet constant. Once deemed if a gap of magnetic ring is small enough, the gap magnetic can be regarded as same as the ring inherent magnetic induced by the current flowing in the conductor. Additionally, based on the Ampere circuital theorem the gap magnetic strength is given by

$$H_g = I/(l_g + u_g l_a/u_a) \tag{2}$$

where l is the polarized light effective optical passing path in the gap, I is the primary current, l_g is the air-gap length, u_g is the air permeability, u_a is the gapped magnetic ring permeability, l_a is the gapped magnetic ring average loop length. As u_g is far less than u_a, Equation (1) is simplified as

$$H_g \approx I/l_g \tag{3}$$

Combined Equation (1) with Equation (3), the Faraday deflection angle θ and the primary current I are given by

$$\theta = V\int Hdl = u_g VIl_c/l_g \tag{4}$$

$$I = l_g\theta/(u_g Vl_c) \tag{5}$$

in which l_c is the crystal length. Set P_i to be the polarizer output light intensity and P_0 to be the analyzer output light intensity. According to the Malus law, the relationship between P_i and P_0 is expressed as

$$P_0 = aP_i\cos^2(\varphi + \theta) \tag{6}$$

where α is the optical attenuation coefficient intensity; φ is the difference optical phase shift between the two output light waves.

Desiring to obtain the maximum output power, φ is set to 45° . If θ is small enough, sin2θ will approximately equal to 2θ. Equation (6) transforms into

$$P_0 = \frac{a}{2}P_i(1 + \frac{2u_g VIl_c}{l_g}) = P_{DC} + P_{AC} \tag{7}$$

*Project supported by the National Natural Science Foundation of China (51177016).

where P_{AC}, the alternate current signal, is expressed as $P_{AC} = (aP_iu_gVIl_c)/l_g$, while P_{DC}, the direct current signal, is expressed as $P_{DC} = (aP_i)/l_g$. Eliminating the fluctuations of optical power by introducing the operation $U=P_{AC}/P_{DC}$. The relationship between U and I is given by

$$U = I \cdot 2u_g VIl_c / l_g \quad (8)$$

3. The Optimal Design of OCT

3.1. The Selection of Magnetic Material

The cold-rolled silicon steel, the permalloy alloy and the amorphous alloy are commonly made on gapped magnetic rings. Their characteristics differences are reflected in the magnetic permeability, the remanence density, the saturation magnetization, the coercive force, the iron loss, the Curie temperature, as well as the magnetostriction coefficient. Considering that the OCT actual operation environment temperature is less than the Curie temperature of magnetic materials, the iron loss and the magnetostriction coefficient are negligible under the 1000A power frequency system, so these three factors, the iron loss, the Curie temperature and the magnetostriction coefficient can be ignored. **Table 1** shows the remainder typical parameters.

Table 1. The comparison of typical parameters of the three materials.

Materials	Typical Parameters			
	Saturation Mag-netiz-ation (T)	Magnetic permeability (*10^4)	Remanence density (T)	Co-erci-ve force (A/m)
amorphous alloy	1.5	0.25	1	2.0
permalloy alloy	0.6	5.8	0.4	4.3
cold-rolled silicon steel	1.7	0.12	1.56	7

Permalloy alloy has the highest initial permeability, the amorphous alloy has a lower one and the cold-rolled silicon steel has a lowest one. Additionally, for closed cores, a higher initial permeability is good for measurement sensitivity. But the gapped magnetic rings are different. As the gap magnetic induction B_g is

$$B_g = \frac{I}{(l_a/u_a + l_g/u_g)} \quad (9)$$

Once the air permeability is 10^4 to 10^5 times bigger than magnetic rings' . Then l_a is 10^2 times larger than l_g, so l_a/u_a is far less than l_g/u_g. Equation (9) is simplified as

$$B_g = u_g I / l_g \quad (10)$$

In this case, the magnetic induction B_g depends on the primary current I, the gap length l_g and the air permeability u_g . That is to say, the permeability u_a is an unimportant factor to the magnetic properties of gapped magnetic rings.

Static field simulations on three magnetic materials shown in **Table 1** are done respectively by using the ANSOFT Maxwell tool. The simulation gapped magnetic ring parameters as follow: the inner radius is 30 mm, the outer radius is 70 mm, the height is 40 mm and, the gap length is 20 mm. Their basic magnetization curve are illustrated in **Figure 1**. When the primary current is less than 4 KA, these three curves are overlapped and the cold-rolled silicon steel one has the longest current linear region. Since normal load current flowing on the conductor is usually under 1 kA level, associating with the linear region length shown in **Figure 1**, cold-rolled silicon steel and amorphous alloy have better magnetic performance compared to permalloy alloy.

The remanence of magnetic concentrator ring is given by[2]:

$$B_r = -f(H) \cdot \frac{u_g l_a}{l_g} = f(H=0) \cdot \frac{u_g l_a}{l_g} \quad (11)$$

where $f(B)$ is a closed core magnetization curve function.

The remanence in the Gapped magnetic ring are depended on the average magnetic path length and the air gap length. These dependencies make the necessity to design a right ring overall size. A smaller remanence has certain distinct advantages to maintain the response characteristics of magnetization, decrease coercive force and reduce hysteresis loss.

Taken together, the norm to select a suitable magnetic material is the one having a small remanence, a small coercive force and a large saturation magnetic flux density. Thereby, amorphous alloy is the most suitable material.

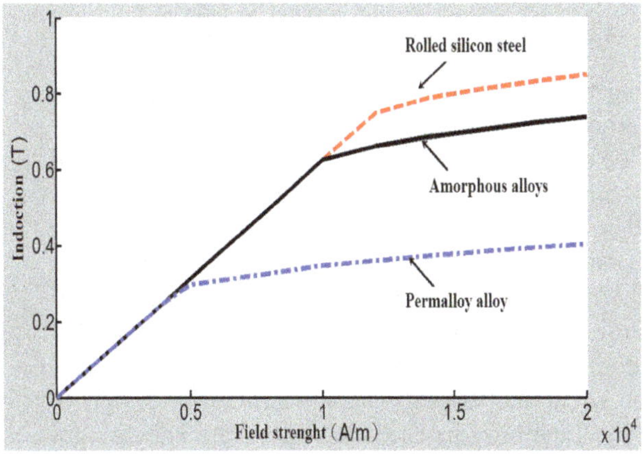

Figure 1. The basic magnetization curve of the three gap magnetic rings.

3.2. The Selection of Magneto-Optical Crystal

Since the Verdet constant varies with the dynamic temperature, the temperature coefficient should be considered during the selection of magneto-optical crystal. When the temperature changes from -25℃ to +80℃, the Verdet constant of FR-5, a paramagnetic material, is reduced by 30% throughout the mutative temperature and YIG that is a ferromagnetic material has 25% irregular variation, while ZF-7 that is an anti-magnetic material only changes 0.79% [3]. Similarly, an anti-magnetic material called MR1 almost has no effect on the Verdet constant in the temperature ranging from -55 to +135℃. Additionally, its Verdet constant is as high as 0.065 - 0.092 min/Oe.cm (@ 632.8 nm) which is helpful to shorten the crystal length and reduce the line birefringence. Therefore, the MR1 is employed in this design.

As the OCT using light to measure the magnetic field surrounding a current carrying conductor has a transfer function with a sine wave characteristics which made the OCT polarization interference be a non-linear portion. An approximate linear relationship between $sin2\theta$ and 2θ introduced in the section 2 can fix this problem when θ is small enough. Ultimately, the OCT accuracy level will determine the maximum Faraday deflection angle and the crystal length. Set X is the difference between l_g and l_c, Equations (5) could be transformed as

$$l_c = \frac{\theta X}{u_g VI - \theta} \tag{12}$$

Assuming the OCT accuracy is a 0.2 class and the primary current is 1000 A, its measurement error is less than 0.2%. Then the maximum Faraday deflection angle θ_{max} is 0.877° . Set the maximum X is 20 mm, (As the prism and polarizer product class size are 7 mm and 2 mm respectively. The other 2mm space is reserved for installation). Putting these data into Equation (12), l is calculated to 16.72 mm. Therefore, the crystal maximum length is 16 mm.

3.3. Methods to Reduce Stress Birefringence

During the process of magneto-optic crystal production and processing or the interaction between the sensor head and the adjacent structure, residual stress will occur and stress birefringence will created simultaneously.

Since expansion coefficient has a function with the temperature when the temperature changes the difference on the materials expansion coefficient will cause temperature gradients owing to the uneven temperature distribution of the sensing head. Then thermal stress is generated within the sensing head and a linear birefringence appears. Line birefringence aliasing on the Faraday polarization angle can undermine the reliability and stability of the OCT [4]. For example, an OCT sensing head has a structure of 70 mm outer diameter, 30 mm inner diameter, 40 mm height and 30 mm air gap length and ZF-7 as the sensing element which is 10 mm long, conducted by a 1000 A primary current, its Faraday rotation angle is 4.81° and its line birefringence angle is about 0.56^0 and it introduces 11.64% measurement error. The method to reduce the linear birefringence is to reduce the magneto-optical material length. For example, using a 200 um thick magneto-optical film that dopes with Ce^{3+} by using the technology of liquid phase epitaxial (LPE) as the sensing element instead of ZF-7 [5], then the Faraday deflection angle is turned to 28.6^0 and its line birefringence angle is 0.0112° only and the corresponding error drops to 0.039%. It turns that the influence of linear birefringence on Faraday deflection angle is effectively weakened.

Further, filling asbestos in sensor head or utilizing materials having the similar expansion coefficient with the magneto-optical material can also free from the temperature variation gradient and weaken the thermal stress birefringence.

3.4. The Optimization Structure of Gapped Magnetic Rings

Since the gapped magnetic ring's inner radius r is determined by the size of a current carrying conductor which is determined by the primary current. Depending on the analysis, the rated current of an OCT designed in this paper is 1000 A, so conductive rod radius is set to 18 mm. Additionally, leaving a 10mm margin space to fill asbestos, the gapped magnetic ring inner radius r is 28 mm. Assuming N is the difference between the inner and outer radius of the gapped magnetic ring. So the outer radius R can be expressed as the sum of r and N. Combing Equation (3) with Equation (11), the gap length not only determines the gap induction, but also affects the remanent density. when the magnetic material is certain, the smaller the gap length, the greater the gap induction and remanence. Since the crystal length is described in selection 3.2, the air gap has a length ranging from 20 mm to 36 mm. Depending on the ANSOFT Maxwell tool, making simulations on three gapped magnetic rings with 25 mm, 30 mm and 35 mm penning gap length under the 1000 A system respectively. The result are illustrated in **Figure 2** and **Figure 3** which show the effect of the difference between a gapped magnetic ring inner and outer radius on the gap induction B_g, as well as on the remanent B_r.

The larger the difference between a gapped magnetic ring's inner and outer radius N, the more uniform gap magnetic field distribution and the smaller the magnetic flux leakage [6]. When N exceeds the length 30 mm, the gap magnetic distribution is completely uniform. Considering to decrease the volume of material and achieve

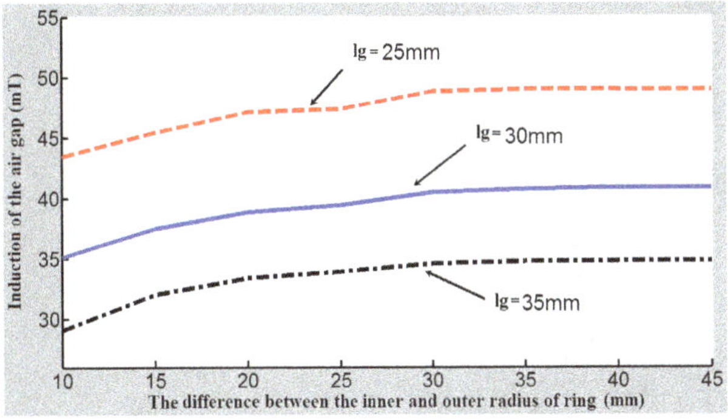

Figure 2. The effect of the difference between a gapped magnetic ring's inner and outer radius on the gap induction.

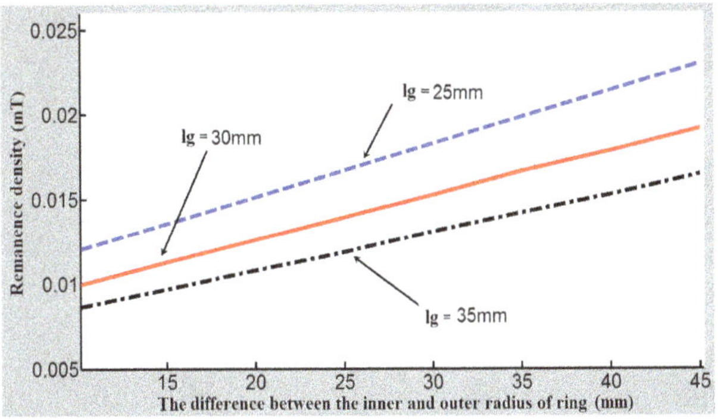

Figure 3. The effect of the difference between a gapped magnetic ring's inner and outer radius on the remanent.

the optimal performance of the OCT, N is set to 30 mm, thereby, the gapped magnetic ring outer radius R is 58 mm. Additionally when N is certain, the longer the gap length, the smaller the remanence density. As the three remanence density curves are at 10^{-2} mT level in **Figure 3**. Comparing to the amorphous alloy saturation magnetic flux density 1.5 T, the remanence density of ring made of amorphous alloy can be ignored. Therefore, under the 1000 A system, the analysis of the OCT transient characteristics can ignore the effects of remanence. Fortunately, a smaller gap length is helpful to obtain a large induced magnetic field, the length of air gap is set to 25 mm, correspondingly the magneto-optical crystal length is set to 5 mm.

4. Conclusions

This paper discusses the characteristics of OCT with gapped magnetic ring, and some considerations to select the magnetic material and the magneto-optical crystal, and reduce the stress birefringence. Depending on the analysis in the paper, the optimized design of an OCT with 1000 A rated primary current is designed by using the ANSOFT Maxwell tools. This optimized sensing head has a structure as follows: 28 mm inner radius, 58 mm outer radius, 15 mm height and 25 mm length gap. Additionally, the magneto-optical crystal choose MR1 with the length of 5 mm. In a temperature range from -40 ℃ to +80℃, the OCT designed can be employed successfully.

REFERENCES

[1] Y. B. Liu, H. B. Li, C. Y. Yu, G. X. Ye and X. Q. Wang, "The Principle, Technology and Application of Electronic Transformer," Science Press, Beijing, 2009, pp. 11-35.

[2] B. B. Afanasyev, "Current Transformer," Mechanical Industry Press, Beijing, 1989, pp. 156-160.

[3] Z. P. Wang, X. Z. Wang and Q. B. Li, "Theoretical Analysis of Temperature Characteristics of a Bulk Glass Optical Current Sensing Element with Polarization-Preserving Layers," *Acta Photonic Sinica*, Vol. 35, No. 6, 2006, pp. 846-849.

[4] Z. P. Wang, Q. Wu, Q. B. Li, Z. J. Huang and J. H. Shi, "Theoretical Analysis of the Effect of the Temperature Features of Linear Birefringence on the Performance of an Optic-glass Current Sensor," *Journal of Harbin Engineering University*, Vol. 26, No. 2, 2005, pp. 272-276.

[5] M. Huang and S. Y. Zhang, "Crystal Growth and Magneto-optical Properties of Ce3+ Doped Rare-earth Iron Garnet," *Chinese Journal of Materials Research*, Vol. 14, No. 4, 2000, pp. 393-396.

[6] Y. Yoshida, S. Kawazoe, K. Ibuki, K. Yamada and N. Ochi, "New Fault Locating System for Air-insulated Substation Using Optical Current Detector," *IEEE Transactions on Power Delivery*, Vol. 7, No. 4, 1992, pp. 1804-1813.
doi: 10.1109/61.156982

Performance Comparison of PIN and APD based FSO Satellite Systems for various Pulse Modulation Schemes in Atmospheric Turbulence

Pooja Gopal[1], V. K. Jain[2], Subrat Kar[2]
[1]Bharti School of Telecom Tech. and Mgmt., IIT Delhi, New Delhi, India
[2]Electrical Engineering Dept., IIT Delhi, New Delhi, India
Email: pooja.gopal@dbst.iitd.ac.in

ABSTRACT

In this paper, the performance of various Pulse Position Modulation (PPM) schemes has been analysed for PIN and APD receivers in the presence of atmospheric turbulence. It is observed that the performance of the APD receiver is always better than that of the PIN receiver as expected. Among the various modulation schemes, the performance of Differential Amplitude PPM (DAPPM) scheme with more number of amplitude levels is better than that of the other schemes for the same single level peak amplitude. Further, the optimum gain of APD receiver does not change substantially for different modulation schemes and turbulent conditions.

Keywords: Free Space Optics; Pulse Modulation; Ground-to-satellite Communications

1. Introduction

Terrestrial Free Space Optical (FSO) systems such as optical fibre backup links, cellular communication back-haul links, multi-campus links, etc. have been performing well right from their emergence. Inter-satellite FSO links have already been established, while the FSO link between a ground station and an orbiting satellite is being extensively reported [1,2].

The major advantages offered by this technology over the conventional Radio Frequency (RF) satellite systems are the following: (i) the size and weight of the payload are critical parameters in any satellite system. These parameters in FSO systems are one third less than that of corresponding RF systems [3], (ii) the beam divergence angle in RF systems is large, which results in large foot-prints. This limits the number of satellites using the same spectrum and also poses a threat to the security of such systems. In contrast to this, the extremely small beam widths and divergence angles of lasers offer links which are basically immune to interference and offer high security, and (iii) there are as yet no restrictions on the bandwidth used in FSO systems. Further, they offer very high data rates which are virtually unconstrained by the carrier frequencies.

Pulse Position Modulation (PPM) is very popular in long distance optical communication systems because of its high power efficiency. Also, due to its high Peak to Average Power Ratio (PAPR), it is resilient to the effects of noise. The bandwidth requirement which increases with the order of the PPM, is directly proportional to the power efficiency i.e., higher order PPM schemes are more power efficient and resilient to noise. Ground-to-satellite optical communications have to undergo the ill effects of atmospheric turbulence, the amount of which is proportional to the strength of turbulence. The pulse broadening limits the rate at which data can be sent through the turbulent channel. Hence, PPM may not be an ideal choice of modulation scheme in all atmospheric conditions because of its high bandwidth requirement. Several variants of PPM scheme have been in use, which offer a trade-off between the power efficiency and bandwidth efficiency. In this paper, the comparison of the Bit Error Rate (*BER*) performance of the various PPM schemes is analysed.

This paper is organized as follows. In Section 2, a brief description of the PPM schemes is given. The Section 3 contains the system model used for subsequent analysis. The methodology for system performance evaluation and the numerical results in graphical form are presented in Section 4. The conclusions of the study are given in Section 5.

2. Modulation Schemes

In Differential Pulse Position Modulation (DPPM), all the empty slots following the pulse in PPM are removed.

This reduces the average symbol length implying improved bandwidth efficiency. Also, there is an inherent symbol synchronization capability as every symbol ends with a pulse. Like PPM and unlike On-Off Keying (OOK), DPPM does not require an adaptive threshold at the receiver.

Differential Amplitude Pulse Position Modulation (DAPPM) is a combination of DPPM and Pulse Amplitude Modulation (PAM). The average number of empty slots following a pulse in DPPM can be reduced by increasing the number of amplitude levels A. This in turn increases the bandwidth efficiency. But, it adds the requirement of having an adaptive threshold, due to the presence of multi-amplitudes. A well designed DAPPM system would require less bandwidth in comparison to OOK, PPM and DPPM systems [2]. It has inherent symbol synchronization capability like DPPM.

3. System Model

The three most reported models for irradiance fluctuations in a turbulent channel are: log-normal, gamma-gamma and negative exponential. Their respective ranges of validity are in the weak, weak-to-strong and saturation regimes. In the region of weak fluctuations, the statistics of the irradiance fluctuations have been experimentally found to obey the log-normal distribution [4]. The probability density function of log-normal distribution is given by

$$p(I)=\frac{1}{\sqrt{2\pi\sigma_I^2}}\frac{1}{I}\exp\left\{-\frac{\left(\ln(I/I_0)-E[I]\right)^2}{2\sigma_I^2}\right\}, I\geq 0 \quad (1)$$

and the scintillation index σ_I^2 is given by the expression

$$\sigma_I^2=\frac{E[I^2]}{E^2[I]}-1 \quad (2)$$

where I is the received field intensity in presence of turbulence and I_0 the received field intensity without the effect of turbulence, σ_I^2 the log-intensity variance and $E[I]$ the mean of log-intensity variance. For the case of strong turbulence, the probability density function is given by the negative exponential distribution

$$p(I)=\frac{1}{I_0}\exp\left\{-\frac{I}{I_0}\right\} \quad (3)$$

The Avalanche Photo Diode (APD) is generally used in long distance optical communications because of the low received power levels. An APD performs better than a PIN diode receiver, when the received power levels are low. A high avalanche gain requires a high reverse bias voltage. The higher gain doesn't imply a better signal to noise ratio (*SNR*) since the performance degrades beyond a certain gain as the effect of noise becomes dominant. Hence the optimum gain of APD for the particular system has to be used. A comparative study of APD receiver vis-à-vis PIN receiver is made in the following section.

4. Methodology for System Performance Evaluation

The number of photons received at the detector, N_s would be a log-normal distributed random variable (in the case of weak turbulence) or a negative exponentially distributed random variable (in the case of strong turbulence). The conditional Bit Error Rate (BER) is then given by

$$P_{b/i}=\frac{1}{2}erfc\left[\frac{Q(N_s)}{\sqrt{2}}\right] \quad (4)$$

where $Q(N_s)$ is the Q-parameter. The unconditional BER is then given by

$$BER=\int_0^\infty P_{b/i}\,p(N_s)\,dN_s \quad (5)$$

where $p(N_s)$ is the probability density function of N_s. Since, N_s is proportional to the received irradiance I, $p(N_s)$ can be determined from eqn. (1) or (3) depending on the level of turbulence. After simplification using the Gauss-Hermite approximation [6], the corresponding BER expressions are obtained. The BER expressions for the different modulation schemes are derived by taking into consideration the respective bandwidth and power requirements. The Symbol Error Rate (SER) expressions are obtained from the respective BER expressions.

The SER expressions for PIN and APD receivers when different modulation schemes are used are obtained from eqs. (1)-(5). The modulation schemes considered are 64-PPM, 64-DPPM, 64-DAPPM (A=2, L=32; A=4, L=16 and A=8, L=8). The numerical results computed from these expressions are shown in **Figures 1-4**. **Figures 1(a)-(b)** and **Figure 2(a)** give the graphs of SER vs. N_s (in dB) for the PIN receiver for σ_{sc}^2 = 0 (no turbulence), 0.5 (low turbulence) and 1 (high turbulence), respectively. The corresponding graphs for the APD receiver are given in **Figures 2(b)**, **Fiugres 3(a)** and **(b)**, respectively.

We observe from **Figure 1(a)** that the performance of DPPM is better than that of PPM. Further, the performance of DAPPM is better than that of DPPM and PPM. The performance of DAPPM becomes still better if the number of levels is increased from 2 to 8. This trend remains the same irrespective of the turbulence level. In case of APD receiver, the comparative performance of different modulation schemes is similar to that of the PIN case. But, the required SNR to obtain a particular SER is much less than that of the PIN receiver.

In **Figures 4(a)** and **(b)**, the variations of SER vs. APD gain are given for low turbulence and high turbulence cases, respectively. It is observed that there is not much difference in the optimum gain for different modulation schemes. The optimum gain for different turbulence conditions is almost same. But, we see that the degradation in performance with increase in gain, beyond the optimum gain is more in the case of high turbulence.

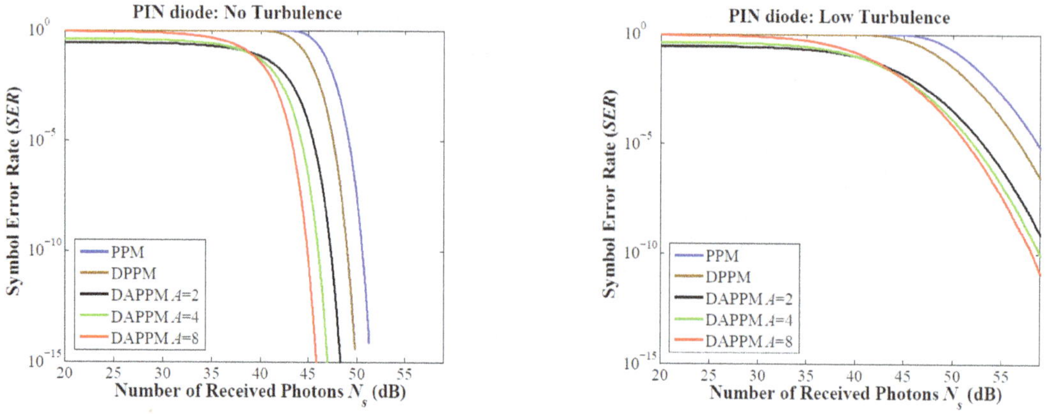

(a) (b)

Figure 1. (a) *SER* vs. N_s in PIN receiver for different modulation schemes without turbulence; (b) *SER* vs. N_s in PIN receiver for different modulation schemes in low turbulence.

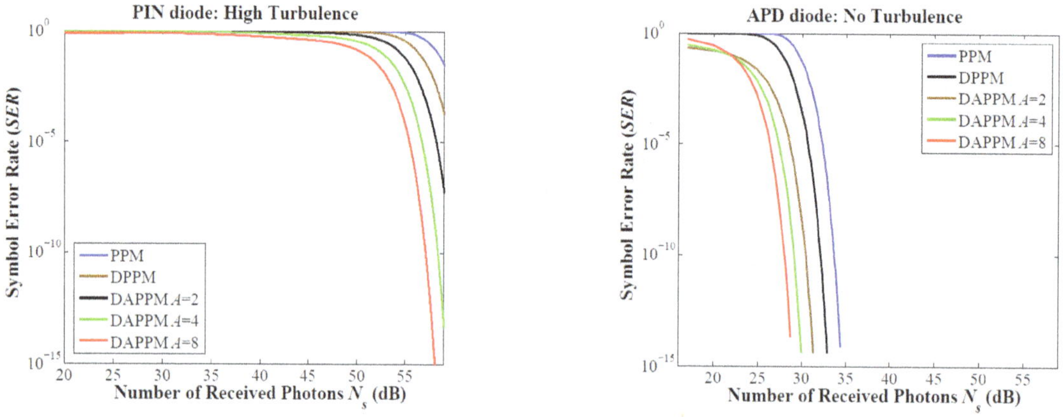

(a) (b)

Figure 2. (a) *SER* vs. N_s in PIN receiver for different modulation schemes in high turbulence; (b) *SER* vs. N_s in APD receiver for different modulation schemes without turbulence.

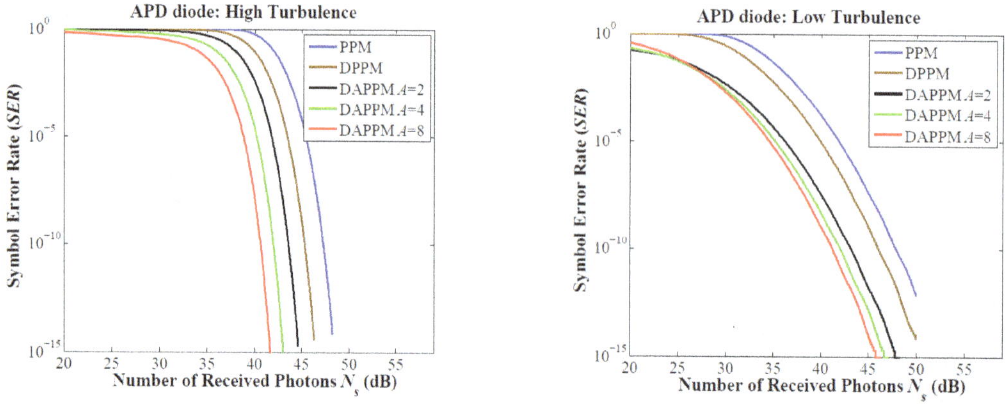

(a) (b)

Figure 3. (a) *SER* vs. N_s in APD receiver for different modulation schemes in low turbulence; (b) *SER* vs. N_s in APD receiver for different modulation schemes in high turbulence.

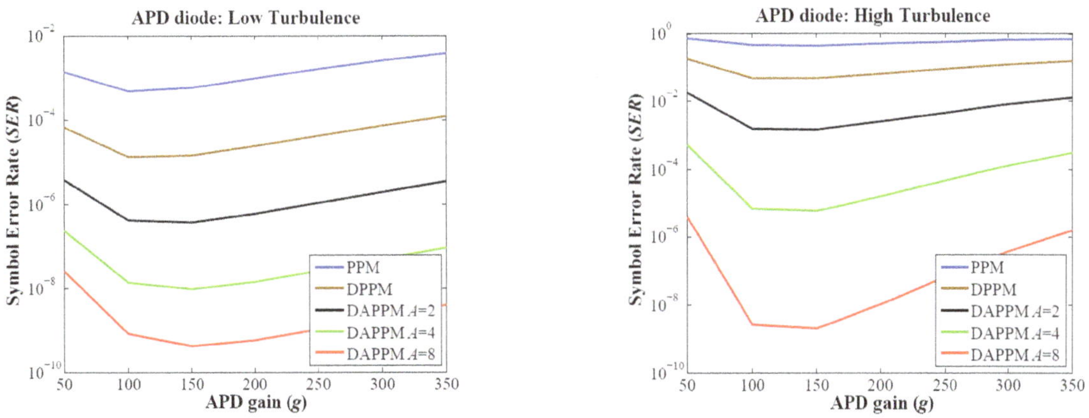

(a) (b)

Figure 4. (a) *SER* vs. *g* in APD receiver for different modulation schemes in low turbulence; (b) *SER* vs. *g* in APD receiver for different modulation schemes in high turbulence.

Table 1. Parameter values in numerical computation.

Parameter	Value
Bit rate ($1/T$)	1 Gbps
Receiver temperature (T_0)	300 K
Average APD gain (g)	150
Load resistance (R_L)	100 Ω
APD noise figure	6.756
Ionization factor	0.02
Order of PPM (M)	64

The numerical values of the parameters used in the computation are given in **Table 1**.

5. Conclusions

The performance of APD receiver is much better than that of corresponding PIN receiver. This is because of the gain factor and also the low received power levels. Hence for satellite communications which involve long distances implying less received power, APD receivers are more suitable. Also, there is no substantial change in the optimum gain for different modulation schemes and atmospheric conditions. The disadvantage is the requirement of high bias voltages for more gains, which will add to the payload.

The better performance of the DAPPM scheme as compared to PPM and DPPM schemes can be attributed to the fact that more number of levels reduce the effective symbol length, which in turn reduces the bit duration and hence inter symbol interference is caused due to turbulence. This also explains why DAPPM schemes with more number of levels perform better.

The degradation of performance with increase in gain beyond an optimum value is due to the fact that the noise is amplified along with the signal and beyond a certain point, the effect of amplified noise is more than that of the signal. Similarly, the degradation is more in the case of strong turbulence because, turbulence causes random variation of received signal which is amplified similar to the noise.

We can conclude that because of the time varying nature of the atmospheric channel, the performance of the link depends on the turbulence conditions. Hence by adaptively changing the modulation, a more robust system performance can be expected.

REFERENCES

[1] H. Kaushal, V. K. Jain and S. Kar, "Performance Improvement with Coding of Free Space Optical Ground to Satellite Link in Atmospheric Turbulence Environment," *AIP Conference Proceedings*, Kerala, 2011.

[2] P. Gopal, V. K. Jain and S. Kar, "Performance Analysis of Ground to Satellite FSO System with DAPPM Scheme in Weak Atmospheric Turbulence," *International Conference on Fibre Optics and Photonics*, Chennai, India, WPo. 43, December 2012.

[3] L. C. Andrews and R. L. Phillips, *Laser Beam Propagation Through Random Media*, 2nd Edition, SPIE Press, Washington, 2005

[4] R. L. Phillips and L. C. Andrews, "Universal Statistical Model for Irradiance Fluctuations in a Turbulent Medium," *Journal of Optical Society of America*, Vol. 72, No. 7, 1982, pp. 864-870. doi:10.1364/JOSA.72.000864

[5] G. P. Agrawal: *Fiber Optic Communication Systems*, Wiley, India, 2002. doi:10.1002/0471221147

[6] K. Kiasaleh, "Performance of APD-based, PPM Free Space Optical Communication Systems in Atmospheric Turbulence," *IEEE Transactions on Communications*, Vol. 53, No. 9, 2005, pp. 1455-1461. doi:10.1109/TCOMM.2005.855009

Permissions

All chapters in this book were first published by Scientific Research Publishing; hereby published with permission under the Creative Commons Attribution License or equivalent. Every chapter published in this book has been scrutinized by our experts. Their significance has been extensively debated. The topics covered herein carry significant findings which will fuel the growth of the discipline. They may even be implemented as practical applications or may be referred to as a beginning point for another development.

The contributors of this book come from diverse backgrounds, making this book a truly international effort. This book will bring forth new frontiers with its revolutionizing research information and detailed analysis of the nascent developments around the world.

We would like to thank all the contributing authors for lending their expertise to make the book truly unique. They have played a crucial role in the development of this book. Without their invaluable contributions this book wouldn't have been possible. They have made vital efforts to compile up to date information on the varied aspects of this subject to make this book a valuable addition to the collection of many professionals and students.

This book was conceptualized with the vision of imparting up-to-date information and advanced data in this field. To ensure the same, a matchless editorial board was set up. Every individual on the board went through rigorous rounds of assessment to prove their worth. After which they invested a large part of their time researching and compiling the most relevant data for our readers.

The editorial board has been involved in producing this book since its inception. They have spent rigorous hours researching and exploring the diverse topics which have resulted in the successful publishing of this book. They have passed on their knowledge of decades through this book. To expedite this challenging task, the publisher supported the team at every step. A small team of assistant editors was also appointed to further simplify the editing procedure and attain best results for the readers.

Apart from the editorial board, the designing team has also invested a significant amount of their time in understanding the subject and creating the most relevant covers. They scrutinized every image to scout for the most suitable representation of the subject and create an appropriate cover for the book.

The publishing team has been an ardent support to the editorial, designing and production team. Their endless efforts to recruit the best for this project, has resulted in the accomplishment of this book. They are a veteran in the field of academics and their pool of knowledge is as vast as their experience in printing. Their expertise and guidance has proved useful at every step. Their uncompromising quality standards have made this book an exceptional effort. Their encouragement from time to time has been an inspiration for everyone.

The publisher and the editorial board hope that this book will prove to be a valuable piece of knowledge for researchers, students, practitioners and scholars across the globe.

List of Contributors

Zainab S. Sadik
Department of Physics, College of Science, University of Baghdad, Baghdad, Iraq

Dhia H. Al-Amiedy
Department of Physics, College of Science for Women, University of Baghdad, Baghdad, Iraq

Amal F. Jaffar
Ministry of High Education & Scientific Research, Foundation of Technical Education, Institute of Medical Technology, Mansour, Iraq

Ansgar Wego
Department of Electrical Engineering, Hochschule Wismar, Wismar, Germany

Gundolf Geske
ASTECH Angewandte Sensortechnik GmbH, Rostock, Germany

Mustapha Remouche, Francis Georges and Patrick Meyrueis
Laboratoire des Systèmes Photoniques, École Nationale Supérieure de Physique de Strasbourg, Université de Strasbourg, Strasbourg, France

Belloui Bouzid
Hafr Al-Batin Community College (HBCC), King Fahd University of Petroleum and Minerals (KFUPM), Dhahran, Saudi Arabia

Xinping Zhang, Shengfei Feng and Tianrui Zhai
College of Applied Sciences, Beijing University of Technology, Beijing, China

Weston Thomas and Christopher Middlebrook
Electrical and Computer Engineering Department, Michigan Technological University, Houghton, USA

Nahid Karimi, Soheil Sharifi and Mousa Aliahmad
Department of Physics, University of Sistan and Baluchestan, Zahedan, Iran

Sneha Salampuria and Manas Banerjee
Department of Chemistry, University of Burdwan, Burdwan, India

Tandrima Chaudhuri
Department of Chemistry, Dr. Bhupendra Nath Dutta Smriti Mahavidyalaya, Hatgobindapur, India

Nallamuthu Ananthi, Umesh Balakrishnan and Sivan Velmathi
Department of Chemistry, National Institute of Technology, Tiruchirappalli, India

Krishna Balakrishna Manjunath and Govindarao Umesh
Department of Physics, National Institute of Technology, Surathkal, India

Hana Mohammed Mousa
Physics Department, Al-Azhar University, Gaza, Palestine

Ali W. Elshaari
Electrical and Electronic Engineering Department, University of Benghazi, Benghazi, Libya

Stefan F. Preble
Microsystems Engineering Department, Rochester Institute of Technology, Rochester, USA

Sergey Borisovich Odinokov and Hike Rafaelovich Sagatelyan
Bauman Moscow State Technical University (BMSTU), Moscow, Russia

Arif Mirjalal Pashayev and Bahadir Guseyn Tagiyev
Institute of Physics, Azerbaijan National Academy of Sciences, Baku, Azerbaijan
National Aviation Academy of Azerbaijan, Baku, Azerbaijan

Said Abush Abushov and Fatma Agaverdi Kazimova
Institute of Physics, Azerbaijan National Academy of Sciences, Baku, Azerbaijan

Ogtay Bahadir Tagiyev
Institute of Physics, Azerbaijan National Academy of Sciences, Baku, Azerbaijan
Lomonosov Moscow State University Baku Campus, Baku, Russia

Mohammad Reza Nasiri-Avanaki
Electronics Department, Azad University of Karaj, Tehran, Iran

Vahid Soleimani and Rohollah Mazrae-Khoshki
Electronic Engineering Department, RAZI University, Kermanshah, Iran

Abdulsalam G. Alkholidi
Faculty of Engineering, Electrical Engineering Department, Sanaa University, Sanaa, Yemen

M. R. A. Moghaddam and H. Ahmad
Photonics Research Center, University of Malaya, Kuala Lumpur, Malaysia

S. W. Harun
Photonics Research Center, University of Malaya, Kuala Lumpur, Malaysia
Department of Electrical Engineering, University of Malaya, Kuala Lumpur, Malaysia

Hisham Imam
National Institute of Laser Enhanced Sciences, Cairo University, Giza, Egypt

Khaled Elsayed, Mohamed A. Ahmed and Rania Ramdan
Materials Science Laboratory (1), Physics Department, Faculty of Science, Cairo University, Giza, Egypt

Hai-Chao Li and Guo-Qin Ge
School of Physics, Huazhong University of Science and Technology, Wuhan, China

Thomas Edward Donaldson Frame and Alexandre Pechev
Surrey Space Centre, University of Surrey, Guildford, UK

Bo Tao, Lianguan Shen, Mujun Li and Jian Zhou
Department of Precision Machinery and Precision Instrumentation, University of Science and Technology of China, Hefei, Anhui, China

Allen Yi
Department of Integrated Systems Engineering, the Ohio State University, Columbus, Ohio, USA

Guozhou Jiang and Ying Mei
College of Educational Information & Technology, Hubei Normal University, Huangshi, China

Xin Zhang, Guohui Yuan and Zhuoran Wang
School of Optoelectronic Information, University of Electronic Science and Technology of China, Chengdu, Sichuan, China

Fernando Andrés Londono Badillo, Jose Antonio Eiras, Flavio Paulo Milton and Ducinei Garcia
Ferroelectric Ceramics Group, Physics Department, Federal University of São Carlos, São Carlos, Brazil

Raied K. Jamal, Mohammed T. Hussein and Abdulla M. Suhail
Department of Physics, College of Science, University of Baghdad, Baghdad, Iraq

Reza Ashrafi, Ming Li and José Azaña
Institut National de la Recherche Scientifique–Énergie, Matériaux et Télécommunications, Montréal, Québec, Canada

Dmitrii Kouznetsov and Makoto Morinaga
Institute for Laser Science, University of Electro-Communications, Tokyo, Japan

Shaojiang Du and Hairan Feng
Physics and Information Engineering Department, Jining University, Qufu, China

Rosmin Elsa Mohan
Amrita Vishwa Vidyapeetham, Kollam, India

M. Sivakumar
Amrita Vishwa Vidyapeetham, Coimbatore, India

K. S. Sreelatha
Govt.Polytechnic College, Kottayam, India

Evgeny V. Chensky
Modern Science Institute, SAIBR, Moscow, Russia

Xiaoyi Su and Qifeng Xu
College of Electrical Engineering and Automation, Fuzhou University, Fuzhou, China

Pooja Gopal
Bharti School of Telecom Tech. and Mgmt., IIT Delhi, New Delhi, India

V. K. Jain and Subrat Kar
Electrical Engineering Dept., IIT Delhi, New Delhi, India

www.ingramcontent.com/pod-product-compliance
Lightning Source LLC
Chambersburg PA
CBHW080527200326
41458CB00012B/4359
* 9 7 8 1 6 3 2 4 0 5 6 4 7 *